Zooplankton Ecology

T0225291

Editors

M. Alexandra Teodósio
and
Ana B. Barbosa

Universidade do Algarve
Centre of Marine Sciences (CCMAR) and
Centre for Marine and Environmental Research (CIMA)
Campus de Gambelas, Faro, Portugal

CRC Press
Taylor & Francis Group
Boca Raton London New York

CRC Press is an imprint of the
Taylor & Francis Group, an **informa** business

A SCIENCE PUBLISHERS BOOK

Cover illustrations
Top, left: zoea larval stage of the decapod *Nepinnotheres pinnotheres* (reproduced by kind courtesy of Joana Cruz).
Top, right: medusa stage of the hydrozoan jellyfish *Odessia maeotica*, with the parasitic isopod *Paragnathia formica* visible at the exumbrella apex (reproduced by kind courtesy of Vânia Baptista).
Bottom, left: unidentified Gobidae fish larvae, with lapilli (anterior) and sagittae (posterior) otoliths visible at the posterior head region (reproduced by kind courtesy of M. Alexandra Teodósio). **Bottom, right**: scanning electron micrograph of the siliceous skeleton of the polycystine radiolarian *Lamprocyclas maritalis* (reproduced by kind courtesy of Stanley A. Kling).

CRC Press
Taylor & Francis Group
6000 Broken Sound Parkway NW, Suite 300
Boca Raton, FL 33487-2742

First issued in paperback 2022

© 2021 by Taylor & Francis Group, LLC
CRC Press is an imprint of Taylor & Francis Group, an Informa business

No claim to original U.S. Government works

Version Date: 20200610

ISBN 13: 978-0-367-62056-1 (pbk)
ISBN 13: 978-1-138-49645-3 (hbk)

DOI: 10.1201/9781351021821

Library of Congress Cataloging-in-Publication Data
Names: Teodósio, M. Alexandra, 1965- editor.
Title: Zooplankton ecology / editors, M. Alexandra Teodósio and Ana B. Barbosa, Centre of Marine Science (CCMAR), University of Algarve, Campus de Gambelas, Faro, Portugal.
Description: First. | Boca Raton : CRC Press, [2021] | Includes bibliographical references and index.
Identifiers: LCCN 2020020520 | ISBN 9781138496453 (hardcover)
Subjects: LCSH: Zooplankton--Ecology. | Zooplankton.
Classification: LCC QL123 .Z635 2021 | DDC 592.177/6--dc23
LC record available at https://lccn.loc.gov/2020020520

Visit the Taylor & Francis Web site at
http://www.taylorandfrancis.com

and the CRC Press Web site at
http://www.routledge.com

Preface

Zooplankton comprise a vast diversity of organisms, from unicellular phagotrophic protists to metazoans, including both non-gelatinous and gelatinous invertebrates, and vertebrate eggs and larval stages. These organisms inhabit all different types of aquatic systems, which cover ca. 72% of the Earth's surface. Zooplankton play a pivotal role in ecosystem functioning, representing a link between phytoplankton and higher trophic levels, including living aquatic resources. Zooplankton also play a fundamental role in global biogeochemical cycles and ocean carbon pump, acting as connectors between ocean surface processes and communities, and the deep benthic ocean realm. Considering the current focus on environmental change and related anthropogenic stressors, it is more crucial than ever to get a comprehensive knowledge on the environmental drivers of zooplankton, in order to able to predict their responses to natural and anthropogenically-induced environmental changes. According to the Intergovernmental Panel on Climate Change (IPCC) projections, by 2100, average temperature will increase by 0.6–2.0°C and pH will decrease by 0.1–0.4 units in the oceans' upper surface layer. Zooplankton are already and will continue facing remarkable changes, including ocean warming, acidification, deoxygenation, and plastification. Global change will likely continue to intensify as the world population continues to grow exponentially. Urban, agricultural, and industrial development will also continue to alter biogeochemical cycles, and possibly increase the frequency of extreme climatic events. Moreover, anthropogenically-mediated introduction of non-native species, namely zooplanktonic life phases (ballast water), facilitated by the absence of physical natural barriers in ocean, may further disrupt food webs and impact commercial fisheries.

The wide diversity of zooplanktonic representatives, illustrated in the book front cover, is associated with multiple organism sizes, functional traits, specific ecological functions, environmental drivers, control modes, environmental stressors and varying methodological approaches, from sampling to quantification. This great complexity has been frequently scattered into textbooks and articles tailored to different audiences. As example, freshwater and marine zooplankton studies have been usually separated, as well as metazoans from phagotrophic protists, bottom-up controls from top-down controls, and even fish larvae are frequently perceived as a zooplankton singularity, at the interface with fisheries biology. Although these dichotomies are understandable, they have fragmented this sub-field of marine ecology.

In this context, this book aims to capture key areas of research on zooplankton ecology, using an integrative approach covering both marine and freshwater zooplankton, including phagotrophic protists and metazoans, and their bottom-up and top-down controls, spatial-temporal distribution patterns and emergent methodological approaches. Overall, the book chapters represent a wide array of studies based on the use of multiple approaches including comprehensive synthesis, observational studies, undertaken at different spatial and temporal scales, manipulative studies, and even sediment analysis, a vibrant imprint of benthic-pelagic coupling and ecosystem connectivity. Omics approaches, that have enabled us to assess the biodiversity at an unparalleled scale, are integrated in several chapters. Most chapters also address the specific impacts of anticipated environmental changes and stressors (e.g., warming, ocean acidification, eutrophication), and conclude with the current challenges, opportunities, and avenues for continued study.

The book is divided into 12 chapters, integrated in three major sections. The first section addresses zooplanktonic organisms, processes and controls, the second section is devoted to zooplankton spatial and temporal distribution patterns and trophic dynamics, and the final section is dedicated to innovative tools and approaches for studying zooplankton ecology.

In the first book section, Chapter 1 provides a comprehensive analysis of zooplankton functional traits. This chapter includes information for both fresh-water and marine zooplankton representatives, and addresses multiple traits (morphological, behavioral, stoichiometric and physiological, life history traits), and highlights lesser studied traits, taxa and habitats. Chapter 2 explores two main interactions between phytoplankton and zooplankton, consumptive and non-consumptive. This chapter emphasizes how chemical communication shapes the interactions between dominant primary producers and consumers, and how zooplankton can act as a bottom-up and top-down control on phytoplankton. Chapter 3 provides a thorough historical overview of the hypothesis behind fish recruitment, considering the dynamic controls of ocean physical, chemical and biological processes on fish larvae. This chapter also explores a specific multidimensional functional trait of fish larvae, behavior, as an integrative modulator of fish recruitment, thus providing a linkage with fisheries biology. Chapter 4 revises the effects of a current stressor, ocean acidification, on zooplankton assemblages, and explores a specific case study over shallow-water CO_2 submarine vents (NE Atlantic). This chapter uses CO_2 vents as a natural surrogate for anticipated ocean acidification, and evaluates its effects on copepod individual fitness and zooplankton structure.

In the second book section, Chapter 5 examines biodiversity patterns of tintinnid ciliates combining complementary approaches, conventional morphology-based microscopy, and DNA sequencing, thus unravelling species crypticity and polymorphism. This chapter explores the distribution patterns of phylogenetic diversity of tintinnids and underlying factors (salinity, depth and latitude), at global, regional, and local scales. Chapter 6 explores the global distribution patterns of other groups of phagotrophic protists, foraminifers and polycystine radiolarians, and the hypothesis that may explain their dissimilarities. The use of sediment samples, in tandem with some of the proposed hypothesis (isothermal submersion), clearly reflect the relevance of connectivity between pelagic and benthic sub-systems. Chapter 7 provides a comparative analysis of zooplankton distribution and trophic dynamics over four highly productive coastal upwelling systems (California, Humboldt, Canary and Benguela Current systems). This chapter also discusses the relevance of zooplankton behavioral strategies for retention and dispersion during upwelling-relaxation events, at small- to mesoscale levels. Chapter 8 evaluates intra- and interannual variability patterns of planktonic assemblages in the Baltic Sea, with emphasis on metazooplankton annual succession, associated functional traits, annual routines and life histories, trophic dynamics, and phenology. The use of this specific case study, a brackish water ecosystem, creates additional opportunities for exploring zooplankton temporal dynamics across salinity and latitudinal gradients. Chapter 9 guides the readers over a set of simple visualization techniques for evaluating and comparing zooplankton intra- and interannual variability patterns and trends across time and space, along with underlying environmental determinants. This chapter explores zooplankton time series, with different methods and lengths, and represents a contribution at the interface between zooplankton distribution patterns and methodological approaches.

In the third book section, Chapter 10 introduces the ecology of the highly diverse gelatinous zooplankton and explores new technological approaches for *in situ* jellyfish monitoring, including acoustic, optical and environmental DNA sensors, installed onto robotic platforms (remotely operated vehicles, autonomous underwater vehicles). This chapter specifically emphasizes the relevance of citizen science-based programs, as a complement to current monitoring efforts, which can globally enhance sampling resolution and promote an engaged ocean literate society. Chapter 11 reviews the main steps in metabarcoding studies of zooplankton, including sampling, DNA extraction, amplification and sequencing, and data analysis. This chapter also provides recommendations for minimizing methodological biases, and obtaining ecologically relevant information (barcode choice, reference database, sequence processing and clustering, and taxonomic assignment). Chapter 12 provides an overview on how omics techniques (metagenomics, metatranscriptomics, metaproteogenomics) are currently fostering the research in zooplankton, contributing to characterize their spatial and temporal distribution patterns, life cycles and ecological functions.

After this brief overview of the book chapters, finally, and above all, we gratefully acknowledge all authors for their effort, talent, and insightful contributions. Overall, this book aims at providing students and researchers an advanced integrative overview of zooplankton ecology. This multi-authored book is not a conventional textbook. Yet, we sincerely hope our book proves useful for our past, current and future students, who have been a constant source of inspiration and the prime motivation to accept the challenge of editing this book. Enjoy …

Faro, February 7, 2020

M. Alexandra Teodósio
Ana B. Barbosa

Contents

Part 3: Advanced Techniques to Study Zooplankton

Part 1

Organisms, Processes and Controls

Chapter 1

Functional Trait Approaches for the Study of Metazooplankton Ecology

Marie-Pier Hébert[1], and *Beatrix E. Beisner[2],**

1.1 A Brief History of Trait-based Approaches and their Application to Aquatic Ecology

Functional trait-based approaches have increased in popularity in ecology since the turn of the millennium. There are a number of advantages to such approaches that explain their widespread adoption. In particular, examining traits present in a community can increase the mechanistic understanding of how environmental factors may influence its structure and how community shifts may in turn influence ecosystem processes (functioning). Moreover, utilizing lists of observed traits across individuals, rather than lists of species present within communities, enables a comparison across regions that may have very different taxonomic composition, while expressed traits that are behavioural, physiological or morphological can sometimes be directly compared as these are common to all life forms. Particularly applicable to plankton communities, the use of functional traits can even enable comparison of marine with freshwater environments, broadening our understanding and predictions of the ecological roles of aquatic organisms (Hébert et al. 2016a). Together, the traits expressed in a community can thus provide an indication of organismal strategies and roles as well as the niche space occupied that can then be compared across habitats and ecosystems (Litchman et al. 2013).

More recent practice describes organisms as a continuous distribution of traits along environmental gradients (Kenitz et al. 2018), a concept forming the core of the emergent field of functional biogeography (Violle et al. 2014). Such an approach can allow for a greater exploration of trait patterns and trade-off relationships (Kiørboe 2011), which can prove useful in predicting novel trait

[1] Department of Biology and Groupe de recherche interuniversitaire en limnologie (GRIL), McGill University, 1205 Dr Penfield Avenue, Montréal, Québec, Canada, H3A 1B1, University of Québec at Montreal, C.P. 8888 Succ. Centre-Ville, Montréal, Québec, Canada H3C 3P8.

[2] Department of Biological Sciences and Groupe de recherche interuniversitaire en limnologie (GRIL), University of Québec at Montreal, C.P. 8888 Succ. Centre-Ville, Montréal, Québec, Canada H3C 3P8.

* Corresponding authors: mphebert4@gmail.com; beisner.beatrix@uqam.ca

assemblages or organismal strategies under different conditions (Chen et al. 2018). In a broader context, investigating trait (spatial) distributions and (temporal) successions can provide mechanistic insight into how organisms may respond to environmental drivers and changes, which can in turn be implemented into scenarios of future global change (e.g., warming responses; Burrows et al. 2019). Using traits rather than taxonomic affiliation also enables the assessment of functional diversity, which can depict more clearly the degree to which organismal roles in ecosystems are complementary or redundant (Hooper et al. 2005).

In aquatic environments, the use of functional traits has been adopted for a variety of organisms, including both marine and freshwater plankton, fish, benthic macroinvertebrates, and macrophytes. Amongst the plankton, functional approaches—generally perceived as the precursor to functional groups—began to be used earlier for phytoplankton, in large part because of the ground-breaking work of Ramón Margalef (1960–70s) and Colin Reynolds (1980–90s) relating "life forms" or "functional morphologies" to environmental conditions (Weithoff and Beisner 2019). As such, phytoplankton ecology is more readily translatable from the pioneering work done earlier in the trait-based ecology of terrestrial plant communities; e.g., Tilman et al. (1997), Walker et al. (1999), Reynolds et al. (2002), Weithoff (2003). Amongst metazoan animal communities, however, functional traits were applied relatively early to aquatic organisms. This is especially the case for stream benthic invertebrates largely because of the well-developed functional groups based on food web position and feeding strategies outlined in the river continuum concept (Cummins and Klug 1979, Vannote et al. 1980). One particularly applied aspect that has emerged from the trait framework in stream benthic ecology is the widely adopted development of bioindicators for habitat quality based on macroinvertebrate tolerance to stress. The early use of benthic macroinvertebrate traits to derive bioindicators likely occurred because this group of organisms is relatively long-lived (integrating information from a few seasons), not particularly motile (reflecting local habitat conditions), and exhibits highly variable response and tolerance to stress (Wallace and Webster 1996), unlike many other aquatic organisms. Fish ecologists were also relatively early adopters of trait-based approaches, mainly focusing on morphological size indicators that influence feeding, swimming and agility (Lankford et al. 2001), with more recent inclusion of metabolic traits to infer the contribution of fish communities to ecosystem processes (Barneche et al. 2014). Much of the earliest uses of traits for aquatic organisms were in relation to the physical-chemical constraints present in aquatic environments; e.g., phytoplankton movement and morphology in relation to lake stratification/mixing (Reynolds 1984), or fish and benthic invertebrate feeding and morphologies in relation to streamflow and dissolved oxygen (Poff et al. 2006, Keck et al. 2014). It is possible that the less obvious effect of lake and ocean physical mixing on the relatively motile zooplankton may have led to a delayed interest in trait-based approaches for this group of organisms. However, the use of chemical elements to characterize organisms at different levels of organization that emerged three decades ago in the form of "biological stoichiometry" can be considered amongst the earliest applications of traits to characterize zooplankton, especially with respect to linking particular taxonomic groups to ecosystem processes (e.g., nutrient recycling; Elser et al. 2000).

Amongst the various organisms that can be classified as metazooplankton, it is for the crustacean component that trait-based approaches have been most commonly used, as this group is globally distributed, very abundant, and a large proportion of zooplankton diversity is known. A first compilation of potential crustacean zooplankton functional traits was done by Barnett et al. (2007) for freshwater species of eastern North America. Since then, Litchman et al. (2013) have further developed a more general conceptual framework within which to consider zooplankton traits, similar to the one previously developed for phytoplankton (Litchman and Klausmeier 2008). Subsequently, Hébert et al. (2016a, b) expanded the original trait compilation of Barnett et al. (2007), including a broader range of crustacean traits more specifically related to ecosystem function in both freshwater and marine environments. Additional trait syntheses for marine species include the databases developed by Benedetti et al. (2015; copepods), Pomerleau et al. (2015; copepods, amphipods, euphausiids, pteropods, ostracods, hydrozoans), and Brun et al. (2016; copepods). Perhaps one of the reasons why there are more trait data compilations for marine species is that there exists more accessible resources documenting species-specific (mostly morphological) information for a wide taxonomic coverage, including Mauchline (1980, 1998) that contains valuable data on the biology of marine calanoids, mysids and euphausiids, or periodically-updated online platforms, such as Razouls et al. (2005–2019) and the World Register of Marine Species (2000–2019). As with phytoplankton (Bruggemann 2011), metazooplankton traits may be phylogenetically constrained to some degree, given that phylogenies were based upon morphological similarities/dissimilarities. The trait synthesis work to date has demonstrated that there are only a few traits for which we have large taxonomic coverage to enable broad application to metazooplankton communities, with trait data being disproportionally more reported for crustaceans relative to other metazooplankton representatives. In particular, quantitative traits that relate to ecosystem processes are more sparse, especially in freshwater ecosystems (Hébert et al. 2017). In this chapter, we will focus on knowledge related to metazooplankton traits, thus not including planktonic phagotrophic protists such as ciliates (see Santoferrara and McManus 2021, this book, Chapter 5), foraminiferans or radiolarians (see Boltovskoy and Correa 2021, this book, Chapter 6).

1.2 Most Commonly used Traits in Metazooplankton

As in other subfields of ecology, functional traits and their relation to community and ecosystem processes have become an active area of research for metazooplankton ecology. Traits are defined as individual-level measurements of behavioural, morphological, molecular, phenological or physiological characters that can influence fitness, while encapsulating information on population, community and ecosystem processes (Table 1.1; Violle et al. 2007). The trait framework can be organized into two main types, dependent on whether organismal traits are examined through the lens of (i) factors determining community structure and assembly processes or (ii) the direct influence of traits on ecosystem processes (Lavorel and Garnier 2002). In the former category, traits are often referred to as "response traits" while in the latter they can be considered as "effect traits". It should be noted, however, that it is the way in which traits are used to address ecological questions that often determines whether they are considered response or effect traits. Some traits are more versatile than others, directly affecting processes at multiple scales and thus falling into more than one trait category (Table 1.1). For example, morphology such as distances between setae that determine the size of phytoplankton prey captured by filter feeding crustacean zooplankton could be considered as a response trait to the feeding preferences of competing zooplankton and to the phytoplankton community present in a habitat, or it could be considered an effect trait influencing

Table 1.1: Integrative framework linking traits to levels of organization and ecological parameters of interest. Some traits may be related to specific ecological parameters and thus operate at particular levels of organization ("mono-type" trait); e.g., response traits such as temperature optima or stress tolerance dictating community composition and succession along environmental gradients. Other, more versatile traits may encapsulate information regarding several or all levels of organization (e.g., body size, growth; *). Adapted from Hébert et al. (2017).

Level of organization and application	Ecological parameters of interest	Commonly used trait type	Examples of traits for metazooplankton
Ecosystem	Ecosystem functioning (properties and processes)	Effect traits	Body size*, growth*, respiration, excretion, clearance rates
Community	Community structure and assembly processes	Response	Body size*, growth*, temperature optima, stress tolerance (e.g., starvation, hypoxia)
Population	Population dynamics	Demographic	Body size*, growth*, generation time, dispersal
Individual	Individual fitness and performance	Life-history	Body size*, growth*, fecundity, individual performance- or survival-related traits

phytoplankton community composition, overall biomass pools or primary productivity. Conversely, there are several "mono-type" traits, i.e., those that can only or primarily be regarded as either a response or effect trait (Table 1.1; Violle et al. 2007). Although some traits may indirectly affect different ecological parameters across levels of organization, "mono-type" traits are typically those that exert direct effects at only one level of organization (Table 1.1; Hébert et al. 2017). For example, amongst the plankton, thermal growth optima would mainly be perceived as a response trait (i.e., dictating community assemblages along temperature gradients; Burrows et al. 2019), while carbon (C) export efficiency is essentially considered as an effect trait (Litchman et al. 2015).

Functional traits can be either qualitative or quantitative in nature, both types being useful in estimates of functional diversity (Section 1.6). Estimating functional diversity indices permits comparison across communities, with most indices allowing for the incorporation of both qualitative and quantitative traits via the use of Gower distances (Petchey and Gaston 2002, 2006, Villéger et al. 2008, Laliberté and Legendre 2010). Several quantitative traits can be considered as "hard" traits (i.e., informative though hard to measure traits) that are often estimated more broadly via "soft" proxy traits (i.e., more readily-measurable correlates of hard traits) that may or may not be quantitative (Hodgson et al. 1999); e.g., several physiological traits are mass-dependent, and their values can sometimes be inferred from body size. For a non-exhaustive list of soft and hard trait associations in zooplankton, see Table 1 in Hébert et al. (2017).

To conceptually represent relations among traits, Litchman et al. (2013) and Hébert et al. (2017) provide frameworks within which the functional traits of metazooplankton can be classified. The Litchman et al. (2013) framework focuses on ecological functions of traits related to metazooplankton feeding, growth/reproduction and survival/mortality. The framework proposed by Hébert et al. (2017) emphasizes inter-related traits based on organismal bioenergetics (energy budget) and considers more specifically potential associations between organismal and ecosystem functions. We combine these perspectives to summarize the general state of knowledge to date for metazooplankton traits, relating them to their primary organismal function, sometimes listing more than one primary function (organismal functions are indicated in capitals in the following outline). Within this framework, qualitative and quantitative traits are classified according to whether they relate primarily to morphology, behaviour, physiology or life history. For most traits, we provide a brief description of how they can be measured and examples of how they can be used as either response or effect traits.

1.2.1 Morphological Traits

Body Size (SURVIVAL/MORTALITY, GROWTH/REPRODUCTION, FEEDING)

Expressed in length or mass, body size is one of the most widely used traits for all groups of metazooplankton. Although most precise if measured at the individual level, taxon-specific means of body size from particular habitats are commonly employed. Body size can act as a proxy for many traits via well-established allometric relationships, especially in poikilothermic organisms for which various biological rates can be reasonably well estimated from body size and ambient temperature (Hébert et al. 2016a). Body size is often considered a "master trait", as it is one of the main soft traits for metazooplankon, providing a value that summarizes many other, harder to measure traits (see Section 1.4), such as physiological traits (e.g., clearance, excretion, growth, respiration; Hébert et al. 2017) or behavioural traits (e.g., body-size dependence of occupied depths in the water column or in diel vertical migration patterns; Ohman and Romagnan 2016).

As a master trait, body size can be used both as a response and effect trait, as it equally integrates environment-induced organismal responses (e.g., to temperature or predation; Patoine et al. 2002, Daufresne et al. 2009) and the influence that organisms can exert on ecosystem processes (e.g., the contribution of community-weighed mass-dependant physiological rates to elemental flow; Hébert et al. 2017). In a multi-trophic context, integrating body size information can provide insight as to how the size structure of food webs may vary along environmental gradients (Kenitz et al. 2018).

Volume to Biomass Ratio (SURVIVAL/MORTALITY, GROWTH/REPRODUCTION, FEEDING)

This trait is one that separates crustacean zooplankton (cladocerans, copepods, mysids, euphausiids) from the rotatorian (rotifers) and gelatinous zooplankton (salps, ctenophores, cnidarians, pteropods, semi-gelatinous chaetognaths) based on body C-density (Litchman et al. 2013). Gelatinous zooplankton tend to have very large volumes relative to mass density compared to crustaceans, resulting in the former having larger body sizes that can increase prey capture rates and reduce their own exposure to predation.

Defence (SURVIVAL/MORTALITY)

Freshwater cladocerans are the crustacean zooplankton with the best-known inducible defence system. For example, many *Daphnia* species are able to develop elongated "helmeted" forms when predators are present. This is an especially effective method to evade gape-limited invertebrate predation. Another example of polymorphism for inducible defense is the ability of some rotifers such as *Brachionus* to develop lateral spines in the presence of predator kairomones; however, given that this is energy-demanding, the expression of such defence structures may not be possible under competition pressure, as observed in *B. calyciflorus* (Yoshida et al. 2003). The high sensitivity and plasticity of such defence-related features makes them good response traits to detect predation signals.

The development of sensory systems is a key adaptation to help zooplankton evade predation. Antipredator sensory modalities include the ability to detect light (photoreception), chemicals (chemoreception) and fluid motion (mechanoreception; Buskey et al. 2002, 2011). Upon the detection of a threat, a response is triggered (e.g., escape behaviour or migration; see sections on Motility and Migration). The development of some neurophysiological features can influence the efficiency of such antipredator systems. For example, in copepods, mechanoreception is used to detect hydrodynamic disturbances created by the motion of predators; once a threat is perceived, copepods must in turn react rapidly, and this response can be faster in species having developed myelinated nerves (Buskey et al. 2002, 2011). The discrepancy in response latencies between myelinate and amyelinate copepod species can have a significant effect on predator avoidance efficiency (and thus survival), with differences in response latencies being greatest across larger copepod species given that nerve signals must be conducted over relatively longer distances (Buskey et al. 2011). In this context, the presence of myelinated nerves, or other analogous neurophysiological features, could be considered as a defense-related trait.

Bioluminescence, primarily present and studied in marine zooplankton, has been considered a warning signal towards predators, aiding in the survival of some taxa of chaetognaths (Thuesen et al. 2010) and gelatinous ctenophores and planktonic cnidaria (Haddock and Case 1999). Bioluminescence may also confer crypsis at depth where there is little ambient light, preventing detection by predators (counterillumination), in some gelatinous zooplankton as well as in euphausiids (Herring 1999, Johnsen 2005). However, there is variation in functional purposes of bioluminescence, and it may also serve in some cases for reproductive purposes (Herring 1999).

Given that most planktonic animals are solitary across all aquatic environments, it is hard to include coloniality as a trait on its own, as has been done for phytoplankton. At least one group of free-swimming rotifers (Conochilidae) has a colonial life-form, which likely helps protect them from predation by gape-limited vertebrate and especially invertebrate predators (Gilbert 1980, Diéguez and Balseiro 1998). Coloniality is relatively more observed in gelatinous zooplankton, such as siphonophores, salps or pyrosomes (Mackie 1986). Defensive reactions can be highly developed in such colonies, but often in the form of escape responses through jet propulsion. For siphonophores and salps, coloniality enables coordinated locomotion, enhancing the control and efficiency of swimming movements (Mackie 1986, Madin 1990). For example, physonect siphonophores use multiple clonal individuals ("nectophores") to propel aggregate colonies, with nectophores from the base of the colony affecting the direction and magnitude of whole-colony movements differently from nectophores situated in the colony apex (Costello et al. 2015).

Transparency (SURVIVAL/MORTALITY)

The degree to which metazooplankton are transparent and therefore less prone to detection by visual predators can also vary across taxa. Gelatinous and rotatorian taxa, with their high volume to biomass ratios and high-water content, may be more transparent (Litchman et al. 2013). On the other hand, normally transparent or translucent crustacean zooplankton may be more pigmented owing to lipid deposition if they are exposed to colder waters, such as those experienced at deeper depths or in winter months (Ohman et al. 1998, Grosbois and Rautio 2018). In the absence of predation threat, crustaceans can also retain protective pigments when exposed to high levels of UV radiation (Hylander et al. 2009). Thus, seasonal variation in this trait is possible. In marine environments, gelatinous zooplankton like salps and pteropods are almost exclusively transparent, while ctenophores show a wide variety of colouration, ranging from almost completely transparent to strongly pigmented.

Food Particle Size and Feeding Apparati (FEEDING)

Food particle size can be expressed as quantitative ranges or bracketed size categories (Barnett et al. 2007). Preferred or selected food size range or category can be defined based on laboratory analyses or *in situ* incubation experiments of single taxa with a range of prey sizes. Mesh size or setulae distance mechanistically extends the prey size range trait but applies only to zooplankton species that filter feed using setulae. The distance between setulae will define the prey size most frequently captured and can thus be considered a trait of that organism or taxon. Other zooplankton may be limited by gape size (e.g., rotifers, chaetognaths) or relatively unlimited such as in the case of cnidarians.

The presence, morphology or composition of particular features associated with feeding apparati can also be indicative of preferred diet. For example, copepods use the gnathobases of their mandibles to grab and crush their prey, and in some species gnathobases can exhibit diet-specific characteristics, such as silica-based tooth-like structures that facilitate the breaking of silicified diatom frustules while protecting the copepod from mechanical damage (Michels et al. 2012, Michels and Gorb 2015). Analogously, in rotifers, certain features of their feeding apparati can reflect diet type or trophic group. For instance, the mastax (i.e., pharynx-like apparatus enabling mastication) can contain calcified jaw-like structures (termed trophi) that highly vary in shape: trophi can take the shape of grinding ridges in suspension feeders, whereas they can be forceps-shaped in carnivorous species, which enhances their grip on prey; in ectoparasitic rotifers, the mastax can serve to grip onto hosts (Barnes 1980).

Together with body size and food web position, morphological traits informing on diet type or food size range can prove useful to determine "who eats whom" in food web models requiring an *a priori* establishment of possible interactions. Taxon-specific information for food particle size in freshwater crustacean zooplankton has been compiled by Barnett et al. (2007). As for using this trait in a response–effect trait framework, it can be used both ways: food size can be mediated by intraguild competition or resource availability within habitats (response trait), but can also be indicative of zooplankton top-down control on phytoplankton standing stocks or community composition (effect trait).

1.2.2 Behavioural Traits

Feeding Mode (FEEDING)

Crustacean zooplankton differ in their foraging tactics, based on swimming behaviour and feeding via passive or raptorial approaches (DeMott and Watson 1991). Cyclopoid copepods are generally raptorial, most often actively capturing motile prey detected via mechanoreception. On the other hand, Calanoid copepods are usually classified as stationary suspension feeders, but in rare cases as current cruisers (e.g., *Epischura lacustris*; DeMott and Watson 1991). Stationary suspension feeders can detect prey at a distance and re-orient their body position or change feeding currents to increase capture rates (Strickler 1982). Current cruisers move continuously while producing a feeding current to passively capture prey.

Most cladocerans exhibit filter feeding, although predatory freshwater species (e.g., *Polyphemus, Bythotrephes,* and *Leptodora* spp.) are raptorial. However, predatory mysids, often present in deep glacial relict lakes, are filter feeders. Freshwater cladoceran filter feeders can be further classified into the *Daphnia* (D-)type, the *Sida* (S-)type and the *Bosmina* (B-)type depending on the degree to which feeding positions are stationary (D and S-types) versus involving active swimming (B-type) (DeMott and Kerfoot 1982). B-type feeders thus resemble copepod current cruisers. D and S-types draw water with particles through their carapace openings, separating particles using thoracic appendages possessing fine filter plates (on different legs to separate the D and S-types) that collect food particles from the feeding current, and drawing them into the food groove. To this, Barnett et al. (2007) have added *Chydorus* (C-)types to represent zooplankton that feed by scraping periphyton from surfaces. As with other feeding-related traits, diversity in feeding modes can reflect environmental constraints (response traits) while informing on the top-down control exerted by zooplankton (effect traits), with some feeding modes being associated with greater feeding rates (stronger influence on resource biomass) or selectivity/specificity (stronger influence on resource community composition).

In part because of the larger diversity of life forms, marine zooplankton show a wide array of prey capture techniques including using tentacles as webs by sit-and-wait predators, ambush predation, filter feeding or prey baiting through hanging sticky droplets on a thread (all techniques used by ctenophores; Haddock 2007). While filter feeding is most commonly observed (cnidarians, ctenophores, euphausiids and salps), raptorial predation through forward darting to capture prey may also be present (e.g., chaetognaths; Hinde 1998). Although more specific, mucous-mesh grazing is also a common feeding mode (pelagic tunicates and pteropods) that implies the capture, consumption and re-packing of very small particles. Erroneously perceived as non-selective filter feeding in the past, the use of adhesive mucous mesh can be prey-specific, with important functional variations in filter mechanisms and hydrodynamics across taxa (Conley et al. 2018).

Food Web Position (FEEDING)

The degree to which zooplankton are herbivorous, carnivorous or omnivorous can also vary. Barnett et al. (2007) classified this trait into 5 categories: pure herbivores, omnivores that are primary herbivorous, omnivores, omnivores that are primarily carnivorous and pure carnivores. This trait is a good example of how trait-based classification may be a means to simplify diversity across groups of zooplankton, where purely carnivorous zooplankton can include representatives of various taxa from marine and freshwater habitats; e.g., ctenophores, cnidarians, amphipods and predatory cyclopoids and cladocerans.

Defining trophic groups can be based on gut content analysis, and in more recent decades on trophic tracers involving stable isotopes (of C and N, most commonly), trace elements, fatty acids, and bulk tissue- or compound-specific stable isotopes (Pethybridge et al. 2018). While most techniques align well when applied to simple diet mixtures, they often tend to diverge with diet complexity

(Nielsen et al. 2017). Biochemical tracers have various advantages (e.g., high specificity, low sample size required) but they can be assimilated over relatively long time scales, providing trophic information that may differ from gut content analyses, for instance. As a result, there is now an increasing number of studies employing complementary approaches to reduce biases of single methods (Schmidt et al. 2006). Through the use of such techniques, trophic groups (based on diet composition, niche width or trophic position) can be characterized for individuals from different environments, and general taxonomic tendencies can be identified. Trophic classification can be applied most easily to adult stages, as ontogenetic niche shifts in feeding preferences by zooplankton may occur in some groups. It should be noted that some taxa may show parasitic feeding (e.g., some cnidaria and at least one ctenophore and many copepod species; Kabata 1982, Haddock 2007, Jiménez-Guri et al. 2007); parasitic feeding could be used as a trait, although many parasitic organisms are not technically free-living and whether they are included will depend on the extent of the community being considered.

As highlighted in the "Food Particle Size and Feeding Apparati" description, food web position is a valuable information to consider to analyze multi-trophic interactions with more realism (Gsell et al. 2016, Kenitz et al. 2018). An appreciable amount of information on food web positions occupied by crustacean zooplankton in both marine and freshwater environments can be found in Hébert et al. (2016b).

Motility (FEEDING, SURVIVAL/MORTALITY)

Metazooplankton differ in their motility traits with some having greater swimming speeds than others based on different movement types. Motility expressed by metazooplankton taxa has much to do with the way in which they capture prey. Thus, raptorial predators such as copepods may demonstrate extremely rapid movements following periods of relative stasis while cruising feeders may show low but consistent velocities. Salps are barrel-shaped planktonic tunicates that move via jet propulsion created when they contract to pump water through their bodies (see coloniality in the section on Defense), from which they also retain phytoplankton food by filter feeding. Zooplankton motility can also reflect predation to which individuals are exposed through their escape response. For example, some rotifers such as *Polyarthra* can swim slowly when feeding, and then can make large jumps (of several times body length) using their "feather-like" appendices when a predator is detected (Gilbert 1985). Similarly, some ctenophores swim through undulating movements, while others appear to have "darting" motions.

Metazooplankton motility traits can be characterized based on the various types of movement mechanisms: swimming using appendages (copepods and cladocerans), rowing (some cnidaria), undulation (ctenophores, some cnidaria) or jet propulsion (salps and some cnidaria). Given the tight link between motility and feeding traits (active/motile versus passive/non-motile feeders), these are often used in conjunction to determine optimal strategies in different physicochemical environments. Along with size-based food preferences, this trait combination can predict predator-prey interactions (e.g., Kenitz et al. 2018).

Migration (SURVIVAL/MORTALITY, GROWTH/REPRODUCTION, FEEDING)

Diel vertical migration (DVM), or diel horizontal migration in shallow freshwater ecosystems (DHM), are ways in which zooplankton can evade predation, but usually at a cost to (diurnal) feeding, especially for herbivorous zooplankton (Burks et al. 2002, Lampert 2005, Williamson et al. 2011, Pierson et al. 2013). Many zooplankton will remain in darker, deeper waters or in refugia provided by macrophytes during daylight hours to avoid predation by visual predators such as fish. At night, they migrate to the upper layer where phytoplankton are more abundant, and also because warmer temperatures from the surface are beneficial for zooplankton egg development. Migration-related traits influence immediate survival but also growth and reproduction via indirect effects on their feeding and egg development.

Euphausiids can show extremely large daily migrations in marine environment. In lakes, it has been shown that the tendency to display DVM may vary between populations and can represent an adaptive response to previous exposure to predators (Pangle and Peacor 2006). DVM is thus a response trait that can indicate current or previous predation pressure. However, active transport through daily migration can also mediate C and nutrient vertical fluxes through differential feeding, respiration, excretion and egestion along depths, especially in marine environments where DVM can sustain mesopelagic food webs (Kelly et al. 2019); hence, migration can also be used as an effect trait.

1.2.3 Stoichiometric and Physiological Traits

Feeding Rate (FEEDING)

Clearance or grazing rates can be considered as a trait that reflects grazing or feeding rates. Clearance rates are estimated as the volume of water from which a zooplankton removes prey per unit time (Gauld 1951). Although responsive to environmental conditions (e.g., temperature, food availability), feeding rate estimates is likely to be an important effect trait, influencing phytoplankton productivity and biomass as well as nutrient and energy transfer. Sloppy feeding observed in copepods can also supply a fair amount of dissolved nutrients and C, which can not only support phytoplankton but also heterotrophic bacterial growth and other aspects of the C cycle (Jumars et al. 1989, Steinberg and Landry 2017).

Respiration/Metabolic Rates (SURVIVAL/MORTALITY)

Basal metabolic rate, which maintains body tissues, is likely to vary between metazooplankton, but it has been little studied to date. As a trait, this rate is likely to influence longevity and other metabolic functions such as excretion. Respiration rates can be estimated, usually on an individual level, by measuring changes in oxygen concentration in a small chamber where the individual plankter resides. Alternatively, electron transport system (ETS) activity can also reflect respiration in zooplankton (Bode et al. 2013). At rest, this rate represents the basal metabolic rate, otherwise respiration rate can vary under different conditions and can be affected by behaviour as well. Respiration rates are primarily used as an effect trait, given the direct impact of oxygen consumption and CO_2 release in the environment. The magnitude of this effect, however, depends on the relative proportion of zooplankton biomass in a given system.

Excretion and Egestion (SURVIVAL/MORTALITY, FEEDING)

These physiological functions release dissolved (excretion) and particulate (egestion; faecal pellets) compounds, contributing to N, phosphorus (P) and C cycling. Typically, only N and P under the forms of ammonia and phosphate are measured in zooplankton excretion; only very rarely other organic compounds such as urea or amino acids are measured, despite their potential role in stimulating heterotrophic bacterial activity (Arístegui et al. 2014). For faecal pallets—one of the primary components of the so-called "marine snow" aggregates found in oceans (Turner and Ferrante 1979)—C content is most often quantified as a measure of C export in marine systems, but it is not rare that nutrient contents and ratios are reported. Although relatively harder to measure, such traits can exert a strong influence on elemental cycling (including nutrient recycling), C sink/burial (mostly in oceans) and other vertical fluxes (when combined with DVM), bacterial activity, and primary productivity, especially in nutrient-poor systems (Frangoulis et al. 2005, Steinberg and Landry 2017, Cavan et al. 2019).

There are multiple examples of how zooplankton excretion and egestion can be used as effect traits, with studies explicitly quantifying the relative contribution of community-weighted traits to overall C, N and P pools and cycling. For instance, Hernández-León et al. (2008) estimated that the total amount of N-ammonia excreted by mesozooplankton in the oceanic upper layer accounted for approximately 1.78 Gt N per year. In their global analysis, they showed that N inputs derived from mesozooplankton excretion decreased from tropical to polar waters, with the largest contribution to photosynthesis observed in the tropical and subtropical areas, due to temperature-dependence. Globally, nutrient recycling by marine mesozooplankton was estimated to be in the range of 12–23% of the requirements for phytoplankton and bacterial production (Hernández-León et al. 2008). Such contributions can, however, increase in importance in smaller scale assessments; e.g., up to 90% in some marine habitats or even exceeding 100% in oligotrophic freshwater systems (Jawed 1973, Villar-Argaiz et al. 2001).

Elemental or Stoichiometric Body Composition (SURVIVAL/MORTALITY, FEEDING)

Organismal elemental ratios (C:N, C:P or N:P, primarily) are increasingly being referred to as 'stoichiometric traits', with likely influences on taxon co-existence, nutrient recycling and growth rate with regards to nutrient limitation (Elser et al. 2000). As opposed to phytoplankton, the stoichiometry of consumers such as zooplankton is homeostatic, making stoichiometric trait values rather "fixed" or static. These ratios capture body requirements and inform on biochemical composition; for example, daphniids are known to have high P requirements, in part due to their greater nucleic acid content (Sterner and Elser 2002). Preferentially storing an element relative to others will influence the stoichiometry of the excretory products (e.g., P-rich daphniids excrete relatively less P), with direct influence on the nutrient balance available to primary producers in the environment. For taxa that are able to select amongst items captured and consumed, their choice of prey can be influenced by their corporal needs.

While species distribution may reflect the signal of habitat filtering, in that some taxa may preferentially be found in habitats that can meet their body requirements, stoichiometric traits primarily operate as effect traits as they dictate diet preferences (for selective feeders), nutrient ratios in excretory products and nutritional value for zooplankton predators. Using stoichiometric traits in food web models can help track C:N:P flow and ratio across trophic levels (Litchman et al. 2013).

Growth Rate (GROWTH/REPRODUCTION)

The maximum growth rate of an organism will determine the degree to which it can respond to environmental variation and will also determine secondary productivity rates of the zooplankton community. Body size will usually be used as a proxy of this rate (often in conjunction with temperature and resource availability). While growing, crustacean zooplankton can moult, with exuviae release contributing to the pool of particulate organic C sinking in the water column (Steinberg and Landry 2017). Organismal growth is a trait that operates at various levels of organization, and can thus be used both as a response trait (e.g., indicator of stressful conditions, such as in nutrient-depleted environments; "growth rate hypothesis"; Sterner and Elser 2002) and as an effect trait, with direct implication for biomass/C stocks comprised in ecosystems (Hébert et al. 2017).

Longevity (SURVIVAL/MORTALITY)

As outlined in the Litchman et al. (2013) framework, longevity is influenced by the degree to which organisms invest in maintenance and repair relative to growth and reproduction. It is also influenced by the degree to which a zooplankter is exposed to predation (which depends on other traits related to predator evasion) and its starvation tolerance. Survivorship, and thus longevity, has not been studied generally enough to date to be a useful trait to distinguish between communities.

Starvation Tolerance (SURVIVAL/MORTALITY)

The ability to survive periods of low prey availability is also likely to differ among metazooplankton, but has not been well studied generally. When present in a community, this response trait may be indicative of higher resistance to stressful conditions, likely capturing signals of habitat filtering. Where it has been examined, it is usually within the context of surviving winter conditions by some freshwater taxa and could therefore be combined with dormancy as an alternative strategy to survive stressful conditions by reducing metabolism (see "dormancy" trait in next section; Ohman et al. 1998). The accumulation of lipid (wax esters) reserves has been identified as a preparation step for winter dormancy in (mostly marine) copepods (Ohman et al. 1998); in freshwaters, however, pre-winter acquisition of algal lipids can support egg production in active overwintering copepods (Schneider et al. 2017). The composition of the lipids stored prior to winter may also be indicative of overwintering strategies; for example, the freshwater cladoceran *Daphnia* retains high levels of polyunsaturated fatty acids to remain active throughout winter (Mariash et al. 2016).

1.2.4 Life History Traits

Reproduction (GROWTH/REPRODUCTION)

Metazooplankton can display diversity in their reproductive traits. Reproductive mode can vary from asexual to sexual reproduction (including hermaphroditism). Taxa reproducing asexually for most of their life cycle include cladocerans, tunicates, jellyfish (cnidaria) and rotifers while copepods and euphausiids typically reproduce sexually (Paffenhöfer and Harris 1976, Gilbert 1983, Lucas 2000, Decaestecker et al. 2009). Hermaphroditism is common to chaetognaths and some gelatinous metazooplankton like many ctenophores (Chiu 1963, Barnes 1980). Asexual taxa may be advantaged by very high population reproductive rates when resources are plentiful (Gilbert 1983, Decaestecker et al. 2009). Production of dormant stages may occur during harsh periods in certain taxa, which can also be considered a trait (Gilbert 1974).

In terms of investment in embryo development and numbers, some taxa are broadcast spawners (e.g., many euphausiid species), usually when large numbers of eggs containing small embryos are released. Meanwhile, at the other end of the parental investment spectrum, some metazooplankton such as other euphausiid species, cladocerans and copepods carry their eggs for longer, providing greater protection to the smaller number of embryos produced (Hirst and Lopez-Urrutia 2006). These differences in parental investment in metazooplankton also lead to the general trade-off between offspring size and number (r versus K strategies in ecology).

Size at Maturity (GROWTH/REPRODUCTION)

This trait is generally traded-off against reproductive investments: a greater size at maturity represents greater investment in growth relative to reproduction. It will vary across metazooplankton taxa, although a comprehensive comparison across the suite of taxa represented has not been done. Some groups appear to be more plastic in their ability to modify size at maturity relative to reproductive investment (e.g., rotifers) than others (e.g., copepods; Litchman et al. 2013).

Dormancy (SURVIVAL/MORTALITY)

There is variation amongst metazooplankton in their ability to produce dormant resting stages, usually in the form of resistant eggs, enabling survival during periods of low food or sub-optimal environmental conditions. In lakes, some copepods and most cladocerans appear to produce such resting eggs. Cladoceran ephippia may be more resistant than copepods in the long-term, as shown by the greater success of resurrection ecology experiments with cladocerans and the apparently greater capacity for dispersal of cladocerans than copepods (Louette and DeMeester 2005). Copepods and euphausiids appear to be the dominant groups possessing this trait in marine environments. Bdelloid rotifers can undergo cryptobiosis, a process by which they can survive desiccation for months at a time in very dry environments. Traits embedding zooplankton tolerance and adaptive strategies to cope with stressful conditions can generally be used as response traits.

1.3 Underrepresented Aspects of Trait-based Studies

1.3.1 Traits

While trait-based studies of metazooplankton have increased over the last two decades, some traits have been more studied than others. This results from the cost of measurement, data scarcity in the literature or the limited ecological relevance of a given trait for a specific study. In a recent systematic review aiming to determine which functional traits were most used to characterize cladoceran, copepod and rotifer communities in inland waters, the most frequently used trait was body size (used in 65% of retained publications), closely followed by feeding-related traits and habitat type (Gomes et al. 2019). Similarly, these traits were among the ones for which Hébert et al. (2016b) found the widest taxonomic coverage across habitats types. In their freshwater trait synthesis, however, Barnett et al. (2007) pointed to greater data paucity of information on cyclopoid copepod feeding. In their meta-analysis of "effect" traits in crustacean zooplankton, Hébert et al. (2016a) reported that physiological traits such as excretion and respiration rates as well as stoichiometric traits characterizing C, N and P corporeal content had overall relatively lower taxonomic coverage. Their contention was that this reflected that effect traits are harder to measure and were thus systematically scarcer in the literature. However, it was noted that marine species were more represented in the entirety of the physiological and stoichiometric trait data compiled by Hébert et al. (2016a). This likely reflects a greater focus of oceanographic studies towards zooplankton metabolism and stoichiometry, including their biogeochemical consequences.

1.3.2 Taxa

Functional trait-based approaches to zooplankton communities tend to focus on crustaceans, likely due to their disproportionally high abundance and widespread geographical coverage, transcending both marine and freshwater habitats. In their review, Gomes et al. (2019) reported that 67.5%, 55% and 37.5% of freshwater studies focused on cladocerans, copepods and rotifers, respectively. Although restricted to freshwaters, studies of rotifer traits are less common than for crustaceans, with data being of limited availability for many species (Gsell et al. 2016). Increasing taxonomic coverage in rotifer traits would favour their incorporation in trait-based studies, especially for food web interaction models aiming at reducing overall complexity (Merico et al. 2009). Analogously, trait information on gelatinous forms of zooplankton are mostly restrained to marine studies, with very few gelatinous taxa found in freshwaters. While the trait framework is also less applied to such groups of larger zooplankton, some of their trait values, relationships and patterns along environmental gradients have been relatively well documented outside of the functional trait literature. For example, bell diameter to mass ratios in gelatinous zooplankton (e.g., the scyphomedusa *Aurelia aurita*) vary along salinity levels, which can influence individual buoyancy and mobility (Hirst and Lucas 1998). Another example would include the vertical distribution patterns of pteropods as a function of aragonite saturation, owing to the high sensitivity of their shell to acidification (Bednaršek and Ohman 2015). Although likely rare in the literature, several physiological traits such as excretion rates have been measured for gelatinous zooplankton (Ikeda 1985; note that this study includes several groups of zooplankton). Thus, for underrepresented zooplankton groups and taxa in trait-based approaches, a key step would be to systematically synthesize this information, as has been done for crustacean species to provide useable databases, but to also identify important lacunae where further trait estimates are needed.

1.3.3 Habitats and Areas

Similar to the way in which some traits may be more represented in marine or freshwater environments, some sub-habitats within ecosystem types may also be differentially studied. For example, limnological studies that focus on lakes and reservoirs often focus much more attention on pelagic species relative to those living in littoral and benthic zones, as shown in Barnett et al. (2007). Similarly, trait information on pelagic metazooplankton from the upper oceanic layer (including coastal areas) tends to be relatively more documented, as opposed to data from the deep-sea or other spatially sparse and more circumscribed habitats (e.g., mangroves, kelp forests). This reveals an important caveat in trait-based studies, given how strongly sub-habitats can constrain trait expression and diversity. For example, in deep waters, marine gelatinous zooplankton have developed particular physio-behavioural strategies to control their luminescence in the absence of light (Haddock and Case 1999); lower temperatures and resource availability also constrain physiological traits, with depth-related declines in respiration and excretion rates being observed in many deep-dwelling crustaceans, such as amphipods, decapods, euphausiids, mysids, and copepods (Ikeda 2014). Furthermore, there are clear knowledge gaps in the geographical coverage at larger scales, such as in the more diverse tropical freshwater areas (de Oliveira Sodré and Bozelli 2019). For instance, according to the review conducted by Gomes et al. (2019), most trait-based studies were performed in Canada and the U.S., followed by Brazil and Italy. While the number of retained publications may be influenced by the constrained methodologies associated with systematic reviews (i.e., use of specific key words or search engines), this trend reflects previously acknowledged biases in trait data towards other parts of the world (but see Rizo et al. 2017 for a synthesis on Asian cladoceran traits).

1.3.4 Seasons

Finally, another less-studied dimension of trait-based ecology in the study of metazooplankton is the temporal coverage of trait expression and variation, especially in areas where aquatic systems exhibit strong seasonality. While most traits are studied during the so-called growing season, the ecological strategies of organisms thriving in less-studied seasons may be better understood through the study of their traits, especially prior to or during colder and harsher (usually winter) months in temperate areas. For instance, traits related to starvation tolerance or energy storage may allow freshwater metazooplankton to overwinter instead of relying on dormancy to survive the winter months; e.g., as observed in *Daphnia* (Mariash et al. 2016).

1.4 Trait Estimates, Correlations and Trade-Offs

1.4.1 Direct Measurement

Metazooplankton traits have been increasingly measured and reported in recent years. According to Gomes et al. (2019), roughly half of the trait-based studies on freshwater zooplankton to date directly measured observations, whereas the other half used literature-based values. Although using published taxon-specific trait estimates instead of species identity in analyses can help uncover relationships that could not be revealed otherwise, using trait values that have been estimated under different environmental conditions than those under study may mask or bias site-specific ecological linkages. This is especially the case for highly-plastic, stress-related or habitat-constrained traits, where trait values can greatly vary even within species. That said, if the species included in a study span various groups of zooplankton that typically express reasonable variability in the traits of interest (e.g., body size across rotifers, copepods, amphipods), using literature-based estimates may still enable the detection of functional patterns. When assessing traits expressed in closely related species or when it is impossible to directly measure traits, using less-variable (i.e., fixed or static) traits from the literature may be more suitable and easily transportable across study systems (e.g., zooplankton body stoichiometry; see Section 1.6 for a study example by Moody and Wilkinson 2019).

1.4.2 *Estimates and Correlations in Quantitative Traits*

There are still many knowledge gaps, especially for quantitative traits in the literature. Thus, generating more data on underrepresented traits and taxa is an obvious recommendation to help move the field forward, as highlighted in Section 3. However, an alternative approach is the use of soft traits or predictive equations to estimate unmeasured trait values based on known relationships (Nock et al. 2016).

Although less mechanistically correlated with precise functions of interest, more readily measurable soft traits can sometimes be used as surrogate metrics to infer hard traits. Because it transcends multiple organismal functions, the most common soft trait is consistent body size. Allometric equations that describe the scaling relationships between organismal size and individual-level biological variables, such as morphological features or metabolic rates, have been extensively developed for zooplankton (Hébert et al. 2017, Arhonditsis et al. 2019). Developing such equations is a powerful means towards generalizing trait relationships, further facilitating the estimation of unknown trait values. A non-exhaustive, though representative, list of morphological, physiological and life-history allometric relationships commonly used for marine and/or freshwater zooplankton is provided in Table 1 from Arhonditsis et al. (2019).

Mass-length equations are often used to infer body mass based on length given that the latter is much easier to measure than the former, enabling the estimation of overall taxon or community biomass when combined with abundance data. Allometric exponents can differ across taxa, including among crustacean taxonomic sub-groups owing to differences in body shapes; e.g., copepods versus daphniid-like cladocerans versus smaller, rounder cladocerans such as Bosminidae or Chydoridae (Hébert et al. 2016a). Such exponents can also vary considerably within genera or species (McCauley 1984); in some cases, they can even be habitat-specific (see Section 1.5; Hébert et al. 2016a). Care should be taken, however, not to extrapolate trait estimates based on size falling outside of the size spectrum covered by established relationships; McCauley et al. (1984) offer an extensive set of species-specific mass-length equations with the associated size ranges of application. A further step in the use of allometry in ecosystem models is to combine mass-length equations and mass-scaled exponents of physiological traits to estimate a community's contribution to a particular ecosystem level process based on organismal body lengths alone (Hébert et al. 2017).

Well-established relationships between body mass and physiological rates can also help estimate hard trait values. There is a wide variety of studies (mostly on marine zooplankton) providing general models to estimate excretion, respiration, growth, fecundity or clearance rates (Ikeda 1985, Hirst and Lampitt 1998, Ikeda et al. 2001, Barnett et al. 2007, Kiørboe and Hirst 2014, Hébert et al. 2016a). For example, Ikeda et al. (2001) have shown that body mass and temperature alone could explain on average 94%, 77% and 51% of the variation in respiration, ammonia, and phosphate excretion rates, respectively, in epipelagic marine copepods along large latitudinal gradients. When corrected for temperature, metabolic or physiological rates theoretically scale with body mass to the ¾ power (or -¼ on a mass-specific basis). This mass-scaled exponent refers to the general power law in ecology (Kleiber 1961, Peters 1983) and has been derived from large datasets on various animal classes. When experimentally measured on a more constrained group of organisms, however, scaling exponents can considerably diverge from theory, and the use of taxon- and process-specific exponents is recommended. For example, Hébert et al. (2016a) reported exponents of 0.70 and 0.84 for the mass dependency of P and N excretion rates, respectively, of crustacean zooplankton; despite the discrepancy, both mass-scaled exponents are within ranges of values observed in other nutrient excretion studies, with exponents for N excretion being typically higher. Furthermore, the use of scaling exponents derived from very large size spectra to make such estimations should also be avoided, as it includes a greater degree of uncertainty (Glazier 2006).

Stoichiometric traits can also be used to estimate other biological parameters. For excretion and egestion, the balance between corporeal and resource N:P ratios is often used to project nutrient ratios in excreted and egested products (Elser and Urabe 1999). When both respiration and excretion rates are measured, the ratio of oxygen (O) respired to N-ammonia excreted (O:N) can also be used as an indicator of the primary substrates (proteins versus lipids) for metabolism, with that of zooplankton from temperate and subpolar areas being typically more lipid-based due to energy storage in colder regions (Hernández-León et al. 2008).

1.4.3 Trade-Offs

Identifying trait trade-offs can help reveal ecological strategies amongst metazooplankton, with regards to fitness optimization under different environmental conditions. Organismal fitness is achieved by allotting energy amongst different fundamental functions: survival, growth, reproduction and feeding. As pointed out in Section 1.2, traits can be classified into these components, and trade-offs may occur between inter-related traits from different categories.

Strong trait trade-offs related to morphology, behaviour, and feeding can be found across multiple zooplankton taxa, with common examples including trade-offs between resource acquisition/ competition versus defence, feeding versus hiding, prey size versus prey selection (Gliwicz 2003, Yoshida et al. 2003, Kiørboe 2011). Some traits can also be related to strategies under stressful or more harsh environmental conditions, with a trade-off that can be imposed between starvation tolerance (survival) and delayed reproduction, as observed in copepods, jellyfish and pteropods (Litchman et al. 2013). Investment in energy storage or tissue repair may enhance longevity, but this leaves less energy for growth and reproduction. More generally, there is a trade-off between growth and reproductive success (Section 1.2). However, trade-offs can also occur between traits related to a single organismal function, such as growth, with investment in growth rate during early ontogenetic stages likely reducing size at maturity- a trade-off that can be modulated by temperature (Gillooly et al. 2002).

Many trade-offs can be associated with mobility given that swimming increases the risk of encountering both prey and mates, but also predators. While searching for prey or mates is beneficial for feeding and reproduction, respectively, swimming can trigger fluid disturbances in marine copepods that may be detected by rheotactic predators, reducing survival (Kiørboe 2008). Performing vertical migration to evade predators also implies lost feeding opportunities (Kiørboe 2011). Such motility-related trade-offs illustrate well how it is impossible for organisms to simultaneously optimize all fundamental activities.

1.5 Variation in Traits between Freshwater and Marine Environments

Examining trait distribution across aquatic ecosystems can provide valuable insight into the drivers shaping community structure. There are inherent differences between traits from marine and freshwater metazooplankton, in large part owing to habitat constraints. One stark contrast is the much greater taxonomic diversity in marine environments compared to freshwaters, which can lead to the quasi-exclusive presence of certain traits in marine ecosystem type (e.g., metazoan bioluminescence). In contrast, seasonality can often be more pronounced in freshwater relative to marine systems, making some traits relatively more important than others for zooplankton life history strategies, such as traits related to reproduction and dormancy. While there are discrepancies in the amount of information that has been reported in marine versus freshwater studies (see Section 1.3), there are trait values and relationships that can vary considerably between ecosystem types.

1.5.1 Trait Values

As reported in Hébert et al. (2016a), marine crustacean zooplankton have significantly larger body sizes, with 2-fold greater body length and 7-fold greater dry mass than freshwater species. Meanwhile, with respect to corporeal composition, although %C and %N body content seem to be similar across marine and freshwater crustaceans, %P and N:P ratios are systematically higher and lower, respectively, in freshwater species. This is in large part due to the known high P requirements in freshwater cladocerans, especially daphniids. Analogously, N:P ratios in the excreted products of freshwater cladocerans, especially for daphniids, are expected to be much higher than any marine taxa (Sterner and Elser 2002). In terms of physiological traits, based on the data compilation provided in Hébert et al. (2016b), mass-scaled rates of respiration and N and P excretion in crustaceans are all higher (3x, 2.5x and 5.5x, respectively) in freshwater representatives, suggesting overall higher metabolic activity in freshwater zooplankton likely due to environmental constraints (e.g., food supply, higher temporally constrained seasons).

1.5.2 Trait Relationships

When comparing allometry in crustaceans, Hébert et al. (2016a) showed that marine species may be heavier at equivalent body length, and may also gain more mass as they increase in length than do those in freshwater habitats. This pattern was not only found across all taxa, but also within taxa, with a 4-fold difference between marine and freshwater calanoids. Because of this, variation in general body shape of dominant taxa (predominantly copepods in marine ecosystems and relatively more cladocerans in freshwaters) cannot explain this pattern, nor can phylogeny exclusively; rather, this could be hypothetically due to habitat-constrained adaptive differences such as the need to control buoyancy in saltwater when performing vertical migration for example. Irrespective of causes, few, if any, studies have compared crustacean body tissue density between these major aquatic ecosystems.

Another striking difference between ecosystem types is the 3-fold higher mass-specific respiration rates in freshwater zooplankton (Hébert et al. 2016a). A similar observation can be made when inspecting the observed versus predicted values of crustacean respiration in the study by Hernández-León and Ikeda (2005) in which observed respiration rates in marine copepods were systematically lower than that predicted by their general model based on body mass and temperature alone. While this divergence in respiration may require further investigation before confirming any pattern, potential ecological explanations could include greater cost associated with osmoregulation or life-history strategies related to differential investment in growth and reproduction. Potentially more productive and seasonally constrained freshwater habitats may induce higher metabolic activity in zooplankton, in contrast to marine environments where food supply can sometimes be consistently scarcer.

1.6 Functional Diversity in Metazooplankton Communities

Once a set of traits can be attributed to a community of plankton, it is possible to investigate the functional ecology of this community in greater detail. The advantage of a trait-based approach is that it brings mechanistic understanding to our examination of communities, including how their ecological role may vary through time and space. For example, Barnett and Beisner (2007) used their trait matrix for eastern North American freshwater zooplankton to examine how functional trait diversity varied along nutrient (P) gradients in lakes. The trait-based approach demonstrated a linear decline in functional diversity with lake enrichment, even though taxonomic diversity (species richness) demonstrated the expected unimodal (hump-shaped) relationship. This difference suggests that oligotrophic lakes, whilst having relatively low species richness, possess communities with very high functional diversity, enabling these communities to harness resources and use survival mechanisms in a variety of ways. On the other hand, nutrient-enriched lakes, also showing low species richness, were made up of relatively similar organisms with respect to their traits, and thus less likely to be resilient to future perturbations to the ecosystem should they arise. This novel hindsight could only be gained through a trait-based approach.

There is now a wide variety of zooplankton traits that have been identified, with trait data sets being increasingly made available, as pointed out in this chapter. However, all traits may not be equally useful to derive functional diversity indices, especially if the goal is to determine relationships between such indices and particular ecological parameters (e.g., related to ecosystem functioning). Therefore, care should be taken when choosing traits to consider in a study (see Section 1.7; Petchey and Gaston 2002). Further, if traits are not directly observed on organisms but are rather literature-based taxon-specific values, static/fixed traits may be more suitable and easily transposable to various study systems given that they may be less responsive to environmental variability (Section 1.4). One example of relatively fixed traits is the stoichiometric composition of zooplankton bodies. For instance, Moody and Wilkinson (2019) used stoichiometric traits compiled by Hébert et al. (2016b) to examine functional and taxonomic diversity variation in lake zooplankton communities along nutrient gradients. In accordance with Barnett and Beisner (2007), they also found that functional dispersion decreased with P-load, but using literature-based, fixed, stoichiometric traits. Their study showed that stoichiometric trait distributions in lake zooplankton shifted with eutrophication, with high N:P ratio organisms increasing in abundance with lake P-levels. Higher dominance of P-poor individuals also indicate that zooplankton communities may have exacerbated the high supply of P in hypertrophic lakes by excreting more P relative to N, suggesting a potential contribution of zooplankton to the eutrophication process in lakes.

Generally, functional diversity has been used in two main contexts. The first is in the context of what has been entitled "assembly rules" in ecology, i.e., what are the patterns in community composition, and what are the environmental factors that affect these. Barnett and Beisner (2007) and Vogt et al. (2013) provide examples of this use of functional trait diversity applied to zooplankton across a landscape of lakes. As discussed in Section 1.2, traits in this context are mainly considered to be response traits with respect to the environment or habitat constraints. The second context is geared towards ecosystem level functional consequences, within which traits are considered to reflect what organisms in a community do in order to affect ecosystem stocks and fluxes (effect traits).

With respect to the response-effect trait framework, the idea is that functional diversity indices capture complementarity and redundancy in a community, and should allow for greater explanatory power with respect to the role of organisms in ecosystems than does the sole number of species (richness), as has frequently been used in diversity–ecosystem functioning relationships. In some cases, species richness and functional diversity can be linearly related (Box 1.1; Fig. 1.1A), while communities that are composed of functionally redundant species exhibit lower functional diversity regardless of the number of species (Box 1.1; Fig. 1.1B), which may result in reduced ecosystem processing if indices are based on effect traits (Box 1.1; Fig. 1.1C versus 1.1D). Particularly relevant in the face of environmental change, greater functional diversity in response traits may allow overall communities to persist despite species loss or sorting, thereby favouring stability (Box 1.1; Fig. 1.2A), whereas functional redundancy in effect traits may exert stabilizing effects on ecosystem processes (Box 1.1; Fig. 1.2B). Functionally equivalent species in terms of their contribution to ecosystem processes (i.e., low to moderate effect trait diversity; functional redundancy) can compensate for the loss of other species (insurance hypothesis) or dampen the effects of species abundance fluctuations (portfolio effect; Box 1.1; Fig. 1.1E versus 1.1F), hence providing greater ecosystem stability (Fig. 1.2B; Rosenfield 2002, Thibaut and Connolly 2013). In this context, assessing the functional

Box 1.1: Relationships between functional diversity, species richness, ecosystem processes and ecosystem stability.

In biodiversity research, it is often assumed that functional diversity is linearly related to ecosystem processing (Fig. 1.1C-D). The linearity of this relationship, however, is independent of that between taxonomic and functional diversity (Fig. 1.1A-B), the latter being entirely dictated by community composition. For a given (fixed) number of species, a set of functionally complementary traits results in greater functional diversity indices (Fig. 1.1A), while functionally redundant traits translate into lower functional diversity indices (Fig. 1.1B), with potentially relatively less ecosystem processing achieved (Fig. 1.1D). Differential relationships between taxonomic and functional diversity (reflecting functional complementarity or redundancy; Fig. 1.1A versus 1.1B, respectively) can be indicative of how species richness may influence ecosystem processes (Fig. 1.1E versus 1.1F, respectively). Regardless of the degree to which ecosystem processes are realized (Y axis in Fig. 1.1C-D-E-F), if the response of a community exposed to environmental changes or stressors is the loss of species, then the supposedly negative impact on ecosystem processes will be greater in communities with high functional complementarity (Fig. 1.1E) relative to those composed of functionally redundant species (Fig. 1.1F). Diminished impacts of species loss on ecosystem processing through functional redundancy, however, implies that functional diversity is evaluated on the basis of effect traits that regulate key ecosystem control functions.

   A common assumption in community ecology is that communities exhibiting a high diversity of response traits may persist under changing or stressful conditions, with assembly mechanisms regulating fluctuations in relative abundances of species based on their response trait optimum ranges. For example, greater functional diversity in a pool of response traits related to thermal affinities or tolerance to starvation or hypoxia may allow community biomass to remain stable despite a change in species relative abundances or richness. From that perspective, greater functional complementarity in response traits may confer greater stability (Fig. 1.2A). Species sorting or loss may, however, be detrimental to ecosystem processing, if key functions can no longer or can only be partly fulfilled due to the loss or reduced presence of some effect traits (e.g., reduced grazing pressure may alter water clarity and thus quality). In such cases, functional redundancy in effect traits may help maintain a certain level of ecosystem processing, with functionally equivalent species compensating for the loss (insurance hypothesis) or lower abundance (portfolio effect) of other species (Rosenfield 2002, Thibaut and Connolly 2013); communities with low or moderate effect trait diversity may thus exert stabilizing effects on ecosystems (Fig. 1.2B). In this context, zooplankton communities with a high diversity of response traits (Fig. 1.2A) but a relatively moderate diversity of effect traits (Fig. 1.2B) may be the most beneficial for the stability and resilience of aquatic ecosystems.

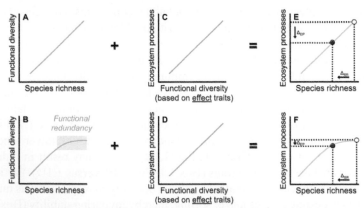

Figure 1.1: Predicted relationships between species richness, functional diversity and ecosystem processes based on biodiversity–ecosystem functioning principles. Panels from the first row (A-D-E) versus the second row (B-D-F) refer to different scenarios of community composition: greater functional complementarity (row 1; A) versus greater functional redundancy (row 2; B). Together, these scenarios indicate how the presumably negative impact of species loss (reducing species richness; transition from the white to the red circle on the horizontal axis; Δ_{SR}) on ecosystem processes (reducing ecosystem processing; transition from the white to the red circle on the vertical axis; Δ_{EP}) can be mediated by functional redundancy in effect trait diversity (B-F; as opposed to A-E) (Adapted from Petchey and Gaston 2006).

Box 1.1 contd. ...

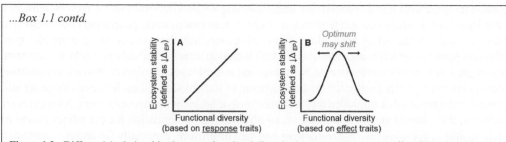

Figure 1.2: Differential relationships between functional diversity in response (A) versus effect (B) traits and ecosystem stability. Ecosystem stability is defined as the magnitude of change in ecosystem processes (Δ_{EP}), with small to no changes in ecosystem processing associated with greater stability. Note that for the relationship between functional effect trait diversity and ecosystem stability (B), the optimum value may be context-dependant, potentially resulting in either hump-shaped, left-skewed or right-skewed relationships.

diversity of species effect traits can not only provide insight into the role of communities in ecosystem functioning, but also how they may confer stability and resilience at higher levels of organization (McNaughton 1977, Hooper et al. 2005).

Over the past decade, a plethora of indices have been developed to estimate functional diversity based on accentuating the various components of diversity (richness and evenness or their combination) to differing degrees. Schleuter et al. (2010) gives an excellent overview of the most commonly used functional trait diversity indices, including those of Villéger et al. (2008): functional richness (FRic), functional evenness (FEve), functional divergence (FDiv) and functional diversity FD. To that, the functional dispersion (FDis) indicator developed by Laliberté and Legendre (2010) can be added. The FD package in R that accompanied the publication of Villéger et al. (2008) can be used to estimate these common diversity indices. In a general sense, functional diversity indices reflect different methods to estimate the way in which the composition of a community statistically "fills" a multivariate trait space: in degree (how much trait-space is covered by different traits; analogous to richness), type (which traits are most represented; analogous to evenness) or both (analogous to diversity).

In addition to constraining the set of traits used to those that reflect the ecological question of interest, another important consideration when calculating functional diversity is the number of traits to include (Petchey and Gaston 2006). Even if multiple traits could be considered relevant to a study, there is a natural limit as including too many traits will reduce any estimate of diversity back to the same as species richness: each species has a unique set of traits when too many are considered. Alternatively, one can try all combinations of traits to search for the best overall combination that corresponds to the gradient of interest, as done in Barnett and Beisner (2007)—analogous to using variable transformations to find the best regression for a study and maximize the variation explained in a dataset. Finally, it is also possible to weigh traits differently such that traits that are highly correlated, for example, can be down-weighted relative to more distinct traits that are not correlated. More general or theoretical publications further elaborate on the steps to follow when using multiple traits in functional diversity research (Lefcheck et al. 2015).

1.7 Remarks and Recommendations for Future Trait-based Approaches

Akin to other subfields of ecology and organism types, the increasing interest in trait-based approaches to describe metazooplankton has fostered a multi-scale framework using species responses and effects within a tropho-dynamic and evolutionary context, with regard to ecological forecast under scenarios of global change (Diaz et al. 2007). The growing body of work on metazooplankton trait distributions along various environmental gradients (e.g., of human activity, or latitude/temperature) has advanced our understanding of how environmental factors influence community assembly processes and feedback with ecosystem functioning, generating knowledge that can then be used to make future predictions. Generating more trait data would improve our ability to quantify and parametrize intra- and inter-specific trait variation, correlation and trade-offs in ecosystem models of static or dynamic environments. Syntheses and assemblages of comprehensive trait databases in recent years have greatly facilitated the development of trait-based studies; however, there are still many gaps in the reported or modelled trait information available for metazooplankton. As previously advocated in this chapter and in other studies, an obvious recommendation for the advancement of this field would be for more trait measurements and data disclosure, especially for under-represented traits, taxa, habitat types and areas. Then, developing, calibrating, and validating modelling techniques to infer trait values based on phylogeny (Bruggemann 2011), known trait relationships or response to particular environmental factors such as temperature (Robson et al. 2018) can help facilitate the implementation of functional traits into zooplankton community characterizations.

As the field of trait-based ecology moves forward, a notional, though not trivial, step that should be reinforced when deriving trait-based community descriptors is to meaningfully consider which traits to incorporate in a study. Although often constrained by data availability or the cost and time associated with acquiring new observations, the selection of traits should primarily rely on study objectives. Identifying which traits are most relevant remain a challenge in ecology; for example, choosing which traits should be included in biodiversity-ecosystem functioning relations can be difficult, given that many traits can interact or synergistically influence different aspects of ecosystem functioning (Hooper et al. 2005). In such cases, using traits that are not directly related to the ecological parameters of interest may introduce bias into analyses, hampering the detection or quantification of ecological linkages. Furthermore, clearly defining goals with respect to the rationale for the choice of traits may facilitate trait selection for future studies addressing similar questions, enabling more direct comparisons of how particular trait assemblages may vary in different contexts.

The integrative nature of trait-based approaches has thus far contributed to consolidate ecological information on metazooplankton from various studies that are largely outside of the functional ecology literature. In addition to permitting comparative assessments of metazooplankton features across types of aquatic environments, the use of traits can also help to gain a more generalized understanding of how whole food webs are structured, from the base of the food web to top consumers. For example, already documented information on diet variety, food size ranges or the frequency distribution of ingested food particles could be included in food web interaction models (Gsell et al. 2016). Incorporating stoichiometric traits of consumers and prey when representing food web interactions can also provide insight as to how elemental stocks are structured and flow across trophic levels (Sterner and Elser 2002). In ecosystem models, the inclusion of feeding and stoichiometric traits, along with physiological trait measurements such as growth, respiration or excretion rates would advance our ability to quantify organismal contributions to ecosystem storage and fluxes. Ultimately then, consideration of zooplankton communities by their functional traits can help better characterize the roles, with respect to other environmental drivers, of organisms in aquatic food webs and in ecosystem processing, including biogeochemical cycling.

References

Arhonditsis, G.B., Y. Shimoda and N.E. Kelly. 2019. Allometric theory: extrapolations from individuals to ecosystems. Encyclopedia of Ecology (2nd Edition), Elsevier 2: 242–255.

Aristegui, J., C.M. Duarte, I. Reche and J.L. Gomez-pinchetti. 2014. Krill excretion boosts microbial activity in the Southern Ocean. PLoS ONE 9: e89391.

Barneche, D., M. Kulbicki, S. Floeter, A. Friedlander, J. Maina and A.P. Allen. 2014. Scaling metabolism from individuals to fish communities at broad spatial scales. Ecol. Lett. 17: 1067–1076.

Barnes, R.D. 1980. Invertebrate Zoology. Philadelphia: W. B. Saunders Coll.

Barnett, A.J. and B.E. Beisner. 2007. Zooplankton biodiversity and primary productivity: explanations invoking resource abundance and distribution. Ecology 88: 1675–1686.

Barnett, A.J., K. Finlay and B.E. Beisner. 2007. Functional diversity of crustacean zooplankton communities: towards a trait-based classification. Freshwater Biol. 52: 796–813.

Bednaršek, N. and M.D. Ohman. 2015. Changes in pteropod distributions and shell dissolution across a frontal system in the California Current System. Mar. Ecol. Prog. Ser. 523: 93–103.

Benedetti, F., S. Gasparini and S.-D. Ayata. 2015. Identifying copepod functional groups from species functional traits. J. Plankton Res. 38: 159–166.

Bode, M., A. Schukat, W. Hagen and H. Auel. 2013. Predicting metabolic rates of calanoid copepods. J. Exp. Mar. Biol. Ecol. 444: 1–7.

Boltovskoy, D. and N.M. Correa. 2021. Planktonic shelled protists (Foraminifera and Radiolaria Polycystina): Global biogeographic patterns in the surface sediments. pp. 119–141. *In*: M.A. Teodósio and A.B. Barbosa [eds.]. Zooplankton Ecology. CRC Press.

Bruggeman, J. 2011. A phylogenetic approach to the estimation of phytoplankton traits. J. Phycol. 47: 52–65.

Brun, P., M.R. Payne and T. Kiørboe. 2016. Trait biogeography of marine copepods—an analysis across scales. Ecol. Lett. 19: 1403–1413.

Burks, R.L., D.M. Lodge, E. Jeppesen and T.L. Lauridsen. 2002. Diel horizontal migration of zooplankton: costs and benefits of inhabiting the littoral. Freshwater Biol. 47: 343–365.

Burrows, M.T., A.E. Bates, M.J. Costello, M. Edwards, G.J. Edgar, C.J. Fox et al. 2019. Ocean community warming responses explained by thermal affinities and temperature gradients. Nat. Clim. Change doi.org/10.1038/s41558-019-0631-5.

Buskey, E.J., P.H. Lenz and D.K. Hartline. 2002. Escape behavior of planktonic copepods to hydrodynamic disturbances: high speed video analysis. Mar. Ecol. Prog. Ser. 235: 135–146.

Buskey, E.J., P.H. Lenz and D.K. Hartline. 2011. Sensory perception, neurobiology, and behavioral adaptations for predator avoidance in planktonic copepods. Adapt. Behav. 20: 57–66.

Cavan, E.L., A. Belcher, A. Atkinson, S.L. Hill, S. Kawaguchi, S. McCormack et al. 2019. The importance of Antarctic krill in biogeochemical cycles. Nat. Commun. 10: 4742.

Chen, B., S.L. Smith and K.W. Wirtz. 2018. Effect of phytoplankton size diversity on primary productivity in the North Pacific: trait distributions under environmental variability. Ecol. Lett. 22: 56–66.

Chiu, S.Y. 1963. On the metamorphosis of the ctenophore *Ocyropsis crystallina* (Rang) from Amoy. Acta Zool. Sinica 15: 10–16.

Conley, K.R., F. Lombard and K.R. Sutherland. 2018 Mammoth grazers on the ocean's minuteness: a review of selective feeding using mucous meshes. Proc. R. Soc. B 285: 20180056.

Costello, J.H., S.P. Colin, B.J. Gemmell, J.O. Dabiri and K.R. Sutherland. 2015. Multi-jet propulsion organized by clonal development in a colonial siphonophore. Nat. Commun. 6: 8158.

Cummins, K.W. and M.J. Klug. 1979. Feeding ecology of stream invertebrates. Annu. Rev. Ecol. Evol. S. 10: 147–172.

Daufresne, M., K. Lengfellner and U. Sommer. 2009. Global warming benefits the small in aquatic ecosystems. P. Natl. Acad. Sci. USA 106: 12788–12793.

Decaestecker, E., L. de Meester and J. Mergeay. 2009. Cyclical parthenogenesis in *Daphnia*: Sexual versus asexual reproduction. pp. 295–316. *In*: Lost Sex: The Evolutionary Biology of Parthenogenesis. Springer.

DeMott, W.R. and K.W. Kerfoot. 1982. Competition among cladocerans: nature of the interaction between *Bosmina* and *Daphnia*. Ecology 63: 1949–1966.

DeMott, W. and M.D. Watson. 1991. Remote detection of algae by copepods: responses to algal size, odours, and motility. J. Plankton Res. 13: 1203–1222.

Díaz, S., S. Lavorel, F. de Bello, F. Quétier, K. Grigulis and M.T. Robson. 2007. Incorporating plant functional diversity effects in ecosystem service assessements. P. Natl. Acad. Sci. USA 104: 20684–20689.

Diéguez, M. and E. Balseiro. 1998. Colony size in *Conochilus hippocrepis*: defensive adaptation to predator size. *In*: E. Wurdak, R. Wallace and H. Segers [eds.]. Rotifera VIII: A Comparative Approach. Developments in Hydrobiology, Vol. 134. Springer, Dordrech.

Elser, J.J. and J. Urabe. 1999. The stoichiometry of consumer–driven nutrient recycling: theory, observations, and consequences. Ecology 80: 735–751.

Elser, J.J., R.W. Sterner, E. Gorokhova, W.F. Fagan, T.A. Markow, J.B. Cotner et al. 2000. Biological stoichiometry from genes to ecosystems. Ecol. Lett. 3: 540–550.

Frangoulis, C., E.D. Christou and J.H. Hecq. 2005. Comparison of marine copepod outfluxes: nature, rate, fate and role in the carbon and nitrogen cycles. Adv. Mar. Biol. 47: 254–309.

Gauld, D.T. 1951. The grazing rate of planktonic copepods. J. Mar. Biol. Ass. U.K. 29: 695–706.

Gilbert, J.J. 1974. Dormancy in rotifers. Trans. Am. Microsc. SOC. 93: 490–513.

Gilbert, J.J. 1980. Observations on the susceptibility of some Protists and rotifers to predation by *Asplanchna Girodi*. *In*: H.J. Dumont and J. Green [eds.]. Rotatoria. Developments in Hydrobiology, Vol. 1 Springer, Dordrecht.

Gilbert, J.J. 1983. Rotifera. pp. 181–209. *In*: K.G. Adiyodi and R.G. Adiyodi [eds.]. Reproduction Biology of Invertebrate: I. Oogenesis, Ovoposition and Oosorption. John Wiley & Sons, New York, NY.

Gilbert, J.J. 1985. Escape response of the rotifer *Polyarthra*: a high-speed cinematographic analysis. Oecologia 66: 322–331.

Gillooly, J.F., E.L. Charnov, G.B. West, V.M. Savage and J.H. Brown. 2002. Effects of size and temperature on developmental time. Nature 417: 70–73.

Glazier, D.S. 2006. The 3/4-power law is not universal: evolution of isometric, ontogenetic metabolic scaling in pelagic animals. BioScience 56: 325–332.

Gliwicz, Z.M. 2003. Between hazards of starvation and risk of predation: The ecology of offshore animals. International Ecology Institute, Nordbünte, Germany.

Gomes, L.F., H.R. Pereira, A.C.A. Missias Gomes, M. Carvalho Vieira, P. Ribeiro Martins, I. Roitman et al. 2019. Zooplankton functional-approach studies in continental aquatic environments: a systematic review. Aquat. Ecol. 53: 191–203.

Grosbois, G. and M. Rautio. 2018. Active and colorful life under lake ice. Ecology 99: 752–754.

Gsell, A., D. Özkundakci, M.P. Hébert and R. Adrian. 2016. Quantifying change in pelagic plankton network stability and topology based on empirical long-term data. Ecol. Indic. 65: 76–88.

Haddock, S.H.D. 2007. Comparative feeding behavior of planktonic ctenophores. Integr. Comp. Biol. 47: 847–853.

Haddock, S. and J. Case. 1999. Bioluminescence spectra of shallow and deep-sea gelatinous zooplankton: ctenophores, medusae and siphonophores. Mar. Biol. 133: 571–582.

Hébert, M.P., B.E. Beisner and R. Maranger. 2016a. A meta-analysis of zooplankton functional traits influencing ecosystem function. Ecology 97: 1069–1080.

Hébert, M.P., B.E. Beisner and R. Maranger. 2016b. A compilation of quantitative functional traits for marine and freshwater crustacean zooplankton. Ecology 97: 1081.

Hébert, M.P., B.E. Beisner and R. Maranger. 2017. Linking zooplankton communities to ecosystem functioning: toward an effect-trait framework. J. Plankton Res. 39: 3–12.

Hernández-León, S. and T. Ikeda. 2005. Zooplankton respiration. pp. 57–82. *In*: P.A. del Giorgio and P. le B. Williams [eds.]. Respiration in Aquatic Ecosystems. Oxford University Press, Oxford, UK.

Hernández-León, S., C. Fraga and T. Ikeda. 2008. A global estimation of mesozooplankton ammonium excretion in the open ocean. J. Plankton Res. 30: 577–585.

Herring, P.J. 1985. Bioluminescence in the Crustacea. J. Crustacean Biol. 5: 557–573.

Hinde, R.T. 1998. The Cnidaria and Ctenophora. pp. 28–57. *In*: D.T. Anderson [ed.]. Invertebrate Zoology. Oxford University Press.

Hirst, A.G. and R.S. Lampitt. 1998. Towards a global model of *in situ* weight-specific growth in marine planktonic copepods. Mar. Biol. 132: 247–257.

Hirst, A.G. and C.H. Lucas. 1998. Salinity influences body weight quantification in the scyphomedusa *Aurelia aurita*: important implications for body weight determination in gelatinous zooplankton. Mar. Ecol. Prog. Ser. 165: 259–269.

Hirst, A. and A. Lopez-Urrutia. 2006. Effects of evolution on egg development time. Mar. Ecol. Prog. Ser. 326: 29–35.

Hodgson, J.G., P.J. Wilson, R. Hunt, J.P. Grime and K. Thompson. 1999. Allocating C-S-R plant functional types: a soft approach to a hard problem. Oikos 85: 282–294.

Hooper, D.U., F.S. Chapin, J.J. Ewel, A. Hector, P. Inchausti, S. Lavorel et al. 2005 Effects of biodiversity on ecosystem functioning: a consensus of current knowledge. Ecol. Monogr. 75: 3–35.

Hylander, S., N. Larsson and L.A. Hansson. 2009. Zooplankton vertical migration and plasticity of pigmentation arising from simultaneous UV and predation threats. Limnol. Oceanogr. 54: 483–491.

Ikeda, T. 1985. Metabolic rates of epipelagic marine zooplankton as a function of body mass and temperature. Mar. Biol. 85: 1–11.

Ikeda, T., Y. Kanno, K. Ozaki and A. Shinada. 2001. Metabolic rates of epipelagic marine copepods as a function of body mass and temperature. Mar. Biol. 139: 587–596.

Ikeda, T. 2014. Respiration and ammonia excretion by marine metazooplankton taxa: synthesis toward a global-bathymetric model. Mar. Biol. 161: 2753–2766.

Jawed, M. 1973. Ammonia excretion by zooplankton and its significance to primary productivity during summer. Mar. Biol. 23: 115–120.

Jimérez-Guri, E., H. Philippe, B. Okamura and P.W. Holland. 2007. *Buddenbrockia* is a Cnidarian Worm. Science 317: 116–118.

Johnsen, S. 2005. The red and the black: Bioluminescence and the color of animals in the deep sea. Integr. Comp. Biol. 45: 234–246.

Jumars, P.A., D.L. Penry, J.A. Baross, M.J. Perry and B.W. Frost. 1989. Closing the microbial loop: dissolved carbon pathway to heterotrophic bacteria from incomplete ingestion, digestion and absorption in animals. Deep-Sea Res. 36: 483–495.

Kabata, Z. 1982. Copepoda (Crustacea) parasitic on fishes: problems and perspectives. Adv. Parasit. 19: 1–71.

Keck, B.P., Z.H. Marion, D.J. Martin, J.C. Kaufman, C.P. Harden, J.S. Schwartz et al. 2014. Fish functional traits correlated with environmental variables in a temperate biodiversity hotspot. PLoS ONE 9: e93237.

Kelly, T.B., P.C. Davison, R. Goericke, M.R. Landy, M.D. Ohman and M.R. Stukel. 2019. The importance of mesozooplankton diel vertical migration for sustaining a mesopelagic food web. Front. Mar. Sci. Doi: 10.3389/fmars.2019.00508.

Kenitz, K.M., A.W. Visser, M.D. Ohman, M.R. Landry and K.H. Andersen. 2018. Community trait distribution across environmental gradients. Ecosystems 22: 968–980.

Kiørboe, T. 2008. Optimal swimming strategies in mate-searching pelagic copepods. Oecologia 155: 179–192.

Kiørboe, T. 2011. How zooplankton feed: mechanisms, traits and trade-offs. Biol. Rev. 86: 311–339.

Kiørboe, T. and A.G. Hirst. 2014. Shifts in mass scaling of respiration, feeding, and growth rates across life-form transitions in marine pelagic organisms. Am. Nat. 183: E118–E130.

Kleiber, M. 1961. The Fire of Life: An Introduction to Animal Energetics. Wiley, New York, USA.

Laliberté, E. and P. Legendre. 2010. A distance-based framework for measuring functional diversity from multiple traits. Ecology 91: 299–305.

Lampert, W. 2005. Vertical distribution of zooplankton: density dependence and evidence for an ideal free distribution with costs. BMC Biol 3: 10 Doi: 10.1186/1741-7007-3-10.

Lankford, T.E., J.M. Billerbeck and D.O. Conover. 2001. Evolution of intrinsic growth and energy acquisition rates. II. Trade-offs with vulnerability to predation in *menidia menidia*. Evolution 55: 1873–1881.

Lavorel, S. and E. Garnier. 2002. Predicting changes in community composition and ecosystem functioning from plant traits: revisiting the Holy Grail. Funct. Ecol. 16: 545–556.

Lefcheck, J.S., V.A. Bastazini and G. Griffin. 2015. Choosing and using multiple traits in functional diversity research. Environ. Conserv. 42: 104–107.

Litchman, E. and C.A. Klausmeier. 2008. Trait-based community ecology of phytoplankton. Annu. Rev. Ecol. Evol. S. 39: 615–639.

Litchman, E., M.D. Ohman and T. Kiørboe. 2013. Trait-based approaches to zooplankton communities. J. Plankton Res. 35: 473–484.

Litchman, E., P. de Teznos Pinto, K.F. Edwards, C.A. Klausmeier, C.T. Kremer and M.K. Thomas. 2015. Global biogeochmical impacts of phytoplankton: a trait-based perspective. J. Ecol. 103: 1384–1396.

Louette, G. and L. de Meester. 2005. High dispersal capacity of cladoceran zooplankton in newly founded communities. Ecology 86: 353–359.

Lucas, C.H. 2000. Reproduction and life history strategies of the common jellyfish, *Aurelia aurita*, in relation to its ambient environment. *In*: J.E. Purcell, W.M. Graham and H.J. Dumont [eds.]. Jellyfish Blooms: Ecological and Societal Importance. Developments in Hydrobiology, Vol. 155. Springer, Dordrecht.

Mackie, G.O. 1986. From aggregates to integrates: physiological aspects of modularity in colonial animals. Phil. Trans. R. Soc. Lond. B 313: 175–196.

Madin, L.P. 1990. Aspects of jet propulsion in salps. Can. J. Zool. 68: 765–777.

Mauchline, J. 1980. The biology of mysids and euphausiids. Adv. Mar. Biol. 18: 373–623.

Mauchline, J. 1998. The biology of calanoid copepods. Adv. Mar. Biol. 33: 1–530.

Mariash, H.L., M. Cusson and M. Rautio. 2016. Fall composition of storage lipids is associated with the overwintering strategy of *Daphnia*. Lipids 52: 83–91.

McCauley, E. 1984. The estimation of the abundance and biomass of zooplankton in samples: a manual on methods for the assessment of secondary production in fresh waters. pp. 228–265. *In*: J.A. Downing and F.H. Rigler [eds.]. A Manual on Methods for the Assessment of Secondary Production in Freshwaters. Rigler, editors. IBP Handbook 17, Second edition. Blackwell Scientific Publications, Oxford, UK.

McNaughton, S.J. 1977. Diversity and stability of ecological communities: A comment on the role of empiricism in ecology. Am. Nat. 111: 515–525.

Merico, A., J. Bruggeman and K. Wirtz. 2009. A trait-based approach for downscaling complexity in plankton ecosystem models. Ecol. Model. 220: 3001–3010.

Michels, J., J. Vogy and S.N. Gorb. 2012. Tools for crushing diatoms—opal teeth in copepods feature a rubber-like bearing composed of resilin. Sci. Rep. 2: 465.

Michels, J. and S.N. Gorb. 2015. Mandibular gnathobases of marine planktonic copepods—feeding tools with complex micro- and nanoscale composite architectures. Beilstein J. Nanotech. 6: 674–685.

Moody, E. and G. Wilkinson. 2019. Functional shifts in lake zooplankton communities with hypereutrophication. Freshwater Biol. 64: 608–616.

Nielsen, J.M., E.L. Clare, B. Hayden, M.T. Brett and P. Kratina. 2017. Diet tracing in ecology: Method comparison and selection. Methods Ecol. Evol. 9: 278–291.

Nock, C.A., R.J. Vogt and B.E. Beisner. 2016. Functional Traits. eLS. John Wiley & Sons, Ltd., Chichester. Doi: 10.1002/9780470015902.a0026282.

Ohman, M.D., A.V. Drits, M.E Clarke and S. Plourde. 1998. Differential dormancy of co-occurring copepods. Deep-Sea Res. Pt II 45: 1709–1740.

Ohman, M.D. and J.B. Romagnan. 2016. Nonlinear effects of body size and optical attenuation on diel vertical migration by zooplankton. Limnol. Oceanogr. 61: 765–770.

Paffenhöfer, G.A. and R.P. Harris. 1976. Feeding, growth and reproduction of the marine planktonic copepod *Pseudo-Calanus Elongatus Boeck*. J. Mar. Biol. Ass. UK 56: 327–344.

Pangle, K.L. and S.D. Peacor. 2006. Behavioral response of *Daphnia mendotae* to the invasive predator *Bythotrephes longimanus* and consequent nonlethal effect on growth. Freshwater Biol. 51: 1070–1078.

Patoine, A., B. Pinel-Alloul and E.E. Prepas. 2002. Influence of catchment deforestation by logging and natural forest fires on crustacean community size structure in lakes of the Eastern Boreal Canadian forest. J. Plankton Res. 24: 601–616.

Petchey, O.L. and K.J. Gaston. 2002. Functional diversity (FD), species richness, and community composition. Ecol. Lett. 5: 402–411.

Petchey, O.L. and K.J. Gaston. 2006. Functional diversity: back to basics and looking forward. Ecol. Lett. 9: 741–758.

Peters, R. 1983. The Ecological Implications of Body Size. Cambridge University Press, NY, USA.

Pethybridge, H.R., C.A. Choy, J.J. Polovina and E.A. Fulton. 2018. Improving marine ecosystem models with biochemical tracers. Annu. Rev. Mar. Sci. 10: 199–228.

Pierson, J.J., B.W. Frost and A.W. Leising. 2013. Foray foraging behavior: seasonally variable, food-driven migratory behavior in two calanoid copepod species. Mar. Ecol. Prog. Ser. 475: 49–64.

Poff, N.L., J.D. Olden, N.K. Vieira, D.S. Finn, M.P. Simmons and B.C. Kondratieff. 2006. Functional trait niches of North American lotic insects: traits-based ecological applications in light of phylogenetic relationships. J. N. Am. Benthol. Soc. 25: 730–755.

Pomerleau, C., A.R. Sastri and B.E. Beisner. 2015. Evaluation of functional trait diversity for marine zooplankton communities in the Northeast subarctic Pacific Ocean. J. Plankton Res. 37: 712–726.

Razouls, C., F. deBovee, J. Kouwenberg and N. Desreumaux. 2005–2019. Diversity and Geographic Distribution of Marine Planktonic Copepods. URL: http://copepodes.obs-banyuls.fr/.

Rizo, E.Z.C., Y. Gu, R.D.S. Papa, H.J. Dumont and B.-P. Han. 2017. Identifying functional groups and ecological roles of tropical and subtropical freshwater Cladocera in Asia. Hydrobiologia 799: 83–99.

Reynolds, C.S. 1984. The Ecology of Freshwater Phytoplankton. Cambridge: Cambridge University Press.

Reynolds, C.S., V. Huszar, C. Kruk, L. Naselli-Flores and S. Melo. 2002. Towards a functional classification of the freshwater phytoplankton. J. Plankton Res. 24: 417–428.

Robson, B.J., G.B. Arhonditsis, M.E. Baird, J. Brebion, K.F. Edwards, L. Geoffroy et al. 2018. Towards evidence-based parameter values and priors for aquatic ecosystem modelling. Environ. Modell. Softw. 100: 74–81.

Rosenfeld, J.S. 2002. Functional redundancy in ecology and conservation. Oikos 98: 156–162.

Santoferrara, L.F. and G.B. McManus. 2021. Diversity and biogeography as revealed by morphologies and DNA sequences: Tintinnid ciliates as an example. pp. 85–118. *In*: M.A. Teodósio and A.B. Barbosa [eds.]. Zooplankton Ecology. CRC Press.

Schleuter, D., M. Daufresne, F. Massol and C. Argillier. 2010. A user's guide to functional diversity indices. Ecol. Monogr. 80: 469–484.

Schneider, T., G. Grosbois, W.F. Vincent and M. Rautio. 2017. Saving for the future: pre-winter uptake of algal lipids supports copepod egg production in spring. Freshwater Biol. 62: 1063–1072.

Schmidt, K., A. Atkinson, K.J. Petzke, M. Voss and P.W. Pond. 2006. Protozoans as a food source for Antarctic krill, *Euphausia superba*: complementary insights from stomach content, fatty acids, and stable isotopes. Limnol. Oceanogr. 51: 2409–2427.

Sodré, E.d.O. and R.L. Bozelli. 2019. How planktonic microcrustaceans respond to environment and affect ecosystem: a functional trait perspective. Inter. J. Aquat. Res. 11: 207–223.

Steinberg, D.K. and M.R. Landry. 2017. Zooplankton and the ocean carbon cycle. Annu. Rev. Mar. Sci. 9: 413–44.

Sterner, R.W. and J.J. Elser. 2002. Ecological Stoichiometry: The Biology of Elements from Molecules to the Biosphere. Princeton University Press, Princeton.

Strickler, J.R. 1982. Calanoid coppepods, feeding currents, and the role of gravity. Science 218: 158–160.

Tilman, D., J. Knops, D. Wedin, P. Reich, P. Ritchie and E. Siemann. 1997. The influence of functional diversity and composition on ecosystem processes. Science 277: 1300–1302.

Thibaut, L.M. and S.R. Connolly. 2013. Understanding diversity-stability relationships: towards a unified model of portfolio effects. Ecol. Lett. 16: 140–150.

Thuesen, E.V., F.E. Goetz and S.H.D. Haddock. 2010. Bioluminescent organs of two deep-sea arrow worms, *Eukrohnia fowleri* and *Caecosagitta macrocephala*, with further observations on bioluminescence in chaetognaths. Biol. Bull. 219: 100–111.

Turner, J.T. and J.G. Ferrante. 1979. Zooplankton fecal pellets in aquatic ecosystems. BioScience 29: 670–677.

Vannote, R.L., G.W. Minshall, K.W. Cummins, J.R. Sedell and C.E. Cushing. 1980. The river continuum concept. Can. J. Fish. Aquat. Sci. 37: 130–137.

Villar-Argaiz, M., J.M. Medina-Sanchez, L. Cruz-Pizarro and P. Carrillo. 2001. Inter-and intra-annual variability in the phytoplankton community of a high mountain lake: the influence of external (atmospheric) and internal (recycled) sources of P. Freshwater Biol. 46: 1017–1034.

Villéger, S., N.W.H. Mason and D. Mouillot. 2008. New multidimensional functional diversity indices for a multifaced framework in functional ecology. Ecology 89: 2290–2301.

Violle, C., M.L. Navas, D. Vile, E. Kazakou, C. Fortunel, I. Hummel et al. 2007. Let the concept of trait be functional. Oikos 116: 882–892.

Violle, C., P.B. Reich, S.W. Pacala, B.J. Enquist and J. Kattge. 2014. The emergence and promise of functional biogeography. P. Natl. Acad. Sci. USA 111: 13690–13696.

Vogt, R.J., P.R. Peres-Neto and B.E. Beisner. 2013. Using functional traits to investigate the determinants of crustacean zooplankton community structure. Oikos 122: 1700–1709.

Walker, B., A. Kinzig and J. Langridge. 1999. Plant attribute diversity, resilience, and ecosystem function: the nature and significance of dominant and minor species. Ecosystems 2: 95–113.

Wallace, J.B. and J.R. Webster. 1996. The role of macroinvertebrates in stream ecosystem function. Annu. Rev. Entol. 41: 115–139.

Weithoff, G. 2003. The concepts of 'plant functional types' and 'functional diversity' in lake phytoplankton–a new understanding of phytoplankton ecology? Freshwater Biol. 48: 1669–1675.

Weithoff, G. and B.E. Beisner. 2019. Measures and approaches in trait-based phytoplankton community ecology–from freshwater to marine ecosystems. Front. Mar. Sci. 6: 40.

Williamson, C.E., J.M. Fischer, S.M. Bollens, E.P. Overholt and J.K. Breckenridge. 2011. Toward a more comprehensive theory of zooplankton diel vertical migration: Integrating ultraviolet radiation and water transparency into the biotic paradigm. Limnol. Oceanogr. 56: 1603–1623.

World Register of Marine Species Editorial Board. 2000–2019. URL: www.marinespecies.org.

Yoshida, T., L.E. Jones, S.P. Ellner, G.F. Fussmann and N.G. Jr Hairston. 2003. Rapid evolution drives ecological dynamics in a predator-prey system. Nature 424: 303–306.

CHAPTER 2

Zooplankton-Phytoplankton Interactions in a Changing World

Maarten Boersma[a,b,*] and *Cédric L. Meunier*[a]

2.1 Introduction

The interaction between zooplankton and phytoplankton[1] is pivotal in determining the transfer of energy and material to higher trophic levels in aquatic systems. It determines the productivity of the ecosystem, and through that, ultimately, the potential use by humans in terms of food production. In the interface between herbivorous zooplankton and their prey, or, depending on one's point of view, microalgae and their predators, it is important to differentiate between consumptive and non-consumptive interactions (Fig. 2.1). Traditionally, the main focus has been on consumptive interactions, and the resulting production of ecosystems, largely emerging from early global programmes such as the International Biological Programme (Boffey 1976), JGOFS (SCOR 1987), GLOBEC (Huntley 1992) and others, which mostly focussed on carbon as the major currency. However, starting in the 1990s, more and more investigators discovered other pathways of interactions, where information through chemical signalling, rather than energy, was transferred between trophic levels (Larsson and Dodson 1993, Lass and Spaak 2003). This sparked two decades of intense research, after which interest weaned to some extent, which could be attributed to the fact that the ecological interactions were well described, but the search for the exact chemical nature of the signalling substances proved much more of a challenge. As both direct consumptive interactions as well as chemical (informative) interactions are important components of the link between primary producers and their consumers, we will discuss both in this chapter. Around the same time as when researchers started to investigate non-trophic interactions, also the focus shifted from carbon (energy) as a single currency towards multiple currencies, both in terms of macronutrients such as nitrogen and phosphorus[2] as well as in

[a] Alfred-Wegener-Institut Helmholtz-Zentrum für Polar- und Meeresforschung, Biologische Anstalt Helgoland, Postfach 180, 27483 Helgoland, Germany.

[b] University of Bremen, FB2, Bremen, Germany; Email: cedric.meunier@awi.de

* Corresponding author: maarten.boersma@awi.de

[1] Phytoplankton and zooplankton are both singular as well as plural, even though zooplanktons exist as well to describe different types of zooplankton. Here, we will consistently use the plural.

[2] It is important to note that phosphorus in both British as well as American spelling is always spelled this way. Phosphorous is not the British spelling for the element, it is an adjective not a noun (Hairston 2003).

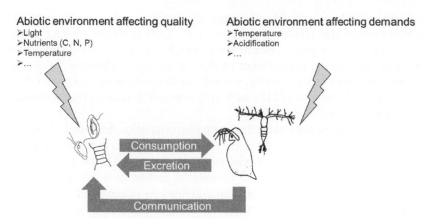

Figure 2.1: Interactions between phytoplankton and zooplankton and the factors that influence this interaction.

terms of different biochemical components of food. This research culminated in the seminal books by Sterner and Elser (2002) on ecological stoichiometry and by Arts and Wainmann (1999) [updated and extended by Arts et al. (2009)] on lipids in aquatic systems. Hence, here we will discuss these aspects as well, first introducing single currency approaches, and extending those two multiple currencies that are now generally accepted to be of importance in the interactions between phytoplankton and zooplankton. Moreover, we will discuss how the currently rapidly changing environment affects these interactions, and finish with an outlook on what the future might bring.

2.2 Consumptive Interactions

Overall, trophic relations within planktonic food webs influence the biomass and dynamics of each trophic level. For example, the abundance of resources (phytoplankton) can drive the dynamics of higher trophic levels (zooplankton). These bottom-up controls have been observed between multiple trophic levels and over large geographic scales. By combining information on the biomass of multiple trophic levels, Ware and Thomson (2005), for example, showed that strong bottom-up trophic linkages exist between phytoplankton, zooplankton, and resident fish along the continental margin of western North America extending to regional areas as small as 10,000 square kilometres. Conversely, the dynamics of zooplankton populations can affect phytoplankton communities, as different zooplankton taxa selectively consume different parts of the phytoplankton size spectrum (Sommer and Sommer 2006). Such selective feeding processes are often primarily driven by size, typically important in filter feeders such as freshwater cladocerans of the genus *Daphnia* (Gliwicz and Lampert 1993), where the size spectrum of the prey is determined by the morphology of the collecting organs (Brendelberger and Geller 1985). They might also be governed by shape, with spines or other structures making some algal species less palatable (Pančić and Kiørboe 2018), or by toxicity (Jungmann and Benndorf 1994). Moreover, mobility of prey, primarily changing encounter rates between predators and their prey (Gerritsen and Strickler 1977), can also result in escape responses, especially with smaller grazers (Jakobsen et al. 2006, Meunier et al. 2013), or by a change in visibility, in those grazers using hydromechanical cues to detect their potential prey (Kiørboe and Visser 1999). As we will see below, many of these attributes of microalgal prey may be affected by abiotic conditions and by the presence of predators.

In the context of this chapter, we will focus mainly on meso-zooplankton as predators and microalgae as prey. We realise that this is a strong simplification, as on the one hand very few mesozooplankters are true herbivores, either by choice (Boersma et al. 2014), or by accident, as filter-feeders essentially select by size only, taking up autotrophs as well as heterotrophs in specific size classes. On the other hand, very few microalgae are true autotrophs, and many rely at least partly on heterotrophic uptake. This mixotrophy (Flynn et al. 2019) is highly relevant in the context of food quality, especially as it allows mixotrophic organisms to switch between autotrophy and heterotrophy depending on the environment (Wilken et al. 2013). However, as the scope of this chapter is mainly on factors of food quality, we will not deal with mixotrophy, and refer the reader to the Horizon paper by Flynn et al. (2019), as well as to excellent reviews by, for example, Stoecker (1998), Jones (2000), or Tittel et al. (2003).

Even though it has been long clear that prey characteristics described above influence the quality of algae as food for zooplankton, the study of the exact factors that determine food quality of algae within and between prey species is relatively new. Early studies on grazing of zooplankton on phytoplankton (Lampert 1977a, b, c, d) focused on carbon as a single currency, and many of the extant models describing planktonic interactions in oceans and lakes still essentially consider carbon as the main currency. So, what are the factors that determine food quality within species? Three main aspects have been mentioned in the literature: (macro) nutrient stoichiometry, morphology, and biochemical composition. As these are interlinked, efforts to identify the most important factor(s) determining food quality have led to much discussion and controversy in the literature. In the following, we will discuss these factors both separately as well in combination.

2.3 Ecological Stoichiometry

Ecological stoichiometry, the study of the balances between different elements, predicts that animal fitness and nutrient recycling are tightly coupled with the nutrient ratios in the food (Sterner and Hessen 1994, Sterner and Elser 2002). In phytoplankton, the balance of carbon and nutrients depends on the availability of dissolved nutrients and light (carbon uptake through photosynthesis) in the environment. Zooplankton consume phytoplankton, and thereby obtain energy and nutrients that was accumulated by the algae. As animals take up food in packages and not as single nutrients, secondary producers often face the problem of nutritional imbalances, as prey items in general do not have the same nutrient requirements and contents as their predators. To cope with sub-optimal quality, consumers need to release much of the nutrients present in excess, while retaining most of the limiting nutrient. More specifically, in the relationship between plants and herbivores, nutrients such as nitrogen and phosphorus are most frequently limiting, as there often is a surplus of carbon in plants (White 1993). Obviously, this carbon is needed as an energy source, and hence will disappear through the food chain, but the corollary of these nutritional imbalances in the food is that many consumers have more problems meeting their nutrient requirements than their energy requirements. At the same time, as a result of the dissipating carbon, organisms feeding at higher trophic levels should normally face food of a higher quality (in terms of nutrient balances) than those feeding lower in the food chain.

Variation in the specific nutrient content of autotrophs is generally larger than the variation in herbivores (Sterner et al. 1998). While there is evidence that other elements such as iron (Chen et al. 2014, Baines et al. 2016, Jeyasingh and Pulkkinen 2019) can influence the interaction between phytoplankton and zooplankton, here, we will focus on the macro nutrients carbon (C), nitrogen (N), and phosphorus (P) as those are the most important ones, and, as a consequence, have been most widely studied. The C:N:P ratio in primary producers shows a very large inter- and intraspecific variation, whereas these ratios are much more constant in primary consumers (Fig. 2.2). In aquatic ecosystems, the issue of stoichiometric nutritional quality and its effect on consumers has received a lot of attention in recent years, as evinced by many publications following the book by Sterner and Elser (2002), and it is now well established that C:N:P ratios are very good descriptors for the functioning of ecosystems.

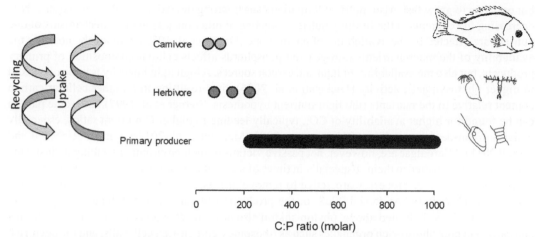

Figure 2.2: Ecological stoichiometry of primary producers, herbivores and carnivores (redrawn after van de Waal and Boersma 2012). Phytoplankton exhibit a wide range of C:P (or C:N) ratios, different heterotrophs such as zooplankton and fish show more constraint stoichiometry, with typically lower values. In this chapter, we will use atomic ratios of elements, as chemical reactions are also described by atom-to-atom interactions, rather than in mass reactions, and ratios are closely linked to the concept of building blocks of different compounds.

2.3.1 Sources of Stoichiometric Variation in Phytoplankton

Phytoplankton obtain their carbon during photosynthesis using dissolved CO_2 and bicarbonate. Hence, both the availability of light as well as of inorganic carbon play important roles in carbon acquisition. Inorganic carbon is converted into carbohydrates, which can be used for synthesis of polysaccharides, and other substances such as into amino acids, fatty acids and nucleic acids. The major pathway of phosphorus uptake is through ortho-phosphates. Nitrogen can be taken through different pathways, most phytoplankton take up nitrate, nitrite or ammonium, some take up organic nitrogen sources such as urea, and several species of cyanobacteria can directly fix nitrogen from N_2. As phytoplankton take up most nutrients singly, the content and ratios of the different elements is highly variable, as indicated above. Moreover, most primary producers even take up nutrients and carbon beyond their needs in so called luxury consumption, storing them in vacuoles or otherwise. This variability in elemental stoichiometry of primary producers is very strong (Sterner et al. 2008), even though Alfred C. Redfield implicitly stated the opposite, posing that the C:N:P ratio of plankton in marine ecosystems was remarkably constraint. This ratio, which was later called the Redfield ratio[3] (106 C:16 N:1 P), is still considered the holy grail of nutrient ratios, but seems to be rather the consequence of bulk samplings and building averages than of biological processes (see Elser et al. 2011, Loladze and Elser 2011). In fact, in different analyses it was shown that the ratio of the macronutrients is highly dependent on, for example, the growth rate of algae (Goldman et al. 1979, Hillebrand et al. 2013). However, the second part of Redfield's observation that the ratio within the living world was very similar to the one in dissolved nutrients is often overlooked, and can be interpreted as proof of how important biological processes are for the chemical composition of global oceans.

As stated above, the elemental composition of phytoplankton is strongly affected by their growth rate and by the availability of resources in the environment. Phytoplankton grown at maximum growth rates have cellular C:N:P ratios resembling Redfield proportions (Hillebrand et al. 2013), whereas slower growing cultures show more variability, indicating that there are many ways to grow slowly,

---

[3] It is important to note that only **the** Redfield ratio (106:16:1) exists, some references refer to any relationship of carbon to nitrogen to phosphorus as **a** Redfield ratio.

but only one to grow fast. Also, nutrient limitation causes strong deviations in phytoplankton C:N:P, decreasing the contents of the limiting nutrient, but also of other ones as uptake mechanisms of one nutrient are affected by the availability of another one (Droop et al. 1982). Obviously, not only the availability of the macronutrients nitrogen and phosphorus affects cellular composition of primary producers, but also the availability of light and carbon sources. A high light intensity is typically linked to higher photosynthetic activity (Dickman et al. 2006), thus resulting in a higher cellular carbon content relative to the nutrients (the light-nutrient hypothesis: Sterner et al. 1997). The same pattern can be found for higher availability of CO_2, typically leading to higher C:nutrient ratios, especially under nutrient-deplete conditions (Urabe et al. 2003, Meunier et al. 2017b, but see Riebesell and Tortell 2011). Microalgae are, however, not passive victims of their environment, taking up just what happens to be presented to them. Especially in times of nutrient limitation and increased carbon/light availability, regulatory processes are active to remove at least parts of the excess carbon through respiration (Hessen and Anderson 2008) or the production of carbon rich polymers (Engel et al. 2014). Similarly, as indicated above, phytoplankton also have the capability to store macronutrients and carbon in phosphorus-rich organelles such as ribosomes, carbon-rich cell walls, and nitrogen-rich chlorophyll, as well as in vacuoles and granules. These storage compounds allow flexible uptake of nutrients and play a key role in decoupling resource availability from growth rate. Hence, both the ability to grow despite non-optimal conditions, as well as luxury uptake mechanisms lead to a large variability in C:N:P ratios in natural environments (Elser and Hassett 1994, Sterner et al. 2008).

2.3.2 *Stoichiometric Homeostasis of Zooplankton*

In contrast to phytoplankton, grazers are much less variable in their stoichiometry (Fig. 2.2). This has consequences, as we will see below, but before we can do this, the concept of homeostasis needs to be introduced. Homeostasis is the maintenance of constant internal conditions, in our case elemental ratios, despite the variability in the external environment (elemental ratios of the food). Homeostasis therefore couples consumers' elemental stoichiometry to that of its food. A large body of literature exists on other regulatory processes, such as of temperature or salinity. Whereas there seems to be a common understanding that, for example, homeothermy has great advantages, as this allows a much broader functioning of organisms despite adverse environmental conditions, no such common understanding exists for stoichiometric homoeostasis. On the one hand, constant internal conditions in terms of nutrients probably allow easier regulatory and information processes (Woods and Wilson 2013), as well as synthesis of biomolecules (Kooijman 1998), on the other hand it puts great demands on the uptake of food in the correct ratios, as it does not allow for luxury uptake or storage. As a result, there is still considerable discussion whether stoichiometric homeostasis is a constraint rather than an adaption. The in-depth analysis of Meunier et al. (2014) suggested that it might be more of a constraint, but more research is certainly needed in this field (Fig. 2.3).

Zooplankton show a fairly strong homeostasis, but large differences exist between individual species (Hessen and Lyche 1991, Gismervik 1997), or even between stage- or age classes within species (Meunier et al. 2016a). Within zooplankton, copepods have a lower phosphorus content compared to cladocerans, whereas the relative variability in carbon and nitrogen content is much lower (Elser et al. 1996). Two groups of explanations exist to explain these differences. First, structural components play a major role in the nutrient stoichiometry of organisms. This becomes most clear when looking at vertebrates, for example, which have a high phosphorus content due to the large amounts of phosphorus present in the skeleton, or alternatively, vascular plants have very high C-contents as a result of the cellulose present in cell walls and supporting structures. In zooplankton, the stoichiometry of gelatinous structures can be reflected in total body stoichiometry (Kutter et al. 2014). Moreover, the biochemical composition of organisms will affect elemental composition. For example, higher investments into proteins will increase the relative nitrogen content, thus increasing the organism N:P ratio, whereas higher investment in nucleic acids will enhance the phosphorus content of an

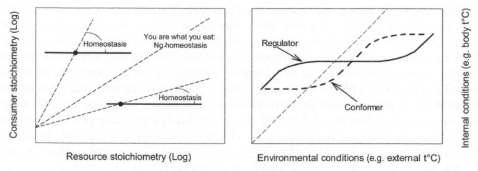

Figure 2.3: Two models of homeostasis. The left panel shows the traditional view on homeostasis, with essentially the slope of the regression line between external and internal conditions indicating the level of homeostasis in an organism (horizontal line complete homeostasis; 1:1 complete lack of homeostasis). The right panel shows the new approach of Meunier et al., indicating that in different ranges of environmental conditions the slopes may be different, and both regulators and conformers have areas in the graph with horizontal lines as well as 1:1 lines (Meunier et al. 2014).

organism, decreasing the organism N:P ratios. The main use for phosphorus in living cells is in nucleic acids, especially RNA (Elser et al. 2003). In fact, the P present in RNA can account for up to 50% of the whole organism phosphorus content. This close link between P and RNA has led to the development of the growth rate hypothesis (Elser et al. 1996, Sterner and Elser 2002), which asserts that RNA is needed in ribosomes for protein synthesis creating new biomass, and thus directly links P, RNA, and growth rate. Organisms with relatively high growth rates have high RNA demands, and hence high P demands, with the resulting high relative phosphorus contents in their body tissue. Thus, slower growing copepods have less RNA than fast growing cladocerans such as *Daphnia*, and as a result have higher N:P ratios. Whereas the general applicability of the growth rate hypothesis is well accepted, especially for relatively homeostatic organisms such as zooplankton (Main et al. 1997), it does not automatically imply complete linearity between RNA, P and growth rate under all circumstances (Acharya et al. 2004, Chícharo and Chícharo 2008). Moreover, care should be taken when applying the hypothesis to non-homeostatic organisms such as microalgae (Flynn et al. 2010).

The elemental composition may even vary with ontogenetic development, for instance in copepods showing lower N:P ratios in fast growing naupliar stages and higher ones for older stages of the same species, which grow slower and invest more in structural mass such as proteins (Bullejos et al. 2013, Meunier et al. 2016a). Thus, we have seen that phytoplankton nutrient composition is highly variable, also within species, strongly influenced by external conditions. At the same time, zooplankton show a lot of variation between species, less between stages within species, and rather little within stages. Because of the fairly strict homeostasis of zooplankton, combined with the lack thereof in algae, there is great scope for a mismatch of nutrient availability and demand, and in the next sections we will investigate the potential consequences of this mismatch. As can be seen from Fig. 2.2, typically algae have higher C:X ratios (X reflecting the macronutrients N or P) than zooplankton. This is not necessarily a mismatch, as carbon is used as an energy source by the zooplankters, and much of the carbon taken up by consumers disappears as CO_2, with growth efficiencies for carbon typically in the range of 30% (Anderson and Hessen 1995, Anderson et al. 2017). However, even taking this into account, carbon is often still available in excess. This excess carbon needs to be dealt with by the herbivore, which can be done in several ways.

2.3.3 *Nutritional Requirements and Zooplankton Feeding Mechanisms*

As different zooplankton species have different nutrient demands, they obviously also have different optimal nutrient ratios in their food. This optimal ratio of food C:X is referred to as the Threshold

Elemental Ratio (TER) (Sterner and Hessen 1994, Frost et al. 2006). As carbon is usually not limiting, the TER was initially defined as the ratio above which food shifts from carbon limited to nutrient limited (i.e., high carbon:nutrient ratios). As a result, the main focus of many studies has been on the effects of high carbon relative to nitrogen or phosphorus foods (DeMott and Gulati 1999, DeMott and Pape 2005). These studies were based on the implicit assumption that, as soon as the stoichiometry of the food was below the TER (on a C:X scale), there were enough nutrients, and the exact nutrient content of the food did not matter anymore. However, new evidence suggests that also food with nutrient ratios clearly below the TER (hence with an excess of nitrogen or phosphorus) is not an optimal food source (Boersma and Elser 2006). This led to the term stoichiometric knife edge for the TER, as essentially on either side of the TER food quality in terms of C:X declines (Elser et al. 2005, 2006, Bullejos et al. 2013, Laspoumaderes et al. 2015).

Most studies investigating the effects of stoichiometric food quality on performances of grazers have found that consuming food with C:X ratios clearly above or below the TER lowers growth and reproduction rates, and increases mortalities of zooplankton (Frost and Elser 2002, Persson et al. 2007, Malzahn and Boersma 2012). One of the interesting questions related to this is of course what are the mechanisms that explain the decreases in animal performance on either side of the TER. Essentially, in the ecological stoichiometry approach the explanation lies in the effort required to expunge excess nutrients rather than in the low availability of the limiting nutrient. This becomes very clear from the early work of Sterner et al. (1993), which shows that growth of the cladoceran *Daphnia obtusa* did not increase when given more low P-food. This means that the absolute amount of phosphorus in the diet is not of interest, but rather the relative amount (Urabe and Sterner 1996). Evidently, keeping homeostasis, and removing excess nutrients or excess carbon is costly, resulting in decreased performance of grazers. When the nutrients have been taken up and assimilated, excretory processes will be needed to void excess material, an issue we will deal with below as we address consumer-driven nutrient recycling. However, there are also other possible mechanisms that prevent the uptake of excess nutrients in the first place. For example, animals may feed selectively on food sources of different quality. This is known for many zooplankton species, both with respect to interspecific differences in food quality (DeMott 1986), as well as to intraspecific differences as a result of differences in nutrient composition (Wootton et al. 2007, Martel 2009, Meunier et al. 2016a). Alternatively, by increasing the feeding effort, and thus enhancing the turnover of food, sufficient amounts of easily accessible nutrients may be acquired by selectively taking up the nutrients from the gut (White 1993), although this may come with additional energetic costs. When carbon is in ample supply, sufficient energy may be available to support this enhanced turnover of food (Plath and Boersma 2001). Other pre-gut processes to balance the intake of different elements are also possible. The fact that, for example, faecal pellet production and composition in copepods is dependent on the composition of the food (Voss 1991) suggests that assimilation efficiencies might change associated with different food sources. Carbon assimilation efficiencies have been reported to be lower when feeding on high carbon food by the copepod *Acartia tonsa* (Kiørboe 1989, Anderson et al. 2017); the same was found by DeMott et al. (1998) for *Daphnia magna* feeding on the green alga *Scendesmus acutus* with different C:P ratios. Interestingly, in the same study DeMott et al. (1998) did not observe changes in P-assimilation efficiencies along the gradient, and neither did Kiørboe (1989) for nitrogen in the C:N gradient. Thus, apparently it is easier to change the assimilation efficiency of carbon, the nutrient usually in ample supply over phosphorus and nitrogen, which are more commonly limiting.

Post-gut mechanisms include selective discharge of the substances in excess, either by respiratory or excretory processes (Hessen and Anderson 2008). In herbivores, this is often carbon, when food stoichiometry in terms of C:X ratios is above the TER, but as indicated above, also the opposite is possible in cases of high N or high P food sources. There are several pathways to void excess material. In the case of carbon this can be either by respiration as CO_2 or by excretion as dissolved organic carbon (DOC). Remarkably little is known about how excretory processes work in zooplankton. In fact, the exact pathways of DOC excretion are still elusive, despite the fact that it is important for the further utilization of the waste products whether excess carbon is released as DOC or CO_2. In the latter case, it can only be used by photoautotrophs, in the former case bacteria and other microbes might use the DOC as their carbon source. Several studies exist showing that, indeed, respiration in zooplankton is higher on high C:P food (Jeyasingh 2007, for *Daphnia*), but Schoo et al. (2013) showed that for the copepod *Acartia tonsa* there is a stage dependence of how excess C is voided. Younger stages mainly respired, whereas older stages excreted excess carbon mainly as DOC. Whereas for some terrestrial invertebrates, wastage respiration or futile cycles have been mentioned as pathways to respire excess carbon (Zanotto et al. 1997, Trier and Mattson 2003), no such mechanisms have been described for zooplankton. In contrast to the studies on the effect of excess carbon on the excretion of carbon, virtually no studies exist on the impact of excess nitrogen and phosphorus on the excretion of these elements. Obviously, theoretical studies (Anderson and Hessen 1995) predict a higher excretion at low C:X levels, but measurements are scarce, and often contradictory (Boersma and Wiltshire 2006, Franco-Santos et al. 2018). Moreover, in measurements, it is very difficult to differentiate between egestion and excretion, and the exact processes are far from completely understood. In any case, as a result of the species and stage-specific relative homeostasis of consumers, in combination with the much larger variation in nutrient stoichiometry of primary producers, there is large scope for selective retention and excretion of nutrients by zooplankters. As a result, the species and stage composition of a zooplankton community drives the availability of nutrients for primary producers (also see below).

2.3.4 *Zooplankton-driven Recycling of Nutrients*

Fast growing cladocerans, younger stages of copepods, as well as heterotrophic dinoflagellates require high amounts of phosphorus, and as a result retain most of the phosphorus they take up (Olsen et al. 1986, Sommer 1992, Meunier et al. 2016a, 2018). As a result, these zooplankton reduce the phosphorus availability in the water (Elser and Urabe 1999). Consequently, phytoplankton become further limited in phosphorus which increases their C:P and N:P ratio. In turn, this will enhance the elemental imbalance between phytoplankton and zooplankton, and thus aggravate nutrient limitation by grazers. Thus, the identity of the dominating zooplankton species may determine whether phosphorus or nitrogen is limiting, and consumer-driven nutrient recycling will enhance or reduce the N:P stoichiometry in the aquatic food web, respectively (Elser et al. 1988, Sommer and Sommer 2006). Zooplankton, however, may also replenish the nutrient availability in the water as a result of so-called sloppy feeding (Lampert 1978). That is, zooplankton break their food in smaller pieces that are not all consumed. Part of the nutrients in the left-over phytoplankton remains dissolved in the water and become available again for phytoplankton growth, while other parts of the cell rests are remineralized by bacteria. This will enhance growth and nutrient content of phytoplankton, and thus facilitate zooplankton grazing. Hence, sloppy feeding may slow down nutrient limitation as a result of direct consumer-driven nutrient recycling. All these recycling processes will not only affect the quality of the living material as food for other zooplankton, but will also strongly impact the quality of the detrital pool as a potential food source (Legendre and Rivkin 2008, 2015, Basedow et al. 2016).

2.4 Biochemical Composition of the Food

Nutrients are not the only constituents that determine the quality of different food items for consumers and the exact nature of the most important food-quality determining factor has caused quite some discussion. Biochemical features of the food are also important factors influencing its quality for grazers (see also Ruess and Müller-Navarra 2019, for an excellent recent review). In this chapter, we want to concentrate mainly on the effects of quality of the food, and we will refrain from discussing the effects of toxic substances produced by algae on survival, growth, and reproduction of zooplankton. We refer readers to the specialised literature, as both for freshwater systems (Carmichael 1992, Dawson 1998, O'Neil et al. 2012) as well as for marine systems (Anderson et al. 2012) a large body of literature exists on the effects of cyanobacteria (mainly freshwater) and dinoflagellates and diatoms (mainly marine) on their grazers.

The contents in essential fatty acids, amino acids, sterols, and vitamins are all important in determining food quality of phytoplankton for zooplankton. There is a large body of literature, investigating, for example, why cyanobacteria, despite their often very high phosphorus content, are not a food source of high quality for herbivorous zooplankton (Brett et al. 2000, von Elert and Wolffrom 2001). The explanation is that (non-toxic) cyanobacteria lack biochemical substances essential for growth in herbivorous zooplankton. Especially the highly unsaturated fatty acids (HUFA), such as eicosapentaenoic acid (EPA) and docosahexaenoic acid (DHA), are essential for many consumers, as they are used in membranes and are important precursors for hormones (Müller-Navarra 1995a, b, Brett and Müller-Navarra 1997). Most animals cannot synthesize these omega-3 fatty acids themselves, and hence are dependent on their food to obtain these substances. Microalgae are the most important producers of essential fatty acids (Ahlgren et al. 1992, Gladyshev et al. 2013), and the complete aquatic food chain depends on the primary producers. Most authors agree that even though potentially present, the process to synthesize EPA is very slow in zooplankton, even though there is some evidence that longer chained highly unsaturated fatty acids can be synthesised from shorter ones through elongation by some zooplankters (Weers et al. 1997, von Elert 2002, Heckmann et al. 2008). Interestingly, Anderson and Pond (2000) speculated that the necessary enzymes do not need to be present because calanoid copepods (and the same could apply to cladocerans) are not normally limited in their EPA uptake (see Brett and Müller-Navarra 1997). Fatty acids and lipids have been studied in great detail (Dalsgaard et al. 2003, Kattner et al. 2007), and it is well established that a lack of essential fatty acids in the food causes the food to be of lower quality, with the resulting reductions in growth and reproduction (von Elert 2004, Wacker and Martin-Creuzburg 2007).

Apart from fatty acids, other biochemicals may also be of importance. A lack of sterols (Martin-Creuzburg and Von Elert 2004) has been described to reduce food quality, and as these are particularly low in concentration in cyanobacteria, this has been invoked as another explanation for the low food quality of cyanobacteria as food for zooplankton. Sterols and fatty acids have also been studied in combination, and several cases have been described where there seems to be a co-limitation (Martin-Creuzburg et al. 2009).

Essential amino acids comprise the third group of biochemical constituents of the food that may determine food quality (Guisande et al. 1999, Nielsen et al. 2015). However, much less is known about threshold levels of essential individual amino acids, a topic that has been well studied for single essential fatty acids in laboratory situations (Müller-Navarra 1995b, Becker and Boersma 2005, Ravet et al. 2012). In fact, it is not completely clear whether the nine amino acids that are essential to humans are also essential to all other vertebrates, and even less known for zooplankton. Furthermore, in many studies on pelagic systems amino acids are used solely as markers (in stable isotope studies) (Nielsen et al. 2015, Eglite et al. 2018), and we are aware of no experimental studies, manipulating single amino acids in algae to investigate the importance of amino acids as essential components of the food for zooplankton.

2.5 Macronutrients versus Biochemicals

What drives zooplankton production in the field? This is not easily answered, as there are several confounding factors that preclude clear conclusions. First, it is important to note, that it is highly unlikely that one size fits all in this case. Freshwater systems will be different from marine ones, and high-nutrient systems will have different limitations than oligotrophic areas. Furthermore, also through the course of a year, limitations may change (Boersma et al. 2001), and hence overarching statements on the importance of nutrient versus biochemical limitations cannot be made. The confounding factors added to this are that most studies investigating food quality effects in natural situations focus on one aspect of nutritional quality only, even though the different quality defining traits are not independent. For example, nutrient-limited algae contain a different fatty acid spectrum when compared to non-limited algae (Weers and Gulati 1997, Boersma 2000), making them potentially less nutritious. Although nutrient-limited algae often contain higher absolute amounts of fatty acids (especially the freshwater ones), the concentrations of highly unsaturated fatty acids, such as EPA or DHA, can be much lower. However, these changes are not identical for all species. Nutrient-limited *Rhodomonas salina*, for example, contain more of the highly unsaturated fatty acids compared to nutrient-replete algae (Malzahn et al. 2010). Thus, as carbon, nitrogen, and phosphorus are the main building blocks of fatty acids, amino acids, and other essential components, there is a close link between the biochemical and elemental composition of primary consumers (Anderson et al. 2004). Moreover, algae may also change their morphology under nutrient limitation (Tillberg et al. 1984). Algal cell walls often become thicker and harder to digest for zooplankters (van Donk et al. 1997), leading to the question whether the nutrient limitation or the structural defences lead to a decrease in food quality. Furthermore, it has been suggested that these digestion resistant algae pass the gut of zooplankters unharmed, and take up nutrients from the zooplankters in the process, but this does not seem to be an important pathway (Boersma and Wiltshire 2006).

Much of the evidence that has been presented in favour of biochemical versus elemental limitations in natural situations is correlational, and concentrating either on elemental limitations or biochemical limitations. Especially in the early days of research on nutritional quality, researchers typically carried out a bio-assay assessing growth or reproduction of a standard animal (mostly *Daphnia* species), and correlated the performance of the animals in these bio-assays to a large set of measurements of seston in the field (Müller-Navarra 1995b, Müller-Navarra et al. 2000, Wacker and von Elert 2001). Those parameters that yielded the highest correlation with the performance parameters were stated to limit growth or reproduction in the field. Correlating a group of sestonic variables with growth rates will always yield one highest correlation, but is this meaningful? The fact that Wacker and von Elert (2001) observed the strongest correlation of *Daphnia* growth with the concentration α-linolenic acid, whereas Müller-Navarra (1995b) reported the highest correlation with EPA, illustrates this point. Hence, although very useful to identify potential factors explaining growth of zooplankton under natural conditions, there is no mechanistic basis to assume that the substance showing the highest correlation coefficient with the zooplankton response is in fact the limiting nutrient (especially as in several cases the differences in correlation coefficients was small). Several authors tried to investigate these limitations in the field experimentally, and carried out addition experiments, adding different substances to natural seston, investigating the response of zooplankters with and without the addition. For example, Park et al. (2003) added a mixture of highly unsaturated fatty acids to lake seston, and observed higher growth and survival of *Daphnia*, whereas Elser et al. (2001) did the same thing with PO_4 additions, and observed a similar pattern: higher growth of *Daphnia* feeding on P-enriched seston (with no changes in the fatty acid composition). Boersma et al. (2001) added both fatty acid emulsions as well as dissolved P, and noted that both additions increased growth of *Daphnia*, especially in the summer months. So experimental evidence of which factor limits growth in nature is equivocal. The experiment of Boersma and Stelzer (2000), who added fatty acid emulsions to mesocosms in a lake, showed that to complicate things even more there was variation in the reaction of the different zooplankton species. Hence, based on this and other studies we need to conclude that the attempt of the last 20 years to find the ultimate answer of what determines food quality for all zooplankton species in all systems throughout all seasons was likely doomed from the start. The link between the elemental composition and the biochemical composition is so tight that separating them is not useful. For example, phosphorus limitation in many algae will both cause an excess of carbon, at the same time change the spectrum of essential components of the food. Both change the quality of the food for the consumer, as the herbivores have to deal with the excess carbon and the scarcity of essential biochemical at the same time. In our future work, we should try to include all the factors, and refrain from advocating one-factor solutions. The approaches of Kooijman (2000) and Sperfeld et al. (2016) dealing with more than one nutrient are certainly the way forward here, and links up perfectly with the work on co-limitations as advocated by Wacker and Martin-Creuzburg (2012) and Sperfeld et al. (2012). Naturally, following Liebig's law of the minimum (von Liebig 1855), a co-limitation is theoretically not possible, as there should always be one nutrient limiting growth at the time. However, interactions between more than one potential resources, for example by one resource being important for the uptake of the other one, may result in resources being co-limiting. Both the increase of the limiting resource, as well as the increase of the resource that is important for the uptake of the limiting resource will increase growth (Sperfeld et al. 2012). Other cases of co-limitation might exist as well, and these have been described in great detail by Sperfeld et al. (2016), even though they stated that few experiments had been carried out to investigate co-limitation of consumers explicitly, and the full extent of how important this phenomenon is in nature is still unclear, and needs much more work.

2.6 Higher Tropic Levels: Predatory Zooplankton and Ichthyoplankton

Most studies on transfer efficiencies between trophic levels in aquatic environments have concentrated on the link between primary producers and herbivores, focusing mainly on algae and microcrustaceans. Interestingly, hardly any work exists investigating nutrient transfer in a stoichiometric context to higher trophic levels like predatory zooplankters and fish (Sterner and George 2000). Moreover, even though a large body of work exists on the importance of, for example, essential fatty acids, on the development of larval fish (Brown et al. 1997, Sargent et al. 1999), most of this work has been done in an aquaculture context (see Paulsen et al. 2014). One of the reasons for this paucity of studies is that zooplankters normally show a large extent of homeostasis with respect to their C:N:P ratios, and as a result also in their biochemistry, with the exception maybe of high-latitude zooplankton which store large amounts of lipids (Aubert et al. (2013), however, show that even then homeostasis is fairly strict when ignoring the lipid reserves). This implies that, independent of the food of the zooplankters, their nutrient content is constant, and thus the effect of food sources of different quality should not affect the quality of the zooplankters as food for higher trophic levels. This has large consequences for the way we view limitations in ecosystems. It implies that one cannot state that a certain system is P- or N- limited, but rather that defined trophic levels, or most likely species, are limited by different factors. If, for example, we state that a P or N limitation in zooplankton does not affect their own nutrient content (or biochemistry), but obviously does affect the production of these organisms, it could well be that the N or P limitation on one level is replaced by a quantity limitation (not enough prey available) on the next trophic level. To date, however, only very few studies have investigated more than two trophic levels at the same time in a food quality context (Malzahn et al. 2007, Boersma et al. 2009, Malzahn et al. 2010, Schoo et al. 2012, Laspoumaderes et al. 2015). Recently, however, it has become clearer that nutrient homeostasis is less strict in herbivorous zooplankton than was previously believed. In fact, differences in C:P ratio of a factor two are not uncommon under different conditions of the animals. This implies that the quality of differently fed zooplankton as food for higher trophic levels might vary. Indeed, there is a growing body of evidence that higher trophic levels may also be affected by quality differences on the primary producer level, and not through changes in food quantity (Lorenz et al. 2019b), but through quality (Boersma et al. 2008).

2.7 Non-Consumptive Interactions: Chemical Communication and Kairomones

Apart from the consumptive interactions described above, other interactions exist between zooplanktonic predators and their algal prey. Zooplankton excrete substances, which are used by algae as cues to indicate the presence of potential predators. Previous research concentrated on the reactions of algae and the potential effects of those reactions on consumptive success of the herbivores, with few attempts and little success to identify the substances excreted by zooplankton (von Elert and Franck 1999, Wiltshire and Lampert 1999). Recently, progress has been made in the identification of zooplankton-released substances which are used as cues by phytoplankton (Yasumoto et al. 2005, Grebner et al. 2019). Excellent reviews have been written on the effects of chemicals excreted by herbivores (Pohnert et al. 2007, Van Donk et al. 2011, Heuschele and Selander 2014, Saha et al. 2019, Pančić and Kiørboe 2018), and we refer the reader to those and the references therein. However, for the purpose of this chapter, we will briefly summarize the reactions of phytoplankton to the presence of grazers. Infochemicals released by grazers are used by the algae to reduce the risk of being grazed upon. Essentially, reactions of unicellular algae can be morphological, physiological, or behavioural. For example, as many grazers are size specific, changing size avoiding the window of highest predation pressure is an effective strategy to reduce mortality. Depending on the identity of the grazer, this might result in an increased individual or colony size, which has been observed between the cladoceran grazer *Daphnia* and their green algal prey *Scenedesmus* (Hessen and van Donk 1993, Lürling 2001), where previously single-celled microalgae formed colonies when exposed to cues released by the grazers. Furthermore, increases in defence-structures such as clumping (Wiltshire et al. 2003) or spines (Hessen and van Donk 1993) have been reported. In contrast, in cases of copepod grazers which prefer larger prey over smaller ones, reductions in colony size or chain length have been observed (Bjaerke et al. 2015, Selander et al. 2019), thus reducing predation pressure. Furthermore, the presence of copepods in the water (i.e., their excreted chemicals) can also increase the concentrations of toxic substances in algal cells (Bergkvist et al. 2008), whereby the ecological relevance of these toxins is still under debate, as in many cases the toxins do not directly affect the copepod grazers, but rather other higher predators such as fish and humans. Induced toxicity of algae to copepods has been reported by Ianora et al. (2004), who reported that oxylipins produced by diatoms in the process of being grazed induced reduced viability of copepod offspring. It is important to note that all of the changes in algal physiology and morphology are defences induced by the presence of grazers. This means that it is implicitly assumed by all researchers that these defences come at a cost, otherwise they would not be absent when no grazers are present. There is a growing body of research on the trade-offs between these defences and, for example, sinking rates, nutrient uptake rates and other important vital rates (Pančić and Kiørboe 2018). Furthermore, there is great scope of eco-evolutionary processes in these interactions (Yoshida et al. 2003, Fussmann et al. 2007, Becks et al. 2010). All of the changes in the algae affect the quality of the algae as food for zooplankton, but mostly because they turn toxic or difficult to ingest. Interestingly, there is very little work on whether grazers induce changes in the quality of the algae in terms of nutritional composition. Wichard et al. (2007) link the predator-induced oxylipin production of diatoms with a decreasing availability of their precursors, the essential fatty acids, and make the connection between food quality in the composition sense. Lürling et al. (1997) observed very small differences in the biochemical composition of induced and non-induced (colonial versus single celled) *Scenedesmus* species, but no other studies exist that link chemical communication processes with food quality. In fact, in their recent review Pančić and Kiørboe (2018) state that being of low nutrient content is an effective defence against being predated upon, but they link this only to changes in morphology as described above. Since changes in physiology and morphology are linked to changes in (bio) chemical composition through the production of different compounds, and different uptake rates of nutrients and light, it is more than likely that these changes affect the quality of the algae as food for zooplankton. Pančić and Kiørboe (2018) approached this to some extent, and computed from data by Lürling et al. (1997) an increase in the relative lipid content between single cells and colonies of *Scenedesmus acutus* from 6.5 to 21%. If, as a result, the protein contents of the cells decrease, this will certainly change the C:N ratio of the algae, with the resulting changes in food quality. However, there are no further studies in the literature investigating this. One could obtain some tentative information on predator-induced changes in food quality if we consider size as the main trait. Changing size has effects on the nutrient stoichiometry of algal cells, with in general smaller cells having lower C:N ratios (Maranon et al. 2013). Thus, a predator-induced change in size may affect the nutritional quality as well. Especially larger cells, with higher C:N ratios of around 10 (Maranon et al. 2013) might be a food of lower quality, not only because they are too large, but also because they contain insufficient amounts of nitrogen.

2.8 Stressors

Thus far, we have taken a fairly static stance in the relationship between herbivorous zooplankton and their prey, but of course environmental conditions may affect either the supply or the demand of important nutritional components. Moreover, as a result of the major change processes currently happening on Earth, these environmental conditions are changing, and the most important factors affecting supply/demand of food are temperature, (ocean) acidification, and eutrophication. Potentially, in future, micro- and nanoplastics will also affect feeding in zooplankton (Cole et al. 2015), even though current particle concentrations in most pelagic environments seem well below those that show impact in experimental studies (Lorenz et al. 2019a).

Changes in the drivers mentioned above might affect the species composition of planktonic communities, through species introductions (Gollasch 2006) and invasions (Jaspers et al. 2018), changes in dominance (Beaugrand et al. 2003), or local extinctions (Giller et al. 2004). Further, and we will not go into detail here, as it is outside the scope of this chapter, especially temperature strongly affects the phenology of ectotherms (Edwards and Richardson 2004, Poloczanska et al. 2013, Chevillot et al. 2017). As the phenology of consumers and producers are expected to react differently to environmental change (in this case mostly warming), mismatch (Cushing 1974) phenomena might occur. Previously closely timed co-occurrences between zooplankton and their prey may be disturbed (Sommer et al. 2012).

What are the direct effects of changes in nutrients, temperature and CO_2 availability on the supply and demand of important nutritional components, as well as on the communication link between herbivores and their prey? Very little is known on the latter, and there is great need to investigate the effects of global change processes on chemical interactions in aquatic systems (Saha et al. 2019).

The effect of eutrophication on the interaction between primary producers and their grazers has been studied in great detail. In short, the eutrophication of the 1960s and 70s in many temperate lakes and coastal seas caused major changes in the community compositions of the primary producers, often leading to persistent cyanobacterial blooms, which as a result of their lack of essential fatty acids and sterols, combined with their potential toxicity, led to strong changes in the zooplankton communities as well (Hartwich et al. 2012). However, also within algal species, and no change in the communities, changes in the nutrient availabilities have the potential to affect planktonic interactions. In many cases, the ongoing oligotrophication of many lakes and several coastal seas has led to changing quality of the primary producers as food for zooplankton, the concurrent lower densities, and the lower production of fish (Eckmann et al. 2006, Boersma et al. 2015).

An increased availability of CO_2 leads to acidification of the water, as CO_2 not only dissolves in water but also dissociates. This decrease in water pH may lead to changing phytoplankton communities (Bach et al. 2017, Bach and Taucher 2019) and additional stress on zooplankton (Riebesell and Tortell 2011). We refer the reader to Campoy et al. (2021, this book, Chapter 4), for an in-depth analysis of the effects of ocean acidification on planktonic communities. Here, we will only consider the indirect effects of acidification, i.e., the higher availability of carbon relative to macronutrients, which may affect food quality. In effect, the increased C availability may increase the C:X ratios in the primary producers (Burkhardt and Riebesell 1997), with the same effects as have been described in the stoichiometry section above (Urabe et al. 2003, Meunier et al. 2017b). Thus, increased CO_2 availability could lead to high C-content in the primary producers, effectively resulting in nutrient limitation and the resulting changes in food quality. Recently, Cotner (2019) introduced the term environmental obesity for this, suggesting that the constant uptake of atmospheric CO_2 might cause changes in the nutrients ratios in ecosystems with unclear consequences. Moreover, Rossoll et al. (2012) reported direct effects of increased CO_2 availability on the fatty acid signatures of the diatom *Thalassiosira pseudonana*, although the exact mechanism for this is unclear. Nevertheless, this change in the fatty acid spectrum reduced the quality of the diatoms as food for the copepod *Acartia tonsa*, as was expected from the lower amount of unsaturated fatty acids in the high CO_2 algae. Hence, there is consensus that acidification lowers the quality of primary producers as food for zooplankters, although in algal communities there may be a lot of dampening of these effects (Urabe and Waki 2009, Algueró-Muñiz et al. 2017), and some unexpected results may be found, as evinced by the finding of Sswat et al. (2018), who observed positive effects on larval herring survival in a mesocosm experiment. Having said this, acidification and increased CO_2 availability are definitely changing the link between grazers and their algal food, in many cases causing a lower efficiency of the flow of matter and energy through this interface.

In a warming environment, both the supply as well as the demands of essential nutritional components of the food may be affected. Again, no information exists on the impacts of warming on the chemical communication pathways between zooplankton and their prey, so we will concentrate again on the consumptive interactions. Let us start with the supply side. Fatty acids are the major component of cell membranes. Unsaturated fatty acids have higher melting points, and consensus is that to maintain membrane fluidity, membranes need more (highly) unsaturated fatty acids at colder temperatures (Hall et al. 2002). Therefore, theoretically, algae grown under lower temperatures would represent a higher quality food for zooplankton. This would imply that warming would yield lower quality food in terms of HUFAs, but as the zooplankton also need less of these to maintain membrane fluidity, the overall impact of warming on this aspect of food quality is still unknown (von Elert and Fink 2018), and needs attention in the future. Since no such mechanistic predictions exist for sterols and amino acids, the effects of temperature in this context have not been studied at all. In the context of ecological stoichiometry, a handful of papers have been published predicting and investigating the effects of temperature change on supply and demand of essential elements. Cross et al. (2015) set the stage by predicting either a high phosphorus demand or a lower one at higher temperatures, depending on the most important mechanisms. If there is a different temperature dependence of anabolic and catabolic processes in ectotherms (Karl and Fischer 2008), and as a result the Q10 of respiratory processes is higher than growth (Larsson and Berglund 2005), then one would expect a higher carbon demand of grazers at higher temperatures. Alternatively, faster growth at higher temperatures might imply higher demands for phosphorus. In geographical studies with fish, herbivory has been reported to be more prevalent at lower latitudes (Floeter et al. 2005), which could be interpreted as a result of a higher demand for carbon at those temperatures, as plants have higher C:X ratios than animals (Moody et al. 2019). Even though the theoretical study of Anderson et al. (2017) suggests that, at least in terms of C:N, there should be small effects of temperature on relative demands of carbon and nutrients, experimental evidence indicates that the TER of many animals is affected (Persson et al. 2011, Malzahn et al. 2016, Zhang et al. 2016), with, in most cases, TER increasing with temperature.

2.9 What Will the Future Bring: The Bigger Picture

Overall, changes in factors affecting the quality of resources and the metabolic requirements of consumers, as well as the biomass of one or more trophic levels, have the potential to disrupt the functioning of planktonic food webs by restructuring trophic interactions. Among the sources of such disturbances, the currently escalating human population growth and its associated activities are arguably the ones having the strongest impact. Before the widespread use of fossil fuels, the world had a population of about 670 million people. By 2019, the world's population had reached

7.7 billion, a more than 10-fold increase in mere 300 years. This major increase in population size, and the coinciding changes in use of energy are responsible for the degradation of many natural environments.

Earth's climate has changed throughout history and most climate changes which occurred in the last 650,000 years are attributed to minor variations in Earth's orbit that modulate the amount of solar energy received by the planet (Otto-Bliesner et al. 2016). However, the current warming trend is of particular significance because it is proceeding at an unprecedented rate. The planet's increasing average surface temperature can be largely explained by carbon dioxide and other human-made emissions into the atmosphere (IPCC 2013). Carbon dioxide emissions have also altered the pH of surface ocean waters. However, human activities not only alter the biogeochemical cycle of carbon, but also that of other nutrients. In the 1800s, developing countries intensively extracted potassium nitrate from mines and imported guano to fertilize their crops. With the depletion of these nitrogen sources, new technological advances were made to convert atmospheric nitrogen into ammonia. Phosphorus levels have also significantly increased because of its use in fertilizers, and other products such as washing powder. While the intensive use of industrial fertilizer significantly enhanced agricultural production in the ninetieth and twentieth centuries, these concentrated nitrogen and phosphorus inputs into the ecosystem have overloaded certain watersheds and caused the eutrophication of many lakes and coastal areas (Elser et al. 2007, Peñuelas et al. 2012).

The already dramatic changes with which zooplankton have to cope will likely continue to intensify as the world population continues to grow. The Intergovernmental Panel on Climate Change (IPCC) has computed different scenarios projecting that, by 2100, temperatures will increase by 0.6–2.0°C and pH will decrease by 0.1–0.4 units in the oceans' upper hundred meters. Further, increasing frequency and intensity of rainfall and storms is expected to decrease coastal waters salinity and the overall water clarity (IPCC 2013). The latter is particularly relevant for phytoplankton-zooplankton interactions as light availability controls the rate at which nutrients are converted into phytoplankton biomass (Cole and Cloern 1987), and as a result the nutritional quality of phytoplankton (Sterner et al. 1997). Urban, agricultural, and industrial development will also continue to alter biogeochemical cycles through nutrient runoffs (Grizzetti et al. 2012, Peñuelas et al. 2012). Zooplankton may also have to cope with lower oxygen availability as the solubility of O_2 is inversely related to temperature and that stratification is predicted to increase with warming (Sarmiento et al. 1998, Bopp et al. 2002, Keeling and Garcia 2002).

2.10 Future Research

The literature has shown that growth, survival, species distribution, and abundance of zooplankton, as well as the interactions between zooplankton and other trophic levels, are all influenced by the above-mentioned changes in environmental conditions (Fabry et al. 2008, Kurihara and Ishimatsu 2008, Richardson 2008, Dam 2013). This highlights the urgent need to apply mitigation strategies to tackle the challenges associated with global change. This work has already started and important mitigation efforts have been in place since a couple of years. Those can be divided into two categories which are integrated at different spatial scales. On the one hand, a number of international agreements were signed to reduce greenhouse gases emissions and limit associated consequences such as warming, acidification, and salinity changes in the ocean. On the other hand, national and regional regulations in Europe and North America have been enforced to reduce, for example, nutrient runoffs. Taking the North Sea as an example of one of the most industrialized marine systems, a succession of conventions and directives (e.g., OSPAR, Nitrates Directive, Marine Strategy Framework Directive) were adopted to reduce nitrogen and phosphorus inputs to areas affected or likely to be affected by eutrophication. As a consequence, nutrient inputs into the North Sea have steadily decreased over the last 20 years, but the nitrogen to phosphorus ratio has steadily increased. This shows that, in spite of the efforts to lower both nutrients, the policies to reduce phosphorus were more successful than those tackling nitrogen (Grizzetti et al. 2012). However, as we have learned above, shifts in the ratios of limiting nutrients are not anodyne as they can alter community composition, size structure, biodiversity, and growth rates of phytoplankton and zooplankton (Sterner and Elser 2002, Frost et al. 2006, Meunier et al. 2016b, 2018).

It is important to note that the success of mitigation strategies tackling greenhouse gas emissions and nutrient outputs strongly rely on scientific knowledge. While studies on the impacts of individual drivers have significantly advanced our understanding of zooplankton ecology and sensitivity (Fabry et al. 2008, Richardson 2008), a more integrative approach, embracing both the fact that many factors have changed simultaneously as well as investigating different organisational levels (individual, population, and community), is needed. This could be achieved by combining different ecological frameworks, such as ecological stoichiometry and trait-based ecology (Meunier et al. 2017a, Hébert and Beisner 2021, this book, Chapter 1), as well as intensifying efforts in relatively novel research topics, such as rapid evolution (Langer et al. 2019, Dam 2013). Despite the urgent need to understand and predict how global change will influence zooplankton, there is still a striking lack of information on multi-driver impacts.

To close this gap, we propose several major topics of research on which future studies should focus. Because plankton distribution, abundance, and composition impact the functioning of entire aquatic ecosystems, future studies should focus on how simultaneous shifts in environmental drivers will alter these key characteristics. To do so, past shifts in the functional structure as well as phenology of planktonic communities could be identified and linked to environmental conditions that have already largely changed over the past decades (Schlüter et al. 2010, Hsieh et al. 2011, Ershova et al. 2015, Winder and Varpe 2021, this book, Chapter 9). In order for climate-related shifts to be linked to productivity on multi-decadal scales, two criteria must be fulfilled: first, a very long and well understood data base is required, and second, the fundamental controlling factors of plankton growth should be determined based on multi-decadal data. Hence, analyses of long-term datasets (O'Brien and Oakes 2021, this book, Chapter 9) should be strengthened to evaluate shifts in zooplankton distribution, abundance, and composition at the ecosystem scale.

Future impacts of interacting drivers on zooplankton individuals, populations, and communities should be investigated using a wide range of experimental approaches. Experiments ranging in scale from single species to large communities and natural comparisons will allow, on the one hand, a proper prediction of future situations given different scenarios, and, on the other hand, the disentanglement of the importance of the different drivers. The effects of global change can be direct and cause physiological changes or indirect and alter trophic interactions. These field, mesocosm, and laboratory experiments should manipulate multiple drivers simultaneously in order to provide a holistic representation of future abiotic conditions in aquatic ecosystems, and to discriminate between direct and indirect processes. This suite of approaches will enable us to study relevant organisational levels (individual, population, and community) and interactions. By providing invaluable data on changes in metabolic rates, population dynamics, and community composition, these experiments will provide the information needed to predict the potential restructuring of plankton communities and the associated shifts in aquatic ecosystem functioning.

In order to understand and predict how zooplankton will be affected by global change, future studies should not only focus on phenotypic responses but also evaluate the evolutionary processes that may enable zooplankton to adapt. Traditionally, evolution has been considered a much slower process than ecological dynamics, but recent studies have found that evolution can be rapid enough to affect ecological processes (Schoener 2011, Ellner 2013, Yamamichi et al. 2015, Fussmann et al. 2007). Although there is unequivocal evidence for genetic adaptation to a variety of environmental drivers, we are still unable to predict whether zooplankton can adapt quickly enough to global environmental change. Hence, future studies should integrate ecological and evolutionary processes within the same timescales to identify how organismal traits related to zooplankton fitness may adapt to changing environmental conditions and how such adaptations may in turn alter zooplankton population and community dynamics.

References

Acharya, K., M. Kyle and J.J. Elser. 2004. Biological stoichiometry of *Daphnia* growth: An ecophysiological test of the growth rate hypothesis. Limnol. Oceanogr. 49: 656–665.

Ahlgren, G., I.B. Gustafsson and M. Boberg. 1992. Fatty acid content and chemical composition of freshwater microalgae. J. Phycol. 28: 37–50.

Algueró-Muñiz, M., S. Alvarez-Fernandez, P. Thor, L.T. Bach, M. Esposito, H.G. Horn et al. 2017. Ocean acidification effects on mesozooplankton community development: Results from a long-term mesocosm experiment. PLOS ONE 12: e0175851.

Anderson, D.M., A.D. Cembella and G.M. Hallegraeff. 2012. Progress in understanding harmful algal blooms: Paradigm shifts and new technologies for research, monitoring, and management. *In*: C.A. Carlson and S.J. Giovannoni [eds.]. Annual Review of Marine Science, Vol. 4. Annual Reviews, Palo Alto.

Anderson, T.R. and D.O. Hessen. 1995. Carbon or nitrogen limitation in marine copepods? J. Plankton Res. 17: 317–331.

Anderson, T.R. and D.W. Pond. 2000. Stoichiometric theory extended to micronutrients: Comparison of the roles of essential fatty acids, carbon, and nitrogen in the nutrition of marine copepods. Limnol. Oceanogr. 45: 1162–1167.

Anderson, T.R., M. Boersma and D. Raubenheimer. 2004. Stoichiometry: Linking elements to biochemicals. Ecology 85: 1193–1202.

Anderson, T.R., D.O. Hessen, M. Boersma, J. Urabe and D.J. Mayor. 2017. Will invertebrates require increasingly carbon-rich food in a warming world? Am. Nat. 190: 725–742.

Arts, M.T. and B.C. Wainmann [eds.]. 1999. Lipids in Freshwater Systems. Springer, New York.

Arts, M.T., M.T. Brett and M.J. Kainz [eds.]. 2009. Lipids in Aquatic Ecosystems. Springer, New York.

Aubert, A.B., C. Svensen, D.O. Hessen and T. Tamelander. 2013. CNP stoichiometry of a lipid-synthesising zooplankton, *Calanus finmarchicus*, from winter to spring bloom in a sub-Arctic sound. J. Mar. Syst. 111: 19–28.

Bach, L.T., S. Alvarez-Fernandez, T. Hornick, A. Stuhr and U. Riebesell. 2017. Simulated ocean acidification reveals winners and losers in coastal phytoplankton. PLOS ONE 12: e0188198.

Bach, L.T. and J. Taucher. 2019. CO_2 effects on diatoms: a synthesis of more than a decade of ocean acidification experiments with natural communities. Ocean Sci. 15: 1159–1175.

Baines, S.B., X. Chen, S. Vogt, N.S. Fisher, B.S. Twining and M.R. Landry. 2016. Microplankton trace element contents: implications for mineral limitation of mesozooplankton in an HNLC area. J. Plankton Res. 38: 256–270.

Basedow, S.L., N.A.L. De Silva, A. Bode and J. Van Beusekorn. 2016. Trophic positions of mesozooplankton across the North Atlantic: estimates derived from biovolume spectrum theories and stable isotope analyses. J. Plankton Res. 38: 1364–1378.

Beaugrand, G., K.M. Brander, J.A. Lindley, S. Souissi and P.C. Reid. 2003. Plankton effect on cod recruitment in the North Sea. Nature 426: 661–664.

Becker, C. and M. Boersma. 2005. Differential effects of phosphorus and fatty acids on *Daphnia magna* growth and reproduction. Limnol. Oceanogr. 50: 388–397.

Becks, L., S.P. Ellner, L.E. Jones and N.G. Hairston. 2010. Reduction of adaptive genetic diversity radically alters eco-evolutionary community dynamics. Ecol. Lett. 13: 989–997.

Bergkvist, J., E. Selander and H. Pavia. 2008. Induction of toxin production in dinoflagellates: the grazer makes a difference. Oecologia 156: 147–154.

Bjaerke, O., P.R. Jonsson, A. Alam and E. Selander. 2015. Is chain length in phytoplankton regulated to evade predation? J. Plankton Res. 37: 1110–1119.

Boersma, M. 2000. The nutritional quality of P-limited algae for *Daphnia*. Limnol. Oceanogr. 45: 1157–1161.

Boersma, M. and C.-P. Stelzer. 2000. Response of a zooplankton community to the addition of unsaturated fatty acids: an enclosure study. Freshwat. Biol. 45: 179–188.

Boersma, M., C. Schöps and E. McCauley. 2001. Nutritional quality of seston for the freshwater herbivore *Daphnia galeata x hyalina*: biochemical versus mineral limitations? Oecologia 129: 342–348.

Boersma, M. and J.J. Elser. 2006. Too much of a good thing: on stoichiometrically balanced diets and maximal growth. Ecology 87: 1335–1340.

Boersma, M. and K.H. Wiltshire. 2006. Gut passage of phosphorus-limited algae through *Daphnia*: do they take up nutrients in the process? Arch. Hydrobiol. 167: 489–500.

Boersma, M., N. Aberle, F.M. Hantzsche, K. Schoo, K.H. Wiltshire and A.M. Malzahn. 2008. Nutritional limitation travels up the food chain. Int. Rev. Hydrobiol. 93: 479–488.

Boersma, M., C. Becker, A.M. Malzahn and S. Vernooij. 2009. Food chain effects of nutrient limitation in primary producers. Mar. Freshwat. Res. 60: 983–989.

Boersma, M., A. Wesche and H.-J. Hirche. 2014. Predation of calanoid copepods on their own and other copepods' offspring. Mar. Biol. 161: 733–743.

Boersma, M., K.H. Wiltshire, S.-M. Kong, W. Greve and J. Renz. 2015. Long-term change in the copepod community in the southern German Bight. J. Sea Res. 101: 41–50.

Boffey, P.M. 1976. International biological program: was It worth the cost and effort? Science 193: 866–868.

Bopp, L., C. Le Quere, M. Heimann, A.C. Manning and P. Monfray. 2002. Climate-induced oceanic oxygen fluxes: Implications for the contemporary carbon budget. Global Biogeochem. Cycles 16: 14.

Brendelberger, H. and W. Geller. 1985. Variability of filter structures in eight *Daphnia* species: mesh sizes and filtering areas. J. Plankton Res. 7: 473–486.

Brett, M.T. and D.C. Müller-Navarra. 1997. The role of highly unsaturated fatty acids in aquatic food web processes. Freshwat. Biol. 38: 483–499.

Brett, M.T., D.C. Müller-Navarra and S.K. Park. 2000. Empirical analysis of the effect of phosphorus limitation on algal food quality for freshwater zooplankton. Limnol. Oceanogr. 45: 1564–1575.

Brown, M.R., S.W. Jeffrey, J.K. Volkman and G.A. Dunstan. 1997. Nutritional properties of microalgae for mariculture. Aquaculture 151: 315–331.

Bullejos, F.J., P. Carrillo, E. Gorokhova, J.M. Medina-Sánchez, E.G. Balseiro and M. Villar-Argaiz. 2013. Shifts in food quality for herbivorous consumer growth: multiple golden means in the life history. Ecology 95: 1272–1284.

Burkhardt, S. and U. Riebesell. 1997. CO_2 availability affects elemental composition (C:N:P) of the marine diatom *Skeletonema costatum*. Mar. Ecol. Prog. Ser. 155: 67–76.

Campoy, A.N., J. Cruz, J. Barcelos e Ramos, F. Viveiros, P. Range and M.A. Teodósio. 2021. Ocean acidification impacts on zooplankton. pp. 64–82. *In*: M.A. Teodósio and A.B. Barbosa [eds.]. Zooplankton Ecology. CRC Press.

Carmichael, W.W. 1992. Cyanobacteria secondary metabolites: the cyanotoxins. J. Appl. Bacteriol. 72: 445–459.

Chen, X., N.S. Fisher and S.B. Baines. 2014. Influence of algal iron content on the assimilation and fate of iron and carbon in a marine copepod. Limnol. Oceanogr. 59: 129–140.

Chevillot, X., H. Drouineau, P. Lambert, L. Carassou, B. Sautour and J. Lobry. 2017. Toward a phenological mismatch in estuarine pelagic food web? PLOS ONE 12: 21.

Chícharo, M.A. and L. Chícharo. 2008. RNA:DNA ratio and other nucleic acid derived indices in marine ecology. Int. J. Mol. Sci. 9: 1453–1471.

Cole, B.E. and J.E. Cloern. 1987. An emperical model for estimating phytoplankton productivity in estuaries. Mar. Ecol. Prog. Ser. 36: 299–305.

Cole, M., P. Lindeque, E. Fileman, C. Halsband and T.S. Galloway. 2015. The impact of polystyrene microplastics on feeding, function and fecundity in the marine copepod *Calanus helgolandicus*. Environ. Sci. Technol. 49: 1130–1137.

Cotner, J.B. 2019. How increased atmospheric carbon dioxide and "The Law of the Minimum" are contributing to environmental obesity. Acta Limnol. Brasil. 31.

Cross, W.F., J.M. Hood, J.P. Benstead, A.D. Huryn and D. Nelson. 2015. Interactions between temperature and nutrients across levels of ecological organization. Global Change Biol. 21: 1025–1040.

Cushing, D.H. 1974. The natural regulation of fish populations. *In*: F.R. Harden Jones [ed.]. Sea Fisheries Research. Paul Elek, London.

Dalsgaard, J., M. St John, G. Kattner, D.C. Müller-Navarra and W. Hagen. 2003. Fatty acid trophic markers in the pelagic marine environment. Adv. Mar. Biol. 46: 225–340.

Dam, H.G. 2013. Evolutionary adaptation of marine zooplankton to global change. Annu. Rev. Mar. Sci. 5: 349–370.

Dawson, R.M. 1998. The toxicology of microcystins. Toxicon 36: 953–962.

DeMott, W.R. 1986. The role of taste in food selection by freshwater zooplankton. Oecologia 69: 334–340.

DeMott, W.R., R.D. Gulati and K. Siewertsen. 1998. Effects of phosphorus-deficient diets on the carbon and phosphorus balance of *Daphnia magna*. Limnol. Oceanogr. 43: 1147–1161.

DeMott, W.R. and R.D. Gulati. 1999. Phosphorus limitation in *Daphnia*: evidence from a long term study of three hypereutrophic Dutch lakes. Limnol. Oceanogr. 44: 1557–1564.

DeMott, W.R. and B.J. Pape. 2005. Stoichiometry in an ecological context: testing for links between *Daphnia* P-content, growth rate and habitat preference. Oecologia 142: 20–27.

Dickman, E.M., M.J. Vanni and M.J. Horgan. 2006. Interactive effects of light and nutrients on phytoplankton stoichiometry. Oecologia 149: 676–689.

Droop, M.R., M.J. Mickelson, J.M. Scott and M.F. Turner. 1982. Light and nutrient status of algal cells. J. Mar. Biol. Assoc. U.K. 62: 403–434.

Eckmann, R., S. Gerster and A. Kraemer. 2006. Yields of European perch from Upper Lake Constance from 1910 to present. Fish. Manage. Ecol. 13: 381–390.

Edwards, M. and A.J. Richardson. 2004. Impact of climate change on marine pelagic phenology and trophic mismatch. Nature 430: 881–884.

Eglite, E., D. Wodarg, J. Dutz, N. Wasmund, G. Nausch, I. Liskow et al. 2018. Strategies of amino acid supply in mesozooplankton during cyanobacteria blooms: a stable nitrogen isotope approach. Ecosphere 9: 20.

Ellner, S.P. 2013. Rapid evolution: from genes to communities, and back again? Funct. Ecol. 27: 1087–1099.

Elser, J.J., M.M. Elser, N.A. Mackay and S.R. Carpenter. 1988. Zooplankton-mediated transitions between N-limited and P-limited algal growth. Limnol. Oceanogr. 33: 1–14.

Elser, J.J. and R.P. Hassett. 1994. A stoichiometric analysis of the zooplankton-phytoplankton interaction in marine and freshwater ecosystems. Nature 370: 211–213.

Elser, J.J., D.R. Dobberfuhl, N.A. Mackay and J.H. Schampel. 1996. Organism size, life history, and N:P stoichiometry. Bioscience 46: 674–684.

Elser, J.J. and J. Urabe. 1999. The stoichiometry of consumer-driven nutrient recycling: theory, observations, and consequences. Ecology 80: 735–751.

Elser, J.J., K. Hayakawa and J. Urabe. 2001. Nutrient limitation reduces food quality for zooplankton: *Daphnia* response to seston phosphorus enrichment. Ecology 82: 898–903.

Elser, J.J., K. Acharya, M. Kyle, J. Cotner, W. Makino, T. Markow et al. 2003. Growth rate-stoichiometry couplings in diverse biota. Ecol. Lett. 6: 936–943.

Elser, J.J., J.H. Schampel, M. Kyle, J. Watts, E.W. Carson, T.A. Dowling et al. 2005. Response of grazing snails to phosphorus enrichment of modern stromatolitic microbial communities. Freshwat. Biol. 50: 1826–1835.

Elser, J.J., J. Watts, J.H. Schampel and J.D. Farmer. 2006. Early Cambrian food webs on a stoichiometric knife-edge? A hypothesis and preliminary data from a modern stromatolite-based ecosystem. Ecol. Lett. 9: 292–300.

Elser, J.J., M.E.S. Bracken, E.E. Cleland, D.S. Gruner, W.S. Harpole, H. Hillebrand et al. 2007. Global analysis of nitrogen and phosphorus limitation of primary producers in freshwater, marine and terrestrial ecosystems. Ecol. Lett. 10: 1135–1142.

Elser, J.J., C. Acquisti and S. Kumar. 2011. Stoichiogenomics: the evolutionary ecology of macromolecular elemental composition. Trends Ecol. Evol. 26: 38–44.

Engel, A., J. Piontek, H.P. Grossart, U. Riebesell, K.G. Schulz and M. Sperling. 2014. Impact of CO_2 enrichment on organic matter dynamics during nutrient induced coastal phytoplankton blooms. J. Plankton Res. 36: 641–657.

Ershova, E.A., R.R. Hopcroft, K.N. Kosobokova, K. Matsuno, R.J. Nelson, A. Yamaguchi et al. 2015. Long-term changes in summer zooplankton communities of the Western Chukchi Sea, 1945–2012. Oceanography 28: 100–115.

Fabry, V., B.A. Seibel, R.A. Feely and J.C. Orr. 2008. Impacts of ocean acidification on marine fauna and ecosystem processes. ICES J. Mar. Sci. 65: 414–432.

Floeter, S.R., M.D. Behrens, C.E.L. Ferreira, M.J. Paddack and M.H. Horn. 2005. Geographical gradients of marine herbivorous fishes: patterns and processes. Mar. Biol. 147: 1435–1447.

Flynn, K.J., J.A. Raven, T.A.V. Rees, Z. Finkel, A. Quigg and J. Beardall. 2010. Is the growth rate hypothesis applicable to microalgae? J. Phycol. 46: 1–12.

Flynn, K.J., A. Mitra, K. Anestis, A.A. Anschutz, A. Calbet, G.D. Ferreira et al. 2019. Mixotrophic protists and a new paradigm for marine ecology: where does plankton research go now? J. Plankton Res. 41: 375–391.

Franco-Santos, R.M., H. Auel, M. Boersma, M. De Troch, C.L. Meunier and B. Niehoff. 2018. Bioenergetics of the copepod *Temora longicornis* under different nutrient regimes. J. Plankton Res. 40: 420–435.

Frost, P.C. and J.J. Elser. 2002. Growth responses of littoral mayflies to the phosphorus content of their food. Ecol. Lett. 5: 232–240.

Frost, P.C., J.P. Benstead, W.F. Cross, H. Hillebrand, J.H. Larson, M.A. Xenopoulos et al. 2006. Threshold elemental ratios of carbon and phosphorus in aquatic consumers. Ecol. Lett. 9: 774–779.

Fussmann, G.F., M. Loreau and P.A. Abrams. 2007. Eco-evolutionary dynamics of communities and ecosystems. Funct. Ecol. 21: 465–477.

Gerritsen, J. and J.R. Strickler. 1977. Encounter probabilities and community structure in zooplankton: a mathematical model. J. Fish. Res. Bd. Can. 34: 73–82.

Giller, P.S., H. Hillebrand, U.G. Berninger, M.O. Gessner, S. Hawkins, P. Inchausti et al. 2004. Biodiversity effects on ecosystem functioning: emerging issues and their experimental test in aquatic environments. Oikos 104: 423–436.

Gismervik, I. 1997. Stoichiometry of some marine planktonic crustaceans. J. Plankton Res. 19: 279–285.

Gladyshev, M.I., N.N. Sushchik and O.N. Makhutova. 2013. Production of EPA and DHA in aquatic ecosystems and their transfer to the land. Prostag. Oth. Lipid M. 107: 117–126.

Gliwicz, Z.M. and W. Lampert. 1993. Body-size related survival of cladocerans in a trophic gradient: an enclosure study. Arch. Hydrobiol. 129: 1–25.

Goldman, J.C., J.J. McCarthy and D.G. Peavey. 1979. Growth-rate influence on the chemical composition of phytoplankton in oceanic waters. Nature 279: 210–215.

Gollasch, S. 2006. Overview on introduced aquatic species in European navigational and adjacent waters. Helgol. Mar. Res. 60: 84–89.

Grebner, W., E.C. Berglund, F. Berggren, J. Eklund, S. Haroadottir, M.X. Andersson et al. 2019. Induction of defensive traits in marine plankton-new copepodamide structures. Limnol. Oceanogr. 64: 820–831.

Grizzetti, B., F. Bouraoui and A. Aloe. 2012. Changes of nitrogen and phosphorus loads to European seas. Global Change Biol. 18: 769–782.

Guisande, C., I. Maneiro and I. Riveiro. 1999. Homeostasis in the essential amino acid composition of the marine copepod *Euterpina acutifrons*. Limnol. Oceanogr. 44: 691–696.

Hairston, N. 2003. Phosphorus: time for us to oust bad spelling. Nature 426: 119–119.

Hall, J.M., C.C. Parrish and R.J. Thompson. 2002. Eicosapentaenoic acid regulates scallop (*Placopecten magellanicus*) membrane fluidity in response to cold. Biol. Bull. 202: 201–203.

Hartwich, M., D. Martin-Creuzburg, K.O. Rothhaupt and A. Wacker. 2012. Oligotrophication of a large, deep lake alters food quantity and quality constraints at the primary producer-consumer interface. Oikos 121: 1702–1712.

Hébert, M.-P. and B.E. Beisner. 2021. Functional trait approaches for the study of metazooplankton ecology. pp. 3–27. *In*: M.A. Teodósio and A.B. Barbosa [eds.]. Zooplankton Ecology. CRC Press.

Heckmann, L.H., R.M. Sibly, M. Timmermans and A. Callaghan. 2008. Outlining eicosanoid biosynthesis in the crustacean *Daphnia*. Front. Zool. 5: 9.

Hessen, D.O. and A. Lyche. 1991. Interspecific and intraspecific variations in zooplankton element composition. Arch. Hydrobiol. 121: 343–353.

Hessen, D.O. and E. van Donk. 1993. Morphological changes in *Scenedesmus* induced by substances released from *Daphnia*. Arch. Hydrobiol. 127: 129–140.

Hessen, D.O. and T.R. Anderson. 2008. Excess carbon in aquatic organisms and ecosystems: Physiological, ecological, and evolutionary implications. Limnol. Oceanogr. 53: 1685–1696.

Heuschele, J. and E. Selander. 2014. The chemical ecology of copepods. J. Plankton Res. 36: 895–913.

Hillebrand, H., G. Steinert, M. Boersma, A.M. Malzahn, C.L. Meunier, C. Plum et al. 2013. Goldman revisited: faster growing phytoplankton has lower N:P and lower stoichiometric flexibility. Limnol. Oceanogr. 56: 2076–2088.

Hsieh, C.H., Y. Sakai, S. Ban, K. Ishikawa, T. Ishikawa, S. Ichise et al. 2011. Eutrophication and warming effects on long-term variation of zooplankton in Lake Biwa. Biogeosciences 8: 1383–1399.

Huntley, M. 1992. GLOBEC—Global Ocean Ecosystem Dynamics. Oceanus 35: 94–99.

Ianora, A., A. Miralto, S.A. Poulet, Y. Carotenuto, I. Buttino, G. Romano et al. 2004. Aldehyde suppression of copepod recruitment in blooms of a ubiquitous planktonic diatom. Nature 429: 403–407.

IPCC. 2013. Climate Change 2013: The Physical Science Basis. Contribution of Working Group I to the Fifth Assessment Report of the Intergovernmental Panel on Climate Change. Cambridge.

Jakobsen, H.H., L.M. Everett and S.L. Strom. 2006. Hydromechanical signaling between the ciliate *Mesodinium pulex* and motile protist prey. Aquat. Microb. Ecol. 44: 197–206.

Jaspers, C., B. Huwer, E. Antajan, A. Hosia, H.H. Hinrichsen, A. Biastoch et al. 2018. Ocean current connectivity propelling the secondary spread of a marine invasive comb jelly across western Eurasia. Global Ecol. Biogeogr. 27: 814–827.

Jeyasingh, P.D. 2007. Plasticity in metabolic allometry: the role of dietary stoichiometry. Ecol. Lett. 10: 282–289.

Jeyasingh, P.D. and K. Pulkkinen. 2019. Does differential iron supply to algae affect *Daphnia* life history? An ionome-wide study. Oecologia 191: 51–60.

Jones, R.I. 2000. Mixotrophy in planktonic protists: an overview. Freshwat. Biol. 45: 219–226.

Jungmann, D. and J. Benndorf. 1994. Toxicity to *Daphnia* of a compound extracted from laboratory and natural Microcystis spp., and the role of microcyctins. Freshwat. Biol. 32: 13–20.

Karl, I. and K. Fischer. 2008. Why get big in the cold? Towards a solution to a life-history puzzle. Oecologia 155: 215–225.

Kattner, G., W. Hagen, R.F. Lee, R. Campbell, D. Deibel, S. Falk-Petersen et al. 2007. Perspectives on marine zooplankton lipids. Can. J. Fish. Aquat. Sci. 64: 1628–1639.

Keeling, R.F. and H.E. Garcia. 2002. The change in oceanic O_2 inventory associated with recent global warming. Proc. Natl. Acad. Sci. USA 99: 7848–7853.

Kiørboe, T. 1989. Phytoplankton growth rate and nitrogen content: implications for feeding and fecundity in a herbivorous copepod. Mar. Ecol. Prog. Ser. 55: 229–234.

Kiørboe, T. and A.W. Visser. 1999. Predator and prey perception in copepods due to hydromechanical signals. Mar. Ecol. Prog. Ser. 179: 81–95.

Kooijman, S.A.L.M. 1998. The synthesizing unit as model for the stoichiometric fusion and branching of metabolic fluxes. Biophys. Chem. 73: 179–188.

Kooijman, S.A.L.M. 2000. Dynamic Energy and Mass Budgets in Biological Systems. Second ed. Cambridge University Press, Cambridge.

Kurihara, H. and A. Ishimatsu. 2008. Effects of high CO_2 seawater on the copepod (*Acartia tsuensis*) through all life stages and subsequent generations. Mar. Pollut. Bull. 56: 1086–1090.

Kutter, V.T., M. Wallner-Kersanach, S.M. Sella, A.L.S. Albuquerque, B.A. Knoppers and E.V. Silva. 2014. Carbon, nitrogen, and phosphorus stoichiometry of plankton and the nutrient regime in Cabo Frio Bay, SE Brazil. Environ. Monit. Assess. 186: 559–573.

Lampert, W. 1977a. Studies on the carbon balance of *Daphnia pulex* as related to environmental conditions. I. Methodological problems of the use of 14C for the measurement of carbon assimilation. Arch. Hydrobiol. (Suppl.) 48: 287–309.

Lampert, W. 1977b. Studies on the carbon balance of *Daphnia pulex* de Geer as related to environmental conditions. II. The dependence of carbon assimilation on animal size, temperature, food concentration and diet species. Arch. Hydrobiol. (Suppl.) 48: 310–335.

Lampert, W. 1977c. Studies on the carbon balance of *Daphnia pulex* de Geer as related to environmental conditions. III. Production and production efficiency. Arch. Hydrobiol. (Suppl.) 48: 336–360.

Lampert, W. 1977d. Studies on the carbon balance of *Daphnia pulex* de Geer as related to environmental conditions. IV. Determination of the "threshold" concentration as a factor controlling the abundance of zooplankton species. Arch. Hydrobiol. (Suppl.) 48: 361–368.

Lampert, W. 1978. Release of dissolved organic carbon by zooplankton grazing. Limnol. Oceanogr. 23: 831–834.

Langer, J.A.F., C.L. Meunier, U. Ecker, H.G. Horn, K. Schwenk and M. Boersma. 2019. Acclimation and adaptation of the coastal calanoid copepod *Acartia tonsa* to ocean acidification: a long-term laboratory investigation. Mar. Ecol. Prog. Ser. 619: 35–51.

Larsson, P. and S.I. Dodson. 1993. Chemical communication in planktonic animals. Arch. Hydrobiol. 129: 129–155.

Larsson, S. and I. Berglund. 2005. The effect of temperature on the energetic growth efficiency of Arctic charr (*Salvelinus alpinus* L.) from four Swedish populations. J. Therm. Biol. 30: 29–36.

Laspoumaderes, C., B. Modenutti, J.J. Elser and E. Balseiro. 2015. Does the stoichiometric carbon:phosphorus knife edge apply for predaceous copepods? Oecologia 178: 557–569.

Lass, S. and P. Spaak. 2003. Chemically induced anti-predator defences in plankton: a review. Hydrobiologia 491: 221–239.

Legendre, L. and R.B. Rivkin. 2008. Planktonic food webs: microbial hub approach. Mar. Ecol. Prog. Ser. 365: 289–309.

Legendre, L. and R.B. Rivkin. 2015. Flows of biogenic carbon within marine pelagic food webs: roles of microbial competition switches. Mar. Ecol. Prog. Ser. 521: 19–30.

Loladze, I. and J.J. Elser. 2011. The origins of the Redfield nitrogen-to-phosphorus ratio are in a homoeostatic protein-to-rRNA ratio. Ecol. Lett. 14: 244–250.

Lorenz, C., L. Roscher, M.S. Meyer, L. Hildebrandt, J. Prume, M.G.J. Loder et al. 2019a. Spatial distribution of microplastics in sediments and surface waters of the southern North Sea. Environ. Pollut. 252: 1719–1729.

Lorenz, P., G. Trommer and H. Stibor. 2019b. Impacts of increasing nitrogen:phosphorus ratios on zooplankton community composition and whitefish (*Coregonus macrophthalmus*) growth in a pre-alpine lake. Freshwat. Biol. 64: 1210–1225.

Lürling, M., H.J. de Lange and E. van Donk. 1997. Changes in food quality of the green alga *Scenedesmus* induced by *Daphnia* infochemicals: biochemical composition and morphology. Freshwat. Biol. 38: 619–628.

Lürling, M. 2001. Grazing-associated infochemicals induce colony formation in the green alga *Scenedesmus*. Protist 152: 7–16.

Main, T.M., D.R. Dobberfuhl and J.J. Elser. 1997. N: P stoichiometry and ontogeny of crustacean zooplankton: a test of the growth rate hypothesis. Limnol. Oceanogr. 42: 1474–1478.

Malzahn, A.M., N. Aberle, C. Clemmesen and M. Boersma. 2007. Nutrient limitation of primary producers affects planktivorous fish condition. Limnol. Oceanogr. 52: 2062–2071.

Malzahn, A.M., F. Hantzsche, K.L. Schoo, M. Boersma and N. Aberle. 2010. Differential effects of nutrient-limited primary production on primary, secondary or tertiary consumers. Oecologia 162: 35–48.

Malzahn, A.M. and M. Boersma. 2012. Effects of poor food quality on copepod growth are dose dependent and non-reversible. Oikos 121: 1408–1416.

Malzahn, A.M., D. Doerfler and M. Boersma. 2016. Junk food gets healthier when it's warm. Limnol. Oceanogr. 61: 1677–1685.

Maranon, E., P. Cermeno, D.C. Lopez-Sandoval, T. Rodriguez-Ramos, C. Sobrino, M. Huete-Ortega et al. 2013. Unimodal size scaling of phytoplankton growth and the size dependence of nutrient uptake and use. Ecol. Lett. 16: 371–379.

Martel, C.M. 2009. Conceptual bases for prey biorecognition and feeding selectivity in the microplanktonic marine phagotroph Oxyrrhis marina. Microb. Ecol. 57: 589–597.

Martin-Creuzburg, D. and E. Von Elert. 2004. Impact of 10 dietary sterols on growth and reproduction of *Daphnia galeata*. J. Chem. Ecol. 30: 483–500.

Martin-Creuzburg, D., E. Sperfeld and A. Wacker. 2009. Colimitation of a freshwater herbivore by sterols and polyunsaturated fatty acids. Proc. R. Soc. B-Biol. Sci. 276: 1805–1814.

Meunier, C.L., K. Schulz, M. Boersma and A.M. Malzahn. 2013. Impact of swimming behaviour and nutrient limitation on predator–prey interactions in pelagic microbial food webs. J. Exp. Mar. Biol. Ecol. 446: 29–35.

Meunier, C.L., A.M. Malzahn and M. Boersma. 2014. A new approach to homeostatic regulation: towards a unified view of physiological and ecological concepts. PLOS ONE 9: e107737.

Meunier, C.L., M. Boersma, K.H. Wiltshire and A.M. Malzahn. 2016a. Zooplankton eat what they need: copepod selective feeding and potential consequences for marine systems. Oikos 125: 50–58.

Meunier, C.L., M.J. Gundale, I.S. Sanchez and A. Liess. 2016b. Impact of nitrogen deposition on forest and lake food webs in nitrogen-limited environments. Global Change Biol. 22: 164–179.

Meunier, C., M. Boersma, R. El-Sabaawi, H. Halvorson, E. Herstoff, D. Van de Waal et al. 2017a. From elements to function: unifying ecological stoichiometry and trait-based ecology. Front. Env. Sci. 5.

Meunier, C.L., M. Algueró-Muñiz, H.G. Horn, J.A.F. Lange and M. Boersma. 2017b. Direct and indirect effects of near-future pCO_2 levels on zooplankton dynamics. Mar. Freshwat. Res. 68: 373–380.

Meunier, C.L., S. Alvarez-Fernandez, A.Ö. Cunha-Dupont, C. Geisen, A.M. Malzahn, M. Boersma et al. 2018. The craving for phosphorus in heterotrophic dinoflagellates and its potential implications for biogeochemical cycles. Limnol. Oceanogr. 63: 1774–1784.

Moody, E.K., N.K. Lujan, K.A. Roach and K.O. Winemiller. 2019. Threshold elemental ratios and the temperature dependence of herbivory in fishes. Funct. Ecol. 33: 913–923.

Müller-Navarra, D.C. 1995a. Biochemical versus mineral limitation in *Daphnia*. Limnol. Oceanogr. 40: 1209–1214.

Müller-Navarra, D.C. 1995b. Evidence that a highly unsaturated fatty acid limits *Daphnia* growth in nature. Arch. Hydrobiol. 132: 297–307.

Müller-Navarra, D.C., M.T. Brett, A.M. Liston and C.R. Goldman. 2000. A highly unsaturated fatty acid predicts carbon transfer between primary producers and consumers. Nature 403: 74–77.

Nielsen, J.M., B.N. Popp and M. Winder. 2015. Meta-analysis of amino acid stable nitrogen isotope ratios for estimating trophic position in marine organisms. Oecologia 178: 631–642.

O'Brien, T.D. and S. Oakes. 2021. Visualizing and exploring zooplankton spatial-temporal variability. pp. 192–224. *In*: M.A. Teodósio and A.B. Barbosa [eds.]. Zooplankton Ecology. CRC Press.

O'Neil, J.M., T.W. Davis, M.A. Burford and C.J. Gobler. 2012. The rise of harmful cyanobacteria blooms: The potential roles of eutrophication and climate change. Harmful Algae 14: 313–334.

Olsen, Y., A. Jensen, H. Reinertsen, K.Y. Børsheim, M. Heldal and A. Langeland. 1986. Dependence of the rate of release of phosphorus by zooplankton on the P:C ratio in the food supply, as calculated by a recycling model. Limnol. Oceanogr. 31: 34–44.

Otto-Bliesner, B.L., E.C. Brady, J. Fasullo, A. Jahn, L. Landrum, S. Stevenson et al. 2016. Climate variability and change since 850 CE: an ensemble approach with the community earth system model. Bull. Am. Met. Soc. 97: 735–754.

Pančić, M. and T. Kiørboe. 2018. Phytoplankton defence mechanisms: traits and trade-offs. Biol. Rev. 93: 1269–1303.

Park, S., M.T. Brett, E.T. Oshel and C.R. Goldman. 2003. Seston food quality and *Daphnia* production efficiencies in an oligo-mesotrophic Subalpine Lake. Aquat. Ecol. 37: 123–136.

Paulsen, M., C. Clemmesen and A.M. Malzahn. 2014. Essential fatty acid (docosahexaenoic acid, DHA) availability affects growth of larval herring in the field. Mar. Biol. 161: 239–244.

Peñuelas, J., J. Sardans, A. Rivas-Ubach and I.A. Janssens. 2012. The human-induced imbalance between C, N and P in Earth's life system. Global Change Biol. 18: 3–6.

Persson, J., M.T. Brett, T. Vrede and J.L. Ravet. 2007. Food quantity and quality regulation of trophic transfer between primary producers and a keystone grazer (*Daphnia*) in pelagic freshwater food webs. Oikos 116: 1152–1163.

Persson, J., M. Wojewodzic, D. Hessen and T. Andersen. 2011. Increased risk of phosphorus limitation at higher temperatures for *Daphnia magna*. Oecologia 165: 123–129.

Plath, K. and M. Boersma. 2001. Mineral limitation of zooplankton: stoichiometric constraints and optimal foraging. Ecology 82: 1260–1269.

Pohnert, G., M. Steinke and R. Tollrian. 2007. Chemical cues, defence metabolites and the shaping of pelagic interspecific interactions. Trends Ecol. Evol. 22: 198–204.

Poloczanska, E.S., C.J. Brown, W.J. Sydeman, W. Kiessling, D.S. Schoeman, P.J. Moore et al. 2013. Global imprint of climate change on marine life. Nat. Clim. Change 3: 919–925.

Ravet, J.L., J. Persson and M.T. Brett. 2012. Threshold dietary polyunsaturated fatty acid concentrations for *Daphnia pulex* growth and reproduction. Inland Waters 2: 199–209.

Richardson, A.J. 2008. In hot water: zooplankton and climate change. ICES J. Mar. Sci. 65: 279–295.

Riebesell, U. and P.D. Tortell. 2011. Effects of ocean acidification on pelagic organisms and ecosystems. *In*: J.P. Gattuso and L. Hansson [eds.]. Ocean Acidification. Oxford University Press, Oxford.

Rossoll, D., R. Bermúdez, H. Hauss, K.G. Schulz, U. Riebesell, U. Sommer et al. 2012. Ocean acidification-induced food quality deterioration constrains trophic transfer. PLoS ONE 7: e34737.

Ruess, L. and D.C. Müller-Navarra. 2019. Essential biomolecules in food webs. Front. Ecol. Evol. 7: 18.

Saha, M., E. Berdalet, Y. Carotenuto, P. Fink, T. Harder, U. John et al. 2019. Using chemical language to shape future marine health. Front. Ecol. Environ. 17: 530–537.

Sargent, J., L. McEvoy, A. Estevez, G. Bell, M. Bell, J. Henderson et al. 1999. Lipid nutrition of marine fish during early development: current status and future directions. Aquaculture 179: 217–229.

Sarmiento, J.L., T.M.C. Hughes, R.J. Stouffer and S. Manabe. 1998. Simulated response of the ocean carbon cycle to anthropogenic climate warming. Nature 393: 245–249.

Schlüter, M.H., A. Merico, M. Reginatto, M. Boersma, K.H. Wiltshire and W. Greve. 2010. Phenological shifts of three interacting zooplankton groups in relation to climate change. Global Change Biol. 16: 3144–3153.

Schoener, T.W. 2011. The newest synthesis: understanding the interplay of evolutionary and ecological dynamics. Science 331: 426–429.

Schoo, K.L., N. Aberle, A.M. Malzahn and M. Boersma. 2012. Food quality affects secondary consumers even at low quantities: an experimental test with larval European lobster. PLoS ONE 7: e33550.

Schoo, K.L., A.M. Malzahn, E. Krause and M. Boersma. 2013. Increased carbon dioxide availability alters phytoplankton stoichiometry and affects carbon cycling and growth of a marine planktonic herbivore. Mar. Biol. 160: 2145–2155.

SCOR. 1987. The Joint Global Ocean Flux Study—background, goals, organisation, and next steps. Report of the International Scientific Planning and Co-ordination Meeting for Global Ocean Flux Studies sponsored by the Scientific Committee on Oceanic Research held at ICSU headquarters, Paris, 17–19 February, 1987.

Selander, E., E.C. Berglund, P. Engstrom, F. Berggren, J. Eklund, S. Hardardottir et al. 2019. Copepods drive large-scale trait-mediated effects in marine plankton. Science Advances 5.

Sommer, U. 1992. Phosphorus-limited *Daphnia*: intraspecific facilitation instead of competition. Limnol. Oceanogr. 37: 966–973.

Sommer, U. and F. Sommer. 2006. Cladocerans versus copepods: the cause of contrasting top–down controls on freshwater and marine phytoplankton. Oecologia 147: 183–194.

Sommer, U., R. Adrian, B. Bauer and M. Winder. 2012. The response of temperate aquatic ecosystems to global warming: novel insights from a multidisciplinary project. Mar. Biol. 159: 2367–2377.

Sperfeld, E., D. Martin-Creuzburg and A. Wacker. 2012. Multiple resource limitation theory applied to herbivorous consumers: Liebig's minimum rule vs. interactive co-limitation. Ecol. Lett. 15: 142–150.

Sperfeld, E., D. Raubenheimer and A. Wacker. 2016. Bridging factorial and gradient concepts of resource co-limitation: towards a general framework applied to consumers. Ecol. Lett. 19: 201–215.

Sswat, M., M.H. Stiasny, J. Taucher, M. Algueró-Muñiz, L.T. Bach, F. Jutfelt et al. 2018. Food web changes under ocean acidification promote herring larvae survival. Nat. Ecol. Evol. 2: 836–840.

Sterner, R.W., D.D. Hagemeier and W.L. Smith. 1993. Phytoplankton nutrient limitation and food quality for *Daphnia*. Limnol. Oceanogr. 38: 857–871.

Sterner, R.W. and D.O. Hessen. 1994. Algal nutrient limitation and the nutrition of aquatic herbivores. Annu. Rev. Ecol. Syst. 25: 1–29.

Sterner, R.W., J.J. Elser, E.J. Fee, S.J. Guildford and T.H. Chrzanowski. 1997. The light: nutrient ratio in lakes: the balance of energy and materials affects ecosystem structure and process. Am. Nat. 150: 663–684.

Sterner, R.W., J. Clasen, W. Lampert and T. Weisse. 1998. Carbon: phosphorus stoichiometry and food chain production. Ecol. Lett. 1: 146–150.

Sterner, R.W. and N.B. George. 2000. Carbon, nitrogen, and phosphorus stoichiometry of cyprinid fishes. Ecology 81: 127–140.

Sterner, R.W. and J.J. Elser. 2002. Ecological Stoichiometry: The Biology of Elements from Molecules to the Biosphere. Princeton University Press, Princeton, NJ.

Sterner, R.W., T. Andersen, J.J. Elser, D.O. Hessen, J.M. Hood, E. McCauley et al. 2008. Scale-dependent carbon: nitrogen: phosphorus seston stoichiometry in marine and freshwaters. Limnol. Oceanogr. 53: 1169–1180.

Stoecker, D.K. 1998. Conceptual models of mixotrophy in planktonic protists and some ecological and evolutionary implications. Eur. J. Protistol. 34: 281–290.

Tillberg, J.E., T. Barnard and J.R. Rowley. 1984. Phosphorus status and cytoplasmic structures in *Scenedesmus* (Chlorophyceae) under different metabolic regimes. J. Phycol. 20: 124–136.

Tittel, J., V. Bissinger, B. Zippel, U. Gaedke, E. Bell, A. Lorke et al. 2003. Mixotrophs combine resource use to outcompete specialists: Implications for aquatic food webs. Proc. Natl. Acad. Sci. USA 100: 12776–12781.

Trier, T.M. and W.J. Mattson. 2003. Diet-induced thermogenesis in insects: A developing concept in nutritional ecology. Environ. Entomol. 32: 1–8.

Urabe, J. and R.W. Sterner. 1996. Regulation of herbivore growth by the balance of light and nutrients. Proc. Natl. Acad. Sci. USA 93: 8465–8469.

Urabe, J., J. Togari and J.J. Elser. 2003. Stoichiometric impacts of increased carbon dioxide on a planktonic herbivore. Global Change Biol. 9: 818–825.

Urabe, J. and N. Waki. 2009. Mitigation of adverse effects of rising CO_2 on a planktonic herbivore by mixed algal diets. Global Change Biol. 15: 523–531.

van de Waal, D.B. and M. Boersma. 2012. Ecological stoichiometry in aquatic ecosystems. *In*: U.-E.J. Committee [ed.]. Encyclopedia of Life Support Systems (EOLSS), Developed under the Auspices of the UNESCO. EOLSS Publishers, Oxford, UK.

van Donk, E., M. Lürling, D.O. Hessen and G.M. Lokhorst. 1997. Altered cell wall morphology in nutrient-deficient phytoplankton and its impact on grazers. Limnol. Oceanogr. 42: 357–364.

Van Donk, E., A. Ianora and M. Vos. 2011. Induced defences in marine and freshwater phytoplankton: a review. Hydrobiologia 668: 3–19.

von Elert, E. and A. Franck. 1999. Colony formation in *Scenedesmus*: grazer-mediated release and chemical features of the infochemical. J. Plankton Res. 21: 789–804.

von Elert, E. and T. Wolffrom. 2001. Supplementation of cyanobacterial food with polyunsaturated fatty acids does not improve growth of *Daphnia*. Limnol. Oceanogr. 46: 1552–1558.

von Elert, E. 2002. Determination of limiting polyunsaturated fatty acids in *Daphnia galeata* using a new method to enrich food algae with single fatty acids. Limnol. Oceanogr. 47: 1764–1773.

von Elert, E. 2004. Food quality constraints in *Daphnia*: interspecific differences in the response to the absence of a long chain polyunsaturated fatty acid in the food source. Hydrobiologia 526: 187–196.

von Elert, E. and P. Fink. 2018. Global warming: testing for direct and indirect effects of temperature at the interface of primary producers and herbivores is required. Front. Ecol. Evol. 6.

von Liebig, J. 1855. Die Grundsätze der Agrikulturchemie. Vieweg, Braunschweig.

Voss, M. 1991. Content of copepod fecal pellets in relation to food-supply in kiel-bight and its effect on sedimentation-rate. Mar. Ecol. Prog. Ser. 75: 217–225.

Wacker, A. and E. von Elert. 2001. Polyunsaturated fatty acids: evidence for non-substitutable biochemical resources in *Daphnia galeata*. Ecology 82: 2507–2520.

Wacker, A. and D. Martin-Creuzburg. 2007. Allocation of essential lipids in *Daphnia magna* during exposure to poor food quality. Funct. Ecol. 21: 738–747.

Wacker, A. and D. Martin-Creuzburg. 2012. Biochemical nutrient requirements of the rotifer *Brachionus calyciflorus*: co-limitation by sterols and amino acids. Funct. Ecol. 26: 1135–1143.

Ware, D.M. and R.E. Thomson. 2005. Bottom-up ecosystem trophic dynamics determine fish production in the northeast Pacific. Science 308: 1280–1284.

Weers, P.M.M. and R.D. Gulati. 1997. Effect of the addition of polyunsaturated fatty acids to the diet on the growth and fecundity of *Daphnia galeata*. Freshwat. Biol. 38: 721–729.

Weers, P.M.M., K. Siewertsen and R.D. Gulati. 1997. Is the fatty acid composition of *Daphnia galeata* determined by the fatty acid composition of the ingested diet. Freshwat. Biol. 38: 731–738.

White, T.C.R. 1993. The Inadequate Environment. Springer Verlag, Berlin.

Wichard, T., A. Gerecht, M. Boersma, S.A. Poulet, K. Wiltshire and G. Pohnert. 2007. Lipid and fatty acid composition of diatoms revisited: Rapid wound-activated change of food quality parameters influences herbivorous copepod reproductive success. ChemBioChem 8: 1146–1153.

Wilken, S., J. Huisman, S. Naus-Wiezer and E. Donk. 2013. Mixotrophic organisms become more heterotrophic with rising temperature. Ecol. Lett. 16: 225–233.

Wiltshire, K.H. and W. Lampert. 1999. Urea excretion by *Daphnia*: A colony-inducing factor in *Scenedesmus*? Limnol. Oceanogr. 44: 1894–1903.

Wiltshire, K.H., M. Boersma and B. Meyer. 2003. Grazer-induced changes in the desmid *Staurastrum*. Hydrobiologia 491: 255–260.

Winder, M. and O. Varpe. 2021. Interactions in plankton food webs: seasonal succession and phenology of Baltic Sea zooplankton. pp. 162–191. *In*: M.A. Teodósio and A.B. Barbosa [eds.]. Zooplankton Ecology. CRC Press.

Woods, H.A. and J.K. Wilson. 2013. An information hypothesis for the evolution of homeostasis. Trends Ecol. Evol. 28: 283–289.

Wootton, E.C., M.V. Zubkov, D.H. Jones, R.H. Jones, C.M. Martel, C.A. Thornton et al. 2007. Biochemical prey recognition by planktonic protozoa. Environ. Microbiol. 9: 216–222.

Yamamichi, M., C.L. Meunier, A. Peace, C. Prater and M.A. Rua. 2015. Rapid evolution of a consumer stoichiometric trait destabilizes consumer-producer dynamics. Oikos 124: 960–969.

Yasumoto, K., A. Nishigami, M. Yasumoto, F. Kasai, Y. Okada, T. Kusumi et al. 2005. Aliphatic sulfates released from *Daphnia* induce morphological defense of phytoplankton: isolation and synthesis of kairomones. Tetrahedron Lett. 46: 4765–4767.

Yoshida, T., L.E. Jones, S.P. Ellner, G.F. Fussmann and N.G. Hairston. 2003. Rapid evolution drives ecological dynamics in a predator-prey system. Nature 424: 303–306.

Zanotto, F.P., S.M. Gouveia, S.J. Simpson, D. Raubenheimer and P.C. Calder. 1997. Nutritional homeostasis in locusts: Is there a mechanism for increased energy expenditure during carbohydrate overfeeding? J. Exp. Biol. 200: 2437–2448.

Zhang, P., B.A. Blonk, R.F. van den Berg and E.S. Bakker. 2016. The effect of temperature on herbivory by the omnivorous ectotherm snail *Lymnaea stagnalis*. Hydrobiologia 812: 147–155.

CHAPTER 3

Leading Hypothesis about the Influence of Temperate Marine Fish Larvae on Recruitment Variability that Shaped Larval Ecology

Pedro Morais

3.1 Introduction

Coastal fishing populations have witnessed through centuries the variability of fisheries with years of rich and extremely poor yields separated by just a few years (Hjort 1914, Bakun 1996). Johan Hjort stated that "*these great fluctuations, irregular as they must at first sight appear, have naturally for many years past occupied the minds of the population along the coast, and innumerable hypotheses and suggestions have been put forward by way of explanation*" (Hjort 1914, pp. 4). However, such sharp fluctuations have occurred even before large-scale commercial fisheries. Therefore, the dynamic interaction between fish and the biological, chemical, and physical oceanographic conditions must play a pivotal role in fish recruitment. Indeed, by the time when Hjort presented the Critical Period Hypothesis, he was already convinced that "*the renewal of the fish stock, as in the case of any stock on land, is dependent upon many factors, all necessary, and all more or less variable*" (Hjort 1914, pp. 203). Therefore, it comes as no surprise that several other hypotheses were proposed after the Critical Period Hypothesis (Hjort 1914). The premise shared by all the hypotheses laid since then is that events happening early on at the life of fish largely dictate recruitment variability. Some hypotheses added nuances to previous hypotheses, while others challenged established dogmas. Some hypotheses focused on revealing or hypothesizing on links between biological processes, while others unveiled the mechanisms linking the biology of fish and the dynamics of physical and chemical oceanographic events and other biota. These hypotheses highlighted either the processes or mechanisms occurring when fish were larvae or already as adults. The timing and duration of events, the spatial scale at which they occur (from millimeters to hundreds of kilometers), and their influence on fish larval stages and adults also vary between hypotheses.

CCMAR – Centre of Marine Sciences, University of Algarve, Campus de Gambelas, 8005-139 Faro, Portugal.
Email: pmorais@ualg.pt

Thus, in this chapter, the processes and mechanisms that influence the mortality of fish larval stages in oceanic and coastal waters will be reviewed by summarizing the findings of the leading hypotheses described since the early 20th century. More recently, the accumulating evidence on the crucial role of fish larvae behavior in modulating the processes and mechanisms regulating fish recruitment adds a new layer of complexity that improves our knowledge of fish larval ecology.

3.2 Theories Explaining Fish Recruitment Variability

3.2.1 *Critical Period Hypothesis by Hjort (1914)*

In his seminal work, Hjort (1914) revealed two aspects about the fluctuation of fisheries. First, Hjort (1914) demonstrated that the year class strength of a cohort influences the fishery yield through subsequent years. This was obvious with the very strong 1904 cohort of Atlantic herring *Clupea harengus* Linnaeus, 1758 which was the strongest cohort of the exploited stock between 1907 and 1919—from ~ 22% (1907) to ~ 70% (1910, 1911) (Hjort 1914, 1926). It is essential to highlight that Hjort (1914) deduced that strong year-class cohorts do not necessarily produce other strong cohorts because of yearly fluctuations on the survival of fish larval stages (Hjort 1914). This leads to the second important finding made by Hjort (1914) which leads to the formulation of the Critical Period Hypothesis and even the basis for a hypothesis set 62 years later, the Match-Mismatch Hypothesis (Cushing and Dickson 1976). So, let us read Hjort's words: "*...later on in the spring (at a time varying probably as to date in different years), that enormous quantities of microscopical plant organisms (diatoms, flagellata, peridinea) suddenly make their appearance, being found in the form of a thick, slimy, odoriferous layer on the silk of the net, which had previously been perfectly clean, containing nothing beyond fish eggs and some few crustaceans. It therefore occurred to me during these last investigations that it should be worth the while to endeavour to ascertain how far the sudden appearance of this extensive growth might be of importance for the continued existence of the young fish larvae. If the time when the eggs of the fish are spawned, and the time of occurrence of this plant growth both be variable, it is hardly likely that both would always correspond in point of time and manner. It may well be imagined, for instance, that a certain—though possibly brief—lapse of time might occur between the period when the young larvae first require extraneous nourishment, and the period when such nourishment is first available. If so, it is highly probable that an enormous mortality would result. It would then also be easy to understand that even the richest spawning might yield but a poor amount of fish, while poorer spawning, taking place at a time more favourable in respect of the future nourishment of the young larvae, might often produce the richest year classes.*" (Hjort 1914, pp. 205).

Overall, the Critical Period Hypothesis—which was not labeled as such by Hjort (1914)—alludes to the transition between endogenous and exogenous feeding of fish larvae, which is deemed as critical for their survival if suitable feeding conditions do not occur during that period. The timing is thus of essence for larvae. However, this timing is also reliant on the timing when adult fish spawn (Hjort 1914). For example, the spring spawning events that occurred in April of 1903 and 1904 determined rather strong year-classes, in opposition to spawning events occurring during late winter —"*the spawning set in so late as to ensure an adequate supply of nourishment for the young larvae at the stage when this was required.*" (Hjort 1914, pp. 206).

3.2.2 *Aberrant Drift Hypothesis by Hjort (1926)*

Hjort adds another layer of complexity to his 1914 hypothesis in 1926. He stated that year-class strength and variability is likely to be determined by the availability of appropriate food for fish larvae when they hatch (i.e., the Critical Period Hypothesis) and advection from nursery grounds. Hjort (1926) states that offshore advection could disable larvae from returning to the continental shelf in time of benefiting from coastal summer productivity. This theory was later labeled as the Aberrant Drift Hypothesis. Hjort also mentions the work of H. H. Gran, who established a relationship between freshwater inflow and coastal productivity and how increased productivity could eventually support a strong year-class (Hjort 1926). However, if a year of higher river inflow would not coincide with a strong year-class, that would mean that advection would have prevailed and led to increased mortality. Hjort refers to the work of O. Sund that examined the temporal variability of precipitation and year-class strength. The findings were remarkable. The years with the lowest precipitation (rain and snow) records from 1895 and 1925 coincided with the strongest year-classes of Atlantic cod *Gadus morhua* Linnaeus, 1758. For example, the very strong year-class of 1904 coincided with the third driest year during this period, which would imply that coastal productivity by itself is insufficient to guarantee larval survival (Hjort 1926). This conundrum led Hjort to set the ground for future research when he wrote, "… *it seems to me for several reasons desirable not to attack this important problem from any preconceived standpoint. On the contrary, the simultaneous investigation of meteorology, hydrography and biology seems the only way to a deeper understanding of the conditions in which the destiny of the spawned ova is being decided*" (Hjort 1926).

3.2.3 *Migration Triangle Hypothesis by Harden Jones (1968)*

In 1968, Harden Jones introduced the Migration Triangle Hypothesis, which postulates that migratory fish—like the Atlantic salmon, Atlantic cod, Atlantic herring, European plaice, and the European eel—use distinct habitats during their life that serve different functions. The notion that fish have to use different habitats during their life is evident in the Aberrant Drift Hypothesis (Hjort 1926), but Harden Jones (1968) added a new layer of complexity when he describes how fish use these habitats during their life and how fish move between some of these areas. So, the first habitat used during a fish life is naturally the spawning area—which may coincide with the hatching ground, or not; a second habitat would serve as a nursery area, and a third habitat—which might include segregated feeding and wintering areas—will be used by the adult stock to develop and mature before reproduction (Harden Jones 1968). Implicitly, the Migration Triangle Hypothesis highlights the importance of ecosystem connectivity for the closure of a population life cycle as well as to the homing (or philopatric) behavior of fish. Another important aspect is about how fish migrate between these three different areas, either swimming or drifting with the current (i.e., denant migration), or the opposite (i.e., contranant migration). The movement between the spawning ground and the nursery area was attributed as being denatant, i.e., swimming, drifting, or migrating with the current. However, it is essential to note that evidence acquired during the first two decades of the 21st century indicate that marine postflexion fish larvae have the sensorial capability to detect coastal nursery areas, and the physical stamina to swim against the direction of the prevailing coastal currents if necessary to then settle in a nursery area (Teodósio et al. 2016). Regarding the migrations of the adult fish towards the spawning area while still in the ocean, there was no supporting evidence to argue in favor of either a denatant or contranant migration towards the latter area (Harden Jones 1968). Finally, the underlying concept of the Migration Triangle Hypothesis has ramifications nowadays in the way some fish stocks are managed, particularly diadromous fish species (Morais and Daverat 2016), when aiming to ensure resilience and stability of fish populations by enhancing the connectivity between spatially-segregated habitats.

3.2.4 *Match-Mismatch Hypothesis by Cushing and Dickson (1976) and Cushing (1990)*

The mechanisms encapsulated by the Match-Mismatch Hypothesis were vaguely described by Hjort (1914). However, the data needed to support it was only recorded in the following decades, specifically those on the timing and variability of fish spawning and plankton production. There are two versions of the Match-Mismatch Hypothesis (Cushing and Dickson 1976, Cushing 1990), with the later version interpreting data on the Atlantic cod, Atlantic herring, European plaice *Pleuronectes platessa* Linnaeus, 1758, and sockeye salmon *Oncorhynchus nerka* (Walbaum, 1792). Overall, the Match-Mismatch Hypothesis clearly emphasizes the importance of fish larval survival for stock variability and the temporal framework of the processes involved, rather than the spatial variability of these processes.

The original Match-Mismatch Hypothesis proposed that the recruitment success of temperate fish, whose reproduction occurs at a relatively fixed time during spring and autumn, depends on the timing of the onset and duration of the primary production blooms (Cushing 1990). The original hypothesis relied on the critical-depth model for primary production developed by Sverdrup (1953), in which primary production occurs during mixed conditions in spring and stratified conditions in autumn at temperate latitudes in the Northeast Atlantic Ocean (Cushing 1990). Please note that current models about phytoplankton dynamics and blooms in the ocean have been proposed, such as the Dilution Recoupling Hypothesis (Behrenfeld 2010), Critical Turbulence Hypothesis (Huisman et al. 1999), and Onset of Stratification Hypothesis (Chiswell 2011). However, this review will not delve into how these new models may interact with the mechanisms controlling fish stock variability. For a full review, consult Behrenfeld and Boss (2018).

The second Match-Mismatch Hypothesis introduces a series of caveats concerning the conditions leading to the onset of primary production blooms in distinct wind regimes, depths, and even latitudes. Cushing (1990) recognized that phytoplankton blooms initiate earlier in enclosed shallow waters and later in deeper waters. In shallower ecosystems, production is controlled by irradiance, whereas for deeper systems, atmospheric drivers of water column stratification and the effect of the exponential attenuation of light over a more extensive mixed layer depth, are also relevant. For example, in offshore environments, strong winds during winter and spring increase mixed layer depth, thus decreasing the average photosynthetic active radiation available for phytoplankton, delay the spring bloom, and increase the chances for a mismatch between fish larvae and their prey. The Match-Mismatch Hypothesis is not applicable during the summer in stratified waters because there are no fluctuations in primary production. Cushing (1990) also proposed that the timing of reproduction is fixed polewards from latitude 40° (as in the original hypothesis), while towards the equator the adult fish will match reproduction with the production of larval food in upwelling areas and oceanic divergences. These two reproductive strategies can be regarded as an evolutionary adaptation of fish to prevailing environmental conditions which when disturbed by "climatic change"—*sensu* Cushing and Dickson (1976) (i.e., variability of interannual/decadal environmental conditions and not climate change as it is used in the early 21st century)—can lead to fluctuations in the abundance of stocks by three or four orders of magnitude (Cushing and Dickson 1976).

3.2.5 *Stable Ocean Hypothesis by Lasker (1978)*

The Stable Ocean Hypothesis establishes a mechanistic link between the ocean's physical conditions and productivity—particularly the availability of functionally-matching food—with fish larval survival and, therefore recruitment variability (Lasker 1978). This hypothesis was formulated after observations made on the Northern anchovy *Engraulis mordax* Girard, 1854 in the Southern California Bight (California, United States).

Lasker (1978) observed that an upwelling event, during early 1975, disrupted the stable oceanographic conditions that promoted and sustained nutritional and functionally-matching prey (30–50 μm in diameter) for first-feeding Northern anchovy larvae, a bloom of the dinoflagellate *Gonyaulax polyedra*. This upwelling event eroded the thermocline and the chlorophyll maximum layer, dispersed the *Gonyaulax polyedra* bloom, while promoting the conditions for a diatom bloom. However, the cell density of this diatom bloom was lower than that required for efficient anchovy larvae feeding, as well as composed by cells smaller than the minimum size threshold for first-feeding larvae to prey on. Overall, this upwelling event promoted a functional mismatch between first-feeding anchovy larvae and their prey despite the substantial increase in primary productivity. If the timing of this upwelling event was delayed, then first-feeding larvae could have benefited from the *Gonyaulax polyedra* bloom. Also, the onset of the upwelling event at a later moment would have incremented secondary production that would have sustained bigger anchovy larvae (Lasker 1978). Definitely, that was an unlucky year for the Northern anchovy larvae.

3.2.6 Stable Retention Hypothesis by Iles and Sinclair (1982)

The Stable Retention Hypothesis also recognizes that events happening early on at the life of fish largely dictate recruitment variability (Iles and Sinclair 1982). This hypothesis was framed around observations made on the Atlantic herring in the northeast and the northwest Atlantic Ocean. Iles and Sinclair (1982) drew attention to the stock structure and spawning behavior of the Atlantic herring—discrete spawning areas and distribution, homing behavior—which roots into the Migration Triangle Hypothesis and specifically into the segregation of habitats used by fish during their life (Harden Jones 1968). The Stable Retention Hypothesis also has reminiscences of the Aberrant Drift Hypothesis, not by acknowledging that the advection of fish larval stages from spawning/nursery areas may decrease fish recruitment (Hjort 1926) but rather by demonstrating that hydrographic features enabling larger retention areas may increase carrying capacity and therefore recruitment. So, larger spawning stocks may take advantage of such retention areas—see Fig. 5 in Iles and Sinclair (1982)—which may minimize the effect of density-dependent factors at an early developmental stage. Finally, Iles and Sinclair (1982) did not find evidence linking productivity and the timely onset of primary and secondary productions to sustain fish larval stages—i.e., the Match-Mismatch Hypothesis—and, thus, to explain the variability of the Atlantic herring recruitment.

Iles and Sinclair (1982) detailed the specific and rather complex hydrographic characteristics of different spawning habitats that promote the retention of the Atlantic herring larval stages in several regions. The oceanographic features observed to facilitate larval retention in tidally energetic seas were the tidally induced anticyclonic gyres in the Georges Bank and Grand Manan, tidally induced anticyclonic gyres formed independently of wind conditions around the Orkney and Shetland Islands, and bottom inshore current with tidally driven centrifugal upwelling in Nova Scotia. In estuarine and adjacent waters, other circulation retention features come into play, as two-layer estuarine circulation in the Norwegian fjords and Gulf of Maine and their adjacent areas, while in the Gulf of Saint Lawrence a geographically fixed gyre promotes larval retention. Similarly, spawning areas located in enclosed regions also promote larval retention, as in the Bras d'Or Lake (Nova Scotia) or Gulf of Riga (Baltic Sea). Interestingly, in regions where hydrographic features do not promote larval retention, larvae may display a behavioral response to favor stationary position when exposed to temperature gradients across transition zones (Iles and Sinclair 1982).

Overall, the Stable Retention Hypothesis proposes that fish recruitment variability depends on the extent of hydrographical features promoting retention of fish larval stages. This process is modulated by the interaction between hydrographical features and behavioral decisions made by fish adults—i.e., selection of spawning ground—and larvae vertical movements to promote retention. This conceptual framework, based on the Atlantic herring, also proposes that stock size is independent of biological parameters as reproduction or growth, and therefore of productivity.

3.2.7 Member/Vagrant Hypothesis by Sinclair and Iles (1987)

The Member/Vagrant Hypothesis is a very complex extension to the Stable Retention Hypothesis. This new hypothesis integrates an evolutionary perspective on the processes and mechanisms dictating recruitment variability under a conceptual framework merging hydrographic and food-chain processes, which can act in a density-dependent or a density-independent manner (Sinclair and Iles 1987). The evolutionary hypothesis framing the Member/Vagrant Hypothesis—and for this, the reason why I consider it an extension to the Stable Retention Hypothesis—is that a species, or a population, spawn in a given area because of a generational imprinting of that area to maximize retention and therefore recruitment.

The Stable Retention Hypothesis was built around the Atlantic herring, while part of the Member/Vagrant Hypothesis includes insights into an array of species with very distinct life histories, including anadromous, catadromous, and marine species—Atlantic cod, American shad *Alosa sapidissima* (Wilson, 1811), Atlantic herring, Atlantic mackerel *Scomber scombrus* Linnaeus, 1758, Atlantic menhaden *Brevoortia tyrannus* (Latrobe, 1802), Atlantic salmon *Salmo salar* Linnaeus, 1758, European eel *Anguilla anguilla* (Linnaeus, 1758), haddock *Melanogrammus aeglefinus* (Linnaeus, 1758), rainbow smelt *Osmerus mordax* (Mitchill, 1814), and striped bass *Morone saxatilis* (Walbaum, 1792) (Sinclair and Iles 1987). Other two species—winter flounder *Pseudopleuronectes americanus* (Walbaum, 1792), yellowtail flounder *Limanda ferruginea* (Storer, 1839)—were included in a paper published in the following year addressing population richness (Sinclair and Iles 1988), one of the three components of the Member/Vagrant Hypothesis (Sinclair and Iles 1987).

The first component of this hypothesis considers that population pattern (i.e., how populations are distributed along the species distribution area) and population richness (i.e., the number of populations of a species) depend on the number and geographical location where a species is capable of completing their life cycle and is determined at the onset of a fish life cycle in relation to specific hydrological conditions (e.g., tidal-induced circulation, river flow, coastal current systems, oceanic currents, oceanic-scale gyres) occurring at particular geographical settings (rivers, estuaries, coastal embayments, banks). The second component suggests that the absolute abundance of a population is a function of the size of the geographical location where a free-crossing population can close their life cycle. Finally, the third component proposes that temporal variability is dictated by intergenerational losses of individuals (vagrancy and mortality) from the areas that will ensure membership within a given population, which can occur at any stage of the life cycle. The losses mentioned in the third premise can either be due to spatial processes (e.g., advection of larval stages; juvenile vagrants; adult vagrants; reproduction failure by not securing a mate or reproducing in the wrong location; offspring that abort or are infertile due to biochemical constraints) or energetic processes (i.e., mortality of larval stages, juveniles, and adults due to predation, disease, starvation, and their interaction; reduced fecundity of adults due to food limitation). Finally, Sinclair and Iles (1987) also proposed that if the regulation of a population is more likely to be controlled by hydrographic-induced losses/mortality then, the importance of density-dependent factors (e.g., food limitation, predation, and/or disease) is minimal to the overall population abundance.

The Member/Vagrant Hypothesis received some criticism after the publication of the book "Marine Populations: An Essay on Population Regulation and Speciation" by Sinclair in 1988. Some of the objections are linked with the excessive reliance on data from the Atlantic herring, insufficient evidence on the role of advection to recruitment variability, exclusion of density-dependent and stabilizing behavior in populations (Zeldis 1989), and the excessive role put on philopatric and homing behavior to explain marine fish migrations (Secor 2015). However, this hypothesis is a stimulating research framework that has received the attention of many studies since its publication.

3.2.8 *The Optimal Environmental Window Hypothesis by Cury and Roy (1989)*

Cury and Roy (1989) proposed the Optimal Environmental Window Hypothesis, which balances the explanation for recruitment variability between the Match-Mismatch Hypothesis (Cushing 1975) and the Stable Ocean Hypothesis (Lasker 1978) depending on the prevailing environmental conditions. The Optimal Environmental Window Hypothesis is based on data from several Clupeiformes, the

European sardine *Sardina pilchardus* (Walbaum, 1792), Madeiran sardinella *Sardinella maderensis* (Lowe, 1838), Pacific sardine *Sardinops sagax* (Jenyns, 1842), Peruvian anchoveta *Engraulis ringens* Jenyns, 1842, and round sardinella *Sardinella aurita* Valenciennes, 1847 (Cury and Roy 1989). This hypothesis expands the Stable Ocean Hypothesis by Lasker (1978), which proposed that oceanographic phenomena, as storms and upwelling, may compromise established abiotic and biotic conditions conducive to successful recruitment, including disruption of food patches and offshore advection. However, Cury and Roy (1989) made a distinction between Ekman- and non-Ekmann-type upwelling events and established the type of relationship between wind mixing and turbulence conditions on recruitment.

Overall, in Ekman-type upwelling regions, the relationship between upwelling intensity and recruitment variability is dome-shaped, showing an optimal environmental window during moderate upwelling conditions. In such situations, fish recruitment may increase because upwelling has a positive effect on primary productivity without causing the offshore advection of fish larval stages—in a certain way, moderate upwelling delivers the adequate food conditions for fish larvae, i.e., Critical Period Hypothesis and Match-Mismatch Hypothesis. However, fish recruitment is generally reduced when weak upwelling conditions prevail because of lower productivity, while strong upwelling conditions cause offshore advection and mixing of the water column despite the overall increase in productivity (Cury and Roy 1989). Cury and Roy (1989) even set two wind speed thresholds, which can provide information on turbulence conditions. One at 5 m s^{-1} that dictates the onset of water column mixing in near-shore areas which breaks phytoplankton patches, and another at 7 m s^{-1}, which sets strong turbulence conditions that may increase offshore transport and decrease primary and secondary productivity. In non-Ekman-type upwelling regions, the relationship between upwelling intensity and recruitment is linear because strong upwelling conditions are not related to wind intensity, i.e., productivity is high, wind-induced turbulence is low, and fish larval off-shore advection may be minimal (Cury and Roy 1989).

3.2.9 Ocean Triads Hypothesis by Bakun (1996)

In 1996, Andrew Bakun wrote a comprehensive book—but rather captivating and clear—that soon after became a cornerstone of contemporary marine population ecology. In this book, "Patterns in the Ocean: Ocean Processes and Marine Population Dynamics", Bakun (1996) brought us the Ocean Triads Hypothesis, or "*The Fundamental Triads*" as he called it then. This hypothesis considers that three processes—enrichment, concentration, and retention—intervene in controlling the mortality of pelagic fish larval stages in coastal ecosystems, consequently dictating fish recruitment variability. This hypothesis is based on data from multiple species, including Clupeiformes (anchovies, sardinellas, sardines), tunas, haddock, and other fish species, but also holoplanktonic and meroplanktonic invertebrates to illustrate specific cases where examples from fish larvae were unavailable.

The enrichment processes include the mechanisms that increase the nutrients that will boost primary and secondary productivity such as seasonal overturn, tidal mixing, upwelling (coastal, shelf-break, equatorial, open-ocean, vortex-driven, stationary thermocline), and coastal runoff. The concentration processes increase the chances of fish larvae to find food. Several concentration mechanims may occur, such as in interface areas (sea surface, thermocline, and nutricline with the often associated chlorophyll-maximum layer), surface fronts, convergent frontal structures (e.g., like those induced by river plumes, oceanic water masses, downwelling, vortex-driven upwelling, or Langmuir circulation), and also by oceanographic mechanisms promoting water column stability or microscale turbulence. Finally, the retention processes include several hydrographic mechanisms—either by itself or coupled with larvae's vertical positioning accomplished by their vertical migration behavior—that promote the transport of fish larval stages to retention areas or retention within such areas. Bakun (1996) highlights a few of those mechanisms, including Taylor columns, shelf-sea fronts, wind-induced oceanographic phenomena, vertical geostrophic shears, bottom boundary layers, Stokes transport (or surface wave transport), and slicks coupled with internal waves.

A great example of the interplay between the different mechanisms integrating the Ocean Triads processes coming into play to promote the recruitment of the European anchovy *Engraulis encrasicolus* (Linnaeus, 1758) larval stages was documented in five basins of the Mediterranean Sea—Adriatic Sea, Aegean Sea, Alboran Sea, Gulf of Lion and the adjacent Catalan coast, and Strait of Sicily (Agostini and Bakun 2002). These regions are home to sizeable European anchovy populations, except the Strait of Sicily, where the combination of the prevailing Ocean Triads mechanisms is unsuited to promote the desirable enrichment, concentration, and retention conditions for a pelagic reproducing fish population. In the Gulf of Lion and adjacent Catalan coast, as well as in the Aegean and Alboran Seas, coupled wind-driven Ekman upwelling and downwelling mechanisms promoted enrichment and concentration. Riverine inputs into the coastal area or intrusion of less saline water (the Atlantic Ocean into the Alboran Sea or the Black Sea into the Aegean Sea) also contributed to enrichment and concentration through surface fronts and stability. In the Aegean Sea and Gulf of Lion, coastal mountains protect the coastal area from intense winds which promotes water column stability (concentration) and retention. In the western Alboran Sea, a clockwise geostrophic gyre transports nutrients and induces a fine-scale convergence zone. In the Adriatic Sea, the enrichment process is mainly caused by riverine inputs since coupled upwelling-downwelling is weak, while concentration and retention processes are favored by the wind-sheltered conditions prompted by the surrounding coastal mountains that induce low turbulence mixing conditions.

3.2.10 *Sense Acuity and Behavior Hypothesis by Teodósio et al. (2016)*

Bakun (1996) clearly highlighted that *"fish larvae are not drift bottles"*; however, most theories deeming to explain fish recruitment variability regarded fish larvae as passive organisms drifting with the currents or limited their behavior to vertical migrations, either to avoid predators, follow prey, or minimize advection. The contribution of fish larvae behavior and swimming capabilities on fish recruitment was formally introduced to help explain fish recruitment with the Sense Acuity and Behavioral Hypothesis, or SAAB Hypothesis (Teodósio et al. 2016). Essentially, the swimming capacity of larvae was considered to be limited to a reduced spatial scale, from a few millimeters to centimeters (Houde 2016), while now it has expanded to the kilometer range with the SAAB Hypothesis (Teodósio et al. 2016). This generalized misconception is probably rooted in the etymology of the word ichthyoplankton in its strictest meaning, i.e., fish (*ichthyo*) drifter (*plankton*).

The SAAB Hypothesis acknowledges that it is imperative to include fish larvae behavior and their swimming capabilities when testing some of the hypotheses mentioned before and to model fish larvae transport in coastal ecosystems (Baptista et al. 2020). Indeed, such models show that recruitment would be lower if fish larvae would drift passively with currents, and that successful recruitment requires the sensory acuity of temperate fish larvae and their behavioral response to estuarine and coastal cues (Teodósio et al. 2016, Baptista et al. 2020).

The SAAB Hypothesis established its two main premises on the observations made on several temperate fish larval species either in nature or under controlled conditions. The species studied are from the families Anguillidae, Atheriniidae, Clupeidae, Congridae, Engraulidae, Gadidae, Percichthyidae, Pleuronectidae, Sciaenidae, Scombridae, Sparidae, and Soleidae (Teodósio et al. 2016). The first premise considers that when postflexion temperate fish larvae are offshore, they possess the necessary sensory organs developed to perform bearing-keeping navigation towards coastal nursery areas guided by the sun compass or the earth's geomagnetic field. The second premise considers that in more nearshore areas, postflexion fish larvae will develop a behavioral response to the cues (odor, sound, visual) coming from coastal nursery areas (coastal lagoons, estuaries, rocky reefs). This behavioral response triggers distinct swimming behaviors that vary according to the intensity of cues present in each coastal area, but that must be conciliated with innate behaviors (nycthemeral, feeding strategies, avoidance of predators) (Fig. 3.1; Teodósio et al. 2016).

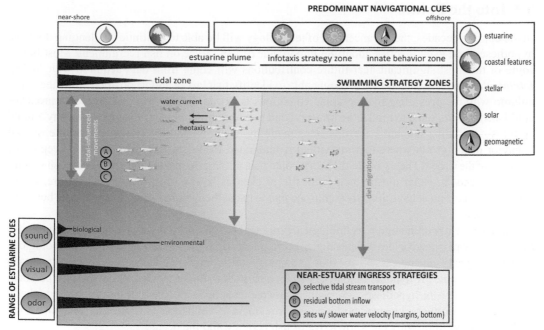

Figure 3.1: Conceptual model proposed by the Sense Acuity and Behavioral Hypothesis (Teodósio et al. 2016) detailing the estuarine (sound, visual, odor) and navigational cues (geomagnetic, solar, stellar, coastal features) used by temperate fish larvae to detect estuarine ecosystems and to navigate towards them, before using active swimming strategies to ingress into these nursery areas. The use of navigational cues varies according to the distance to the estuary, as well as the range of the estuarine cues. Swimming strategies also vary according to the distance to the estuary. The larvae may use an innate or an infotaxis strategy when away from estuarine cues, while under the influence of estuarine cues, larvae will use rheotaxis coupled with directional swimming along the estuarine cue concentration gradient. Swimming towards the estuary may be combined with daily vertical migrations, which will be superimposed with tidal-induced movements once larvae are near the estuary. Here, the larvae may use an array of strategies to ingress into the estuary, or to maintain stationary position and to aggregate. Reprinted from: Teodósio, M.A., E. Wolanski, C.B. Paris and P. Morais. 2016. Biophysical processes leading to the ingress of temperate fish larvae into estuarine nursery areas. Estuar. Coast. Shelf Sci. 183: 187–202, © 2016 Elsevier Inc., with permission from Elsevier.

Despite the small size of temperate fish larvae, their swimming capabilities and stamina are truly remarkable. For example, in laboratory trials, one white seabream *Diplodus sargus* (Sparidae) Linnaeus, 1758 larvae swam continuously for ten days, without being fed, for a total of 86.5 km, while the maximum value of critical swimming speed recorded for this species reached 25.1 body lengths s^{-1} (Baptista et al. 2019), which is the same range of a cheetah (25.8 body lengths s^{-1}) but still far behind the peregrine falcon (235 body lengths s^{-1}) or marine bacteria (200 body lengths s^{-1}) (Seymour and Stoecker 2018). However, white seabream larvae swimming performance surpass those of the fastest fish species, the sailfish *Istiophorus platypterus* (Istiophoridae) (Shaw, 1792), which reaches an approximate maximum dislocation of 10.2 body lengths s^{-1} when swimming at a maximum speed of 110 km h^{-1}. Even human top-athletes performances are easily surpassed by fish larvae. For example, when Usain Bolt broke the 100 meters world record, he moved at 5.354 body lengths s^{-1}, while Eliud Kipchoge moved at 3.52 body lengths s^{-1} when he broke the marathon record. Human performance in water is even worst- the fastest human swimmer, César Cielo only swam at 1.226 body lengths s^{-1}. Although some might consider these facts just as fun facts, there is accumulating evidence that minimal behavioral responses by larvae to environmental cues, which are possible due to their acute sense and swimming capabilities, have meaningful differences on their transport and should not be neglected (Rossi et al. 2019, Downie et al. 2020, Baptista et al. 2020). This is particularly important for the development of individual-based models merged into hydrographic models aiming at inferring recruitment areas, predicting recruitment variability, and supporting fishery policies.

3.3 Into the Future

In the coming decades, the advancement of technology will be able to continue challenging dogmas, hypotheses, and even theories about the importance of fish larvae on fish recruitment. Most likely, some of the most significant near-future contributions will come from data acquired with *In Situ Ichthyoplankton Imaging Systems*—being able to identify fish larvae and its position along the water column with the prevailing hydrographic conditions is extraordinary. New conceptual approaches could also be integrated to model the transport and behavior of larvae. For example, in many animal systems—including humans—the decisions made by an individual are influenced by the overall behavior of the group where that individual is integrated to. When transposing this concept—the Wisdom of the Crowd Theory or Herd Theory—into fish, it is impossible to forget that many fish species gather in schools, either during their adult life or even as larvae and juveniles. Therefore, future transport models must have this aspect into account to make them more similar to the natural behavior of fish larvae.

Independently of the future technological progress brought into fish larval ecology research, it is the ecological perspective into the question of recruitment variability—autoecological, demecological, or synecological—and necessarily the time and spatial scales associated with each ecological approach that will determine the hypothesis to be tested for a particular model species. Also, and on a personal note, I do not view the hypotheses described in this chapter as absolute trues nor that a given hypothesis is better than the others. I instead prefer considering them as valid tools to frame our research efforts in the pursuit of improving our knowledge of fish ecology and the ocean.

References

Agostini, V.N. and A. Bakun. 2002. Ocean triads' in the Mediterranean Sea: physical mechanisms potentially structuring reproductive habitat suitability (with example application to European anchovy, *Engraulis encrasicolus*). Fish. Oceanogr. 11: 129–142.

Bakun, A. 1996. Patterns in the Ocean: Ocean Processes and Marine Population Dynamics. California Sea Grant College System/NOAA/Centro de Investigaciones Biológicas del Noroeste, La Paz, Mexico.

Baptista, V., P. Morais, J. Cruz, S. Castanho, L. Ribeiro, P. Pousão-Ferreira et al. 2019. Swimming abilities of temperate pelagic fish larvae prove that they may control their dispersion in coastal areas. Diversity 11(10): 185.

Baptista, V., F. Leitão, P. Morais, M.A. Teodósio and E. Wolanski. 2020. Modelling the ingress of a temperate fish larva into a nursery coastal lagoon. Estuar. Coast. Shelf Sci. Doi: 10.1016/j.ecss.2020.106601 (in press).

Behrenfeld, M.J. 2010. Abandoning Sverdrup's critical depth hypothesis on phytoplankton blooms. Ecology 91: 977–989.

Behrenfeld, M.J. and E.S. Boss. 2018. Student's tutorial on bloom hypotheses in the context of phytoplankton annual cycles. Glob. Change Biol. 24: 55–77.

Boss, E. and M. Behrenfeld 2010. *In situ* evaluation of initiation of the North Atlantic phytoplankton bloom. Geophys. Res. Lett. 37: L18603.

Chiswell, S.M. 2011. Annual cycles and spring blooms in phytoplankton: don't abandon Sverdrup completely. Mar. Ecol. Progr. Ser. 443: 39–50.

Cury, P. and C. Roy. 1989. Optimal environmental window and pelagic fish recruitment success in upwelling areas. Can. J. Fish. Aquat. Sci. 46: 670–680.

Cushing, D.H. and R.R. Dickson. 1976. The biological response in the sea to climatic changes. Adv. Mar. Biol. 14: 1–122.

Cushing, D.H. 1990. Plankton production and year-class strength in fish populations: an update of the match/mismatch hypothesis. Adv. Mar. Biol. 26: 249–293.

Downie, A., B. Illing, A. Faria and J. Rummer. 2020. Swimming performance of marine fish larvae: review of a universal trait under ecological and environmental pressure. Rev. Fish Biol. Fisher. Doi: 10.1007/s11160-019-09592-w.

Harden Jones, F.R. 1968. Fish Migration. Edward Arnold Ltd., London.

Hjort, J. 1914. Fluctuations in the great fisheries of northern Europe viewed in the light of biological research. Cons. Perm. Int. Explor. Mer. 20: 228.

Hjort, J. 1926. Fluctuations in the year classes of important food fishes. ICES J. Mar. Sci. 1: 5–38.

Houde, E.D. 2016. Recruitment variability. pp. 98–187. *In*: T. Jakobsen, M.J. Fogarty, B.A. Megrey and E. Moksness [eds.]. Fish Reproductive Biology: Implications for Assessment and Management. 2nd Edition. Wiley-Blackwell, U.K.

Huisman, J., P. van Oostveen and F.J. Weissing. 1999. Critical depth and critical turbulence: Two different mechanisms for the development of phytoplankton blooms. Limnol. Oceanogr. 44: 1781–1787.

Iles, T.D. and M. Sinclair. 1982. Atlantic herring: stock discreteness and abundance. Science 215: 627–633.

Lasker, R. 1978. The relation between oceanographic conditions and larval anchovy food in the California Current: identification of factors contributing to recruitment failure. Rapp. p.-v. Réun.-Cons. Int. Explor. Mer. 173: 212–230.

Morais, P. and F. Daverat. 2016. Definitions and concepts related to fish migration. pp. 14–19. *In*: P. Morais and F. Daverat [eds.]. An Introduction to Fish Migration. CRC Press, Boca Raton, Florida, USA.

Rossi, A., M. Levaray, C. Paillon, E.D.H. Durieux, V. Pasqualini and S. Agostini. 2019. Relationship between swimming capacities and morphological traits of fish larvae at settlement stage: a study of several coastal Mediterranean species. J. Fish Biol. 95: 348–356.

Secor, D.H. 2015. Migration Ecology of Marine Fishes. Johns Hopkins University Press, USA.

Seymour, J.R. and R. Stoecker. 2018. The Ocean's microscale: A microbe's view of the sea. pp. 289–344. *In*: J.M. Gasol and D.L. Kirchman [eds.]. Microbial Ecology of the Oceans. Third Edition. John Willey & Sons, Hoboken NJ, USA.

Sinclair, M. and T.D. Iles. 1987. Population regulation and speciation in the oceans. C.M. 1987/Mini No. 3. Mini-Symposium on Recruitment Processes in Marine Ecosystems. International Council for the Exploration of the Sea. Denmark. 22 p.

Sinclair, M. and T.D. Iles. 1988. Population richness of marine fish species. Aquat. Living Resour. 1(1): 71–83.

Sverdrup, H.U. 1953. On conditions for the vernal blooming of phytoplankton. ICES J. Mar. Sci. 18: 287–295.

Teodósio, M.A., E. Wolanski, C.B. Paris and P. Morais. 2016. Biophysical processes leading to the ingress of temperate fish larvae into estuarine nursery areas. Estuar. Coast. Shelf Sci. 183: 187–202.

Zeldis, J. 1989. Book review: Marine populations: an essay on population regulation and speciation. New Zeal. J. Mar. Fresh. 23: 605–606.

CHAPTER 4

Ocean Acidification Impacts on Zooplankton

*Ana N. Campoy,[1] Joana Cruz,[2] Joana Barcelos e Ramos,[3] Fátima Viveiros,[4] Pedro Range[5] and M. Alexandra Teodósio[2],**

4.1 Introduction

Global change is and will continue impacting biodiversity, as many studies have already documented. Rising atmospheric CO_2 is alleviated by oceanic uptake, since atmosphere and surface ocean exchange CO_2, but it also modifies the ocean carbonate system towards decreased carbonate ion concentrations and a corresponding decline in seawater pH. This process is known as ocean acidification (OA) and has a direct effect on plankton, namely calcifying organisms, such as coccolithophores, foraminifers, corals, molluscs and crustaceans, with consequences for the entire marine ecosystem (see review by Reibesell and Tortell 2011).

Zooplankton is a key component of aquatic communities and play a pivotal role in the structure and functioning of marine planktonic food webs as a major link between pelagic primary producers and planktivorous. The effect of OA on the fitness of individual zooplanktonic species has been reported by many studies mostly developed under laboratory conditions. Reported individual responses have been used to predict community scale responses but few studies exist including natural acidified zones in important oceanographic systems (see Boyd et al. 2018).

In this context, the present chapter starts with a general description of the process of OA and how it impacts oceanic systems worldwide, by modifying their physicochemical characteristics and

[1] Department of Marine Biology, Catholic University of the North, Larrondo 1281, Coquimbo, Chile; Center for Advanced Study of Arid Zones (CEAZA), Av. Bernardo Ossandón 877, C.P. 1781681, Coquimbo, Chile.
[2] Centre of Marine Science (CCMAR), Universidade do Algarve, FCT, Campus de Gambelas, 8005-139, Faro, Portugal.
[3] Institute for Agricultural and Environmental Research and Technology of the Azores, University of Azores, Angra do Heroísmo, Açores, Portugal.
[4] IVAR – Research Institute for Volcanology and Risks Assessment, University of the Azores, Ponta Delgada, Portugal.
[5] Environmental Science Center (ESC), Qatar University, P.O. Box 2713, Doha, Qatar.
Emails: anavcampoy@gmail.com; jmcruz@ualg.pt; joana.b.ramos@uac.pt; Maria.FB.Viveiros@azores.gov.pt; Prange@qu.edu.qa
* Corresponding author: mchichar@ualg.pt

influencing the biota. Then, we focus on reviewing the effects of OA on zooplankton and present the potential of natural shallow-water CO_2 vents as *in situ* laboratories for OA studies. This is exemplified with a case study from the North Atlantic (Azores islands) where the effect of elevated CO_2 on mesozooplankton assemblages is shown and the suitability of this area for future studies is assessed.

Field experiments at sites with naturally elevated CO_2 conditions are described as analogues for investigating the effect of future dissolved CO_2 levels on marine organisms and ecosystems. Beyond individual responses, the abundance, diversity and structure of mesozooplankton among sites with different CO_2 concentrations have been found to differ in the North Atlantic. Experiments in this area also provide the first *in situ* evidence of a significant decrease of Egg Production Rate (EPR) of copepods under near future CO_2 levels. These results are highly relevant to improve the predictions on zooplankton responses to climate change stressors, including OA. Nevertheless, given that most OA stress studies have focused on the effects of short-term exposure, the long-term multigenerational exposure to multiple stressors (e.g., increased pCO_2 and food shortage) is a priority to understand the adaptation capacity of common species and how zooplankton communities will shift (Kroeker et al. 2020).

4.2 The Marine Mesozooplankton in a Changing World

Exponential population growth during the nineteenth and twentieth centuries has led to an unavoidable transformation of the natural world. Unfortunately, humans have been exerting a negative direct impact on terrestrial and aquatic biodiversity through increased greenhouse gas emissions, land use change (agriculture, deforestation) and overfishing among others. Specifically, the atmospheric concentration of greenhouse gases, mainly represented by carbon dioxide (CO_2), has increased at an unprecedented rate during this period (Hoegh-Guldberg et al. 2014, Pörtner et al. 2014).

During the last 800,000 years, atmospheric CO_2 concentrations have fluctuated between 180 ppmv (parts per million per volume) in colder glacial and 290 ppmv in warmer interglacial periods (Ciais et al. 2013). Nevertheless, since the second half of the 18th century, when the Industrial Revolution started, energy production by burning of fossil fuels and cement production has led to an increase in atmospheric CO_2 levels to 408.5 ppmv in 2019 (www.co2.earth). Atmospheric CO_2 levels are predicted to continue increasing for at least the next century, and unless emissions are substantially reduced, levels will exceed 1,000 ppm by 2100 (Waldbusser and Salisbury 2014). Continued emission of greenhouse gases will also increase the natural greenhouse effect, causing atmosphere and ocean warming, diminishing snow and ice, increasing the sea level (Collins et al. 2013) and modifying the environmental properties of the ocean realm, not only temperature but also pH, water column oxygenation and food supply (Sweetman et al. 2017).

4.2.1 Ocean Acidification

Oceans cover about 71% of Earth's surface (Costello et al. 2010) and they play a vital role in ameliorating the effect of climate change on terrestrial life by virtue of a continuous interchange of gases with the atmosphere. As CO_2 increases in the atmosphere, the flux to the ocean also increases, heightening the partial pressure of CO_2 in seawater and consequently affecting the carbonate system of the surface ocean. Dissolved CO_2 in the oceans exists in three main inorganic forms known as dissolved inorganic carbon (DIC), whose relative proportions are ~ 0.5% aqueous CO_2, ~ 89% bicarbonate (HCO_3^-) and ~ 11% carbonate ions (CO_3^{2-}) (Waldbusser and Salisbury 2014). The equilibrium between these three forms of DIC keeps the seawater pH more or less stable, but when large amounts of CO_2 are dissolved, the equilibrium is altered, decreasing the carbonate ions and lastly reducing the pH (Raven et al. 2005), which also prevents the formation of carbonate minerals.

Thus, pH has reduced about 0.1 pH units over the past century, an equivalent to a ca. 25% increase in acidity (Havenhand 2012), with the greatest reduction occurring at high latitudes (Cramer et al. 2014). According to the Representative Concentration Pathways (RCPs) in the IPCC AR5, ocean pH is expected to decline between 0.14 unit (RCP2) and 0.43 unit (RCP8.5) until the end of the century (Hoegh-Guldberg et al. 2014). Despite this, without the oceans acting as a carbon sink, the concentration of CO_2 in the atmosphere would be 55% higher (Fabry et al. 2008).

This process, known as ocean acidification (OA), has direct and indirect effects on the entire marine ecosystem. One of the most important uses of carbonate in the ocean is the formation of calcium carbonate ($CaCO_3$) or limestone structures, such as in coral skeletons, pearls, and the shells or analogues of coccolithophores, foraminiferans, pteropods, bivalves and a variety of other organisms known as calcifiers (Caldeira and Wickett 2003, Raven et al. 2005, Fabry et al. 2008). Decreasing CO_3^{2-} derived from dissolved CO_2 also decreases the saturation state of $CaCO_3$ (Waldbusser and Salisbury 2014), hindering the formation of these structures. $CaCO_3$ solubility decreases with depth due to decreasing temperature, increasing pressure and increasing amounts of dissolved CO_2 until the so called "saturation horizon", where seawater is undersaturated and will tend to dissolve. This saturation horizon also becomes shallower under OA conditions (Waldbusser and Salisbury 2014).

Despite being the most directly affected and more widely reported, calcification is only one of the physiological parameters affected by OA. Marine organisms can experience physiological stress due to an increase in CO_2 (hypercapnia) and/or a decrease in pH in the surrounding environment (Dupont and Thorndyke 2009), the consequences of which range from undetectable to very severe and certainly depend on the exposure time. Different life cycle stages also show different degrees of susceptibility, with reproductive and early life stages being considered particularly vulnerable (Kurihara and Ishimatsu 2008, Dupont and Thorndyke 2009, Fitzer et al. 2012). Impacts include decreased growth rate, reduced reproductive output, disrupted respiratory and nervous system function and increased susceptibility to predators and disease, all of which could produce ripple effects through food webs and ecosystems. Evidence from the geological record shows that previous periods of intense OA, e.g., at the end of the Paleocene, coincided with mass extinction events, including zooplankton species (Zachos et al. 2005, Jackson 2010).

4.2.2 *Mesozooplankton Responses to Ocean Acidification*

Zooplankton is a key component of aquatic communities (Chan et al. 2008) and plays a pivotal role in the structure and functioning of marine planktonic food webs (Debes et al. 2008, Zervoudaki et al. 2013), such as a major link between pelagic primary producers and planktivorous (Vehmaa et al. 2013). It is not surprising that responses to OA are very variable, given the wide spectrum of organisms that are part of the zooplankton. The responses can be negative, neutral or even positive and appear to be species-specific, even in closely related species (Troedsson et al. 2013). Holozooplanktonic (organisms that expend their entire life cycle in the plankton) $CaCO_3$ producers include foraminifera (shells of different materials including calcite and aragonite), euthecosomatous pteropods (shells of aragonite), heteropods (tropical and subtropical oceans, shells of aragonite not always present) and gymnosomes (shells of aragonite cast off at metamorphosis) (Fabry et al. 2008). It is expected that these groups will be affected under acidified conditions, by reducing calcification. Broadcast spawning invertebrates are particularly vulnerable to ocean acidification because fertilization and larval development occur in the water column. Lecithotrophic larvae (10% of marine benthic invertebrates) may be better competitors and less affected since they spend less time in plankton than planktotrophic larvae (60–90% of marine organisms), which feed on exogenous sources. Any sub-lethal reductions in rate of development and larval size may also have significant consequences for the survival of marine larvae because prolonged larval life phase and delayed settlement may lead to a concomitant increase in the likelihood of predation (Ross et al. 2011). Larvae of benthic calcifying organisms will respond with a reduction in calcification rate.

There are a relevant number of published studies regarding OA effects, but few of them have reported effects under near future conditions (Zervoudaki et al. 2013), and even less using an ecosystem approach (Lischka et al. 2017) or a long-term exposure of the organisms (Thor and Dupont 2015). Instead, most studies are short-term assays under controlled conditions focused on very restricted groups. These are not always relevant to predict climate impacts on ecosystems (e.g., using unrealistic pH values and/or acid-based acidification without correcting carbonates and bicarbonates) or ecologically realistic (e.g., single species cultures) conditions (Dupont and Thorndyke 2009).

Copepods typically account for about 55–95% of zooplankton biomass and are relevant herbivores, transferring carbon captured at lower trophic levels to higher trophic levels (Longhurst 1985). Thus, any change in their populations can affect the growth and reproduction of higher trophic organisms (Zhang et al. 2011). In addition, the timing and intensity of copepod reproduction is considered to be essential for survival of fish larvae, since several pelagic fish stocks feed on copepods during their entire lifetime, and their individual growth as well as stock production is highly affected by copepod availability (Debes et al. 2008). Thus, knowledge of how they cope with environmental stressors as well as any potential effect on copepod productivity and population structure (Cripps et al. 2014) is important for understanding how the aquatic ecosystem as a whole responds (Chan et al. 2008, Kurihara and Ishimatsu 2008).

Given their important role in marine food webs, copepods require and, in fact have received, much attention in OA studies. Many previous studies have assessed the OA effect on copepods, the vast majority under controlled conditions (which sometimes are not representative of natural environment), with low exposure times (few days), under extreme pH/CO_2 concentrations and focused on a single or few species. Since copepods are non-calcifying organisms (i.e., they possess a chitinous exoskeleton rather than calcium carbonate shell), and they are not directly affected by the decreases in carbonate saturation state (Caldeira and Wickett 2003, Raven et al. 2005, Fabry et al. 2008, Kurihara 2008, Riebesell and Tortell 2011). OA has been shown to have mainly sub-lethal effects on copepods (Zervoudaki et al. 2013), affecting physiology (Thor and Dupont 2015) and behaviour (Li and Gao 2012), but also survival (Zhang et al. 2011). It has also been observed that early life-history stages may be the most sensitive to CO_2-induced OA (Dupont and Thorndyke 2009). Thus, early developmental and reproductive stages are important to understand the effect of OA at a population level (Kurihara 2008), since they can be sensitive to CO_2 concentrations that will not affect the survival of adults (Kurihara et al. 2004).

Sub-lethal responses have been shown to vary considerably between copepod species, particularly with regard to reproductive success (Cripps et al. 2014). Previous studies have demonstrated a decrease of the Egg Production Rate (EPR), under different levels of elevated CO_2, relative to ambient levels: at 1,000 ppm in *Acartia tonsa* (Cripps et al. 2014) and *Pseudocalanus acuspes* (Thor and Dupont 2015); at 2,000 ppm in *Centropages tenuiremis* (Zhang et al. 2011); at 5,000 ppm in *Acartia erythraea* (Kurihara et al. 2004) and *Acartia spinicauda* (Zhang et al. 2011) and at 10,000 ppm in *Acartia steueri* (Kurihara et al. 2004), *Calanus sinicus* (Zhang et al. 2011) and *Centropages typicus* (McConville et al. 2013).

Other studies have shown that when zooplankton is exposed to increased pCO_2 over multiple generations, it can acclimate or even adapt to OA conditions, but this effect is not well understood yet and has only been documented for few species. Shifts toward smaller brood sizes and adults (Fitzer et al. 2012) have been reported. Physiological acclimation also includes transgenerational effects, such as epigenetic changes exemplified by DNA methylation and/or histone modification. In contrast, an adaptive response, which is due to changes in the genetic structure of a population (i.e., changes in allele frequencies) is likely to be sustained for a long time, regardless of the presence of the stressor (Tsui and Wang 2005). This adaptive potential has also been suggested for some species (Pedersen et al. 2014).

4.3 Natural CO$_2$ Vents to Study Ocean Acidification

A more integrated approach for OA studies may be achieved in environments with naturally high concentrations of CO$_2$, such as submarine volcanic vents (Hall-Spencer et al. 2008). These *in situ* laboratories entail an opportunity to conduct field experiments at sites with naturally-elevated CO$_2$ conditions, over timescales of months to decades and more, converting them in potentially useful analogues for investigating the effects of expected CO$_2$ levels on marine organisms and ecosystems (Hall-Spencer et al. 2008, Boyd et al. 2018). The existence of shallow-water submarine volcanic vents is well known off volcanic islands and provinces (Cardigos et al. 2005), occurring over a wide depth range, from the intertidal to the abyss. The deepest active hot vent known so far with associated fauna is the Ashadze field located at 4000–4100 m depth on the Mid-Atlantic Ridge (Tarasov et al. 2005). Within submersed CO$_2$ vents, it is important to distinguish between deep-sea hydrothermal vents and shallow-water vents. The first are hot springs that form chimney-like structures close to mid-ocean ridges and represent a distinguish ecosystem with low diversity and usually high endemism and biomass whose energy source comes from chemosynthesis (Cardigos et al. 2005, Won 2006, Levin et al. 2016). By contrast, shallow-water vents extend from the intertidal to more than 200 m depth (Couto et al. 2015) and can be surrounded by a more diverse fauna dependent on their location.

While previous studies on shallow-water volcanic vents consider a great diversity of groups, like bacteria (Brinkhoff et al. 1999, Cardigos et al. 2005, Kerfahi et al. 2014), benthic invertebrates (Gamenick et al. 1998, Cardigos et al. 2005, Hall-Spencer et al. 2008b, Calosi et al. 2013, Pettit et al. 2013), algae (Cardigos et al. 2005, Hall-Spencer et al. 2008b), fishes (Cardigos et al. 2005) or seagrass (Hall-Spencer et al. 2008b, Arnold et al. 2012), only a few recent studies take into account these environments to study possible effects on zooplankton communities (Kâ and Hwang 2011, Smith et al. 2016). This opens a new door of opportunities at many levels, moving beyond single-species studies under so restricted conditions for the zooplankton community, where the interactions within and between species and with the environment cannot be addressed.

Any study performed in these environments requires a previous description of the physical-chemical characteristics of the vent site, to define the areas with CO$_2$ emissions and where the degassing activity ceases. Daily variations can appear around the main points of emission (Kerrison et al. 2011) as well as other gases and sulphides. Accordingly, analyses of gas composition, temperature, salinity and alkalinity acquire special relevance in order to assess the aptitude of these locations to act as natural laboratories for testing ocean acidification impacts (Barry et al. 2010).

4.3.1 Shallow-Water Volcanic CO$_2$ Vents of the North Atlantic: Azores as a Case Study

The Azores archipelago, composed of nine volcanic islands, is located in the mid-Atlantic ridge. Active submarine volcanic areas with emissions of above 98% of CO$_2$ (Viveiros et al. 2016) are present. Specifically, one off the island of São Miguel (Ribeira Quente) and one off the island of Faial (Espalamaca) (Fig. 4.1) were the target to study the effects of OA on the zooplanktonic community in the important ocean biogeochemical province "North Atlantic Subtropical Gyre".

A proper analysis of the characteristics of each venting site is crucial on vent-associated community studies, including gas composition, the geology of the site and other physicochemical features, to ensure which are the factors influencing differences among communities. In the study case of the Azores islands, three sampling sites with decreasing degassing activity were chosen in each island, in order to have a gradient of CO$_2$ for comparison. These presented similar bottom geology and depth conditions in each island: a strong degassing area (High CO$_2$), a transitional site, where the carbonate chemistry of the seawater was still affected by the emissions, but with no obvious degassing (Low CO$_2$) and a reference site (Ref), unaffected by the CO$_2$ emissions (Ref < Low CO$_2$

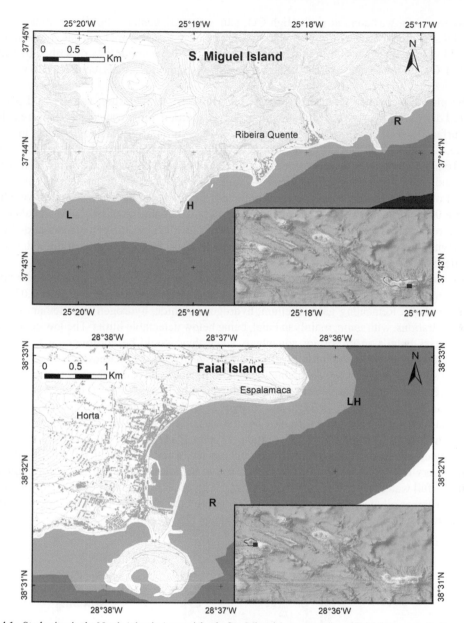

Figure 4.1: Study sites in the North Atlantic Azores islands, São Miguel (upper panel) and Faial (lower panel): R – reference site, L – Low CO_2, and H – High CO_2 (Reference < Low CO_2 < High CO_2). The colour gradient denotes the 30 m and 50 m bathymetric contours.

< High CO_2). Test dives and pH measurements with a CTD equipped with a pH probe were used to define these sites. They are also located near the coast and extending beyond the intertidal (Couto et al. 2015), which facilitate the logistics for sampling and experimental studies.

Conditions varied substantially between the two islands, with a maximum depth of 10 meters in São Miguel and 37 meters in Faial. Given their proximity to the Faial-Pico channel, the sites in Faial were strongly exposed to tidal currents. The venting sites in both islands also differed in terms of their geological characteristics, since some of the emissions in São Miguel are warm while in Faial, they are always cold. According to these characteristics, they can be denominated as hot vents and cold seeps, respectively (Tarasov et al. 2005).

Gas sampling was done at each High CO_2 site, using Giggenbach bottles (bottles filled with NaOH 4N and under vacuum). Three replicates were collected in each area and gas composition was analysed by titration and gas chromatography, according to the methodology described by Caliro et al. (2015). Gas emission rates were measured using inverted plastic funnels, connected to volumetric flasks.

The physical-chemical characteristics of the seawater (temperature, salinity and pH) were measured daily, *in situ*, using a YSI6000 multiprobe. In addition, three water samples were collected at the surface (0–2 m) and three near the bottom (6–10 m in São Miguel and 25–37 m in Faial) at each site, for subsequent laboratory analyses. Total pH (pH_t) was corrected with a TRIS seawater buffer. Total alkalinity (TA) was obtained through potentiometric titration, following Dickson et al. (2003), and using a Metrohm Titrino Plus 848 equipped with a 869 Compact Sample Changer, and corrected with certified Reference material supplied by A. Dickson. Seawater samples were filtered through a 0.2 µm membrane and measured within 48 h. Carbonate chemistry was calculated from temperature, salinity, silicate, phosphate, multiprobe and glass sensor pHt values and TA using CO2sys with the equilibrium constants determined by Mehrbach et al. (1973) as refitted by Dickson and Millero (1987).

CO_2 accounted for more than 99 molar% in São Miguel, while at Faial CO_2 accounted for between 98 and 99 molar%. Nitrogen, argon and oxygen followed with values ranging between 0.02 molar% and 1.27 molar%. Remaining gases (helium, hydrogen sulphide, hydrogen and methane) had very low concentrations, with some, mainly in Faial, being below detectable limits. The low concentration of toxic gases entails an important advantage, since they would be confounding factors in the interpretation of the results. Accordingly, from the perspective of the composition of gas emissions, both these sites can be considered suitable for long-term studies on the effects of OA.

The Reference sites showed a similar and stable pH mostly above 8 units (Figs. 4.2A and 4.2D) in both islands, but differences in depth and currents are reflected in how the CO_2 bubbling from the seabed is influencing the chemical variables. Maximum sampled depth in São Miguel was 10.65 m in the Reference site, 9.71 in the Low CO_2 site and 6.97 in the High CO_2 site. At the Low CO_2 site in São Miguel, lower values of pH are found close to the surface, while at the High CO_2 site the degassing activity is very recognizable close to the bottom and at the surface, not having a continuous effect through the full water column (Figs. 4.2B and 4.2C). Lower pH values are below 7 units in the High CO_2 site, whereas all measurements are above 7.5 in the Low CO_2 site, but mean values indicate a decrease from 8.03 (Low CO_2) to 7.86 (High CO_2) units (Table 4.1).

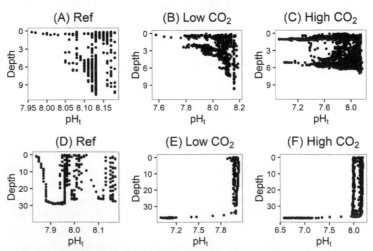

Figure 4.2: pH variation along the full water column in São Miguel (A–C) and Faial (D–F) vents, for the three sites sampled: Ref – Reference < Low CO_2 < High CO_2. Notice that the depth scale is the same on each island (A–C and D–F) only for comparison, maximum depth at each site is detailed in the text. Note also differences in x-scale between sites.

Table 4.1: Seawater carbonate chemistry variables (mean ± SE, n = 3) at each site (Ref-Reference < Low-Low CO_2 < High-High CO_2) in São Miguel and Faial. The parameters represented are: temperature (Temp), salinity (Sal), pH (total scale), total alkalinity (TA), CO_2 (total scale), CO_2 partial pressure, bicarbonate (HCO_3^-), carbonate (CO_3^{2-}), CO_2, saturation state for calcite (ΩCa) and aragonite (ΩAr). These correspond to the entire water column in São Miguel and the deeper 10 m (upper row) or the entire water column (lower row) in Faial.

Local	Site	Temp (°C)	Sal	pH_t	TA (µmol/kg)	TCO_2 (µmol/kg)	pCO_2 (µatm)	HCO_3^- (µmol/kg)	CO_3^{2-} (µmol/kg)	ΩCa	ΩAr
São Miguel	Ref	20.13 ± 0.01	36.93 ± 0.00	8.10 ± 0.00	2371.47 ± 0.07	2056.80 ± 0.27	347.80 ± 0.45	1827.08 ± 0.43	218.65 ± 0.18	5.15 ± 0.00	3.36 ± 0.00
	Low	20.06 ± 0.01	36.98 ± 0.00	8.03 ± 0.00	2380.92 ± 0.00	2104.04 ± 1.02	430.63 ± 2.69	1895.06 ± 195.25	195.25 ± 0.63	4.60 ± 0.01	3.00 ± 0.01
	High	20.54 ± 0.00	36.94 ± 0.00	7.86 ± 0.00	2397.89 ± 0.06	2199.16 ± 1.42	818.95 ± 10.07	2021.84 ± 1.92	151.51 ± 0.77	3.57 ± 0.02	2.33 ± 0.01
Faial	Ref	19.58 ± 0.07	37.56 ± 0.02	8.10 ± 0.00	2371.70 ± 0.00	2056.85 ± 1.34	347.74 ± 2.80	1827.84 ± 2.09	217.83 ± 0.83	5.10 ± 0.02	3.32 ± 0.01
		20.14 ± 0.04	37.65 ± 0.01	8.10 ± 0.00	2371.70 ± 0.00	2050-46 ± 0.85	345.94 ± 1.59	1817.50 ± 1.33	222.00 ± 0.53	5.20 ± 0.01	3.39 ± 0.01
	Low	18.02 ± 0.08	37.54 ± 0.01	7.51 ± 0.04	2345.2 ± 0.00	2286.11 ± 17.81	2009.71 ± 117.73	2133.19 ± 23.13	84.64 ± 9.24	1.98 ± 0.22	1.29 ± 0.14
		19.91 ± 0.05	37.62 ± 0.01	8.04 ± 0.01	2345.20 ± 0.00	2044.58 ± 6.31	594.01 ± 38.00	1811.93 ± 8.37	213.05 ± 3.34	4.99 ± 0.08	3.26 ± 0.05
	High	17.86 ± 0.03	37.47 ± 0.01	7.84 ± 0.02	2353.59 ± 0.11	2162.46 ± 9.94	1430.79 ± 105.41	1954.25 ± 10.88	159.35 ± 4.33	3.73 ± 0.10	2.42 ± 0.07
		19.58 ± 0.05	37.56 ± 0.01	8.02 ± 0.01	2353.48 ± 0.08	2069.69 ± 4.52	744.14 ± 44.22	1839.47 ± 5.16	205.37 ± 2.06	4.81 ± 0.05	3.14 ± 0.03

In Faial, the maximum depth oscillated from 37.35 m in the High CO_2 site to 37.27 m in the Low CO_2 site and 29.33 m in the Reference site. The situation was very similar between the Low and High CO_2 sites, where CO_2 peaks close to the seabed and the effect is totally dissipated during the first 5 m, because of the proximity among sites together with the greater depth and strong tidal currents (Figs. 4.2E and 4.2F). Even though lower values of pH are reached in the High CO_2 site of this Island (6.81), mean pH values resulted equal for the full water column, but not for the deeper 10 m, particularly at the Low CO_2 site (minimum value of 7.21) (Table 4.1).

Seawater temperature and salinity did not vary substantially among sites in both islands. Whereas pH values varied correspondingly to the CO_2 partial pressure, TA remained constant among sites. CO_2 partial pressure in São Miguel showed a gradient within the range of the Representative Concentration Pathways in the latest IPCC assessment report (Hoegh-Guldberg et al. 2014, Pörtner et al. 2014), varying between 936 ppm (RCP8.5) and 421 ppm (RCP2.6) CO_2. Accordingly, results at the High CO_2 site can be considered as representatives of the more extreme scenario at the end of this century. This corresponded to a pH decrease from 8.06 to 7.75 (0.31 units), which is also within the range predicted by 2100 (0.3/0.4 units) (Dupont and Thorndyke 2009, Fitzer et al. 2012, Havenhand 2012).

Conditions in São Miguel, including shallow depth, absence of strong currents, easy access and constant emissions of almost pure CO_2, make this island an extraordinary location for further studies of the OA effect on Atlantic communities. On the other hand, studies carried out in Faial will require deeper analysis and reconsideration of the objectives. Effectively, emissions are above 98.6% of CO_2, avoiding the possible toxic effect of other gases, and lower pH values follow a gradient from the Reference to the Low and High CO_2 sites. Nevertheless, the location of the degassing site in the middle of the Faial-Pico channel, at a depth of almost 40 m, contributes to dissipate the effect of the emissions beyond the first 5 m closer to the bottom, with no significant differences at the surface among the three sites. This type of lateral advective processes and short residence time is seen as limiting the use of shallow-water vents as natural laboratories to evaluate AO effects on planktonic organisms. The need for biogeochemical characterization of these sites, to ensure that they are robust analogues for the future acidified ocean, has been previously emphasized (Burrell et al. 2015).

4.3.2 *Mesozooplankton Assemblages on CO_2 Vents*

Mesozooplankton was sampled using oblique tows with a WP2 net Ø60 cm and 200 μm mesh (São Miguel) or 500 μm mesh (Faial), for 10 minutes and at approximately 2 knots. Five tows per station were done at São Miguel (3–7 July 2014) and three in Faial (10–15 July 2014). Immediately after sampling, mesozooplankton was preserved in 4% borax buffered formaldehyde. One additional haul was done at each location with a modified cod end (without a mesh) in order to minimize damage to the organisms kept alive for manipulative experiments described in the next section (Egg Production Rate).

Patterns of variation in zooplanktonic assemblage structure along the CO_2 gradient in each island were analysed by multivariate statistical methods. Each Island (São Miguel and Faial) was considered independently, and two orthogonal factors were considered in the analyses: Site was considered a fixed factor with three levels corresponding to areas with different volcanic CO_2 emissions (Low CO_2 and High CO_2) and control areas, without emissions (Reference); Date was considered a random factor with five levels in São Miguel and three levels in Faial. Statistical differences among assemblages were tested using Permutational multivariate analysis of variance (PERMANOVA) on Bray Curtis similarities with untransformed abundances. The similarity percentages routine (SIMPER) was used to examine the contribution of each taxon to average dissimilarities between sample groups. Ordination by non-metric multidimensional scaling (nMDS) was used to visualize patterns in the biological dataset. Abundance (N), number of taxa (S), diversity (d) and evenness (J′) were calculated for the entire assemblage and for the dominant phyla. Univariate analyses of variance (ANOVA) were used to test for statistical differences in the taxa highlighted by the SIMPER routine, abundances, number of taxa and diversity indices. Differences between means were considered statistically significant for p < 0.05. Homogeneity of variances was previously tested with the Bartlett test, and pair-wise tests for group means were done *a posteriori* on significant effects. Evenness (J′) was obtained with the DIVERSE routine on PRIMER 6 statistical (PRIMER - E. Plymouth Marine Laboratory). Other analyses and data representation were done using R (R Core Team 2019).

Organisms from nine different phyla were identified, with a total of 71 different taxa, 45 in São Miguel and 61 in Faial. *Pelagia noctiluca* was collected only in the High CO_2 site of both islands. Given the mesh sizes used (200 or 500 µm), the most abundant zooplankton groups (e.g., phagotrophic protists) were clearly under-sampled. A separation on functional groups was also effectuated for the main phyla. As described in other studies (Sun et al. 2010, Shi et al. 2015), classification of zooplankton functional groups can be based on zooplankton size, food preferences, trophic functionality, interactions between one another or relationships with higher trophic levels. In this case, the crustaceans, dominant zooplankton group in terms of abundance in São Miguel, were separated according to their size (small: < 2 mm, medium-sized: 2–5 mm, and large: > 5 mm crustaceans). Cnidaria, Mollusca, fishes (Chordata), Tunicata (Chordata) and Radiozoa were differentiated as independent functional groups, while Annelida, Chaetognatha, Echinodermata and Foraminifera were pooled as "Others" since their abundances were too low in both islands.

São Miguel

Copepods were the most abundant component of zooplankton, the main species being calanoids as *Paracalanus parvus* or *Centropages typicus*. Along with cladocerans as *Evadne spinifera*, they lead to the dominance of the phylum Arthropoda. This and Chordata presented a significantly greater abundance at the High CO_2 site, relative to the Low CO_2 and Reference sites (Arthropoda—F value = 17.81, p = 0.0003; Chordata—F value = 9.98, p = 0.003). The total abundance also differed (F value = 22.94, p = 0) (Fig. 4.3A). Patterns in terms of richness, diversity (Shannon index) and evenness among sites varied inversely to the CO_2 gradient, although no significant differences among sites were observed for any particular phylum or the total taxa (results not shown).

Permutational multivariate analysis of variance (PERMANOVA) showed that composition of the zooplanktonic assemblages differed significantly between the three sites considered (p = 0.001). Non-metric multi-dimensional scaling (nMDS-Bray-Curtis similarities) analyses were done to illustrate the distribution of the main species highlighted by the Similarity percentage analysis (SIMPER). The main contributors (> 5%) to the differences among sites were the cladoceran *Evadne spinifera* (37.90% Ref-Low, 50.23% Ref-High, 44.25% Low-High), the calanoid copepod *Paracalanus parvus* (30.83% Ref-Low, 22.93% Ref-High, 29.66% Low-High), Cirripedia nauplii (7.92% Ref-Low) and Radiozoa (10.22% Ref-High, 10.22% Low-High). The nMDS (Fig. 4.4A) shows the distribution of the sample sites. Dimension 1 on the nMDS separated the samples of the High CO_2 site from the Reference and Low CO_2 sites, except for one sample of Reference and Low CO_2. Dimension 2 divided the Reference and Low CO_2 samples, except for a sample in the Reference site. Radiozoa, *Paracalanus parvus* and *Evadne spinifera* seem more related to High CO_2 conditions, while Cirripedia nauplii are closer to the Reference conditions.

Univariate analysis of variance for the abundance of the species highlighted by the SIMPER routine showed significant differences in the abundance of *Evadne spinifera* among sites (F value = 7.562, p = 0.0075), while for *P. parvus*, Cirripedia nauplii and Radiozoa abundances there were no significant differences. Previous studies have reported increased abundances of calanoid copepods under elevated $p\text{CO}_2$ conditions, probably as an indirect effect of increased food availability, due to enhanced primary productivity and phytoplankton biomass (Taucher et al. 2017, Algueró-Muñiz et al. 2019).

Small crustacean dominated over medium and large crustaceans, being representative of the assemblage (Reference – 86.07%, Low CO_2 – 86.49%, High CO_2 – 82.20%). Analysis of variance (ANOVA) on the abundances of functional groups showed significant differences among sites for small

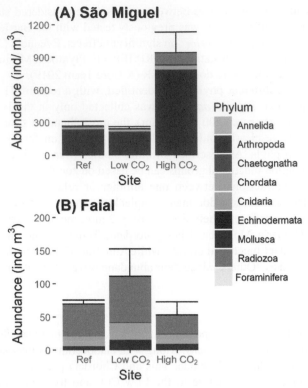

Figure 4.3: Abundances for each zooplankton phylum (individuals m⁻³, mean ± SE) on the three sampling sites (Ref – Reference < Low CO_2 < High CO_2) in São Miguel (A) and Faial (B) vents.

crustacean (F value = 16.4, p = 0.00037) and Tunicata (F value = 10.47, p = 0.00233). Oikopleuridae was the main representative of tunicates, and its increased abundance goes in accordance with previous studies concerning gelatinous zooplanktonic components. A positive correlation between the appendicularian *Oikopleura dioica* and pCO_2 was already observed in a mesocosm experiment (Troedsson et al. 2013), consistent with hypotheses concerning increased gelatinous zooplankton in future oceans. Thus, these results and subsequent studies (Winder et al. 2017, Bouquet et al. 2018, Algueró-Muñiz et al. 2019) suggest that appendicularians will play an important role in marine pelagic communities and vertical carbon transport under projected OA and elevated temperature scenarios. These species ingest smaller-sized particles than copepods, by-passing the microbial loop by directly transferring heterotrophic prokaryotes and nanoplankton to higher trophic levels, and they are vectors of global vertical carbon flux through trapping of prey in their frequently discarded "houses" (Troedsson et al. 2013, Conley et al. 2018). *Pelagia noctiluca* was not included in the analysis but its occurrence only in the High CO_2 sites on both islands does not seem fortuitous. Future sampling campaigns in these areas including a task aimed at these macrozooplankton components can bring more specific and useful information.

Radiozoa had low representation in the Reference and Low CO_2 sites, and higher abundance in the High CO_2 site, where it was the third most abundant group. There, high CO_2 concentration in the water may have enhanced the photosynthetic activity of endosymbiotic microalgae associated with Radiozoans, thus explaining their increased abundance. Similarly, phytoplankton could be benefited by the CO_2 emissions, and indirectly explain the higher zooplankton abundance (Schippers et al. 2004). Cirripedia nauplii appeared mainly associated with the Reference site, but other studies focusing on the larval stage of barnacles do not report this effect. For example, *Amphibalanus amphitrite* showed the early life phases unaffected under a pH of 7.4 (McDonald et al. 2009), and a population

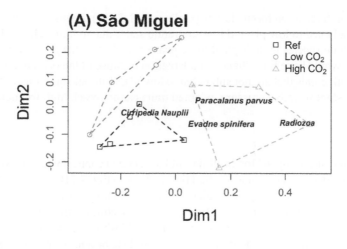

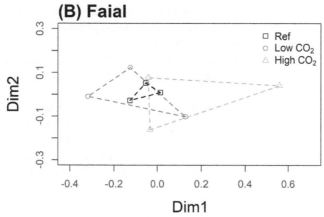

Figure 4.4: Non-metric multi-dimensional scaling (nMDS-Bray-Curtis similarities) analyses of zooplankton assemblages at each of the three sampling sites (Ref – Reference < Low CO_2 < High CO_2) in São Miguel (A) and Faial (B).

model of *Semibalanus balanoides* using empirical data showed that a decrease on pH from 8.2 to 7.8 only produces a significant effect below a critical temperature (Findlay et al. 2010). Although the planktonic life stages of Cirripedia usually lack a calcium carbonate external body, the early developmental stages of many marine species are suspected to be most sensitive to OA (Dupont and Thorndyke 2009, Kroeker et al. 2010, Cripps et al. 2014).

Paracalanus parvus and *Evadne spinifera*, the dominant species of the zooplankton assemblage, seem to be quite resilient under high CO_2 conditions. Most of the previous studies in copepods showed no effect on the survival of these organisms under controlled conditions of decreased pH. Detrimental effects of CO_2 were only detected during the early stages, like the egg production and the hatching success. Havenhand (2012), in his study on the Baltic Sea key functional groups, concluded that copepods will be resilient to near-future OA (≤ 1000 µatm CO_2). According to this, the extreme pCO_2 levels used in experiments elicited little or no response, and the relatively rapid generation times of copepods confer a high potential for adaptability. Nevertheless, as Mayor et al. (2012) concluded, there is a context-dependency that highlights the need for cautious interpretation and application of data from individual climate-change studies. Copepods, as a group, may be well equipped to deal with the chemical changes associated with OA; however, long-term exposures examining the synergistic effects of OA with other climate stressors, particularly warming on population viability and success, are yet to be conducted (Weydmann et al. 2012, Kroeker et al. 2017). Moreover, there are examples of invertebrates, such as corals, adapted to acidic water in shallow vents, but in areas with high retention for the larvae (Golbuu et al. 2016). It was suggested that the corals of Nikko Bay have adapted to low pH and low saturation state, due to high self-seeding and isolation. Notably, invertebrate species in their planktonic phase near CO_2 vents in other regions, such as the ones in our study, would have little chance of adaptation because of the open nature of the gene pool and the mixing with corals that are not subjected to low pH. Without some form of genetic isolation and without consistent selective pressure, adapted traits cannot reach high frequencies.

Faial

Radiozoa dominated mesozooplankton in all sites of Faial, and in contrast with São Miguel, the smallest values of total abundance were found in the High CO_2 site (Fig. 4.3B). No significant differences were found in the abundance of any particular phylum or the total taxa.

In general, there were no significant differences in the richness, diversity and evenness patterns among sites in Faial. The only exception was the diversity of Arthropoda, which decreased significantly along the CO_2 gradient (F value = 8.35, p = 0.02). The distribution of relative abundances among groups was more balanced. The most abundant group was Radiozoa (Reference – 69.33%, Low CO_2 – 62.10%, High CO_2 – 53.47%), while Cnidaria had a good representation compared to São Miguel (Reference – 11.56%, Low CO_2 – 15.0 %, High CO_2 – 15.08%).

PERMANOVA showed that composition of the zooplanktonic assemblages did not differ between sites (p = 0.502). The non-metric multi-dimensional scaling (nMDS-Bray-Curtis similarities) shows the distribution of the sample sites (Fig. 4.4B), where no differentiation can be made. This reveals the unstable conditions in Faial, which are manifested among the different sampling days. Since the CO_2 concentration is completely dissolved in the vicinity of the emissions, and its effect is not discernible in the remainder water column, this area is less suitable to test the OA effect on both planktonic and pelagic communities than the sites in São Miguel. In the same line, the differences in the diversity of arthropods among sites cannot be justified with the CO_2 levels considering that these organisms are distributed along the water column.

4.4 Ocean Acidification Effect on Planktonic Copepods Reproduction

Mesozooplankton was collected using oblique tows with a WP2 net Ø60 cm with a modified cod (without a mesh) in order to minimize damage to the organisms. Immediately after collection, the live zooplankton samples were pooled (no pre-selection was used), evenly distributed among the incubation chambers and placed by divers 1 m above the sea bottom using a line tied to a buoy (São Miguel) or a concrete anchor (Faial). Each chamber was 38 cm high and 10 cm in diameter, supported by a PVC frame and divided inside with a 500 µm mesh in a superior part with 30 cm and an inferior part with 8 cm, in order to allow eggs fall to avoid predation (Fig. 4.5). Around the chambers, a 50 µm mesh maintained the eggs inside while allowing exchange of seawater and smaller food particles. A total of 18 chambers were deployed in each island and exposed to the previously described conditions at each site. The incubations lasted 24 h and 72 h (three replicates per site and time). Due to logistical constraints, only the 72 h incubations could be retrieved in Faial (six replicates per site).

At the end of the incubations, the chambers were retrieved, and the content was immediately preserved in 1 ugol solution. The dominant copepod species were sorted and counted under a stereomicroscope. Total adult copepod females, eggs and nauplii emitted in each chamber were considered for determining the EPR, according to the following equation:

$$EPR = ((eggs+nauplii)/(adult\ female))/(Incubation\ time)$$

Previous studies calculated the egg hatching rate of different copepod species at *in situ* conditions (Andersen and Nielsen 1997). Escaravage and Soetaert (1993) define hatching as the moment when at least 50% of the eggs have hatched, and Andersen and Nielsen (1997) consider a 100% hatching

Figure 4.5: Incubation chambers designed for quantification of copepod Egg Production Rate.

criterion. Here, egg production rate was defined as the number of eggs released and nauplii hatched in relation to the number of females in the sample. The nauplii produced were taken into consideration because the hatching of eggs can occur during the time of the experiment so, otherwise, the egg production would be underestimated (Andersen and Nielsen 1997). In other methodologies, the capture and transport of copepods, as well as the change of the natural conditions like temperature or feeding, may cause a variation in the natural rates. This is the reason why incubation chambers were prepared, and the experiments were carried out in the field, with freshly caught copepods.

As the CO_2 gradient among sites affecting the incubation chambers was very similar between islands, results were analysed together. The EPR variation was tested through a two-way analysis of variance with CO_2 level and Incubation as orthogonal factors and EPR as dependent variable. Island and incubation time were merged in a single independent variable with three levels: São Miguel 24 and 72 h (SM-24 h, SM-72 h) and Faial 72 h (F-72 h). Differences between means were considered statistically significant for $p < 0.05$. Pairwise tests for group means were done *a posteriori* on significant effects using the Student-Newman-Keuls method. All parameters were initially tested for normality (Kolmogorov–Smirnov test) and homogeneity of variance (Bartlett's test). These analyses were performed in R (R Core Team 2019).

All individuals inside the chambers were identified as Calanoida (*Paracalanus* sp. and *Clausocalanus* sp.) in São Miguel, and Calanoida (*Centropages* sp.) and non-identified Harpacticoida species in Faial. The EPR showed a decreasing trend, from the Reference to the Low CO_2 and the High CO_2 sites in both islands (Fig. 4.6).

Results showed that differences among sites (Reference, Low CO_2 and High CO_2) were statistically significant ($p = 0.010$). Differences in *Incubation* were also statistically significant ($p \leq 0.001$), while there was not a statistically significant interaction between *Site* and *Incubation* ($p = 0.255$). Thus, decreasing EPR with the CO_2 gradient is independent of the incubation time and location (island).

The pairwise multiple comparison procedure showed that differences between the Low and High CO_2 sites were not significant ($p = 0.261$) but only between the Reference and Low CO_2 ($p = 0.031$) or High CO_2 ($p = 0.009$) sites. Significant differences for *Incubation* arose between SM-24 h and SM-72 h ($p \leq 0.001$) and SM-24 h and F-72 h ($p \leq 0.001$), but not for SM-72 h and F-72 h ($p = 0.209$).

The EPR significantly decreased under acidification conditions in São Miguel and Faial. Also, the decreasing EPR independent of the island, as shown in the ANOVA, demonstrates the suitability of the sites in Faial for these experiments. This effect was also independent from the incubation time (24 and 72 h in São Miguel), most probably because hatching occurred mainly during the first 24 h, and therefore an incubation of 72 h would not be necessary to detect the effect of CO_2. This information is relevant for future studies focused on EPR on similar systems. *Acartia tonsa* also show a decreasing EPR after 24 h (Cripps et al. 2014), but other species that have shown a response in previous studies were under a longer exposure.

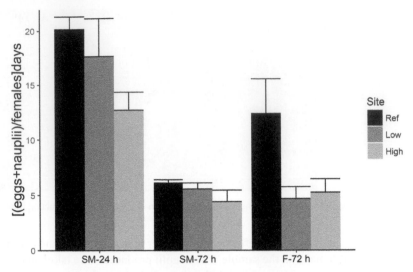

Figure 4.6: Copepod Egg Production Rate [(eggs+nauplii) female^{-1} day^{-1}] at each site, for different incubation periods (Ref – Reference < Low – Low CO_2 < High CO_2) of São Miguel (SM-24 h and SM-72 h) and Faial (F-72 h) (mean ± SE). Incubation experiments were carried out during July 2014.

In previous studies, the egg production rate and/or hatching rate were tested under controlled conditions and there was an effect of decreasing pH. Most of them were obtained for unrealistic pH values under 7, which are clearly beyond any existing scenarios for OA in the foreseeable future. Only Cripps et al. (2014) found a decreasing trend on *Acartia tonsa* for a pH of 7.8 and Thor and Dupont (2015) on *Pseudocalanus acuspes* for a pH of 7.75. Conditions of higher pH showed no significant effect. Considering that mean pH values in this study were between 7.51 and 8.10 (Table 4.1), this is one of the few studies that reports a significant decrease in EPR under realistic near future conditions, and the first one that report it *in situ* and for natural assemblages of zooplankton.

The studies by Smith et al. (2017, 2016) on tropical waters are the only evidence of how zooplankton communities are structured on natural CO_2 seeps. These are reef-associated communities where coral bleaching near the seeps alter the quality of the habitat, so CO_2 is exerting an indirect effect on the zooplanktonic community. This study reveals a different behaviour in temperate communities, where the substrate consists of rock and sand, which does not vary significantly between different sites gradually exposed to CO_2 and, therefore, the effects may be directly related to changes in the water chemistry.

4.5 Conclusions and Future Challenges

This chapter sheds some light to OA impacts on zooplankton communities, especially on early developmental and reproductive stages. Moreover, incubation chambers resulted in an adequate methodology to evaluate real ecophysiological impacts on zooplankton in the naturally acidic ocean areas. The case study presented evidence that is extremely important to further investigate zooplankton's sensitivity to global change, understanding the mechanisms affected and shifts in community composition as the first step to determine potential feedbacks to climate. Therefore, the study of the impact of OA on zooplankton occurring in the North Atlantic Subtropical Gyre was timely, since most studies address species in artificial conditions or at the most ecosystems from nutrient-rich coastal areas. This study was carried out in an open water system without evident retention areas leading to genetic isolation and consistent selective pressure. However, under the continuous exposure to increased pCO_2, zooplankton can acclimate or even adapt to OA when they are exposed over multiple generations. Thus, it is important to address which process regulates the effects of short-term and long-term exposures of zooplankton to OA in natural acidified conditions. Moreover, given that most OA stress studies have focused on the effects of short-term exposure (shorter than a single generation), experiments using adults might have underestimated the damaging effects of OA and the long-term multigenerational exposure to multiple stressors (e.g., increased pCO_2 and food shortage) will be required. Particularly, omics-based technologies (e.g., genomics, proteomics, and metabolomics) are helpful to better understand the underlying processes behind biological responses (e.g., survival, development, and offspring production) at the mechanistic level, which will improve our predictions of the responses of zooplankton to climate change stressors including OA (Dahms et al. 2018, Wang et al. 2018).

Acknowledgements

This work was part of the project MOFETA (EXPL/MAR-EST/0604/2013), funded by the Portuguese Foundation for Science and Technology (FCT). This study received Portuguese national funds from FCT - Foundation for Science and Technology, through the projects EXPL/MAR-EST/0604/2013 and UIDB/04326/2020. Sampling was done under the permit CCPJ015/2014, issued by the Regional Government of the Azores. The authors acknowledge the contribution of the following team members (in alphabetical order): Catarina Silva, Francesca Gallo, Francisco Leitão, Gustavo Martins, Hugo Parra, Jason Hall-Spencer, João Monteiro, Lucia Rodriguez, Marina Carreiro e Silva, Marta Martins, Ruben Couto. We are also grateful to "Bombeiros Voluntários de Vila Franca do Campo" and the boat crews of "Toninha Pintada", "Cachalote I", "Pintado" and "Águas Vivas".

References

Algueró-Muñiz, M., G.H. Henriette, S. Alvarez-Fernandez, C. Spisla, N. Aberle, L.T. Bach et al. 2019. Analyzing the impacts of elevated-CO_2 levels on the development of a subtropical zooplankton community during oligotrophic conditions and simulated upwelling. Front. Mar. Sci. 6: 61.

Andersen, C.M. and T.G. Nielsen. 1997. Hatching rate of the egg-carrying estuarine copepod *Eurytemora affinis*. Mar. Ecol. Prog. Ser. 160: 283–289.

Arnold, T., C. Mealey, H. Leahey, A.W. Miller, J.M. Hall-Spencer, M. Milazzo et al. 2012. Ocean acidification and the loss of phenolic substances in marine plants. PloS One 7(4): e35107.

Barry, J.P., J. Hall-Spencer and T. Tyrrell. 2010. *In situ* perturbation experiments: natural venting sites, spatial/temporal gradients in ocean pH, manipulative *in situ* p(CO₂) perturbations. pp. 123–136. *In*: U. Riebese [ed.]. Guide to Best Practices in Ocean Acidification Research and Data Reporting. Luxembourg: Publications Office of the European Union.

Bouquet, J., C. Troedsson, A. Novac, M. Reeve, A.K. Lechtenbörger, W. Massart et al. 2018. Increased fitness of a key appendicularian zooplankton species under warmer, acidified seawater conditions. PloS One 13(1): e0190625.

Boyd, P.W., S. Collins, S. Dupont, K. Fabricius, J.-P. Gattuso, J. Havenhand et al. 2018. Experimental strategies to assess the biological ramifications of multiple drivers of global ocean change—A review. Global Change Biology 24: 2239–2261.

Brinkhoff, T., S.M. Sievert, J. Kuever and G. Muyzer. 1999. Distribution and diversity of sulfur-oxidizing *Thiomicrospira* spp. at a shallow-water hydrothermal vent in the Aegean Sea (Milos, Greece). Appl. Environ. Microbiol. 65(9): 3843–3849.

Caldeira, K. and M.E. Wickett. 2003. Anthropogenic carbon and ocean pH. Nature 425: 365.

Caliro, S., F. Viveiros, G. Chiodini and T. Ferreira. 2015. Gas geochemistry of hydrothermal fluids of the S. Miguel and Terceira Islands, Azores. Geochim. Cosmochim. Acta 168: 43–57.

Calosi, P., S.P.S. Rastrick, M. Graziano, S.C. Thomas, C. Baggini, H.A. Carter et al. 2013. Distribution of sea urchins living near shallow water CO_2 vents is dependent upon species acid-base and ion-regulatory abilities. Mar. Pollut. Bull. 73(2): 470–484.

Cardigos, F., A. Colaço, P.R. Dando, S.P. Ávila, P.M. Sarradin, F. Tempera et al. 2005. Shallow water hydrothermal vent field fluids and communities of the D. João de Castro Seamount (Azores). Chem. Geol. 224(1-3): 153–168.

Chan, E.M., A.M. Derry, L.A. Watson and S.E. Arnott. 2008. Variation in calanoid copepod resting egg abundance among lakes with different acidification histories. Hydrobiologia 614(1): 275–284.

Ciais, P., C. Sabine, G. Bala, L. Bopp, V. Brovkin, J. Canadell et al. 2013. Carbon and other biogeochemical cycles. pp. 465–570. *In*: Climate Change 2013: The Physical Science Basis: Working Group I Contribution to the Fifth Assessment Report of the Intergovernmental Panel on Climate Change. Cambridge: Cambridge University Press.

Collins, M., R. Knutti, J. Arblaster, J.-L. Dufresne, T. Fichefet, P. Friedlingstein et al. 2013. Long-term climate change: projections, commitments and irreversibility. pp. 1029–1136. *In*: G.-K. Stocker, T.F.D. Qin, V. Bex, P.M. Midgley Plattner, M. Tignor, S.K. Allen et al. [eds.]. Climate Change 2013: The Physical Science Basis. Working Group I Contribution to the Fifth Assessment Report of the Intergovernmental Panel on Climate Change. Cambridge University Press, Cambridge, United Kingdom and New York, NY, USA.

Conley, K.R., F. Lombard and K.R. Sutherland. 2018. Mammoth grazers on the ocean's minuteness: a review of selective feeding using mucous meshes. Proc. R. Soc. B 285: 20180056.

Costello, M.J., A. Cheung and N. De Hauwere. 2010. Surface area and the seabed area, volume, depth, slope, and topographic variation for the world's seas, oceans, and countries. Environ. Sci. Technol. 44(23): 8821–8828.

Couto, R.P., A.S. Rodrigues and A.I. Neto. 2015. Shallow-water hydrothermal vents in the Azores (Portugal). J.I.C.Z.M. 15(4): 495–505.

Cramer, W., G.W. Yohe, M. Auffhammer, C. Huggel, U. Molau, M.A.F. Da Silva Dias et al. 2014. Detection and attribution of observed impacts. pp. 979–1037. *In*: C.B. Field, V.R. Barros, D.J. Dokken, K.J. Mach, M.D. Mastrandrea, T.E. Bilir et al. [eds.]. Climate Change 2014: Impacts, Adaptation, and Vulnerability. Part A: Global and Sectoral Aspects. Contribution of Working Group II to the Fifth Assessment Report of the Intergovernmental Panel on Climate Change. 979–1037. Cambridge, United Kingdom and New York, NY, USA: Cambridge University Press.

Cripps, G., P. Lindeque and K. Flynn. 2014. Parental exposure to elevated pCO_2 influences the reproductive success of copepods. J. Plankton Res. 36(5): 1165–1174.

Dahms, H.-U., N.V. Schizas, R.A. James, L. Wang and J.-S. Hwang. 2018. Marine hydrothermal vents as templates for global change scenarios. Hydrobiologia 818: 1–10.

Debes, H., K. Eliasen and E. Gaard. 2008. Seasonal variability in copepod ingestion and egg production on the Faroe shelf. Hydrobiologia 600(1): 247–265.

Dickson, A.G. and F.J. Millero. 1987. A comparison of the equilibrium constants for the dissociation of carbonic acid in seawater media. Deep Sea Res. I 34(10): 1733–1743.

Dupont, S. and M.C. Thorndyke. 2009. Impact of CO_2-driven ocean acidification on invertebrates early life-history— What we know, what we need to know and what we can do. Biogeosci. Discuss. 6(2): 3109–3131.

Escaravage, V. and K. Soetaert. 1993. Estimating secondary production for the brackish westerschelde copepod population *Eurytemora affinis* (Poppe) combining experimental data and field observations. Cah. Biol. Mar. 34(2): 201–214.

Fabry, V.J., B.A. Seibel, R.A. Feely and J.C. Orr. 2008. Impacts of ocean acidification on marine fauna and ecosystem processes. ICES J. of Mar. Sci. 65: 414–432.

Findlay, H.S., M.T. Burrows, M.A. Kendall, J.I. Spicer and S. Widdicombe. 2010. Can ocean acidification affect population dynamics of the barnacle *Semibalanus balanoides* at its southern range edge? Ecology 91(10): 2931–2940.

Fitzer, S.C., G.S. Caldwell, A.J. Close, A.S. Clare, R.C. Upstill-Goddard and M.G. Bentley. 2012. Ocean acidification induces multi-generational decline in copepod naupliar production with possible conflict for reproductive resource allocation. J. Exp. Mar. Biol. Ecol. 418-419: 30–36.

Gamenick, I., M. Abbiati and O. Giere. 1998. Field distribution and sulphide tolerance of *Capitella capitata* (Annelida: Polychaeta) around shallow water hydrothermal vents off Milos (Aegean Sea). A new sibling species? Mar. Biol. 130(3): 447–453.

Golbuu, Y., M. Gouezo, H. Kurihara, L. Rehm and E. Wolanski. 2016. Long-term isolation and local adaptation in Palau's Nikko Bay help corals thrive in acidic waters. Coral Reefs 35(3): 909–918.

Hall-Spencer, J.M., R. Rodolfo-Metalpa, S. Martin, E. Ransome, M. Fine, Suzanne M. Turner et al. 2008. Volcanic carbon dioxide vents show ecosystem effects of ocean acidification. Nature 454: 96–99.

Havenhand, J.N. 2012. How will ocean acidification affect Baltic Sea ecosystems? An assessment of plausible impacts on key functional groups. Ambio 41(6): 637–644.

Hoegh-Guldberg, O., R. Cai, E.S. Poloczanska, P.G. Brewer, S. Sundby, K. Hilmi et al. 2014. The Ocean. pp. 1655–1731. *In*: V.R. Barros, C.B. Field, D.J. Dokken, M.D. Mastrandrea, K.J. Mach, T.E. Bilir et al. [eds.]. Climate Change 2014: Impacts, Adaptation, and Vulnerability. Part B: Regional Aspects. Contribution of Working Group II to the Fifth Assessment Report of the Intergovernmental Panel of Climate Change. Cambridge, United Kingdom and New York, NY, USA: Cambridge University Press.

Jackson, J. B.C. 2010. The future of the oceans past. Philos. Trans. Royal Soc. B: Biological Sciences 365: 3765–3778.

Kâ, S. and J.-S. Hwang. 2011. Mesozooplankton distribution and composition on the northeastern coast of Taiwan during Autumn: Effects of the Kuroshio current and hydrothermal vents. Zoological Studies 50: 155–163.

Kerfahi, D., J.M. Hall-Spencer, B.M. Tripathi, M. Milazzo, J. Lee and J.M. Adams. 2014. Shallow water marine sediment bacterial community shifts along a natural CO_2 gradient in the Mediterranean Sea off Vulcano, Italy. Microb. Ecol. 67(4): 819–828.

Kerrison, P., J.M. Hall-Spencer, D.J. Suggett, L.J. Hepburn and M. Steinke. 2011. Assessment of pH variability at a coastal CO_2 vent for ocean acidification studies. Estuar. Coast. Shelf Sci. 94(2): 129–137.

Kroeker, K.J., R.L. Kordas, R.N. Crim and G.G. Singh. 2010. Meta-analysis reveals negative yet variable effects of ocean acidification on marine organisms. Ecol. Lett. 13(11): 1419–1434.

Kroeker, K.J., R.L. Kordas and C.D.G. Harley. 2017. Embracing interactions in ocean acidification research: confronting multiple stressor scenarios and context dependence. Biology Letters 13: 20160802.

Kroeker, K.J., L.E. Bell, E.M. Donham, U. Hoshijima, S. Lummis, J.A. Toy et al. 2020. Ecological change in dynamic environments: Accounting for temporal environmental variability in studies of ocean change biology. Global Change Biology 26: 54–67.

Kurihara, H., S. Shimode and Y. Shirayama. 2004. Effects of raised CO_2 concentration on the egg production rate and early development of two marine copepods (*Acartia steueri* and *Acartia erythraea*). Mar. Pollut. Bull. 49(9-10): 721–727.

Kurihara, H. 2008. Effects of CO_2-driven ocean acidification on the early developmental stages of invertebrates. Mar. Ecol. Prog. Ser. 373: 275–284.

Kurihara, H. and A. Ishimatsu. 2008. Effects of high CO_2 seawater on the copepod (*Acartia tsuensis*) through all life stages and subsequent generations. Mar. Pollut. Bull. 56(6): 1086–1090.

Levin, L.A., A.R. Baco, D.A. Bowden, A. Colaco, E.E. Cordes, M.R. Cunha et al. 2016. Hydrothermal vents and methane seeps: rethinking the sphere of influence. Front. Mar. Sci. 3: 72.

Li, W. and K. Gao. 2012. A marine secondary producer respires and feeds more in a high CO_2 ocean. Mar. Pollut. Bull. 64(4): 699–703.

Lischka, S., L.T. Bach, K.-G. Schulz and U. Riebesell. 2017. Ciliate and mesozooplankton community response to increasing CO_2 levels in the Baltic Sea: insights from a large-scale mesocosm experiment. Biogeosciences 14(2): 447–466.

Longhurst, A.R. 1985. The structure and evolution of plankton communities. Progress in Oceanography 15(1): 1–35.

Mayor, D.J., N.R. Everett and K.B. Cook. 2012. End of century ocean warming and acidification effects on reproductive success in a temperate marine copepod. Journal of Plankton Research 34(3): 258–262.

McConville, K., C. Halsband, E.S. Fileman, P.J. Somerfield, H.S. Findlay and J.I. Spicer. 2013. Effects of elevated CO_2 on the reproduction of two calanoid copepods. Marine Pollution Bulletin 73(2): 428–434.

McDonald, M.R., J.B. McClintock, C.D. Amsler, D. Rittschof, R.A. Angus, B. Orihuela et al. 2009. Effects of ocean acidification over the life history of the barnacle *Amphibalanus amphitrite*. Mar. Ecol. Prog. Ser. 385: 179–187.

Mehrbach, C., C.H. Culberson, J.E. Hawley and R.M. Pytkowicx. 1973. Measurement of the apparent dissociation constants of carbonic acid in seawater at atmospheric pressure 1. Limnology and Oceanography 18 (6): 897–907.

Pedersen, S.A., O.J. Håkedal, I. Salaberria, A. Tagliati, L.M. Gustavson, B.M. Jenssen et al. 2014. Multigenerational exposure to ocean acidification during food limitation reveals consequences for copepod scope for growth and vital rates. Environ. Sci. Technol. 48(20): 12275–12284.

Pettit, L.R., M.B. Hart, A.N. Medina-Sánchez, C.W. Smart, R. Rodolfo-Metalpa, J.M. Hall-Spencer et al. 2013. Benthic foraminifera show some resilience to ocean acidification in the northern gulf of California, Mexico. Marine Pollution Bulletin 73(2): 452–462.

Pörtner, H.-O., D. Karl, P.W. Boyd, W. Cheung, S.E. Lluch-Cota, Y. Nojiri et al. 2014. Ocean systems. pp. 411–484. *In*: C.B. Field, V.R. Barros, D.J. Dokken, K.J. Mach, M.D. Mastrandrea, T.E. Bilir et al. [eds.]. Climate Change 2014: Impacts, Adaptation, and Vulnerability. Part A: Global and Sectoral Aspects. Contribution of Working Group II to the Fifth Assessment Report of the Intergovernmental Panel of Climate Change. Cambridge, United Kingdom and New York, NY, USA: Cambridge University Press.

Raven, J., K. Caldeira, H. Elderfield, O. Hoegh-Guldberg, P. Liss, U. Riebesell et al. 2005. Ocean acidification due to increasing atmospheric carbon dioxide. The Royal Society Policy Document. Vol. 12/05. London.

R Core Team. 2019. R: a language and environment for statistical computing. Vienna, Austria: R Foundation for Statistical Computing. Retrieved from https://www.r-project.org/.

Riebesell, U. and P. D. Tortell. 2011. Effects of ocean acidification on pelagic organisms and ecosystems. pp 99–121. *In*: J.-P. Gattuso and L. Hansson [eds.]. Ocean Acidification. Oxford, UK: Oxford University Press.

Ross, Pauline M., L. Parker, W.A. O'Connor and E.A. Bailey. 2011. The impact of ocean acidification on reproduction, early development and settlement of marine organisms. Water 3(4): 1005–1030.

Schippers, P., M. Lürling and M. Scheffer. 2004. Increase of atmospheric CO_2 promotes phytoplankton productivity. Ecol. Lett. 7(6): 446–451.

Shi, Y.Q., S. Sun, G.T. Zhang, S.W. Wang and C.L. Li. 2015. Distribution pattern of zooplankton functional groups in the Yellow Sea in June: a possible cause for geographical separation of giant jellyfish species. Hydrobiologia 754(1): 43–58.

Smith, J.N., G. De'ath, C. Richter, A. Cornils, J.M. Hall-Spencer and K.E. Fabricius. 2016. Ocean acidification reduces demersal zooplankton that reside in tropical coral reefs. Nat. Clim. Change 6(12): 1124–1129.

Smith, J.N., C. Richter, K.E. Fabricius and A. Cornils. 2017. Pontellid copepods, *Labidocera* spp., affected by ocean acidification: a field study at natural CO_2 seeps. PloS One 12(5): e0175663.

Sun, S., Y. Huo and B. Yang. 2010. Zooplankton functional groups on the continental shelf of the Yellow Sea. Deep-Sea Res. II 57(11-12): 1006–1016.

Sweetman, A.K., A.R. Thurber, C.R. Smith, L.A. Levin, C. Mora, C.-L. Wei et al. 2017. Major impacts of climate change on deep-sea benthic ecosystems. Elem. Sci. Anth. 5: 4.

Tarasov, V.G., A.V. Gebruk, A.N. Mironov and L.I. Moskalev. 2005. Deep-sea and shallow-water hydrothermal vent communities: two different phenomena? Chem. Geol. 224(1-3): 5–39.

Taucher, Jan, M. Haunost, T. Boxhammer, L.T. Bach, M. Algueró-Muñiz and U. Riebesell. 2017. Influence of ocean acidification on plankton community structure during a winter-to-summer succession: an imaging approach indicates that copepods can benefit from elevated CO_2 via indirect food web effects. PloS One 12(2): e0169737.

Thor, P. and S. Dupont. 2015. Transgenerational effects alleviate severe fecundity loss during ocean acidification in a ubiquitous planktonic copepod. Global Change Biol. 21(6): 2261–2271.

Troedsson, C., J.-M. Bouquet, C.M. Lobon, A. Novac, J. Nejstgaard, S. Dupont et al. 2013. Effects of ocean acidification, temperature and nutrient regimes on the appendicularian *Oikopleura dioica*: a mesocosm study. Mar. Biol. 160(8): 2175–2187.

Tsui, M.T.K. and W.X. Wang. 2005. Multigenerational acclimation of *Daphnia magna* to mercury: relationships between biokinetics and toxicity. Environ. Toxicol. Chem. 24(11): 2927–2933.

Vehmaa, A., H. Hogfors, E. Gorokhova, A. Brutemark, T. Holmborn and J. Engström-Öst. 2013. Projected marine climate change: effects on copepod oxidative status and reproduction. Ecol. Evol. 3(13): 4548–4557.

Viveiros, F., L. Moreno, M. Carreiro-Silva, R.P. Couto, C. Silva, P. Range et al. 2016. Volcanic gas emissions offshore of São Miguel and Faial Islands (Azores Archipelago). Lisboa: 4ª Jornadas de Engenharia Hidrográfica.

Waldbusser, G.G. and J.E. Salisbury. 2014. Ocean acidification in the coastal zone from an organism's perspective: multiple system parameters, frequency domains, and habitats. Annu. Rev. Mar. Sci. 6: 221–247.

Wang, M., C.B. Jeong, Y.H. Lee and J.S. Lee. 2018. Effects of ocean acidification on copepods. Aquat. Toxicol. 196: 17–24.

Weydmann, A., J.E. Søreide, S. Kwasniewski and S. Widdicombe. 2012. Influence of CO_2-induced acidification on the reproduction of a key arctic copepod calanus glacialis. J. Exp. Mar. Biol. Ecol. 428: 39–42.

Winder, M., J.-M. Bouquet, J.R. Bermúdez, S.A. Berger, T. Hansen, J. Brandes et al. 2017. Increased appendicularian zooplankton alter carbon cycling under warmer more acidified ocean conditions. Limnol. Oceanogr. 62(4): 1541–1551.

Won, Y.-J. 2006. Deep-sea hydrothermal vents: ecology and evolution. J. Ecol. Environ. 29(2): 175–83.

Zachos, J.C., U. Röhl, S.A. Schellenberg, A. Sluijs, D.A. Hodell, D.C. Kelly et al. 2005. Rapid acidification of the ocean during the Paleocene-Eocene thermal maximum. Science 308(5728): 1611–1615.

Zervoudaki, S., C. Frangoulis, L. Giannoud and E. Krasakopoulou. 2013. Effects of low pH and raised temperature on egg production, hatching and metabolic rates of a Mediterranean copepod species (*Acartia clausi*) under oligotrophic conditions. Mediterr. Mar. Sci. 15(1): 74–83.

Zhang, D., S. Li, G. Wang and D. Guo. 2011. Impacts of CO_2-driven seawater acidification on survival, egg production rate and hatching success of four marine copepods. Acta Oceanol. Sin. 30(6): 86–94.

Part 2
Spatial-Temporal Distribution Patterns and Trophic Dynamics

CHAPTER 5

Diversity and Biogeography as Revealed by Morphologies and DNA Sequences
Tintinnid Ciliates as an Example

Luciana F. Santoferrara[1,*] and *George B. McManus*[2]

5.1 Introduction

Quantifying zooplankton diversity was initially focused on discovering species and their distributions in the word ocean, as catalogued in early monographs (Haeckel 1887, Sars 1903, Brandt 1906–1907, Lohmann 1908). Later studies have contributed to our understanding of diverse marine processes, such as species dispersal and water circulation patterns (van der Spoel and Heyman 1983, Chiba et al. 2013). The progression from cataloguing diversity to understanding its role in marine ecology has more recently been accompanied by a transition from morphological species examination to molecular methods that allow distinctions to be made at every taxonomic level, from phyla down to populations.

Among zooplankton, attempts to differentiate species have frequently run into difficulties when based only on morphology, even among some of its best-known members, the copepods. For example, at one time it was nearly impossible to differentiate *Pseudocalanus* spp. in boreal and temperate coastal ecosystems based on microscopy alone (Corkett and McLaren 1978, Frost 1989). The development of assays based on DNA sequencing allowed for both faster and more precise identifications within this species complex, revealing important ecological differences (Bucklin et al. 1998, Grabbert et al. 2010). Also based on DNA methods, the ubiquitous estuarine copepod *Acartia tonsa* actually presents species-level genetic divergences in Atlantic and Pacific specimens (Caudill and Bucklin 2004), and shows populations adapted to low- or medium-salinity in Chesapeake Bay (Chen and Hare 2008).

Application of molecular techniques in diversity studies has become common, but combination with morphological taxonomy is important in order to link DNA data to the large body of information accumulated over nearly two centuries of plankton studies. When molecular and morphological methods are compared, they often reveal the presence of cryptic (morphologically-identical forms

[1] University of Connecticut, Department of Ecology and Evolutionary Biology and Department of Marine Sciences, 1 University Place, Stamford, CT 06901, USA.
[2] University of Connecticut, Department of Marine Sciences, 1080 Shennecossett Rd., Groton, CT 06340, USA.
   Email: george.mcmanus@uconn.edu
* Corresponding author: luciana.santoferrara@uconn.edu

that are genetically distinct) or polymorphic (genetically-identical forms with different morphologies) species (McManus and Katz 2009, Bucklin et al. 2010, Dolan 2016). This is one of the reasons that has led to the extensive practices of DNA "barcoding" and "metabarcoding" (see Box 5.1). For a

---

Box 5.1: DNA barcoding and metabarcoding are complementary methods.

DNA barcoding aims at delineating and classifying species based on the sequence of a short, standardized gene marker (Hebert et al. 2003). The requisites for a good barcoding protocol include marker and primer universality, ease of DNA sequencing, the presence of a barcode gap (see Fig. 5.3A) and voucher deposition. However, it is very difficult to fulfill all these criteria simultaneously for organisms as different as those present in plankton. For multicellular organisms, such as copepods, a small piece of taxonomically-unimportant tissue from a voucher specimen can be used for DNA extraction and sequencing, with the most common marker being the mitochondrial cytochrome oxidase *c* subunit I gene (COI; Bucklin et al. 2010). For protists, such as ciliates, COI presents many limitations and instead a region of the nuclear small-subunit ribosomal RNA gene (SSU rDNA) is used as a pre-barcode (Pawlowski et al. 2012). While SSU rDNA is widely used for phylogenetics, even its most variable regions (such as V4 and V9) may be too conserved to accurately delineate species (Fig. 5.3). Thus, complementary, group-specific barcodes are proposed, such as the D1-D2 region of the large-subunit ribosomal RNA gene (LSU rDNA) for ciliates (Santoferrara et al. 2013, Stoeck et al. 2014a), or parts of the ITS region (the 5.8S rRNA gene combined with the internally transcribed spacer regions 1 and 2) for dinoflagellates (Litaker et al. 2007). Another challenge in protist metabarcoding is that the availability of voucher cells depends on culturing success. For uncultured species or specimens isolated directly from the environment, the cell is destroyed during DNA extraction and thus diagnostic images are acceptable as depository material (Pawlowski et al. 2012). Regardless of the organism, once the barcode is amplified by polymerase chain reaction (PCR), it may or may not be inserted into a plasmid for subsequent Sanger sequencing, with products that usually range from 400 to 1800 nucleotides depending on the marker. These barcodes have formed a large body of sequences linked to species names that are available in public repositories (e.g., GenBank), which in turn are a reference for new identifications of isolated specimens or the members of natural communities. For some markers and lineages, curated reference databases are being built (Guillou et al. 2013, del Campo et al. 2018).

DNA metabarcoding aims at identifying and tracking taxa in natural communities. Total DNA from an environmental sample is used to build a library of barcodes via PCR amplification and sequencing. These libraries used to be cloned and Sanger-sequenced (López-García et al. 2001, Doherty et al. 2007), but are now analyzed by parallel, high-throughput sequencing (e.g., Illumina), with the hypervariable region V4 or V9 of SSU rDNA as the most common markers (Amaral-Zettler et al. 2009, Stoeck et al. 2009). These markers are easily amplified, are short enough for current technologies (e.g., Illumina), contain enough taxonomic information at least above the species level, and count with a growing number of reference sequences for comparison. Metabarcode reads go through several bioinformatic procedures for quality filtering, grouping into Operational Taxonomic Units (OTUs) based on a sequence similarity cut-off (generally 97–100%), taxonomic identification and biodiversity analyses (reviewed by Santoferrara 2019). Note that sequences obtained with the early forms of metabarcoding (clone libraries) are usually longer than 500 nucleotides and are available in GenBank as "uncultured organisms". Instead, high-throughput metabarcodes are usually shorter than 500 nucleotides and are obtained in the order of thousands to millions per sample, so they are deposited in special repositories such as the NCBI Sequence Read Archive (SRA).

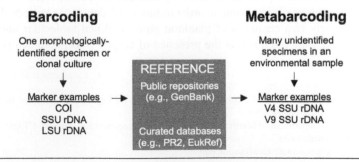

given morphospecies (e.g., a single copepod picked from a plankton net tow or a newly established protist culture), the barcode sequence (often from a nuclear ribosomal RNA gene or a mitochondrial cytochrome oxidase gene) can serve to verify identity. Making a library of such sequences from an environmental DNA sample now provides standard metrics for diversity of molecular operational taxonomic units (OTUs), which many times can be linked to named species. Such libraries have not only filled out the picture of diversity and biogeography for known organisms but have also revealed previously-unknown or poorly known clades (López-García et al. 2001, Slapeta et al. 2006, Kim et al. 2011). A number of recent and ongoing projects that have used metabarcoding on global, regional and local scales are greatly expanding our knowledge of plankton diversity and distribution (de Vargas et al. 2015, Villarino et al. 2018).

Understanding the diversity and biogeography of zooplankton is a key to oceanographic processes, including advection in the great ocean gyres, mesoscale horizontal mixing, and vertical displacement from motility or sinking. Larger organisms such as gelatinous zooplankton and fish appear to have endemism, and changes in species ranges due to climate change or anthropogenic introductions have been documented (Shiganova et al. 2001, Graham et al. 2003, Boero et al. 2009). Such changes can disrupt food webs and impact commercial fisheries. On the other hand, at least some zooplankton species, especially in microzooplankton, may not be dispersal-limited and thus their distributions may not be affected by accidental introductions or other dispersal-related factors (Fenchel and Finlay 2006). But because most of these smaller zooplankton (even most unicellular forms) have sexual phases in their life cycles, they fit the "biological" species definition (Sonneborn 1957) and may thus be subject to reproductive isolation. Therefore, documenting species distributions in a way that is not reliant on ambiguous species delineations or increasingly rare taxonomic expertise provides direct evidence of dispersal limitation and reproductive isolation, or lack thereof. This reinforces the idea that molecular identifications, in concert with morphology, can serve to illuminate diverse aspects of plankton ecology.

The goal of this chapter is to illustrate assessment of diversity and biogeography in an important group of microzooplanktonic protists, the order Tintinnida (Spirotrichea, Ciliophora). Tintinnids have both a long history of morphology-based study and several recent examples of molecular surveys, thus representing a good example on how the rich classical literature can be reconciled with emerging DNA data. The interest in tintinnids as model organisms goes beyond characterizing their diversity, and several aspects of their biology and roles in marine ecosystems have been recently compiled (Dolan et al. 2013a). Tintinnids and other spirotrichs, the naked choreotrichs and oligotrichs, are the most diverse and abundant pelagic ciliates (Pierce and Turner 1992). Quantitatively, therefore, tintinnids are an important component of the consumers in plankton and provide a link to higher trophic levels such as omnivorous copepods (reviewed by McManus and Santoferrara 2013).

One feature that sets tintinnids apart is that the cell is attached to a lorica, a loose-fitting proteinaceous shell (Fig. 5.1A), which can be hyaline (clear) or agglomerated with mineral or biogenic particles. Like other protist zooplankton groups such as foraminiferans and radiolarians, the possession of a hard structure makes the tintinnids collectible via plankton nets, and ensures recognition even when poorly preserved. Thus, the cataloguing of tintinnid diversity began early in the history of plankton studies, and convenient illustrated keys and monographs to help identifications were available long before they were available for other unicellular zooplankton (Kofoid and Campbell 1929, 1939, Marshall 1969). However, the apparent simplicity of lorica-based identifications presents many challenges, including artificial classification and unclear species delineations (Alder 1999, Agatha and Strüder-Kypke 2013).

The limitations of a lorica-based taxonomy have led to complementation with other data sources, notably genetic information. In common with most ciliates, tintinnids have germline micronuclei that are capable of sexual recombination and highly polyploid somatic macronuclei that propagate asexually (Lynn 2008). Because they have many copies of the ribosomal RNA genes, which may number in the thousands (Gong et al. 2013), tintinnids are amenable to single-cell barcoding for

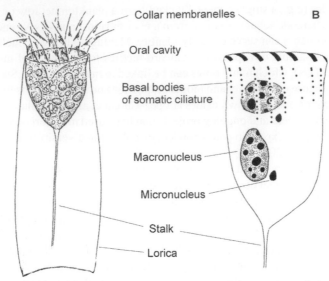

Figure 5.1: A tintinnid example, *Dartintinnus alderae*. (A) specimen observed *in vivo* (ventral view). The cell attaches to the posterior wall of the hyaline (clear) lorica by a contractile stalk. The collar membranelles expand beyond the lorica or can retract inside. (B) dorsal view after protargol staining; basal bodies (where cilia emerge from the cell surface) and nuclei are revealed. Adapted from: Smith, S., W. Song, N. Gavrilova, A. Kurilov, W. Liu, G. McManus and L. Santoferrara. 2018. *Dartintinnus alderae* n. g., n. sp., a minute brackish tintinnid (Ciliophora, Spirotrichea) with dual-ended lorica collapsibility. J. Eukaryot. Microbiol. 65: 400–411, © 2017 The Author(s) Journal of Eukaryotic Microbiology © 2017 International Society of Protistologists), with permission from John Wiley and Sons.

comparisons of morphological and DNA sequence variations with no need for clonal cultures (Snoeyenbos-West et al. 2002, Bachy et al. 2012, Xu et al. 2012, Santoferrara et al. 2013, 2015). Single specimens, which usually range from about 20 to 200 µm in diameter, can be picked onto a microscope slide, photographed to document lorica morphology, and then placed in buffer for DNA extraction and sequencing. Thus, individual concordance between shape and DNA barcodes can be assessed, just as in larger organisms. Collections of matched DNA sequences and morphologies have helped form the basis for linking the emerging blizzard of DNA metabarcoding data to what is known from microscope-based studies (Bachy et al. 2013, Santoferrara et al. 2016a).

We have recently retrieved all the tintinnid DNA barcode sequences available in GenBank (National Center for Biotechnology Information, NCBI) and revised these data in terms of classification and phylogenetics (Santoferrara et al. 2017), species delineation (Santoferrara et al. 2013, 2015) and global diversity and distribution (Santoferrara et al. 2018). We have integrated molecular and morphological information as much as possible, and also compared both approaches using DNA metabarcoding and microscopy to evaluate tintinnid assemblages (Santoferrara et al. 2014, 2016a). This chapter synthesizes our work and other literature in order to critically analyze these complementary approaches and to identify future directions in diversity studies of ciliates and other zooplankton.

5.2 Setting the Foundation for Studies of Diversity and Biogeography

Estimates of diversity and distribution are determined by the criteria used to catalog organisms (e.g., morphology, DNA sequences), the target taxonomic level (e.g., family, genus, species), and the overall reliability of taxon classification. Taxonomy sets the basis to track organisms and, along with phylogenetics, is necessary to understand not only the patterns, but also the ecological and evolutionary processes that determine how diversity is generated, maintained and distributed. Tintinnids have a long history in morphological taxonomy and relatively recent efforts for integration of DNA sequences, which need consideration before focusing on their diversity and biogeography.

5.2.1 Taxonomy and Phylogenetics

The first tintinnid species was described almost 250 years ago (Müller 1779). Since then, tintinnid taxonomy can be very broadly summarized in three periods. The first period was based almost exclusively on the lorica shape and size, and led to the description of almost all of ~ 1,000 known species (Ehrenberg 1832, Claparède and Lachmann 1858, Daday 1887, Brandt 1906–1907, Laackmann 1907, 1910, Jörgensen 1924, Kofoid and Campbell, 1929, 1939, Balech 1968, Hada 1970). This period of discoveries was facilitated by the fact that the lorica both endures plankton net collection and is conspicuous enough for identification by conventional microscopy, even though it was early recognized that overlooking the cell features would create an artificial taxonomy (Brandt 1906–1907, Entz Jr. 1909).

The second period in tintinnid taxonomy incorporated cytological and ultrastructural analyses. Some early studies considered cell features visible *in vivo* (Fig. 5.1A), such as the conspicuous oral membranelles (Fol 1881). But cytology became much more informative with stains (especially silver proteinate, or protargol) that also reveal the somatic ciliature, nuclei and morphogenesis details (Foissner and Wilbert 1979, Snyder and Brownlee 1991, Choi et al. 1992, Petz et al. 1995, Agatha and Riedel-Lorjé 2006, Agatha 2008, 2010, Agatha and Tsai 2008; Fig. 5.1B). Also, electron microscopy started to further reveal diagnostic characters such as the lorica texture (Agatha et al. 2013), extrusome-like 'capsules' and associated organelles (Laval-Peuto and Barria de Cao 1987) and, more recently, the ultrastructure of somatic kinetids (Gruber et al. 2019). Cytological and ultrastructural data are key for tintinnid taxonomy (Laval-Peuto and Brownlee 1986, Laval-Peuto 1994), but remain rare due to the laborious procedures and (often unachievable) cultivation required.

The third and current period has incorporated DNA sequences, mainly of the small-subunit ribosomal RNA gene (SSU rDNA). Initial molecular studies were based on lorica-identified specimens and were facilitated by the fact that sequences could be obtained from single or few specimens, many times picked directly from environmental samples (Snoeyenbos-West et al. 2002, Strüder-Kypke and Lynn 2003, 2008, Bachy et al. 2012, Santoferrara et al. 2012, 2017, Xu et al. 2013, Zhang et al. 2017). These sequences caused a change of paradigm in tintinnid taxonomy (for example, revealing the non-monophyly of hyaline and hard-agglomerated taxa), although the heterogeneous quality of the obtained data and interpretations have also caused confusion (see examples noted by Agatha and Strüder-Kypke 2014, Santoferrara et al. 2017). More reliably but sparsely, recent descriptions and redescriptions have incorporated lorica, cell and DNA sequencing data (Kim et al. 2010, Agatha and Strüder-Kypke 2012, Saccà et al. 2012, Ganser and Agatha 2018, Gruber et al. 2018, Smith et al. 2018).

The lorica is still the foundation of tintinnid taxonomy (it is the only link to virtually all the original descriptions), but integration of more detailed phenotypic characters and genetic information is key for classification and phylogenetics (Santoferrara et al. 2016b). Reliable cytological data and DNA sequences exist for only about 3% and 10% of the known species, respectively, but have allowed for significant improvements in the classification of genera and families (Agatha and Strüder-Kypke 2007, 2013, 2014, Santoferrara et al. 2017). Tintinnids are currently classified into 77 genera and 14 families (Table 5.1). Following a recent trend of taxonomic rearrangements, and given that most genera and some families still need cytological and/or genetic investigation (Table 5.1), these taxonomic levels are expected to undergo additional merging or splitting once the required data are collected (Santoferrara et al. 2016b).

Current phylogenetic inferences (Santoferrara et al. 2017) also suggest the need for additional taxonomic rearrangements (Fig. 5.2): the sequential arrangement of four well-supported, monophyletic families (Tintinnidiidae, Tintinnidae, Eutintinnidae and Favellidae) is followed by an unresolved polytomy of at least nine families (some appearing so closely related that they may need synonymization) intertwined with thirteen *incertae sedis* lineages (some appearing so divergent that they may represent multiple families). The most dramatic case is "*Tintinnopsis*", an artificial genus with about 160 species united by a hard, fully agglomerated lorica, but now known to include at least five distinct patterns of somatic ciliature (Agatha and Studer-Kypke 2007, 2013, 2014, Gruber et al.

Table 5.1: Classification of the tintinnid ciliates.

Taxa	Taxa (continued)
Ascampbelliellidae Corliss 1960 (3 genera)	*Buschiella* Corliss 1960
Ascampbelliella Corliss 1960[1, 3]	*Canthariella* Kofoid and Campbell 1929[1]
Incertae sedis: Luxiella Lecal 1953	*Clevea* Balech 1948
Incertae sedis: Niemarshallia Corliss 1960	*Daturella* Kofoid and Campbell 1929[1]
Cyttarocylididae Kofoid and Campbell 1929 (2 genera)	*Epicranella* Kofoid and Campbell 1929
Cyttarocylis Fol 1881[1, 3]	*Odontophorella* Kofoid and Campbell 1929
Petalotricha Kent 1881[1, 3]	*Ormosella* Kofoid and Campbell 1929[1]
Dictyocystidae Haeckel 1873 (6 genera)	*Proamphorella* Kofoid and Campbell 1939
Codonaria Kofoid and Campbell 1929[1, 3]	*Prostelidiella* Kofoid and Campbell 1939
Codonella Haeckel 1873[1, 3]	*Rhabdosella* Kofoid and Campbell 1929
Codonellopsis Jörgensen 1924[1, 2, 3]	*Salpingacantha* Kofoid and Campbell 1929[1, 3]
Dictyocysta Ehrenberg 1854[1, 3]	*Salpingella* Jörgensen 1924[1, 3]
Incertae sedis: Laackmanniella Kofoid and Campbell 1929[1, 2, 3]	*Salpingelloides* Campbell 1942
	Steenstrupiella Kofoid and Campbell 1929[1, 3]
Incertae sedis: Wangiella Nie 1934	*Stelidiella* Kofoid and Campbell 1929
Epiplocylididae Kofoid and Campbell 1939 (3 genera)	*Tintinnus* Schrank 1803
Epicancella Kofoid and Campbell 1929[1]	Tintinnidiidae Kofoid and Campbell 1929 (3 genera)
Epiplocylis Jörgensen 1924[1, 3]	*Antetintinnidium* Ganser and Agatha 2019[1, 2, 3]
Epiplocyloides Hada 1938[1, 3]	*Membranicola* Foissner, Berger and Schaumburg 1999[2]
Eutintinnidae Bachy et al. 2012 (2 genera)	*Tintinnidium* Kent 1881[1, 2, 3]
Dartintinnus Smith and Santoferrara in Smith et al. 2017[1, 2, 3]	Undellidae Kofoid and Campbell 1929 (7 genera)
	Amplectella Kofoid and Campbell 1929[1]
Eutintinnus Kofoid and Campbell 1939[1, 2, 3]	*Amplectellopsis* Kofoid and Campbell 1929
Favellidae Kofoid and Campbell 1929 (1 genus)	*Cricundella* Kofoid and Campbell 1929
Favella Jörgensen 1924[1, 2, 3]	*Parundella* Jörgensen 1924[1, 3]
Nolaclusiliidae Sniezek et al. 1991 (1 genus)	*Proplectella* Kofoid and Campbell 1929[1]
Nolaclusilis Snyder and Brownlee 1991[1, 2]	*Undella* Daday 1887[1, 3]
Ptychocylididae Kofoid and Campbell 1929 (4 genera)	*Undellopsis* Kofoid and Campbell 1929[1]
Cymatocylis Laackmann 1910[1, 2, 3]	Xystonellidae Kofoid and Campbell 1929 (5 genera)
Protocymatocylis Kofoid and Campbell 1929	*Parafavella* Kofoid and Campbell 1929[1, 3]
Ptychocylis Brandt 1896[1, 3]	*Spiroxystonella* Kofoid and Campbell 1939
Wailesia Kofoid and Campbell 1939	*Xystonella* Brandt 1906[1, 3]
Rhabdonellidae Kofoid and Campbell 1929 (7 genera)	*Xystonellopsis* Jörgensen 1924[1]
Epirhabdonella Kofoid and Campbell 1939	*Incertae sedis*: *Dadayiella* Kofoid and Campbell 1929[1, 3]
Metacylis Jörgensen 1924[1, 3]	
Protorhabdonella Jörgensen 1924[1, 3]	*Incertae sedis*:
Pseudometacylis Balech 1968	*Acanthostomella* Jörgensen 1927[1, 3]
Rhabdonella Brandt 1906[1, 3]	*Climacocylis* Jörgensen 1924[1, 3]
Rhabdonellopsis Kofoid and Campbell 1929[1]	*Codonopsis* Kofoid and Campbell 1939
Schmidingerella Agatha and Strüder-Kypke 2012[1, 2, 3]	*Helicostomella* Jörgensen 1924[1, 3]
Stenosemellidae Campbell 1954 (1 genus)	*Leprotintinnus* Jörgensen 1900[1, 3]
Stenosemella Jörgensen 1924[1, 2, 3]	*Poroecus* Cleve 1902[1]
Tintinnidae Claparède and Lachmann 1858 (21 genera)	*Rhizodomus* Strelkow and Wirketis 1950[1, 2, 3]
Albatrossiella Kofoid and Campbell 1929[1]	*Rotundocylis* Kufferath 1950
Amphorellopsis Kofoid and Campbell 1929[1, 3]	*Stylicauda* Balech 1951[1, 3]
Amphorides Strand 1928[1, 3]	*Tintinnopsis* Stein 1867[1, 2, 3]
Brandtiella Kofoid and Campbell 1929[1]	*Nomen inquirendum*: *Coxliella* Brandt 1906[3]
Bursaopsis Kofoid and Campbell 1929	

Based on Agatha and Strüder-Kypke (2013) and Santoferrara et al. (2017). This update adds two recently established genera and places *Acanthostomella* as *incertae sedis* (as shown in Fig. 5.2, this genus is no longer supported in Ascampbelliellidae and its affiliation needs confirmation with cytological and ultrastructural data; Santoferrara et al. 2018). Genera [1] reported at least four times by two different author teams (updated from Dolan and Pierce 2013), [2] with published protargol-staining data (excluding *Codonella cratera*, which needs reassignment; Gruber et al. 2018) and [3] with DNA sequences in GenBank.

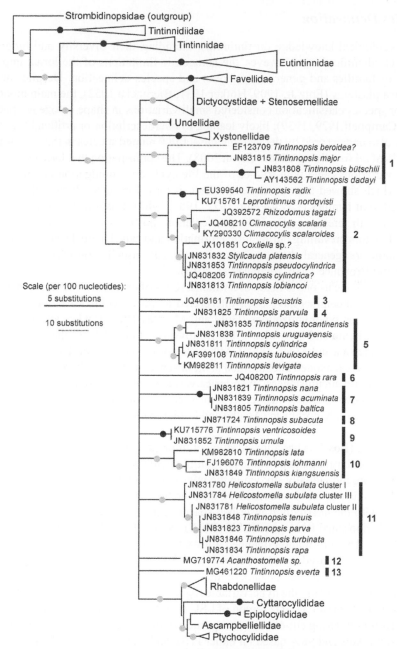

Figure 5.2: Phylogenetic tree inferred from tintinnid SSU rDNA sequences. Families are collapsed, while *incertae sedis* lineages (1 to 13) are expanded and enumerated as reported by Santoferrara et al. (2017). The reference alignment produced by Santoferrara et al. (2017) was updated, re-aligned with MAFFT (Katoh and Standley 2013) and used for maximum likelihood inferences with RAxML under the GTR model with a Γ model of rate heterogeneity and a proportion of invariable sites (Stamatakis 2014). Node support values inferred from 10,000 bootstraps are represented with grey or black circles (50–99% and 100% support, respectively; nodes with support < 50% were eliminated).

2018) and twelve SSU rDNA clades or branches (Fig. 5.2). Some of these clades also include hyaline genera (e.g., *Helicostomella*), to which agglomerated specimens resemble if grown in particle-free cultures (Santoferrara et al. 2017), and that may end up hosting "*Tintinnopsis*" species once more data are available.

5.2.2 *Species Delineation*

The bulk of ecological knowledge on tintinnids, including their diversity and biogeography, is based on lorica identifications. However, the taxonomic limitations of the lorica impact both the classification of families and genera (see above) and species delineation. Despite early warnings about the lorica plasticity (Entz Jr. 1909, Hofker 1931, Biernacka 1952), the main monographs still used today for species classification considered small variations in shape or size as species-specific (Kofoid and Campbell 1929, 1939), likely leading to overdescription or artificial "splitting" in the group. This is also underscored by the fact that the ~ 1,000 named species in this one order represent more than 10% of all free-living ciliates (Lynn 2008). The confirmation that loricae can have marked intra-specific variations came with studies of the life cycle and conjugation in cultures of *Favella ehrenbergii*, which showed that the same species can form loricae so distinct that they had been assigned to different families (Laval-Peuto 1977, 1981, 1983). It is now accepted that many lorica features (e.g., length, presence of spirals, denticulation, posterior process) are influenced by the environmental factors prevailing during lorica formation and can change during the life cycle, while only a few characters (general shape, opening diameter) are usually conserved at the species level (Laval-Peuto and Brownlee 1986, Agatha et al. 2013).

Applying these conclusions to field observations, series of loricae with continuous morphological variations or small anomalies have been interpreted as intraspecific variations, thus leading to suggestions of species synonymizations (Gold and Morales 1975, Bakker and Phaff 1976, Davis 1981, 1985, Laval-Peuto and Brownlee 1986, Boltovskoy et al. 1990, Williams et al. 1994, Alder 1995, 1999, Santoferrara and Alder 2009a). Although many synonyms probably exist, most of these "lumped" species have remained unconfirmed in the absence of clonal cultures or alternative species-specific traits. In these circumstances, even lorica variants with known ultrastructure and cytology (data so sparsely known that they lack criteria for species delineation) are admittedly inconclusive proofs of intraspecific polymorphism (Agatha and Riedel-Lorjé 2006, Agatha and Tsai 2008, Agatha 2008, 2010).

DNA barcoding (Box 5.1) has allowed study of intra- and inter-specific lorica variability. Given the difficulties in establishing clonal cultures, multiple field specimens (alive or fixed with non-acid Lugol's solution) are individually studied in the microscope and then used for DNA extraction and sequencing of genetic markers less conserved than the SSU rDNA, such as parts of the 5.8S rRNA gene combined with the internally transcribed spacer regions 1 and 2 (ITS1-5.8S rDNA-ITS2) or the large-subunit ribosomal RNA gene (LSU rDNA) (Xu et al. 2012, Kim et al. 2013, Santoferrara et al. 2013, 2015). The information obtained allows for pairwise comparisons of genetic variation within and among lorica-based morphospecies (Fig. 5.3). Based on the tintinnid sequences available in GenBank for different markers, the bulk of intraspecific variation is lower than 1% for SSU rDNA (full length or considering only the hypervariable regions V4 or V9), 1.5% for the ITS1-5.8S rDNA-ITS2 region and 0.6% for the D1-D2 region of LSU rDNA (Santoferrara et al. 2013, 2015). At such threshold levels (or clustering cut-offs), the proportion of differentiated morphospecies is 72% based on SSU rDNA markers and 98% based on either ITS1-5.8S rDNA-ITS2 or the LSU rDNA marker (Santoferrara et al. 2013, 2015). More recently, Jung et al. (2018) sequenced the mitochondrial cytochrome oxidase c subunit I gene (COI) in one genus and found a high intraspecific variation (up to 9%), along with multiple haplotypes inconsistent with morphology and/or geographical origin. Although none of the markers tested so far show a 'barcode gap' (none of them is perfect for species delineation; Fig. 5.3), the available data supports the D1-D2 region of LSU rDNA as the best DNA barcode in ciliates (Santoferrara et al. 2013, 2015, Stoeck et al. 2014a).

5.2.3 *Crypticity and Polymorphism*

Although lorica morphology and molecular markers correlate relatively well in tintinnid species (> 70% concordance, as detailed above; Santoferrara et al. 2013, 2015), examples of both polymorphism

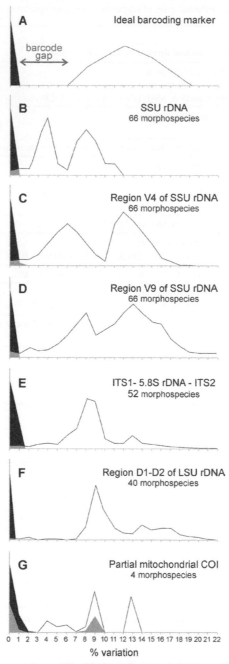

Figure 5.3: Distribution of genetic variation within (black) and among (white) species based on different gene regions. Grey areas represent overlap between both distributions. (A) an ideal marker for species delineation. (B–G) schematic representations of tintinnid data from Santoferrara et al. (2013, 2015) and Jung et al. (2018). The number of morphospecies analyzed in each case is indicated. Frequencies (*y* axes) are not drawn to scale.

and crypticity exist (Table 5.2; Fig. 5.4). Polymorphism here refers to the same species producing either clearly distinct lorica types or series of loricae with continuous variation in shape and size. The first type of polymorphism is exemplified by *Favella ehrenbergii*, which produces a protolorica after cell division (with a single, anterior spiral and a posterior process) and a fully spiraled replacement paralorica at later stages (Laval-Peuto 1981, 1983; Fig. 5.4A-B). Specimens with these distinct loricae

Table 5.2: Different cases of crypticity and polymorphism in tintinnids.

Taxa	Evidence	References
Intra-species polymorphism: distinct loricae formed by the same species		
Favella ehrenbergii	Clonal cultures; conjugation Clonal cultures; intraspecific variation, SSU rDNA < 0.4% Intraspecific variation, SSU rDNA = 0.1%, LSU rDNA = 0%	Laval-Peuto 1981, 1983 Kim et al. 2010 Santoferrara et al. 2013
Schmidingerella arcuata	Cytology; lorica ultrastructure Clonal cultures	Agatha and Strüder-Kypke 2012 This study (Fig. 5.2C-D)
Intra-species polymorphism: loricae with continuous variation formed by the same species		
Cymatocylis convallaria, C. calyciformis, C. drygalskii	Cytology (except *C. drygalskii*) Intraspecific variation, SSU rDNA = 0%, ITS = 0%, LSU rDNA = 0%	Petz et al. 1995 Kim et al. 2013
Favella ehrenbergii	Intraspecific variation, SSU rDNA = 0.1%, LSU rDNA = 0%	Santoferrara et al. 2013
Stenosemella pacifica	Cytology; lorica ultrastructure Intraspecific variation, SSU rDNA = 0%, LSU rDNA = 0.1%	Agatha and Tsai 2008 Santoferrara et al. 2013
Tintinnopsis acuminata	Intraspecific variation, SSU rDNA = 0%, LSU rDNA = 0%	Santoferrara et al. 2013, 2017
Tintinnopsis parvula	Cytology; lorica ultrastructure Intraspecific variation, SSU rDNA = 0.1%, LSU rDNA = 0%	Agatha 2010 Santoferrara et al. 2013
Intra-species polymorphism and inter-species crypticity/pseudocrypticity: loricae with continuous variation within and between species		
Helicostomella spp. (one or two morphospecies split in three clades)	Intraclade variation, SSU rDNA = 0%, ITS < 0.1%, LSU rDNA < 0.2% Interclade variation, SSU rDNA < 0.3%, ITS = 1.2–4.5%, LSU rDNA = 0.4–4.9%	Santoferrara et al. 2015 Xu et al. 2012
Parafavella spp. (five morphospecies split in two clades)	Intraclade variation, SSU rDNA = 0%, ITS = 0.2%, LSU rDNA = 0% Interclade variation, SSU rDNA = 0%, ITS = 2.0%, LSU rDNA = 1.3%	Jung et al. 2018 Santoferrara et al. 2017
Tintinnopsis butschlii, T. major *T. butschlii, T. major, T. dadayi, T. everta*	Intraspecific variation (within each), SSU rDNA = 0%, LSU rDNA = 0% Interspecific variation (between both), SSU rDNA = 3.3%, LSU rDNA = 4.1% Interspecific variation (among the four), SSU rDNA = 0.5–8.1%	Santoferrara et al. 2013 Strüder-Kypke and Lynn 2003 Gruber et al. 2018
Tintinnopsis parvula, T. rapa	Intraspecific variation (within each), SSU rDNA < 0.1%, LSU rDNA = 0% Interspecific variation (between both), SSU rDNA = 3.2%, LSU rDNA = 10.2%	Santoferrara et al. 2013

Only examples confirmed with relevant evidence (clonal cultures, conjugation, DNA sequences) are included; availability of additional data (cytology, ultrastructure) is mentioned. ITS refers to the ITS1-5.8S rDNA-ITS2 region.

are found in clonal cultures (Laval-Peuto 1981), have sexual exchange (Laval-Peuto 1983) and have almost identical SSU and LSU rDNA sequences (Kim et al. 2010, Santoferrara et al. 2013). Such protoloricae and paraloricae are also seen in cultures of *Schmidingerella arcuata* (Fig. 5.4C-D). The second type of polymorphism is probably present in most tintinnids and has been confirmed in some species with continuous lorica variation paired with identical DNA barcode sequences (Table 5.2; Fig. 5.4E; Santoferrara et al. 2013).

Intra-specific polymorphism with continuous lorica variation had been suspected for a long time, but the incorporation of more detailed morphological studies and DNA sequencing also led to somewhat unexpected results: intra-species polymorphism can overlap with inter-species crypticity

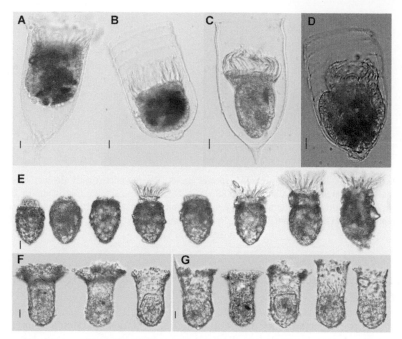

Figure 5.4: The lorica presents limitations for tintinnid taxonomy. Some polymorphic species form distinct loricae during the life cycle: *Favella ehrenbergii* protolorica, or *favella* form (A) and paralorica, or *coxliella* form (B). *Schmidingerella arcuata* forms a protolorica (C) and a paralorica (D) similar to those of *Favella*, but both genera differ in ultrastructure, cytology and genetic markers. Some polymorphic species form loricae with continuous variation in shape and size, such as *Stenosemella pacifica* (E). In some cases, lorica polymorphism and crypticity overlap, such as in *Tintinnopsis butschlii* (F) and *Tintinnopsis major* (G). Scale bar = 10 µm. Panels A, C and E–G adapted from: Santoferrara, L.F., G.B. McManus and V.A. Alder. 2013. Utility of genetic markers and morphology for species discrimination within the order Tintinnida (Ciliophora, Spirotrichea). Protist 164: 24–36, © 2011 Elsevier GmbH, with permission from Elsevier.

(Table 5.2; Santoferrara et al. 2013, 2015). Crypticity (or pseudo-crypticity) here refers to genetically distinct species with identical (or almost identical) loricae. For example, several *Tintinnopsis* morphospecies with loricae difficult to differentiate by conventional microscopy present genetic distances that disprove conspecificity (Table 5.2; Fig. 5.4F-G). *Helicostomella* morphospecies are divided in three clades based on ITS1-5.8S rDNA-ITS2 and LSU rDNA sequences, despite almost invariant lorica diameter and overlapping length fluctuations in field and cultured specimens (Table 5.2; Xu et al. 2012, Santoferrara et al. 2015). Some of these cases of crypticity or pseudo-crypticity are likely the result of looking at only basic lorica features, as more detailed morphological study (e.g., cytology, ultrastructure) may actually reveal phenotypic differences. For example, specimens previously classified as *Favella*, with loricae only differing by a small subapical bulge but with clear cytological, ultrastructural and genetic differences, are now assigned to the genus *Schmidingerella* (Fig. 5.4C-D) and a different family (Agatha and Strüder-Kypke 2012).

   While DNA sequences can be very useful in species delineation, this is a far from perfect criterion and should be used with caution in taxonomy. Species identical in one or more markers can differ distinctly in alternative genes, as observed in some *Tintinnopsis* species with identical SSU rDNA but different LSU rDNA (Santoferrara et al. 2013). This can be even more problematic above the species level. For example, the morphologically distinct *Petalotricha* and *Cyttarocylis* (Fig. 5.5) were recently synonymized based on identical SSU rDNA and ITS1-5.8S rDNA-ITS2 (Bachy et al. 2012), but these genera differ by 1.8% in LSU rDNA (Santoferrara et al. 2017). In this case, the molecular divergence in LSU rDNA and the fact that lorica differences are not confirmed as intra-taxon polymorphism (Dolan 2016) suggest caution in potential species and genera synonymizations until more features are studied and a unified diagnosis can be provided.

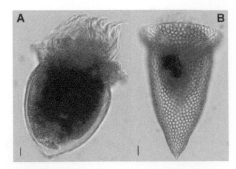

Figure 5.5: Genetic markers do not always resolve species limits. The morphologically distinct *Petalotricha* (A) and *Cyttarocylis* (B) have identical SSU rDNA and ITS1-5.8S rDNA-ITS2 sequences, but differ by almost 2% in LSU rDNA. Scale bar = 10 μm. Adapted from: Santoferrara, L.F., V.A. Alder and G.B. McManus. 2017. Phylogeny, classification and diversity of Choreotrichia and Oligotrichia (Ciliophora, Spirotrichea). Mol. Phylog. Evol. 112: 12–22, © 2017 Elsevier Inc., with permission from Elsevier.

5.2.4 *Importance of Reference Databases: Linking Morphological and Genetic Data*

Despite the ever-improving taxonomy and remaining uncertainties in species delineation, we need practical ways to track taxa for understanding the patterns and processes that define zooplankton diversity and distribution. Increasingly, this involves DNA sequencing (Box 5.1). In the case of tintinnids, the recent improvements in taxonomic classification and phylogenic relationships outlined above are crucial pillars to build an accurate reference for environmental metabarcoding surveys. As for other organisms, such a reference is populated with sequences obtained worldwide and that are available in public repositories, for example GenBank. Although it has recently been concluded that GenBank is a reliable resource for metazoan identifications based on COI (Leray et al. 2019), it is well-known that this repository should be used with caution. For example, publicly-available SSU rDNA sequences of planktonic ciliates are plagued with issues, such as misidentifications, insufficient or nonexistent published data to confirm identifications and outdated taxonomic labels (Santoferrara et al. 2017). Another issue is that many sequences in GenBank are labeled as an "uncultured organism" (usually obtained by the early forms of metabarcoding; see Box 5.1) and not only can include methodological artifacts (e.g., chimeric sequences) but also do not carry any taxonomic information for identification of newly-obtained sequences. These kinds of issues and the need for reliable ways to classify environmental sequences have motivated community efforts that combine taxonomic and phylogenetic expertise for curation and annotation of GenBank sequences (linked to species names or not). An example product is the Protist Ribosomal Reference database (PR[2]; Guillou et al. 2013) and its recently integrated, taxon-specific EukRef databases (del Campo et al. 2018, Boscaro et al. 2018). The backbone provided by traditional systematics continues to be a key to link the ever-increasing collection of environmental DNA sequences with meaningful diversity at the taxonomic, ecological and functional levels.

5.3 Global Diversity and Biogeography of Tintinnid Taxa

Because of the lorica, tintinnid taxa are among the microzooplankton groups that have the richest body of literature on occurrence and distribution. Synthesis of hundreds of lorica-based reports has indicated consistent global patterns of diversity and biogeography (Pierce and Turner 1993, Dolan and Pierce 2013). More recently, DNA sequence data have emerged as a new opportunity to explore traditional views that may be obscured by the limitations of a lorica-based taxonomy (e.g., synonyms, cryptic species; see above). However, rather than replacement, we emphasize that the combination of traditional lorica information and modern DNA data is the most informative and robust approach for studying tintinnid diversity and distribution. In this section, we integrate both lines of evidence to illustrate patterns of particular taxa at the global level. Smaller-scale assemblage reports are exemplified here to support these global patterns but are considered in more detail in the next section.

5.3.1 Global Species Richness and Phylogenetic Diversity

The real number of tintinnid species is currently unknown. Out of the almost 1,000 described species, many are based on minute lorica details or do not show discontinuity in lorica shapes and sizes (Jörgensen 1924, Kofoid and Cambell 1929, 1939). This has led to several proposals for species "lumping" (Bakker and Phaff 1976, Boltovskoy et al. 1990, Santoferrara and Alder 2009a). However, very few cases of potential synonyms have been confirmed with cultures (Laval-Peuto 1981) or multiple identical DNA barcodes (Kim et al. 2013), while examples of cryptic species continue to appear (Xu et al. 2012, Santoferrara et al. 2013, 2015). The existence of both polymorphic, genetically cohesive species as well as cryptic, genetically distinct species (Table 5.2) indicates that the classical views of "lumpers" and "splitters" are actually complementary. This also stresses the importance of relying on original descriptions, rather than monographs that may have changed species limits (Kofoid and Campbell 1929, 1939), for accurate microscope-based identifications (Santoferrara et al. 2016b).

There is no criterion that can unequivocally tell us the real diversity of tintinnid species: clearly the lorica-based species delineations cannot do this, but neither can the DNA barcodes tested so far given that all of them show some overlap among the relatively few sequenced morphospecies (Fig. 5.3). Among the explored markers, SSU rDNA delineates species incompletely (72% morphospecies differentiated at a 1% similarity cut-off, see above; Santoferrara et al. 2013), but has the largest number of sequences available in global repositories and remains the most widely-used marker in phylogenetic analyses and diversity estimates by environmental sequencing (Box 5.1). In a recent study (Santoferrara et al. 2018), we retrieved all of the tintinnid SSU rDNA sequences available in GenBank from surveys around the world, including both sequences linked to a species name and unidentified environmental sequences (only those longer than 500 bp, so mostly from Sanger-sequenced clone libraries; Box 5.1). After a stringent quality filter, we obtained 870 tintinnid sequences (average length = 1573 nucleotides). The compiled sequences came from 50 published studies and 150 geographical locations (Fig. 5.6A), biased towards a few sites that have been sampled more intensively (neritic waters off the Northeast U.S. and China; Mediterranean Sea). Most samples were obtained in epipelagic waters and on single dates. The sequences were grouped into 126 phylotypes defined as monophyletic groups of sequences that are more than 99% similar. This criterion has been practical for delimitation of operational taxonomic units (OTUs) based on full or partial SSU rDNA sequences (Bachy et al. 2013, Santoferrara et al. 2013), but it is admittedly too conservative to estimate species richness. Despite limitations in spatiotemporal coverage and marker resolution, our dataset (subsequently referred as to "GenBank survey") is useful to explore the global phylogenetic diversity (next paragraph) and biogeography (rest of this section) of tintinnid families and genera.

Most of the tintinnid phylotypes sequenced so far can be linked to morphologically-defined taxa (Fig. 5.6B). Even phylotypes that are composed exclusively of unidentified environmental sequences are generally clustered within known families. A notable exception is a clade that may correspond to the un-sequenced Nolaclusiliidae, which is represented only by sequences obtained in brackish waters (as expected based on its known distribution; Sniezek et al. 1991, Snyder and Brownlee 1991) and has a sister relationship to Eutintinnidae (as expected based on cell features; Agatha and Strüder-Kypke 2013). However, between the putative Nolaclusiliidae and the known Eutintinnidae there is a distinct branch of unpredictable affiliation (GenBank sequence EU333101, from open marine waters), which could represent a known lineage that needs reassignment or a novel lineage. Most of the few environmental phylotypes with genus-level divergence probably correspond to known genera for which identified specimens have not yet been sequenced. Considering that about

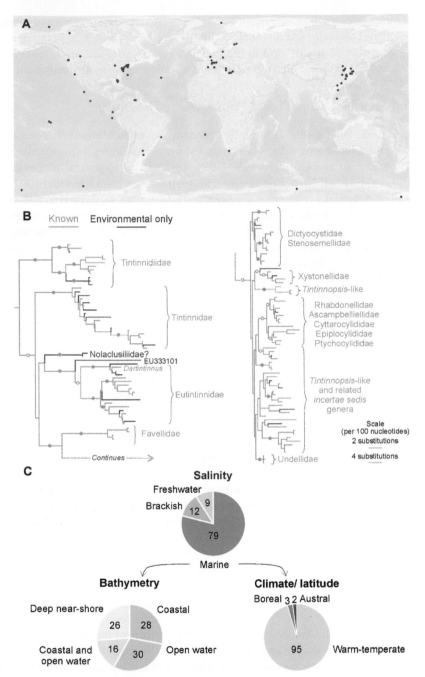

Figure 5.6: Global diversity and distribution of tintinnid phylotypes based on SSU rDNA sequences: (A) geographical origin of the 870 tintinnid sequences retrieved from GenBank; (B) phylogenetic tree based on 126 phylotypes (monophyletic groups of sequences that are more than 99% similar); phylotypes are differentiated into those that include at least one named species and those that include only unidentified environmental sequences. The tree was built as described in Fig. 5.2 (with node support values of 50–89% and 90–100% represented with white or grey circles, respectively); (C) proportion (%) of phylotypes distributed based on salinity (freshwater: < 0.5; brackish: 0.5–30; marine: > 30), bathymetry (coastal: < 50 m in inner shelves; open water: > 50 m in outer shelves, slopes and oceans; deep near-shore sites: on narrow shelves reaching > 50 m depth at less than 2 km from shore) and climate/latitude (warm-temperate; boreal: Arctic and Subarctic; austral: Antarctic and Subantarctic). Panel A adapted from: Santoferrara, L.F., E. Rubin and G.B. McManus. 2018. Global and local DNA (meta)barcoding reveal new biogeography patterns in tintinnid ciliates. J. Plankton Res. 40: 209–221, © 2018 The Author(s), with permission from Oxford University Press.

half of the acknowledged genera have not been sequenced from morphologically-identified isolates yet (38 out of 77 genera; Table 5.1), and only 12 of them pass the condition of being reported at least four times by two different author teams (Table 5.1, excluding *Nolaclusilis*), it is likely that many of the genera not yet barcoded are either genuinely rare (thus rarely sampled, even by environmental sequencing) or invalid (synonyms of sequenced genera).

Tintinnids likely include fewer undescribed taxa than their aloricate relatives (Santoferrara et al. 2017), but even this well-known clade shows some potential for novel diversity. For example, within Eutintinnidae, a clade sister to *Eutintinnus* would be entirely environmental if *Dartintinnus* (the most recently discovered tintinnid; Smith et al. 2018) was not placed there (Fig. 5.6B). This proves the importance of meta-analyses such as our GenBank survey, which may reveal insights unappreciated in dozens of individual publications. These efforts show the potential of environmental sequencing for pointing at novel taxa awaiting formal description, as it has happened for other ciliates (Orsi et al. 2012) and planktonic groups such as dinoflagellates (Guillou et al. 2008) and stramenopiles (Massana et al. 2004). It is also important to explore environments that have not been commonly surveyed even by traditional microscopy and that can hide potentially novel tintinnids, such as those recently reported in mesopelagic waters (Dolan et al. 2019). Finally, detailed taxonomic re-investigation of many well-known taxa is also crucial to fully clarify the global diversity of tintinnid taxa (Ganser and Agatha 2019).

5.3.2 Global Distribution Patterns: Salinity

Salinity is one of the main determinants in the distribution of aquatic species, with the marine-freshwater dichotomy as one of the better-known biogeographic patterns in plankton (Logares et al. 2009, Boxshall and Jaume 2000), including ciliates (Forster et al. 2012). For tintinnids, current evolutionary hypotheses based on ciliary patterns and SSU rDNA suggest that they originated from a marine ancestor (Agatha and Strüder-Kypke 2013) and transitioned into freshwater only rarely and recently (Bachy et al. 2012).

Based on our GenBank survey (Santoferrara et al. 2018), we evaluated the global distribution of tintinnid phylotypes (Fig. 5.6C), using environmental metadata retrieved from the same repository or the publications associated with each sequence. Sequences categorized based on the salinity of the sampled waters showed the expected prevalence of marine phylotypes. Nine percent of freshwater phylotypes are included in the family Tintinnidiidae (which includes known freshwater species in *Tintinnidium* and *Membranicola*; Foissner and Wilbert 1979, Foissner et al. 1999) or have scattered affiliation (two *Tintinnopsis* and one unidentified Undellidae; Bachy et al. 2012, Santoferrara et al. 2013). Some phylotypes (12%) were detected only in brackish waters, including *Dartintinnus* and the putative *Nolaclusilis*, which were described (Snyder and Brownlee 1991, Smith et al. 2017) and subsequently reported (Sniezek et al. 1991, Stoecker et al. 2000, Gavrilova and Dolan 2007, Gavrilova and Dovgal 2016, Li et al. 2019) only under this salinity regime.

The restriction of certain tintinnid lineages to brackish waters (Santoferrara et al. 2018, Li et al. 2019) agrees with recent reports for other planktonic organisms. For example, bacterial and protistan communities differ significantly between brackish waters and adjacent fresh- and marine waters in both the stable salinity gradient along the Baltic Sea (Herlemann et al. 2011, Dupont et al. 2014, Hu et al. 2016, Celepli et al. 2017) and dynamic estuaries of North America (Crump et al. 2004, Hewson et al. 2014). To test the hypothesis that some tintinnid phylotypes are restricted to brackish waters, we took advantage of the sensitive DNA metabarcoding approach (Box 5.1) to analyze assemblages in a 5 to 36 salinity gradient from the tidal portion of a river to adjacent Northwest Atlantic waters (Santoferrara et al. 2018). This local-scale analysis confirmed that some tintinnid phylotypes (at least one within each of the genera *Dartintinnus*, *Eutintinnus*, *Tintinnopsis* and the putative *Nolaclusilis*) are found almost exclusively in estuarine waters (Santoferrara et al. 2018). This is further supported for *Dartintinnus* by no matches against the metabarcoding dataset obtained in several marine locations around the world during the Tara Oceans expedition (we could not test the other phylotypes because they do not overlap or are not differentiated with the genetic marker used in Tara Oceans; de Vargas et al. 2015).

The ecophysiological basis for brackish-water restriction in certain tintinnids remains unexplored, with adaptations to specific salinity ranges as possible determinants. Still, other variables that correlate with the salinity gradient in land-margin environments could influence the observed pattern, such as the enrichment with prey that depend on terrestrially-derived nutrients. For example, the genus *Rhizodomus* has a coastal distribution but it is usually more abundant in estuaries and lagoons, apparently associated with eutrophication rather than salinity (Saccà and Giuffrè 2013). Experimentation on salinity tolerance and other factors potentially linked to this distribution pattern is needed.

5.3.3 Global Distribution Patterns: Bathymetry and Climate/Latitude

Marine tintinnids show global distribution patterns that match the biogeographies of other planktonic organisms, such as foraminiferans and dinoflagellates (Pierce and Turner 1993, Dolan and Pierce 2013). In their compilation of almost 300 lorica-based studies in 1,800 marine locations worldwide, Dolan and Pierce (2013) divided tintinnid genera into five categories: cosmopolitan (from Artic to Antarctic, not restricted to nearshore waters), neritic (from Artic to Antarctic, but restricted to nearshore waters), warm-temperate (both coastal and open waters, but not at polar and sub-polar latitudes), boreal (both coastal and open waters, Arctic and Subarctic), and austral (both coastal and open waters, Antarctic and Subantarctic). These patterns are supported by our GenBank survey (Santoferrara et al. 2018), except for cosmopolitanism that could not be assessed due to data limitations (Fig. 5.6A).

Based on our GenBank survey, most marine phylotypes were reported in either coastal (28%) or open waters (30%), while only 16% of the phylotypes were recorded under both regimes (Fig. 5.6C). This corresponds with well-known coastal genera (e.g., *Favella*, *Helicostomella*; Dolan and Pierce 2013), although our results confirm that there are also genera restricted to open waters (e.g., *Xystonella*; Alder 1999, Santoferrara et al. 2018). The differentiation between coastal and open water phylotypes agrees with analyses across gradients from coast to ocean: on wide or relatively wide shelves of the Southwest Atlantic, East China Sea and Northwest Atlantic (about 600, 500 and 150 km wide, respectively), tintinnid assemblages change considerably between the 50- and 100-m isobaths based on microscope counts (Santoferrara and Alder 2012, Li et al. 2016) or metabarcoding (Santoferrara et al. 2016a, 2018). Such bathymetry pattern, however, becomes less clear in deep near-shore sites where the continental shelf drops abruptly and there is a constant influence of adjacent open waters: in the much-sampled Bay of Villefranche (where 26% of phylotypes were sampled; Fig. 5.6C), a mix of coastal and open water taxa are detected. These molecular findings (Bachy et al. 2012, 2013) agree with morphology-based studies in the same location (Dolan 2017) and in other narrow shelves of the Mediterranean (Sitran et al. 2007, 2009).

Our GenBank survey also supported biographic patterns related with climate/latitude. Most phylotypes were obtained in warm-temperate environments (Fig. 5.6C). Still, the small proportion of phylotypes from known austral or boreal genera were only detected in their expected range: *Cymatocylis* and *Laackmanniella* were detected only in Antarctica; *Ptychocylis* and *Parafavella* were detected in the Artic and towards temperate waters, but always in the northern hemisphere (Santoferrara et al. 2018). More sensitive metabarcoding surveys did not detect austral genera in the northern hemisphere (Bachy et al. 2013, Santoferrara et al. 2016a). This supports the well-known restriction of these austral or boreal genera based on microscopy reports (Pierce and Turner 1993, Alder 1999, Dolan et al. 2012, 2017, Dolan and Pierce 2013). A more gradual latitudinal gradient that reflects the poleward diversity decrease known for many terrestrial and aquatic organisms (Gaston and Spicer 2003) is also seen for tintinnid morphospecies (Dolan et al. 2016).

The bathymetry and climate/latitude distribution patterns known for many groups of planktonic protists are typically attributed to the contrasting ecological features of these realms (Forster et al.

2012). Factors that can affect these patterns are oceanographic phenomena such as surface or deep currents, upwelling and mesoscale eddies, which can result in occasional dispersal of tintinnid taxa outside their normal range (Balech 1972, Boltovskoy and Alder 1992, Kato and Taniguchi 1993, Kim et al. 2012). However, these occasional expatriations usually result in rare cells (or their DNA) detected outside their range, which suggest that environmental selection limits colonization (growth) in the new habitat (Santoferrara et al. 2016a, 2018). This is consistent with the idea that effective dispersal is at least partially limited for tintinnids and other microzooplankton, in comparison with the potentially high dispersal proposed for smaller marine microbes based on their large populations, metabolic flexibility and apparent lack of barriers in the ocean (Weisse 2008).

5.3.4 DNA Sequences and Cosmopolitanism

Whether species show ubiquitous or restricted distributions in space and time has been a central debate in microbial biogeography, including microzooplankton, for years (Finlay 2002, Foissner 2008). Now, the coexistence of both patterns is well-supported (Fontaneto 2011, van der Gast 2015). As shown in the previous subsection, both lorica- and DNA-based surveys detect tintinnid lineages restricted to different environments. Cosmopolitan genera are also evident from the abundant lorica data (Dolan and Pierce 2013), but as mentioned above, this cannot be tested with our GenBank survey (Fig. 5.6A).

Recently, the Tara Oceans metabarcoding project studied plankton (0.8–2,000 µm) in 334 samples around the world (de Vargas et al. 2015). These authors identified 381 cosmopolitan OTUs, of which 12 correspond to ciliates, seven to oligotrichs and choreotrichs and two to tintinnids, *Tintinnopsis* and *Salpingella* (http://taraoceans.sb-roscoff.fr/EukDiv/#content5). These are only two examples out of the 13 tintinnid genera known as cosmopolitans (Dolan and Pierce 2013); the others (e.g., *Eutintinnus*) probably remained undetected in some Tara Oceans samples given the broad taxonomic goal of the project. Upon re-analysis of the same Tara Oceans data, however, one of the cosmopolitan OTUs was reassigned to *Laackmanniella* (Gimmler et al. 2016). This is probably a methodological artifact due to the short metabarcode region sequenced (V9) and the broad similarity threshold used for OTU clustering (97%), as this approach cannot distinguish between the austral genus *Laackmanniella* and several widely distributed taxa (e.g., *Tintinnopsis, Codonellopsis* and *Stenosemella*) that are genetically closely-related (Santoferrara et al. 2018). Caution is thus needed when conclusions on cosmopolitanism are based on molecular data that lump potentially endemic and commonly widespread taxa.

Given that neither the lorica-based taxonomy nor the current DNA barcodes can unambiguously delineate species, we refrain from detailed assessment of biogeography at this taxonomic level, but only conclude that current data do not support the existence of cosmopolitan tintinnid species. Based on lorica records, the most commonly reported and widespread morphospecies (e.g., *Amphorides quadrilineata* and *Steenstrupiella steenstrupii*) are not found in some environments (Dolan and Pierce 2013). Some distinct morphospecies have highly similar (> 99%) SSU rDNA in distant locations, such as *Protorhabdonella curta* or *Antetintinnidium mucicola* from the Northwest Atlantic and Pacific oceans (Santoferrara et al. 2013, Xu et al. 2013, Zhang et al. 2017), but broader geographical coverage and less conserved markers are needed to assess cosmopolitanism versus different, locally-adapted species. On the other hand, some cases suggestive of endemism exist based on lorica records (e.g., *Codonellopsis ecaudata* in the Indian and central Pacific oceans; Dolan and Pierce 2013) or both lorica and SSU rDNA data (e.g., *Metacylis angulata* in estuarine and neritic waters of the northeast USA; Santoferrara et al. 2017).

5.3.5 Overview of Distribution Patterns in a Phylogenetic Context

Our GenBank survey indicates that environmentally-restricted taxa are not monophyletic, but interspersed in the phylogenetic tree (Fig. 5.7; Santoferrara et al. 2018). Nor are there lorica features that clearly characterize restricted taxa, except that tintinnids with loricae that agglutinate mineral

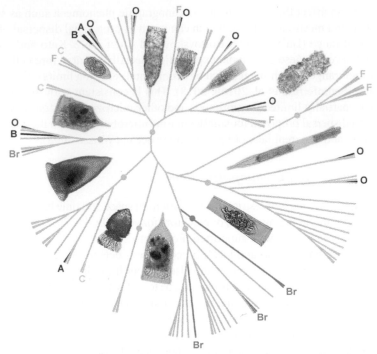

Figure 5.7: Tintinnid ciliates restricted by salinity, bathymetry or climate/ latitude are not phylogenetically or morphologically cohesive. Different representation of the tree shown in Fig. 5.6B, indicating phylotypes restricted to freshwater (F), brackish water (Br), marine coasts (C), open sea (O), boreal latitudes (B) or austral latitudes (A). Note that proportions in Fig. 5.6C cannot be estimated directly from here (for more information, see Santoferrara et al. 2018).

particles are usually circumscribed to nearshore waters that are shallow and turbulent enough to provide a constant supply of such particles (Dolan and Pierce 2013). The fact that certain hyaline taxa are also restricted to estuarine and coastal environments could be related to cyst formation as an important part of their life cycle; these taxa would also depend on shallow and turbulent waters in order to pass from sedimented cysts into swimming vegetative stages (Dolan and Pierce 2013, Kamiyama 2013). Cysts may also explain the detection of tintinnid DNA in extreme environments, such as hypersaline or deep anoxic waters and hydrothermal fields (Santoferrara et al. 2018). Tintinnid cysts are known to survive adverse conditions such as anoxia and high sulfide concentration (Kamiyama 2013) but, although these resistant stages certainly favor dispersal into favorable environments, it is unlikely (or at least unknown) that the vegetative cells survive in extreme environments.

The heterogeneous phylogenetic resolution, diversity and biogeography within and among tintinnid lineages (Fig. 5.7) raise questions about the processes that promote their diversification and determine their spatial distributions. As detailed below, ecological studies have found contemporary environmental selection (Sitran et al. 2009, Santoferrara et al. 2016a) or random dispersal (Dolan et al. 2007) as the main structuring mechanisms of local tintinnid assemblages. Instead, the phylogenetic structure and spatial distribution of tintinnid lineages (Fig. 5.7) suggest that various evolutionary processes have shaped their global diversity and biogeography. For example, the austral versus boreal isolation of the closely related *Cymatocylis* and *Ptychocylis* may be explained by allopatric diversification, while these and several other morphologically distinct taxa inferred as polytomies in the tintinnid phylogenetic tree may reflect a rapid radiation (Figs. 5.2 and 5.7). These aspects remain highly speculative but raise interesting questions about when and why these macroevolutionary events happened. A recent time-calibrated multigene phylogeny estimated that ciliates (1) originated about 1.1 billion years ago, (2) have had low extinction rates and (3) their fastest speciation rates corresponded to the spirotrichean clade that today includes tintinnids (as well as aloricate choreotrichs, oligotrichs and hypotrichs) about 570 million years ago, during the Ediacaran-Cambrian transition, but (4) tintinnids originated only about 263 million years ago, during the Paleozoic-Mesozoic transition (Fernandes and Schrago 2019). Tintinnid loricae actually form most of the ciliate fossil record needed to calibrate these types of evolutionary studies, although the few examples that can be unambiguously assigned to tintinnids are less than 200 million years old, from the Jurassic (Lipps et al. 2013, Dunthorn et al. 2015). The morphological, genetic and ecological factors as well as the evolutionary processes underlying tintinnid diversification are still unknown.

5.4 Diversity and Distribution of Regional and Local Tintinnid Assemblages

The diversity, taxonomic composition and abundance of tintinnid assemblages have been studied in multiple locations around the world, either with a group-specific focus or as part of microzooplankton. We do not attempt to review the huge body of information produced by dozens of microscopy studies and, more recently, by DNA metabarcoding. Instead, we summarize how both methods are complementing or providing new insights, as well as their strengths and limitations.

5.4.1 *The Legacy of Microscope Surveys*

Most of what we know about tintinnid ecology derives from microscopy studies. Observation in the inverted microscope of a known, settled volume of fixed sample (Utermöhl 1958) allows for morphospecies identification, quantification of their abundances and measurement of lorica dimensions for biomass estimates (Verity and Langdon 1984). Microscopy is thus still highly-used for spatial and temporal tintinnid surveys (Feng et al. 2018, Liang et al. 2018, Al-Yamani et al. 2019, Anjusha et al. 2018, Diociaiuti et al. 2019, Wang et al. 2019) and has provided many insights on assemblage structure and distribution. For example, lorica-based analyses have shown changes in abundance, diversity and/or taxonomic composition of tintinnid assemblages across estuarine gradients (Dolan and Gallegos 2001, Godhantaraman and Uye 2003, Urrutxurtu 2004, Li et al. 2019), coast-to-ocean gradients (Santoferrara and Alder 2012, Li et al. 2016), oceanic water masses (Alder 1999, Thompson et al. 1999, Modigh et al. 2003) and vertical profiles down to aphotic waters (Alder and Boltovskoy 1993, Dolan et al. 2019). Assemblages also change with seasons in temperate coasts (Kamiyama and Tsujino 1996, Modigh and Castaldo 2002, Bojanić et al. 2012) and with rain patterns in tropical areas (Godhantaraman 2002), but appear as temporally-stable in open waters (Dolan and Pierce 2013).

Microscopy has also informed us about the factors and processes that affect tintinnid assemblages. Changes in diversity and/or abundance often correlate with abiotic factors such as temperature and salinity (examples in previous paragraph) and biotic factors such as chlorophyll concentration (Santoferrara and Alder 2012), phytoplankton diversity (Stoecker et al. 2000, Dolan et al. 2002) and predation by copepods (Dolan and Gallegos 2001). Processes such as currents and eddies may homogenize distant assemblages (Boltovskoy and Alder 1992, Kim et al. 2012), while boundaries between water masses (fronts) can have major effects in isolating adjacent assemblages (Alder 1999) and generating hotspots of diversity and abundance (Thompson and Alder 2005, Santoferrara 2008, Santoferrara and Alder 2009b). Also associated with diversity maxima are temporal transitions from winter to summer in temperate coastal systems (Dolan and Pierce 2013). Overall structuring of tintinnid assemblages has been explained by environmental selection at a coastal site (Sitran et al. 2009) or random dispersal in open waters (Dolan et al. 2007, 2009). Finally, potential alterations in tintinnid assemblages and species distributions due to climate change, ballast water transport, aquaculture and other anthropogenic changes have been reported (Pierce et al. 1997, Gavrilova and Dolan 2007, Saccà and Giuffrè 2013).

5.4.2 *Environmental DNA Methods: Advantages and Challenges*

For the past 15 years, assemblages of tintinnids and related planktonic ciliates (aloricate choreotrichs and oligotrichs) have been studied with molecular methods such as DNA fingerprinting (Tamura et al. 2011, Grattepanche et al. 2014a, 2015, Tucker et al. 2017) and metabarcoding (Box 5.1), the latter initially done by Sanger-sequenced clone libraries (Doherty et al. 2007, 2010, Bachy et al. 2014) and more recently with high-throughput sequencing (Bachy et al. 2013, Santoferrara et al. 2014, 2016a, 2018, Gimmler et al. 2016, Grattepanche et al. 2016, Onda et al. 2017, Zhao et al. 2017). Environmental DNA metabarcoding is now a standard method to study plankton diversity and distribution, and it allows for simultaneous analysis of ciliates and many other distinct lineages in a single effort (Stoeck et al. 2010, de Vargas et al. 2015). This reduces the need for lineage-specific sample preservation and taxonomic expertise, as well as processing times and overall costs. Additional advantages include less ambiguity in the identification of cryptic and polymorphic taxa, as well as a higher sensitivity for rare (low-abundance) taxa not detected in the microscope (Bachy et al. 2013, Santoferrara et al. 2016a).

Like all methods, metabarcoding also has disadvantages. Metabarcode reads are usually clustered into OTUs, which are expected to define taxonomically-meaningful units and to group intraspecific variants and sequencing errors together with the valid sequences. The inventory of OTUs and their frequencies are typically used as proxies of the taxa present in a sample and their relative abundances. However, results can change markedly based on sampling strategies, markers, lab protocols and bioinformatic procedures (reviewed by Santoferrara 2019). In addition to these technical issues, OTUs also present biological limitations. The boundaries between intra- and interspecific sequence variation are unknown for most taxa and can be additionally blurred by artifactual sequences (Bachy et al. 2013, Decelle et al. 2014, Grattepanche et al. 2014b), which prevents equating OTUs and species. Another problem is that the number of gene copies can vary by orders of magnitude among closely-related taxa and even within a single specimen (Gong et al. 2013, Biard et al. 2017), which can distort relative abundances (Medinger et al. 2010, Santoferrara et al. 2014).

Tintinnids and other plankton have served as models to evaluate metabarcoding accuracy (Table 5.3; Fig. 5.8). Compiling studies that have compared this method with microscopy counts or that have sequenced artificially-assembled samples of known composition indicates that, even after careful optimization of field, lab and bioinformatic procedures, some metabarcoding estimates are not reliable. While metabarcoding is proven as a sensitive and accurate method for the detection and identification of diverse taxa, their relative abundances may appear as strongly biased (Table 5.3; Fig. 5.8). Data interpretations also need to consider what is actually being measured with metabarcoding. For example, we know that OTU richness does not equate to species richness in tintinnids and other protists (Santoferrara and McManus 2017, Caron and Hu 2019) and that many OTUs may remain only identified into broad taxonomic categories (e.g., phylum or class; de Vargas et al. 2015. Still, the use of tightly calibrated bioinformatics in combination with accurate and complete reference databases enables tracking of species across samples at least as accurately as with the microscope (Santoferrara 2019).

The limitations mentioned above clearly affect the interpretation of assemblage structure within a sample, but are expected to have a smaller effect for tracking relative changes in diversity and distribution across space and/or time (Creer et al. 2016). This is exemplified by a detailed comparison of tintinnid assemblages as defined by microscopy and metabarcoding across an environmental gradient (Fig. 5.9A; Santoferrara et al. 2016a). In this study, morphospecies and OTUs captured similar patterns of spatial distribution: three different assemblages in inshore, offshore and deep-water sample groups (Fig. 5.9B). As expected, within each assemblage there was only partial agreement between morphological and molecular estimates (Fig. 5.9C), probably due to: (1) the known limitations of metabarcoding for estimations of relative abundance (previous paragraphs); (2) the lack of reference sequences for most of the deep-water morphospecies, which thus remained unlinked to metabarcodes (grey in Fig. 5.9C); and (3) inaccuracies in microscopy quantifications, especially for small and cryptic *Salpingella* species (red-purple palette in Fig. 5.9C; see also next subsection).

Table 5.3: Metabarcoding of tintinnids and other plankton evaluated by either microscopy of samples collected simultaneously (Mi) or sequencing of artificially-assembled mock samples of known composition (MS).

Target	Test	Marker	Detections	Identities	Relative abundances	References
Tintinnids	Mi	v4 SSU rDNA or ITS	82% (9/11 genera)	> 70%	Similar	Bachy et al. 2013
Tintinnids	Mi	v2-v3 SSU rDNA	77% (17/22 species)	> 70%	Correlated*	Santoferrara et al. 2016a
Ciliates	Mi	v2-v3 SSU rDNA	86% (6/7 families)	> 70%	Different	Santoferrara et al. 2014
Ciliates	Mi	v4 SSU rDNA	65% (11/17 genera)	46%	Different	Stoeck et al. 2014b
Ciliates	Mi	v9 SSU rDNA	–	> 70%	Correlated* (biomass > density)	Pitsch et al. 2019
Protists	Mi	v3 SSU rDNA	–	–	Different	Medinger et al. 2010
Protists	Mi	v4 or v9 SSU rDNA	–	> 70%	Different	Piredda et al. 2017
Copepods	MS	d2 LSU rDNA	97% (32/33 species)	> 90%	Correlated (biomass)*	Hirai et al. 2017
Copepods	Mi	d2 LSU rDNA	91% (21/23 families)	> 70%	Correlated (biomass)*	Hirai et al. 2017
Zooplankton	MS	v4 SSU rDNA	95% (58/61 species)	> 90%	–	Flynn et al. 2015
Zooplankton	Mi	v1-v3 SSU rDNA	78% (14/18 taxa)	> 70%	Different	Lindeque et al. 2013
Zooplankton	Mi	v4 SSU rDNA or COI	75% (42/56 taxa)	> 70%	Correlated (biomass)*	Clarke et al. 2017
Zooplankton	Mi	d1-d2 LSU rDNA and COI	66% (40/61 taxa)	> 70%	Correlation n.s. (density, biomass)	Harvey et al. 2017
Zooplankton	Mi	COI	76% (52/68 species)	> 70%	Correlated (biomass)*	Yang et al. 2017
Zooplankton	Mi	v9 SSU rDNA and COI	74% (40/54 species)	93%	–	Stefanni et al. 2018
Zooplankton	Mi	v9 SSU rDNA	–	–	Correlated*	Bucklin et al. 2019

For works testing multiple metabarcoding strategies, only optimized results are included. Detections = proportion of taxa included in a MS or detected also by Mi (a slash / means 'out of'). Taxonomic identities = proportion of correctly identified taxa. Relative abundances = it is stated if results were similar, different or correlated compared to MS composition or Mi quantifications; results refer to density, unless stated otherwise; if statistically tested, results significant at $p < 0.05$ (*) or non-significant (n.s.) are indicated. Dashes (–) mean 'not evaluated'. For full name of metabarcoding markers, see Box 5.1. Adapted from Santoferrara (2019).

Still, when morphospecies and OTUs could be linked, it was clear that not only taxonomy but also niche-related variables changed across assemblages: morphospecies with decreasing lorica size dominated from inshore to offshore to deep waters (Fig. 5.9B, C). A correlation between the diameter of the lorica aperture and prey size is well-known (Dolan 2010, Dolan et al. 2013b) and we found that proxies for food resources (chlorophyll *a* concentration and ratios of phytoplankton size-fractions) were the best predictors (over predator biomass, temperature and other environmental factors) of morphology- and OTU-based assemblage variation (Santoferrara et al. 2016a). This agrees with the idea that environmental selection, in particular food partitioning, is an important mechanism for assemblage structuring in planktonic ciliates (Sitran et al. 2009, Claessen et al. 2010, Wickham et al.

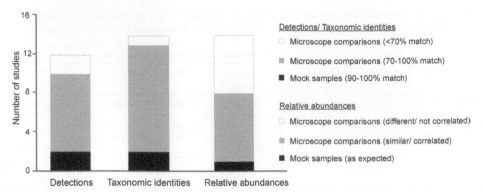

Figure 5.8: Evaluation of plankton metabarcoding by either microscope comparisons or sequencing of artificially-assembled mock samples of known composition. Black and grey areas denote agreement; white areas denote disagreement (details in Table 5.3). For detections and taxonomic identities, most studies found matching results at arbitrary thresholds of 90% (for mock samples) and 70% (for microscope comparisons; a lower value is used due to usual limitations in standardization of sample volumes and completeness of reference databases). For relative abundances (density or biomass), only about half of the studies found similar or correlated results. Adapted from Santoferrara (2019).

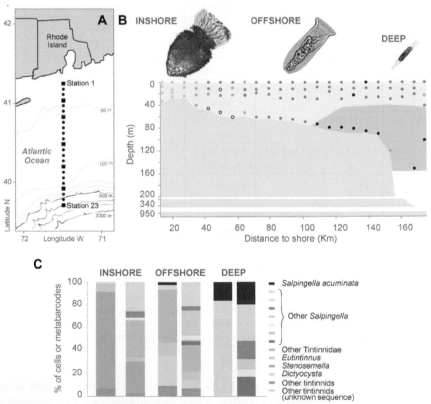

Figure 5.9: Both microscopy and metabarcoding differentiate tintinnid assemblages across a coast-to-ocean gradient in the Northwest Atlantic: (A) geographical location of stations sampled (four depths each) and analyzed by both methods (square) or only by microcopy (circle); (B) sample groupings based on several multivariate analyses that separately considered the occurrence or relative abundance of morphospecies or molecular operational taxonomic units (see details in Santoferrara et al. 2016b). Each dot represents a sample; grey or white dots indicate samples with variable results or non-detections, respectively. The pictures show representative taxa in each sample group: *Stenosemella*, *Amphorides* and *Salpingella* in inshore, offshore and deep samples, respectively; and (C) assemblage structure changed markedly among sample groups, based on relative abundance of either cells (left bars) or metabarcode reads (right bars). Adapted from Santoferrara et al. (2016a).

2011). Again, to explore not only patterns, but also the processes that structure plankton communities, a combination of morphological and DNA sequence data (by parallel microscopy and/or with reliable reference databases) is paramount.

5.4.3 Unmasking Cryptic Diversity

One of the most exciting findings of metabarcoding data is that a large diversity of poorly-known or unknown lineages is hidden even within relatively well-studied groups, such as foraminiferans (Lecroq et al. 2011, Morard et al. 2019). For tintinnids, we found that the OTU diversity and relative abundance in a coast-to-ocean gradient was dominated by *Salpingella* lineages almost undetected in the microscope (Fig. 5.9C; Santoferrara et al. 2016a). The small proportion of *Salpingella* cells observed in the microscope corresponded to the relatively large *S. acuminata* (with similar proportions and distribution based on both microscope and metabarcoding; Fig. 5.9C) and few specimens of smaller morphospecies (diameter 4–14 μm) that were probably underestimated, could not be barcoded (and thus remained unlinked to metabarcodes), and in some cases could not be linked to described species (Santoferrara et al. 2016a). Instead, a carefully-tested bioinformatic analysis of metabarcodes (including the now common practice of analyzing OTUs clustered at 100% similarity) allowed us to detect at least three closely-related *Salpingella* variants that are differentially distributed across shelf waters in the Northwest Atlantic (prevailing in the photic zone offshore, below the photic zone offshore, or everywhere) and that may be thus exploiting different ecological niches (Fig. 5.10; Santoferrara et al. 2016a). Hidden diversity within this well-known genus is also supported in the Mediterranean, where part of the *Salpingella* diversity detected by metabarcoding does not correspond with microscopy counts (Bachy et al. 2013).

One reason for underestimation of *Salpingella* in the microscope might be that some members of this genus (morphologically and genetically similar to the ones detected by Santoferrara et al. 2016a) can attach to large diatoms that may fully hide the tintinnid and make them difficult to detect (Vincent et al. 2018). Interestingly, the epibiotic association between *Salpingella* and diatoms involves living cells and it is widely distributed, thus suggesting a mutualist interaction that may confer advantages such as predation avoidance for the tintinnid and enhanced motility for the diatom (Vincent et al. 2018). Other examples of potential mutualism with diatoms (Armbrecht et al. 2017, Gómez 2007) or cyanobacteria (Foster et al. 2006), as well as parasitism by dinoflagellates (Coats et al. 2012), suggest an important role of biotic interactions other than the most commonly studied feeding and predation in tintinnid abundance and distribution. Symbiosis and parasitism are now seen to prevail in plankton, with surprisingly high sequence abundance and diversity of known photosymbiotic hosts (e.g., Acantharia and Collodaria radiolarians) and parasites of tintinnids and other microzooplankton (e.g., Syndiniales and other dinoflagellates), which suggest that these interactions have been underestimated in planktonic food webs (de Vargas et al. 2015).

We observed distinct distributions of multiple, closely-related tintinnid OTUs not only spatially, but also temporally (Fig. 5.10). Using single-cell barcoding in a coastal site, we found that *Helicostomella* shows temporal alternation of three closely-related molecular clades (two of which have also been reported in a coastal site of the Pacific; Xu et al. 2012). The three *Helicostomella* variants are almost identical in lorica opening diameter (a parameter that correlates with prey size in tintinnids; Dolan 2010) and almost never co-occur, thus suggesting that competitive exclusion and niche parameters other than food size separate these clades temporally (Santoferrara et al. 2015). Alternation of these forms may be related to complex interaction of other seasonally-influenced niche factors such as food type and quantity, temperature and stratification/turbulence (Santoferrara and Alder 2009a). These factors influence periodic encystment and excystment, one of the mechanisms that lead to the periodic disappearance and reappearance, respectively, of *Helicostomella* and other coastal tintinnids from the plankton (Kamiyama 2013).

Cases like *Salpingella* and *Helicostomella* support the idea that some planktonic morphospecies are actually assemblages of cryptic species restricted temporally and/or spatially due to adaptation

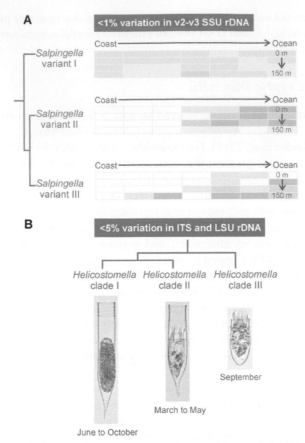

Figure 5.10: Two examples of closely-related species or populations with potentially different ecological niches. (A) *Salpingella* variants with different horizontal and vertical distributions across a coast-to-ocean gradient in the Northwest Atlantic (July 2012), as detected by metabarcoding. For each variant, shading intensity indicates sequence proportions increasing from non-detection (white) to its maximum (darkest). (B) *Helicostomella* variants with different temporal distribution in a coastal site of Long Island Sound, U.S.A. (between 2010 and 2013), as detected by single-cell barcoding. The variants actually overlap in lorica size based on the examination of multiple specimens (Table 5.2). Data from Santoferrara et al. (2015, 2016a). Ciliate images reprinted from: Santoferrara, L.F., M. Tian, V.A. Alder and G.B. McManus. 2015. Discrimination of closely related species in tintinnid ciliates: New insights on crypticity and polymorphism in the genus *Helicostomella*. Protist 166: 78–92, © 2014 Elsevier GmbH, with permission from Taylor & Francis.

to different ecological niches (de Vargas et al. 2004). It is still unclear whether they represent different species or populations and, most importantly, if this small genetic divergence translates into ecophysiological differences. For these and other taxa that are difficult to culture, single-cell genomics and transcriptomics are promising tools to explore species boundaries and functional differentiation.

5.5 Conclusions and Future Directions

While traditional ways to study microzooplankton (microscopy, cultivation) will continue to improve, molecular methods are advancing at a rapid pace and will continue to impact assessments of diversity and biogeography. For example, some newer high-throughput sequencing methods produce reads much longer than those provided by previous technologies, and now allow to sequence the entire ribosomal RNA operon (from SSU to LSU rDNA) for improved taxonomic and phylogenetic resolution of environmental metabarcodes (Jamy et al. 2020).

Complementation of taxonomic and functional diversity is starting to be possible by quantifying libraries of expressed protein-coding genes via metatranscriptomics (Carradec et al. 2018).

Understanding what functional genes are actually expressed in the environment, in conjunction with measurements of environmental conditions, will permit quantification of functional roles and differentiation of community members that may be merely present from those that comprise the most active members of zooplankton assemblages. This will allow studying ecophysiology directly within natural communities, although efforts for isolating and cultivating organisms will remain relevant to populate much needed functional reference databases (with taxonomically and functionally annotated protein sequences) and for experimental hypothesis testing.

Other promising approaches for studying taxonomic and functional diversity in microzooplankton are single-cell genomics and transcriptomics, which enable us to target specimens isolated either from cultures or directly from natural assemblages. These methods make it feasible to explore the repertoire of functional genes and to quantify their expression in individual cells that have been identified and documented by microscopy. These 'omics' methods are also relevant to the issue of cryptic diversity, as comparisons of whole genomes may resolve species identity at a much higher resolution, compared to barcoding. This is especially true for the ciliates, as closely-related but reproductively isolated species may make identical transcripts from genes that are scrambled (transcribed and assembled from multiple, differently-ordered coding regions) in the micronucleus (Gao et al. 2015, Maurer-Alcalá et al. 2018).

We are the beneficiaries of two centuries of careful studies on zooplankton diversity. For the tintinnid microzooplankton in particular, quantification of distributions and morphological details has given us a rich heritage on which genomic and transcriptomic methods can build. As diversity studies move from a taxonomic focus to a functional one, it will be crucial to continue to incorporate this heritage with the staggering growth in molecular information provided by newer technologies.

Acknowledgments

This work was supported by the National Science Foundation (OCE1924527) and the University of Connecticut.

References

Agatha, S. and J.C. Riedel-Lorjé. 2006. Redescription of *Tintinnopsis cylindrica* Daday, 1887 (Ciliophora: Spirotricha) and unification of tintinnid terminology. Acta Protozool. 45: 137–151.

Agatha, S. and M.C. Strüder-Kypke. 2007. Phylogeny of the order Choreotrichida (Ciliophora, Spirotricha, Oligotrichea) as inferred from morphology, ultrastructure, ontogenesis, and SSrRNA gene sequences. Eur. J. Protistol. 43: 37–63.

Agatha, S. 2008. Redescription of the tintinnid ciliate *Tintinnopsis fimbriata* Meunier, 1919 (Spirotricha, Choreotrichida) from coastal waters of Northern Germany. Denisia 23: 261–272.

Agatha, S. and S.-F. Tsai. 2008. Redescription of the tintinnid *Stenosemella pacifica* Kofoid and Campbell, 1929 (Ciliophora, Spirotricha) based on live observation, protargol impregnation, and scanning electron microscopy. J. Eukaryot. Microbiol. 55: 75–85.

Agatha, S. 2010. Redescription of *Tintinnopsis parvula* Jörgensen, 1912 (Ciliophora: Spirotrichea: Tintinnina), including a novel lorica matrix. Acta Protozool. 49: 213–234.

Agatha, S. and M.C. Strüder-Kypke. 2012. Reconciling cladistic and genetic analyses in choreotrichid Ciliates (Ciliophora, Spirotricha, Oligotrichea). J. Eukaryot. Microbiol. 59: 325–350.

Agatha, S. and M.C. Strüder-Kypke. 2013. Systematics and evolution of tintinnid ciliates. pp. 42–84. *In*: J.R. Dolan, D.J.S. Montagnes, S. Agatha, D.W. Coats and D.K. Stoecker [eds.]. The Biology and Ecology of Tintinnid Ciliates: Models for Marine Plankton. Wiley-Blackwell, Oxford, UK.

Agatha, S., M. Laval-Peuto and P. Simon. 2013. The tintinnid lorica. pp. 17–41. *In*: J.R. Dolan, D.J.S. Montagnes, S. Agatha, D.W. Coats and D.K. Stoecker [eds.]. The Biology and Ecology of Tintinnid Ciliates: Models for Marine Plankton. Wiley-Blackwell, Oxford, UK.

Agatha, S. and M.C. Strüder-Kypke. 2014. What morphology and molecules tell us about the evolution of Oligotrichea (Alveolata, Ciliophora). Acta Protozool. 53: 77–90.

Alder, V.A. and D. Boltovskoy. 1993. The ecology of larger microzooplankton in the Weddell-Scotia Confluence area: horizontal and vertical distribution patterns. J. Mar. Res. 51: 323–344.

Alder, V.A. 1995. Ecología y sistemática de Tintinnina (Protozoa, Ciliata) y microzooplancteres asociados de aguas antárticas. Ph.D. Thesis, Universidad de Buenos Aires, Argentina.

Alder, V.A. 1999. Tintinnoinea. pp 321–384. *In*: D. Boltovskoy [ed.]. South Atlantic Zooplankton. Backhuys Publishers, Leiden, Netherlands.

Al-Yamani, F., R. Madhusoodhanan, V. Skryabin and T. Al-Said. 2019. The response of microzooplankton (tintinnid) community to salinity related environmental changes in a hypersaline marine system in the northwestern Arabian Gulf. Deep Sea Res. Pt. II 166: 151–170.

Amaral-Zettler, L.A., E.A. McCliment, H.W. Ducklow and S.M. Huse. 2009. A method for studying protistan diversity using massively parallel sequencing of V9 hypervariable regions of small-subunit ribosomal RNA genes. PLoS ONE 4: e6372.

Anjusha, A., R. Jyothibabu and L. Jagadeesan. 2018. Response of microzooplankton community to the hydrographical transformations in the coastal waters off Kochi, along the southwest coast of India. Continental Shelf Res. 167: 111–124.

Armbrecht, L.H., R. Eriksen, A. Leventer and L. Armand. 2017. First observations of living sea-ice diatom agglomeration to tintinnid loricae in East Antarctica. J. Plankton Res. 39: 795–802.

Bachy, C., F. Gómez, P. López-García, J.R. Dolan and D. Moreira. 2012. Molecular phylogeny of tintinnid ciliates (Tintinnida, Ciliophora). Protist 163: 873–887.

Bachy, C., J.R. Dolan, P. López-García, P. Deschamps and D. Moreira. 2013. Accuracy of protist diversity assessments: Morphology compared with cloning and direct pyrosequencing of 18s rRNA genes and its regions using the conspicuous tintinnid ciliates as a case study. ISME J. 7: 244–255.

Bachy, C., D. Moreira, J.R. Dolan and P. López-García. 2014. Seasonal dynamics of free-living tintinnid ciliate communities revealed by environmental sequences from the north-west Mediterranean Sea. FEMS Microbiol. Ecol. 87: 330–342.

Bakker, C. and W.J. Phaff. 1976. Tintinnida from coastal waters of the S. W. Netherlands 1. The genus *Tintinnopsis* Stein. Hydrobiologia 50: 101–111.

Balech, E. 1968. Algunas especies nuevas o interesantes de Tintinnidos del Golfo de Mexico y Caribe. Revista del Museo Argentino de Ciencias Naturales Bernard Rivadavia e Instituto Nacional de Investigacion de las Ciencias Naturales. Hidrobiologia 2: 165–197.

Balech, E. 1972. Los Tintinnidos indicadores de afloramientos de aguas (Cilliata). Physis XXXI: 519–528.

Biard, T., E. Bigeard, S. Audic, J. Poulain, A. Gutierrez-Rodriguez, S. Pesant et al. 2017. Biogeography and diversity of Collodaria (Radiolaria) in the global ocean. ISME J. 11: 1331–1344.

Biernacka, I. 1952. Studies on the reproduction of some species of the genus *Tintinnopsis* Stein. Annales Universitatis Marie Curie Sklodowska Lublin Polonia Sectio C 6: 211–247.

Boero, F., M. Putti, E. Trainito, E. Prontera, S. Piraino and T.A. Shiganova. 2009. First records of *Mnemiopsis leidyi* (Ctenophora) from the Ligurian, Thyrrhenian and Ionian Seas (Western Mediterranean) and first record of *Phyllorhiza punctata* (Cnidaria) from the Western Mediterranean. Aquat. Inv. 4: 675–680.

Bojanić, N., O. Vidjak, M. Šolić, N. Krstulović, I. Brautović, S. Matijević et al. 2012. Community structure and seasonal dynamics of tintinnid ciliates in Kaštela Bay (middle Adriatic Sea). J. Plankton Res. 34: 510–530.

Boltovskoy, D., E.O. Dinofrio and V.A. Alder. 1990. Intraspecific variability in Antarctic tintinnids: the *Cymatocylis affinis/convallaria* species group. J. Plankton Res. 12: 403–413.

Boltovskoy, D. and V.A. Alder. 1992. Microzooplankton and tintinnid species-specific assemblage structures: patterns of distribution and year-to-year variations in the Weddell Sea (Antarctica). J. Plankton Res. 14: 1405–1423.

Boscaro, V., L.F. Santoferrara, Q. Zhang, E. Gentekaki, M.J. Syberg-Olsen, J. del Campo et al. 2018. EukRef-Ciliophora: a manually curated, phylogeny-based database of small subunit rRNA gene sequences of ciliates. Environ. Microbiol. 20: 2218–2230.

Boxshall, G.A. and D. Jaume. 2000. Making waves: The repeated colonization of fresh water by copepod crustaceans. Adv. Ecol. Res. 31: 61–79.

Brandt, K. 1906–1907. Die Tintinnodeen der Plankton-Expedition. Ergebnisse der Plankton-Expedition der Humboldt-Stiftung Systematischer Teil 3: 1–499 + 70 plates.

Bucklin, A., A.M. Bentley and S.P. Franzen. 1998. Distribution and relative abundance of *Pseudocalanus moultoni* and *P. newmani* (Copepoda: Calanoida) on Georges Bank using molecular identification of sibling species. Mar. Biol. 132: 97–106.

Bucklin, A., B. Ortman, R. Jennings, L. Nigro, C. Sweetman, N. Copley et al. 2010. A "Rosetta Stone" for metazoan zooplankton: DNA barcode analysis of species diversity of the Sargasso Sea (Northwest Atlantic Ocean). Deep Sea Res. Pt. II 57: 2234–2247.

Bucklin, A., H. Yeh, J. Questel, D. Richardson, B. Reese, N. Copley et al. 2019. Time-series metabarcoding analysis of zooplankton diversity of the NW Atlantic continental shelf. ICES J. Mar. Sci. 76: 1162–1176.

Caron, D. and S. Hu. 2019. Are we overestimating protistan diversity in nature? Trends Microbiol. 27: 197–205.

Carradec, Q., E. Pelletier, C. da Silva, A. Alberti, Y. Seeleuthner, R. Blanc-Mathieu et al. 2018. A global ocean atlas of eukaryotic genes. Nat. Commun. 9: 373.

Caudill, C. and A. Bucklin. 2004. Molecular phylogeography and evolutionary history of the estuarine copepod, *Acartia tonsa*, on the Northwest Atlantic coast. Hydrobiologia 511: 91–102.

Celepli, N., J. Sundh, M. Ekman, C. Dupont, S. Yooseph, B. Bergman et al. 2017. Meta-omic analyses of Baltic Sea cyanobacteria: diversity, community structure and salt acclimation. Environ. Microbiol. 19: 673–686.

Chen, G. and M. Hare. 2008. Cryptic ecological diversification of a planktonic estuarine copepod, *Acartia tonsa*. Mol. Ecol. 17: 1451–1468.

Chiba, S., E. Di Lorenzo, A. Davis, J. Keister, B. Taguchi, Y. Sasai et al. 2013. Large-scale climate control of zooplankton transport and biogeography in the Kuroshio-Oyashio Extension region. Geoph. Res. Let. 40: 5182–5187.

Choi, J.K., D.W. Coats, D.C. Brownlee and E.B. Small. 1992. Morphology and Infraciliature of three species of *Eutintinnus* (Ciliophora; Tintinnina) with guidelines for interpreting protargol-stained Tintinnine ciliates. J. Eukaryot. Microbiol. 39: 80–92.

Claessens, M., S.A. Wickham, A.F. Post and M. Reuter. 2010. A paradox of the ciliates? High ciliate diversity in a resource-poor environment. Mar. Biol. 157: 483–494.

Claparède, E. and J. Lachmann. 1858. Etudes sur les infusoires et les rhizopodes. Mem. Inst. Natn. Genev. 5: 1–260.

Clarke, L.J., J. Beard, K. Swadling and B. Deagle. 2017. Effect of marker choice and thermal cycling protocol on zooplankton DNA metabarcoding studies. Ecol. Evol. 7: 873–883.

Coats, D.W., T.R. Bachvaroff and C.F. Delwiche. 2012. Revision of the family Duboscquellidae with description of *Euduboscquella crenulata* n. gen., n. sp. (Dinoflagellata, Syndinea), an intracellular parasite of the ciliate Favella panamensis Kofoid and Campbell. J. Eukaryot. Microbiol. 59: 1–11.

Corkett, C.J. and I.A. McLaren. 1978. The biology of *Pseudocalanus*. pp. 2–231. *In*: F.S. Russell and M. Yonge [eds.]. Advances in Marine Biology. Academic Press, New York, USA.

Creer, S., K. Deiner, S. Frey, D. Porazinska, P. Taberlet, W.K. Thomas et al. 2016. The ecologist's field guide to sequence-based identification of biodiversity. Meth. Ecol. Evol. 7: 1008–1018.

Crump, B., C. Hopkinson, M. Sogin and J. Hobbie. 2004. Microbial biogeography along an estuarine salinity gradient: combined influences of bacterial growth and residence time. Appl. Environ. Microbiol. 70: 1494–1505.

Daday, E. von. 1887. Monographie der familie der Tintinnodeen. Mittheilungen aus der Zool. Station zu Neapel 7: 473–591.

Davis, C.C. 1981. Variation of lorica shape in the genus *Ptychocylis* (Protozoa, Tintinnina) in relation to species identification. J. Plankton Res. 3: 433–443.

Davis, C.C. 1985. *Acanthostomella norvegica* (Daday) in insular Newfoundland waters, Canada (Protozoa: Tintinnina). Int. Rev. Ges. Hydrobiol. 70: 21–26.

de Vargas, C., A. Sáez, L. Medlin and H. Thierstein. 2004. Super-Species in the calcareous plankton. pp. 271–298. *In*: H. Thierstein and J. Young [eds.]. Coccolithophores. Springer, Berlin, Heidelberg.

de Vargas, C., S. Audic, N. Henry, J. Decelle, F. Mahé, R. Logares et al. 2015. Eukaryotic plankton diversity in the sunlit ocean. Science 348: 1261605.

del Campo, J., M. Kolisko, V. Boscaro, L.F. Santoferrara, R. Massana, L. Guillou et al. 2018. EukRef: phylogenetic curation of ribosomal RNA to enhance understanding of eukaryotic diversity and distribution. PLoS Biol. 16: e2005849.

Decelle, J., S. Romac, E. Sasaki, F. Not and F. Mahé. 2014. Intracellular diversity of the V4 and V9 regions of the 18S rRNA in marine protists (Radiolarians) assessed by high-throughput sequencing. PLoS ONE 9: e104297.

Diociaiuti, T., F.B. Aubry and S.F. Umani. 2019. Vertical distribution of microbial communities abundance and biomass in two NW Mediterranean Sea submarine canyons. Prog. Oceanog. 175: 14–23.

Doherty, M., B.A. Costas, G.B. McManus and L.A. Katz. 2007. Culture-independent assessment of planktonic ciliate diversity in coastal Northwest Atlantic waters. Aquat. Microb. Ecol. 48: 141–154.

Doherty, M., M. Tamura, B.A. Costas, M. Ritchie, G.B. McManus and L.A. Katz. 2010. Ciliate diversity and distribution across an environmental and depth gradient in Long Island Sound, USA. Environ. Microbiol. 12: 886–898.

Dolan, J.R. and C. Gallegos. 2001. Estuarine diversity of tintinnids (planktonic ciliates). J. Plankton Res. 23: 1009–1027.

Dolan, J.R., H. Claustre, F. Carlotti, S. Plounevez and T. Moutin. 2002. Microzooplankton diversity: relationship of tintinnid ciliates with resources, competitors and predators from the Atlantic Coast of Morocco to the Eastern Mediterranean. Deep-Sea Res. Pt. I 49: 1217–1232.

Dolan, J.R., M. Ritchie and J. Ras. 2007. The "neutral" community structure of planktonic herbivores, tintinnid ciliates of the microzooplankton, across the SE tropical Pacific Ocean. Biogeosci. 4: 297–310.

Dolan, J.R., M. Ritchie, A. Tunin-Ley and M. Pizay. 2009. Dynamics of core and occasional species in the marine plankton: Tintinnid ciliates in the north-west Mediterranean Sea. J. Biogeogr. 36: 887–895.

Dolan, J.R. 2010. Morphology and ecology in tintinnid ciliates of the marine plankton: Correlates of lorica dimensions. Acta Protozool. 49: 235–344.

Dolan, J.R., R.W. Pierce, E.J. Yang and S.Y. Kim. 2012. Southern Ocean biogeography of tintinnid ciliates of the marine plankton. J. Eukaryot. Microbiol. 59: 511–519.

Dolan, J.R. and R.W. Pierce. 2013. Diversity and distributions of tintinnids. pp. 214–243. *In*: J.R. Dolan, D.J.S. Montagnes, S. Agatha, D.W. Coats and D.K. Stoecker [eds.]. The Biology and Ecology of Tintinnid Ciliates: Models for Marine Plankton. Wiley-Blackwell, Oxford, UK.

Dolan, J.R., D.J.S. Montagnes, S. Agatha, D.W. Coats and D.K. Stoecker. 2013a. The Biology and Ecology of Tintinnid Ciliates. Wiley-Blackwell, Oxford, UK.

Dolan, J.R., M.R. Landry and M.E. Ritchie. 2013b. The species-rich assemblages of tintinnids (marine planktonic protists) are structured by mouth size. ISME J. 7: 1237–1243.

Dolan, J.R. 2016. Planktonic protists: little bugs pose big problems for biodiversity assessments. J. Plankton Res. 38: 1044–1051.

Dolan, J.R., E.J. Yang, S.-H. Kang and T.S. Rhee. 2016. Declines in both redundant and trace species characterize the latitudinal diversity gradient in tintinnid ciliates. ISME J. 10: 2174–2183.

Dolan, J.R. 2017. Historical trends in the species inventory of tintinnids (ciliates of the microzooplankton) in the Bay of Villefranche (NW Mediterranean Sea): Shifting baselines. Eur. J. Protistol. 57: 16–25.

Dolan, J.R., R.W. Pierce and E.J. Yang. 2017. Tintinnid ciliates of the marine microzooplankton in Arctic Seas: a compilation and analysis of species records. Polar Biol. 40: 1247–1260.

Dolan, J.R., M. Ciobanu and L. Coppola. 2019. Past President's address: Protists of the mesopelagic and a bit on the long path to the deep sea. J. Eukaryot. Microbiol. 66: 966–980.

Dunthorn, M., J. Lipps, J.R. Dolan, M. Abboud-Abi Saab, E. Aescht, C. Bachy et al. 2015. Ciliates—Protists with complex morphologies and ambiguous early fossil record. Mar. Micropaleontol. 119: 1–6.

Dupont, C., J. Larsson, S. Yooseph, K. Ininbergs, J. Goll, J. Asplund-Samuelsson et al. 2014. Functional tradeoffs underpin salinity-driven divergence in microbial community composition. PLoS ONE 9: e89549.

Ehrenberg, C.G. 1832. Über die Entwickelung und Lebensdauer der Infusionsthiere; nebst ferneren Beiträgen zu einer Vergleichung ihrer organischen Systeme. Abhandlungen der Königlichen Akademie Wissenschaften zu Berlin, Physikalische Klasse 1831: 1–154, pls I–IV.

Entz, G. Jr. 1909. Studien über organisation und biologie der tintinniden. Arch. Protisten. 15: 93–226.

Fenchel, T. and B.J. Finlay. 2006. The diversity of microbes: resurgence of the phenotype. Philos. Trans. R. Soc. B 361: 1965–1973.

Feng, M., C. Wang, W. Zhang, G. Zhang, H. Xu, Y. Zhao et al. 2018. Annual variation of species richness and lorica oral diameter characteristics of tintinnids in a semi-enclosed bay of western Pacific. Estuar. Coast. Shelf Sci. 207: 164–174.

Fernandes, N.M. and C.G. Schrago. 2019. A multigene timescale and diversification dynamics of Ciliophora evolution. Mol. Phylog. Evol. 139: 106521.

Finlay, B. 2002. Global dispersal of free-living microbial eukaryote species. Science 296: 1061–1063.

Flynn, J.M., E.A. Brown, F.J.J. Chain, H.J. MacIsaac and M.E. Cristescu. 2015. Toward accurate molecular identification of species in complex environmental samples: testing the performance of sequence filtering and clustering methods. Ecol. Evol. 5: 2252–2266.

Foissner, W. and N. Wilbert. 1979. Morphologie, Infraciliatur und ökologie der limnischen Tintinnina: *Tintinnidium fluviatile* Stein, *Tintinnidium pusillum* Entz, *Tintinnopsis cylindrata* Daday und *Codonella cratera* (Leidy) (Ciliophora, Polyhymenophora). J. Protozool. 26: 90–103.

Foissner, W., H. Berger and J. Schaumburg. 1999. Identification and ecology of limnetic plankton ciliates. Informationsberichte des Bayer. Landesamtes fur Wasserwirtschaft 3/99. Bayer. Landesamt fur Wasserwirtschaft, Munich, Germany.

Foissner, W. 2008. Protist diversity and distribution: Some basic considerations. Biodivers. Conserv. 17: 235–242.

Fol, H. 1881. Contribution to the knowledge of the family Tintinnodea. Ann. Magaz. Nat. Hist. 7: 237–250.

Fontaneto, D. 2011. Biogeography of Microscopic Organisms. Is Everything Small Everywhere? Cambridge University Press, New York, USA.

Forster, D., A. Behnke and T. Stoeck. 2012. Meta-analyses of environmental sequence data identify anoxia and salinity as parameters shaping ciliate communities. Syst. Biodiv. 10: 277–288.

Foster, R.A., E.J. Carpenter and B. Bergman. 2006. Unicellular cyanobionts in open ocean dinoflagellates, radiolarians, and tintinnids: ultrastructural characterization and immuno-localization of phycoerythrin and nitrogenase. J. Phycol. 42: 453–463.

Frost, B. 1989. A taxonomy of the marine calanoid copepod genus *Pseudocalanus*. Can. J. Zool. 67: 525–551.

Ganser, M.H. and S. Agatha. 2019. Redescription of *Antetintinnidium mucicola* (Claparède and Lachmann, 1858) nov. gen., nov. comb. (Alveolata, Ciliophora, Tintinnina). J. Eukaryot. Microbiol. 66: 802–820.

Gao, F., S. Roy and L.A. Katz. 2015. Analyses of alternatively processed genes in ciliates provide insights into the origins of scrambled genomes and may provide a mechanism for speciation. MBio 6: e01998–14.

Gaston, K.J. and J.I. Spicer. 2003. Biodiversity: An Introduction. Blackwell Publishing, Oxford, UK.

Gavrilova, N. and J.R. Dolan. 2007. A note of species lists and ecosystem shifts: Black Sea tintinnids, ciliates of the microzooplankton. Acta Protozool. 46: 279–288.

Gavrilova, N.A. and I.V. Dovgal. 2016. Tintinnid ciliates (Spirotrichea, Choreotrichia, Tintinnida) of the Black Sea: recent invasions. Protistol. 10: 91–96.

Gimmler, A., R. Korn, C. de Vargas, S. Audic and T. Stoeck. 2016. The Tara Oceans voyage reveals global diversity and distribution patterns of marine planktonic ciliates. Sci. Rep. 6: 33555.

Godhantaraman, N. 2002. Seasonal variations in species composition, abundance, biomass and estimated production rates of tintinnids at tropical estuarine and mangrove waters, Parangipettai' region, southeast coast of India. J. Mar. Syst. 36: 161–171.

Godhantaraman, N. and S. Uye. 2003. Geographical and seasonal variations in taxonomic composition, abundance and biomass of microzooplancton across a brackish-water lagoonal system of Japan. J. Plankton Res. 25: 465–482.

Gold, K. and E. Morales. 1975. Seasonal changes in lorica sizes and the species of Tintinnida in the New York Bight. J. Protozool. 22: 520–528.

Gómez, F. 2007. On the consortium of the tintinnid *Eutintinnus* and the diatom *Chaetoceros* in the Pacific Ocean. Mar. Biol. 151: 1899–1906.

Gong, J., J. Dong, X. Liu and R. Massana. 2013. Extremely high copy numbers and polymorphisms of the rDNA operon estimated from single cell analysis of oligotrich and peritrich ciliates. Protist 164: 369–379.

Grabbert, S., J. Renz, H.-J. Hirche and A. Bucklin. 2010. Species-specific PCR discrimination of species of the calanoid copepod *Pseudocalanus, P. acuspes* and *P. elongatus*, in the Baltic and North Seas. Hydrobiologia 652: 289–297.

Graham, W.M., D. Martin, D. Felder, V. Asper and H. Perry. 2003. Ecological and economic implications of a tropical jellyfish invader in the Gulf of Mexico. pp. 53–69. *In*: J. Pederson [ed.]. Marine Bioinvasions: Patterns, Processes and Perspectives. Springer, Dordrecht, Netherlands.

Grattepanche, J.D., L. Santoferrara, J. Andrade, A. Oliverio, G.B. McManus and L.A. Katz. 2014a. Distribution and diversity of oligotrich and choreotrich ciliates assessed by morphology and by DGGE in temperate coastal waters. Aquat. Microb. Ecol. 71: 211–221.

Grattepanche, J.D., L.F. Santoferrara, G.B. McManus and L.A. Katz. 2014b. Diversity of diversity: Conceptual and methodological differences in biodiversity estimates of eukaryotic microbes as compared to bacteria. Trends Microbiol. 22: 432–437.

Grattepanche, J.D., L.F. Santoferrara, G.B. McManus and L.A. Katz. 2015. Distinct assemblage of planktonic ciliates dominates both photic and deep waters on the New England shelf. Mar. Ecol. Prog. Ser. 526: 1–9.

Grattepanche, J.D., L. Santoferrara, G. McManus and L. Katz. 2016. Unexpected biodiversity of ciliates in marine samples from below the photic zone. Mol. Ecol. 25: 3987–4000.

Gruber, M.S., M. Strüder-Kypke and S. Agatha. 2018. Redescription of *Tintinnopsis everta* Kofoid and Campbell 1929 (Alveolata, Ciliophora, Tintinnina) based on taxonomic and genetic analyses—Discovery of a new complex ciliary pattern. J. Eukaryot. Microbiol. 65: 484–504.

Gruber, M.S., A. Mühlthaler and S. Agatha. 2019. Ultrastructural studies on a model tintinnid-*Schmidingerella meunieri* (Kofoid and Campbell, 1929) Agatha and Strüder-Kypke, 2012 (Ciliophora). I. Somatic kinetids with unique ultrastructure. Acta Protozool. 57: 195–214.

Guillou, L., M. Viprey, A. Chambouvet, R. Welsh, A. Kirkham, R. Massana et al. 2008. Widespread occurrence and genetic diversity of marine parasitoids belonging to Syndiniales (Alveolata). Environ. Microbiol. 10: 3349–3365.

Guillou, L., D. Bachar, S. Audic, D. Bass, C. Berney, L. Bittner et al. 2013. The Protist Ribosomal Reference database (PR^2): a catalog of unicellular eukaryote Small Sub-Unit rRNA sequences with curated taxonomy. Nucleic Acids Res. 41: D597–D604.

Hada, Y. 1970. The protozoan plankton of the Antarctic and Subantarctic seas. JARE Sci. Rep. Ser. E 31: 1–51.

Haeckel, E. 1887. Report on the Radiolaria. Report on the Scientific Results of the Voyage of H.M.S. Challenger. Zoology, Volume XVIII. Eyre and Spottiswoode, Edinburgh, London.

Harvey, J.B.J., S.B. Johnson, J.L. Fisher, W.T. Peterson and R.C. Vrijenhoek. 2017. Comparison of morphological and next generation DNA sequencing methods for assessing zooplankton assemblages. J. Exp. Mar. Biol. Ecol. 487: 113–126.

Hebert, P., A. Cywinska, S. Ball and J. deWaard. 2003. Biological identifications through DNA barcodes. Proc. R. Soc. Lon. B 270: 313–321.

Herlemann, D., M. Labrenz, K. Jürgens, S. Bertilsson, J. Waniek and A. Andersson. 2011. Transitions in bacterial communities along the 2000 km salinity gradient of the Baltic Sea. ISME J. 5: 1571.

Hewson, I., E. Eggleston, M. Doherty, D. Lee, M. Owens, J. Shapleigh et al. 2014. Metatranscriptomic analyses of plankton communities inhabiting surface and subpycnocline waters of the Chesapeake Bay during oxic-anoxic-oxic transitions. Appl. Environ. Microbiol. 80: 328–338.

Hirai, J., S. Nagai and K. Hidaka. 2017. Evaluation of metagenetic community analysis of planktonic copepods using Illumina MiSeq: Comparisons with morphological classification and metagenetic analysis using Roche 454. PLoS One 12: e0181452.

Hofker, J. 1931. Studien über Tintinnoidea. Archiv für Protistenkunde 75: 315–402.

Hu, Y., B. Karlson, S. Charvet and A. Andersson. 2016. Diversity of pico- to mesoplankton along the 2000 km salinity gradient of the Baltic Sea. Front. Microbiol. 7: 679.

Jamy, M., R. Foster, P. Barbera, L. Czech, A. Kozlov, A. Stamatakis et al. 2020. Long-read metabarcoding of the eukaryotic rDNA operon to phylogenetically and taxonomically resolve environmental diversity. Mol. Ecol. Res. 20: 429–443.

Jörgensen, E. 1924. Mediterranean Tintinnids. Report on the Danish Oceanographical Expeditions 1908–10 to the Mediterranean and adjacent Seas. 2 J.3. (Biology): 1–110.

Jung, J.-H., J. Moon, K.-M. Park, S. Kim, J.R. Dolan and E.J. Yang. 2018. Novel insights into the genetic diversity of *Parafavella* based on mitochondrial CO1 sequences. Zool. Scripta 47: 743–755.

Kamiyama, T. and M. Tsujino. 1996. Seasonal variation in the species composition of tintinnid ciliates in Hiroshima Bay, the Seto Inland Sea of Japan. J. Plankton Res. 18: 2313–2327.

Kamiyama, T. 2013. Comparative biology of tintinnid cysts. pp. 171–185. *In*: J.R. Dolan, D.J.S. Montagnes, S. Agatha, D.W. Coats and D.K. Stoecker [eds.]. The Biology and Ecology of Tintinnid Ciliates: Models for Marine Plankton. Wiley-Blackwell, Oxford, UK.

Katoh, K. and D. Standley. 2013. MAFFT multiple sequence alignment software version 7: improvements in performance and usability. Mol. Biol. Evol. 30: 772–780.

Kato, S. and A. Taniguchi. 1993. Tintinnid ciliates as indicator species of different water masses in the western North Pacific Polar Front. Fish. Oceanogr. 2: 166–174.

Kim, E., J. Harrison, S. Sudek, M. Jones, H. Wilcox, T. Richards et al. 2011. Newly identified and diverse plastid-bearing branch on the eukaryotic tree of life. Proc. Natl. Acad. Sci. USA 108: 1496–1500.

Kim, S.Y., E.J. Yang, J. Gong and J.K. Choi. 2010. Redescription of *Favella ehrenbergii* (Claparède and Lachmann, 1858) Jörgensen, 1924 (Ciliophora: Choreotrichia), with phylogenetic analyses based on small subunit rRNA gene sequences. J. Eukaryot. Microbiol. 57: 460–467.

Kim, S.Y., J.K. Choi, J.R. Dolan, H.C. Shin, S. Lee et al. 2013. Morphological and ribosomal DNA-based characterization of six Antarctic ciliate morphospecies from the Amundsen Sea with phylogenetic analyses. J. Eukaryot. Microbiol. 60: 497–513.

Kim, Y.-O., K. Shin, P.-G. Jang, H.-W., Choi, J.-H., Noh, E.-J. Yang et al. 2012. Tintinnid species as biological indicators for monitoring intrusion of the warm oceanic waters into Korean coastal waters. Ocean Sci. J. 47: 161–172.

Kofoid, C.A. and A.S. Campbell. 1929. A conspectus of the marine and fresh-water Ciliata belonging to the suborder Tintinnoinea, with descriptions of new species principally from the Agassiz Expedition to the eastern tropical Pacific 1904–1905. Univ. Calif. Publ. Zool. 34: 1–403.

Kofoid, C.A. and A.S. Campbell. 1939. The Ciliata: The Tintinnoinea. Bull. Mus. Comp. Zool. Harv. 84: 1–473.

Laackmann, H. 1907. Antarktische Tintinnen. Zoologischer Anzeiger 31: 235–239.

Laackmann, H. 1910. Die Tintinnodean der Deutschen Sudpolar-expedition 1901–1903, 11: 340–396.

Laval-Peuto, M. 1977. Reconstruction d'une lorica de forme Coxliella par le trophonte nu de *Favella ehrenbergii* (Ciliata, Tintinnina). C. R. Acad. Sc. Paris 284: 547–550.

Laval-Peuto, M. 1981. Construction of the lorica in Ciliata Tintinnina. *In vivo* study of *Favella ehrenbergii*: variability of the phenotypes during the cycle, biology, statistics, biometry. Protistologica 17: 249–272.

Laval-Peuto, M. 1983. Sexual reproduction in *Favella ehrenbergii* (Ciliophora, Tintinnina). Taxonomical implications. Protistologica 19: 503–512.

Laval-Peuto, M. and D.C. Brownlee. 1986. Identification and systematics of the Tintinnina (Ciliophora): Evaluation and suggestions for improvement. Ann. Inst. Océanogr. Paris 62: 69–84.

Laval-Peuto, M. and M.S. Barria de Cao. 1987. Les capsules, extrusomes caracteristiques des Tintinnina (Ciliophora), permettent une classification evolutive des genres et des familles du sous-ordre. Ile Re´un. Scientif. GRECO 88, Trav. C.R.M. 8: 53–60.

Laval-Peuto, M. 1994. Classe des Oligotrichea Bütschli, 1887. Ordre des Tintinnida Kofoid et Campbell, 1929. pp. 181–219. *In*: P. de Puytorac [ed.]. Traité de Zoologie. Anatomie, systématique, biologie. Volume II. Infusoires ciliés. 2. Systématique. Masson, Paris.

Lecroq, B., F. Lejzerowicz, D. Bachar, R. Christen, P. Esling, L. Baerlocher et al. 2011. Ultra-deep sequencing of foraminiferal microbarcodes unveils hidden richness of early monothalamous lineages in deep-sea sediments. Proc. Natl. Acad. Sci. USA 108: 13177–13182.

Leray, M., N. Knowlton, S.-L. Ho, B. Nguyen and R. Machida. 2019. GenBank is a reliable resource for 21st century biodiversity research. Proc. Natl. Acad. Sci. USA 116: 22651.

Li, H., Y. Zhao, X. Chen, W. Zhang, J. Xu, J. Li et al. 2016. Interaction between neritic and warm water tintinnids in surface waters of East China Sea. Deep Sea Res. Pt. II 124: 84–92.

Li, H., C. Wang, C. Liang, Y. Zhao, W. Zhang, G. Grégori and T. Xiao. 2019. Diversity and distribution of tintinnid ciliates along salinity gradient in the Pearl River Estuary in southern China. Estuar. Coast. Shelf Sci. 226: 106268.

Liang, C., H. Li, Y. Dong, Y. Zhao, Z. Tao, C. Li et al. 2018. Planktonic ciliates in different water masses in open waters near Prydz Bay (East Antarctica) during austral summer, with an emphasis on tintinnid assemblages. Pol. Biol. 41: 2355–371.

Lindeque, P.K., H. Parry, R.A. Harmer, P.J. Somerfield and A. Atkinson. 2013. Next generation sequencing reveals the hidden diversity of zooplankton assemblages. PLoS One 8: e81327.

Lipps, J.H., T. Stoeck and M. Dunthorn. 2013. Fossil tintinnids. pp. 186–197. *In*: J.R. Dolan, D.J.S. Montagnes, S. Agatha, D.W. Coats and D.K. Stoecker [eds.]. The Biology and Ecology of Tintinnid Ciliates: Models for Marine Plankton. Wiley-Blackwell, Oxford, UK.

Litaker, W.R., M. Vandersea, S. Kibler, K. Reece, N. Stokes, F. Lutzoni et al. 2007. Recognizing dinoflagellate species using its rdna sequences. J. Phycology 43: 344–355.

Logares, R., J. Bråte, S. Bertilsson, J. Clasen, K. Shalchian-Tabrizi and K. Rengefors. 2009. Infrequent marine–freshwater transitions in the microbial world. Trends Microbiol. 17: 414–422.

Lohmann, H. 1908. Untersuchungen zur Festellung des vollstandigen Gehalten des Meeres an Plankton. Wiss. Meeresuntes., N F, Abt Kiel 10: 129–370.

López-García, P., F. Rodriguez-Valera, C. Pedros-Alio and D. Moreira. 2001. Unexpected diversity of small eukaryotes in deep-sea Antarctic plankton. Science 409: 603–607.

Lynn, D.H. 2008. The Ciliated Protozoa. Characterization, Classification, and Guide to the Literature. Springer, Verlag Dordrecht.

Marshall, S.M. 1969. Protozoa, order Tintinnia. Conseil International pour l'Exploration de la Mer, Fiches d'Identification de Zooplancton, fiches 117–127.

Massana, R., J. Castresana, V. Balagué, L. Guillou, K. Romari, A. Groisillier et al. 2004. Phylogenetic and ecological analysis of novel marine stramenopiles. Appl. Environ. Microbiol. 70: 3528–3534.

Maurer-Alcalá, X.X., Y. Yan, O. Pilling, R. Knight and L.A. Katz. 2018. Twisted tales: Insights into genome diversity of ciliates using single-cell 'omics. Genome Biol. Evol. evy133-evy133.

McManus, G.B. and L.A. Katz. 2009. Molecular and morphological methods for identifying plankton: what makes a successful marriage? J. Plankton Res. 31: 1119–1129.

McManus, G.B. and L. Santoferrara. 2013. Tintinnids in microzooplankton communities. pp. 198–213. *In*: J.R. Dolan, D.J.S. Montagnes, S. Agatha, D.W. Coats and D.K. Stoecker [eds.]. The Biology and Ecology of Tintinnid Ciliates: Models for Marine Plankton. Wiley-Blackwell, Oxford, UK.

Medinger, R., V. Nolte, R.V. Pandey, S. Jost, B. Ottenwälder, C. Schlötterer et al. 2010. Diversity in a hidden world: potential and limitation of next-generation sequencing for surveys of molecular diversity of eukaryotic microorganisms. Mol. Ecol. 19: 32–40.

Modigh, M. and M. Castaldo. 2002. Variability and persistence in tintinnid assemblages at a Mediterranean coastal site. Aquat. Microb. Ecol. 28: 299–311.

Modigh, M., S. Castaldo, M. Saggiomo and I. Santarpia. 2003. Distribution of tintinnid species from 42° N to 43° S through the Indian Ocean. Hydrobiologia 503: 251–262.

Morard, R., N. Vollmar, M. Greco and M. Kucera. 2019. Unassigned diversity of planktonic foraminifera from environmental sequencing revealed as known but neglected species. PLoS ONE 14: e0213936.

Müller, O.F. 1779. Zoologia Danica seu animalium Daniae et Norvegiae rariorum ac minus notorum, descriptiones et historia. Vol. I Explicationi iconum fasciculi primi, eiusdem operis inservieris. Havniae et Lipsiae, Sumtibus Weygandinis. p. ix +124.

Onda, D., E. Medrinal, A. Comeau, M. Thaler, M. Babin and C. Lovejoy. 2017. Seasonal and interannual changes in ciliate and dinoflagellate species assemblages in the Arctic Ocean (Amundsen Gulf, Beaufort Sea, Canada). Front. Mar. Sci. doi.org/10.3389/fmars.2017.00016.

Orsi, W., V. Edgcomb, J. Faria, W. Foissner, W. Fowle, T. Hohmann et al. 2012. Class Cariacotrichea, a novel ciliate taxon from the anoxic Cariaco Basin, Venezuela. Int. J. Syst. Evol. Microbiol. 62: 1425–1433.

Pawlowski, J., S. Audic, S. Adl, D. Bass, L. Belbahri, C. Berney et al. 2012. CBOL protist working group: Barcoding eukaryotic richness beyond the animal, plant, and fungal kingdoms. PLoS Biol. 10: e1001419.

Petz, W., W. Song and N. Wilbert. 1995. Taxonomy and ecology of the ciliate fauna (Protozoa, Ciliophora) in the endopelagial and pelagial of the weddell Sea, Antartica. Stapfia 40: 1–223.

Pierce, R.W. and J.T. Turner. 1992. Ecology of planktonic ciliates in marine food webs. Rev. Aquat. Sci. 6: 139–181.

Pierce, R.W. and J.T. Turner. 1993. Global biogeography of marine tintinnids. Mar. Ecol. Prog. Ser. 94: 11–26.

Pierce, R.W., J.T. Carlton, D.A. Carlton and J.B. Geller. 1997. Ballast water as a vector for tintinnid transport. Mar. Ecol. Prog. Ser. 149: 295–297.

Piredda, R., M.P. Tomasino, A.M. D'erchia, C. Manzari, G. Pesole, M. Montresor et al. 2017. Diversity and temporal patterns of planktonic protist assemblages at a Mediterranean Long Term Ecological Research site. FEMS Microbiol. Ecol. 93: fiw200.

Pitsch, G., E.P. Bruni, D. Forster, Z. Qu, B. Sonntag, T. Stoeck et al. 2019. Seasonality of planktonic freshwater ciliates: are analyses based on v9 regions of the 18S rRNA gene correlated with morphospecies counts? Front. Microbiol. Doi: 10.3389/fmicb.2019.00248.

Saccà, A., M. Strüder-Kypke and D. Lynn. 2012. Redescription of *Rhizodomus tagatzi* (Ciliophora: Spirotrichea: Tintinnida), based on morphology and small subunit ribosomal RNA gene sequence. J. Eukaryot. Microbiol. 59: 218–231.

Saccà, A. and G. Giuffrè. 2013. Biogeography and ecology of *Rhizodomus tagatzi*, a presumptive invasive tintinnid ciliate. J. Plankton Res. 35: 894–906.

Santoferrara, L. 2008. Estudio sistemático y ecológico de los ciliados (Protista, Ciliophora) planctónicos del Mar Argentino y Pasaje Drake. Ph.D. Thesis, Universidad de Buenos Aires, Argentina.

Santoferrara, L. and V. Alder. 2009a. Morphological variability, spatial distribution and abundance of *Helicostomella* species (Ciliophora, Tintinnina) in relation to environmental factors (Argentine shelf; 40–55° S). Sci. Mar. 73: 701–716.

Santoferrara, L. and V. Alder. 2009b. Abundance trends and ecology of planktonic ciliates of the south-western Atlantic (35–63° S): A comparison between neritic and oceanic environments. J. Plankton Res. 31: 837–851.

Santoferrara, L.F. and V.A. Alder. 2012. Abundance and diversity of tintinnids (planktonic ciliates) under contrasting levels of productivity in the Argentine Shelf and Drake Passage. J. Sea. Res. 71: 25–30.

Santoferrara, L., G. McManus and V. Alder. 2012. Phylogeny of the order Tintinnida (Ciliophora, Spirotrichea) inferred from small and large subunit rRNA genes. J. Eukaryot. Microbiol. 59: 423–426.

Santoferrara, L.F., G.B. McManus and V.A. Alder. 2013. Utility of genetic markers and morphology for species discrimination within the order Tintinnida (Ciliophora, Spirotrichea). Protist 164: 24–36.

Santoferrara, L.F., J.D. Grattepanche, L.A. Katz and G.B. McManus. 2014. Pyrosequencing for assessing diversity of eukaryotic microbes: Analysis of data on marine planktonic ciliates and comparison with traditional methods. Environ. Microbiol. 16: 2752–2763.

Santoferrara, L.F., M. Tian, V.A. Alder and G.B. McManus. 2015. Discrimination of closely related species in tintinnid ciliates: New insights on crypticity and polymorphism in the genus *Helicostomella*. Protist 166: 78–92.

Santoferrara, L.F., J.D. Grattepanche, L.A. Katz and G.B. McManus. 2016a. Patterns and processes in microbial biogeography: do molecules and morphologies give the same answers? ISME J. 10: 1779–1790.

Santoferrara, L., C. Bachy, V. Alder, J. Gong, Y.-O. Kim, A. Saccà et al. 2016b. Updating biodiversity studies in loricate protists: the case of the tintinnids (Alveolata, Ciliophora, Spirotrichea). J. Eukaryot. Microbiol. 63: 651–656.

Santoferrara, L.F. and G.B. McManus. 2017. Integrating the dimensions of biodiversity in choreotrichs and oligotrichs of marine plankton. Eur. J. Protistol. 61: 323–330.

Santoferrara, L.F., V.A. Alder and G.B. McManus. 2017. Phylogeny, classification and diversity of Choreotrichia and Oligotrichia (Ciliophora, Spirotrichea). Mol. Phylog. Evol. 112: 12–22.

Santoferrara, L.F., E. Rubin and G.B. McManus. 2018. Global and local DNA (meta)barcoding reveal new biogeography patterns in tintinnid ciliates. J. Plankton Res. 40: 209–221.

Santoferrara, L.F. 2019. Current practice in plankton metabarcoding: optimization and error management. J. Plankton Res. 41: 571–582.

Sars, G.O. 1903. An Account of the Crustacea of Norway: Copepoda. Calanoida. The Bergen Museum, Bergen.

Shiganova, T., Z. Mirzoyan, E. Studenikina, S. Volovik, I. Siokou-Frangou, S. Zervoudaki et al. 2001. Population development of the invader ctenophore *Mnemiopsis leidyi*, in the Black Sea and in other seas of the Mediterranean basin. Mar. Biol. 139: 431–445.

Sitran, R., A. Bergamasco, F. Decembrini and L. Guglielmo. 2007. Temporal succession of tintinnids in the northern Ionian Sea, Central Mediterranean. J. Plankton Res. 29: 495–508.

Sitran, R., A. Bergamasco, F. Decembrini and L. Guglielmo. 2009. Microzooplankton (tintinnid ciliates) diversity: coastal community structure and driving mechanisms in the southern Tyrrhenian Sea (Western Mediterranean). J. Plankton Res. 31: 153–170.

Slapeta, J., P. López-García and D. Moreira. 2006. Global dispersal and ancient cryptic species in the smallest marine eukaryotes. Mol. Biol. Evol. 23: 23–29.

Smith, S., W. Song, N. Gavrilova, A. Kurilov, W. Liu, G. McManus et al. 2018. *Dartintinnus alderae* n. g., n. sp., a minute brackish tintinnid (Ciliophora, Spirotrichea) with dual-ended lorica collapsibility. J. Eukaryot. Microbiol. 65: 400–411.

Sniezek, J., G. Capriulo, E. Small and A. Russo. 1991. *Nolaclusilis hudsonicus* n. sp. (Nolaclusiliidae n. fam.) a bilaterally symmetrical tintinnine ciliate from the lower Hudson River estuary. J. Protozool. 38: 589–594.

Snoeyenbos-West, O., T. Salcedo, G. McManus and L. Katz. 2002. Insights into the diversity of choreotrich and oligotrich ciliates (Class: Spirotrichea) based on genealogical analyses of multiple loci. Int. J. Syst. Evol. Microbiol. 52: 1901–1913.

Snyder, R.A. and D.C. Brownlee. 1991. *Nolaclusilis bicornis* n. g., n. sp. (Tintinnina: Tintinnidiidae): a tintinnine ciliate with novel lorica and cell morphology from the Chesapeake Bay estuary. J. Protozool. 38: 583–589.

Sonneborn, T.M. 1957. Breeding systems, reproductive methods and species problems in protozoa. pp. 155–324. *In*: E. Mayr [ed.]. The Species Problem. American Association for the Advancement of Science, Washington DC, USA.

Stamatakis, A. 2014. RAxML version 8: a tool for phylogenetic analysis and post-analysis of large phylogenies. Bioinformatics 30: 1312–1313.

Stefanni, S., D. Stanković, D. Borme, A. de Olazabal, T. Juretić, A. Pallavicini et al. 2018. Multi-marker metabarcoding approach to study mesozooplankton at basin scale. Sci. Rep. 8: 12085.

Stoeck, T., A. Behnke, R. Christen, L. Amaral-Zettler, M. Rodriguez-Mora, A. Chistoserdov et al. 2009. Massively parallel tag sequencing reveals the complexity of anaerobic marine protistan communities. BMC Biology 7: 72.

Stoeck, T., D. Bass, M. Nebel, R. Christen, M. Jones, H. Breiner et al. 2010. Multiple marker parallel tag environmental DNA sequencing reveals a highly complex eukaryotic community in marine anoxic water. Mol. Ecol. 19: 21–31.

Stoeck, T., E. Przybos and M. Dunthorn. 2014a. The D1-D2 region of the large subunit ribosomal DNA as barcode for ciliates. Mol. Ecol. Res. 14: 458–468.

Stoeck, T., H.-W. Breiner, S. Filker, V. Ostermaier, B. Kammerlander and B. Sonntag. 2014b. A morpho-genetic survey on ciliate plankton from a mountain lake pinpoints the necessity of lineage-specific barcode markers in microbial ecology. Environ. Microbiol. 16: 430–444.

Stoecker, D.K., K. Stevens, J. Gustafson and E. Daniel. 2000. Grazing on *Pfiesteria piscicida* by microzooplankton. Aquat. Microb. Ecol. 22: 261–270.

Strüder-Kypke, M.C. and D.H. Lynn. 2003. Sequence analyses of the small subunit rRNA gene confirm the paraphyly of oligotrich ciliates sensu lato and support the monophyly of the subclasses Oligotrichia and Choreotrichia (Ciliophora, Spirotrichea). J. Zool. 260: 87–97.

Strüder-Kypke, M.C. and D.H. Lynn. 2008. Morphological versus molecular data—Phylogeny of tintinnid ciliates (Ciliophora, Choreotrichia) inferred from small subunit rRNA gene sequences. Denisia 23: 417–424.

Tamura, M., L.A. Katz and G.B. McManus. 2011. Distribution and diversity of oligotrich and choreotrich ciliates across an environmental gradient in a large temperate estuary. Aquat. Microb. Ecol. 64: 51–67.

Thompson, G., V. Alder, D. Boltovskoy and F. Brandini. 1999. Abundance and biogeography of tintinnids (Ciliophora) and associated microzooplancton in the Southwestern Atlanctc Ocean. J. Plankton Res. 21: 1265–1298.

Thompson, G.A. and V.A. Alder. 2005. Patterns in tintinnid species composition and abundance in relation to hydrological conditions of the southwestern Atlantic during austral spring. Aquat. Microb. Ecol. 40: 85–101.

Tucker, S.J., G.B. McManus, L.A. Katz and J.-D. Grattepanche. 2017. Distribution of abundant and active planktonic ciliates in coastal and slope waters off New England. Fron. Microbiol. 8.

Urrutxurtu, I. 2004. Seasonal succession of tintinnids in the Nervión River estuary, Basque Country, Spain. J. Plankton Res. 26: 307–314.

Utermöhl, H. 1958. Zur vervolkommung der quantitativen phytoplankton-methodik. Mitteilung Internationale Vereinigung fuer Theoretische unde Amgewandte Limnologie 9: 1–38.

van der Gast, C.J. 2015. Microbial biogeography: The end of the ubiquitous dispersal hypothesis? Environ. Microbiol. 17: 544–546.

van der Spoel, S. and R. Heyman. 1983. A comparative Atlas of Zooplankton. 1–186. Bunge, Utrecht.

Verity, P. and C. Langdon. 1984. Relationship between lorica volume, carbon, nitorgen, and ATP content of tintinnids in Narragansett Bay. J. Plankton Res. 6: 859–868.

Villarino, E., J. Watson, B. Jönsson, J.M. Gasol, G. Salazar, S. Acinas et al. 2018. Large-scale ocean connectivity and planktonic body size. Nat. Comm. 9: 142.

Vincent, F.J., S. Colin, S. Romac, E. Scalco, L. Bittner, Y. Garcia et al. 2018. The epibiotic life of the cosmopolitan diatom *Fragilariopsis doliolus* on heterotrophic ciliates in the open ocean. ISME J. 12: 1094–1108.

Wang, C., Z. Xu, C. Liu, H. Li, C. Liang, Y. Zhao et al. 2019. Vertical distribution of oceanic tintinnid (Ciliophora: Tintinnida) assemblages from the Bering Sea to Arctic Ocean through Bering Strait. Pol. Biol. 42: 2105–2117.

Weisse, T. 2008. Distribution and diversity of aquatic protists: an evolutionary and ecological perspective. Biodivers. Conserv. 17: 243–259.

Wickham, S.A., U. Steinmair and N. Kamennaya. 2011. Ciliate distributions and forcing factors in the Amundsen and Bellingshausen seas (Antarctic). Aquat. Microb. Ecol. 62: 215–230.

Williams, R., H. McCall, R.W. Pierce and J.T. Turner. 1994. Speciation of the tintinnid genus *Cymatocylis* by morphometric analysis of the loricae. Mar. Ecol. Progr. Ser. 107: 263–272.

Xu, D., P. Sun, M.K. Shin and Y.O.K. Kim. 2012. Species boundaries in tintinnid ciliates: A case study—morphometric variability, molecular characterization, and temporal distribution of *Helicostomella* species (Ciliophora, Tintinnina). J. Eukaryot. Microbiol. 59: 351–358.

Xu, D., P. Sun, A. Warren, J. Noh, D. Choi and M.K. Shin. 2013. Phylogenetic investigations on ten genera of tintinnid ciliates (Ciliophora: Spirotrichea: Tintinnida), based on small subunit ribosomal RNA gene sequences. J. Eukaryot. Microbiol. 60: 192–202.

Yang, J., X. Zhang, Y. Xie, C. Song, Y. Zhang, H. Yu et al. 2017. Zooplankton community profiling in a eutrophic freshwater ecosystem-Lake Tai Basin by DNA metabarcoding. Sci. Rep. 7: 1773.

Zhang, Q., S. Agatha, W. Zhang, J. Dong, Y. Yu and N. Jiao. 2017. Three rDNA loci-based phylogenies of tintinnid ciliates (Ciliophora, Spirotrichea, Choreotrichida). J. Eukaryot. Microbiol. 64: 226–241.

Zhao, F., S. Filker, K. Xu, P. Huang and S. Zheng. 2017. Patterns and drivers of vertical distribution of the ciliate community from the surface to the abyssopelagic zone in the Western Pacific Ocean. Front. Microbiol. 8: 2559.

CHAPTER 6

Planktonic Shelled Protists (Foraminifera and Radiolaria Polycystina)

Global Biogeographic Patterns in the Surface Sediments

Demetrio Boltovskoy[1],* and *Nancy M. Correa*[2]

6.1 Introduction

The Radiolaria Polycystina (or Polycystinea) and the Foraminifera (included recently in the Supergroup Rhizaria, Clade Retaria; Adl et al. 2018) are marine, free-living protists. All radiolarians are planktonic, whereas the Foraminifera include both planktonic and benthic species (the present chapter is restricted to the former only).

The two groups share several traits, including their widespread distribution throughout the World Ocean, habitat, mode of life, size, trophic status, possession of a mineral shell (amorphous silica in Radiolaria, calcium carbonate in Foraminifera), and both often host symbiotic algae (Anderson 1983, Hemleben et al. 1989, Boltovskoy et al. 2017, Schiebel and Hemleben 2017). The possession of hard, skeletal structures makes them major contributors to open-ocean sediments, although carbonate deposits (chiefly represented by planktonic Foraminifera and Coccolithophorida), the dominant biogenic sediments, are scarce or absent below 3,500–5,000 m because calcium carbonate dissolves faster than it accumulates at these depths, whereas siliceous oozes (formed largely by diatoms and radiolarians), are less widespread, but their distribution is not affected by depth (Berger 1974). The presence of the remains of both protists is very common in open-ocean sediments worldwide, albeit their proportions and absolute abundance vary substantially (Lisitzin 1971a, b). This circumstance has fostered the use of Foraminifera and Radiolaria from sediment samples for biogeographic (surface sediments) and paleoceanographic (surface and downcore sediments) investigations (De Wever et al. 2001, Schiebel and Hemleben 2017).

[1] Facultad de Ciencias Exactas y Naturales, Instituto de Ecología, Genética y Evolución de Buenos Aires (IEGEBA), Universidad de Buenos Aires-CONICET, 1428, Buenos Aires, Argentina.
[2] Servicio de Hidrografía Naval (Ministerio de Defensa) and Escuela de Ciencias del Mar (Instituto Universitario Naval), Av. Montes de Oca 2124, 1271, Buenos Aires, Argentina.
Email: ncorrea59@gmail.com
* Corresponding author: boltovskoy@gmail.com

Aside from the composition of their shells, a major difference between planktonic Foraminifera and Radiolaria is their extant species richness. Living Foraminifera comprise ca. 50 morphospecies (Schiebel and Hemleben 2017), whereas for Radiolaria around 400–800 species are thought to exist (Lazarus et al. 2015, Boltovskoy et al. 2017). For the latter, however, the taxonomy of many forms is still questionable (Lazarus et al. 2015), and only for ~ 200 species the nomenclature is reasonably stable allowing reliable identifications across publications. It should be stressed that these figures are based on morphological traits only (i.e., morphospecies). Molecular methods suggest that actual numbers of species (or cryptic species) are considerably higher for both groups (de Vargas et al. 2004, Sierra et al. 2013, de Vargas et al. 2015, Biard et al. 2015, Ishitani et al. 2014, Darling et al. 2017), but the information available is still too scarce and fragmentary to allow its use in large-scale biogeographic studies (see below).

The biogeography of pelagic oceanic organisms is a key feature that underpins our understanding of the ecology, paleoecology and evolution of marine biota. In this chapter, we compare the worldwide biogeographic patterns of two protist taxa extensively used in paleoceanographic investigations. The radiolarian data used are a subset of the information analyzed previously by Boltovskoy and Correa (2016), but restricted to sedimentary materials only. Our main goal is contrasting the partitions based on these two groups, as well as with those found for other pelagic organisms and process-oriented features, and interpreting them in the light of ecological constrains, post-mortem mechanisms and methodological approaches.

6.2 Materials

Data used for this study were extracted from two worldwide compilations, the ForCenS foraminiferal database of Siccha and Kucera (2017) (based on PANGAEA and NOAA data repositories uploaded between 1999 and 2017), and the surface sediment information from the WoRaDD database of Boltovskoy et al. (2010) (30 publications between 1967 and 2008) (Fig. 6.1). For the definition of discrete biogeographic regions (Domains), from ForCenS we used 32 taxonomic categories, excluding several multispecies entries, species often ignored in routine counts, and species with > 14% of missing datapoints. In total, 4205 surface sediment samples were employed. From WoRaDD we selected the 139 species present in ≥ 10% of the surface sediment samples and accounting for ≥ 1% of the assemblage in at least one sample (total samples used: 1150). Biogeographic Domains were derived on the basis of the sample groups defined by cluster analyses (Unweighed Pair Group Method with Arithmetic Means, UPGMA) using Morisita's (Morisita 1959) quantitative similarity index (based on the percentages of the species recorded in each sample). This approach has been shown to adequately synthesize compositional relationships (Boltovskoy and Correa 2016). Cluster cutoff values were defined on the basis of visual inspections of the resulting groupings with a view to maximize intra-cluster similarity and congruency in the geographic distribution of the samples involved. Characterization of dominant species of each biogeographic Domain was performed by means of Nodal Analysis using sample groups and individual species (rather than species group) intersections, a straightforward and intuitive technique for interpreting which species are chiefly responsible for the partitions indicated by the cluster analysis (Boltovskoy and Riedel 1987, Hartwell and Claflin 2005). Species composition and environmental differences between the biogeographic areas defined (sample clusters) were subsequently tested with an ANOSIM (Analysis of Similarities) non-parametric post-hoc test (Clarke 1993), Spearman (non-parametric) correlation coefficient, and Kruskal-Wallis ANOVA and Bonferroni-corrected pairwise Mann-Whitney comparisons. Assessments of latitude-diversity trends were evaluated with LOESS (locally weighted scatterplot smoothing) regressions. For all species used, a Mean Weighed Sea Surface Temperature (MWSST), the SST where the species peaks in abundance, was calculated as follows: MWSST = $[\sum X_i * SST_i]/\sum X_i$, where X_i is the % of species X in sample i, and SST_i is the corresponding mean annual SST. The species assumed to best characterize each Domain are those that, on average, are ≥ two times (Foraminifera)

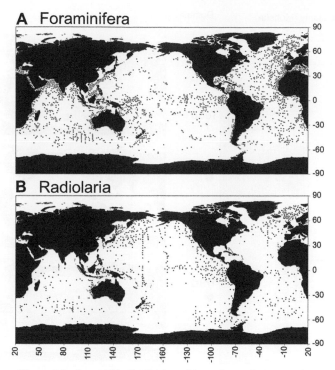

Figure 6.1: Geographic distribution of the surface sediment samples.

or ≥ three times (Radiolaria) more abundant (in % of the overall taxocoenosis; absolute abundances are very seldom assessed in routine counts) in the given Domain than in the entire collection. These cutoff values were established with a view to minimize the numbers of cosmopolitan taxa and species with unclear distributional patterns.

Mean sea surface temperature (SST) data are from Boyer et al. (2013) (mean data for 1955–2012 at 0 m averaged over 5° × 5° squares). Mean annual primary productivity (PP) data are from the Ocean Productivity database (www.science.oregonstate.edu/ocean.productivity) (means for 1997–2012 based on the Vertically Generalized Production Model averaged over 5° × 5° squares). Diversity values are based on Shannon-Wiener's (log base e) expression (Shannon and Weaver 1963); evenness values are according to Pielou (1975). All numerical analyses were performed with the aid of the PAST (Hammer et al. 2001) and PRIMER (Clarke and Warwick 2001) programs.

6.3 Biogeographic Domains

6.3.1 *Foraminifera*

Foraminiferal data show a coherent pattern of four main Domains, three of which can be further subdivided into smaller units (Fig. 6.2). Differences in the specific composition of the seven units identified are statistically highly significant ($p < 0.01$ for all pairwise contrasts, ANOSIM analysis).

The Polar Domain (P), encompassing 339 samples, is split geographically into two well-defined areas: a southern band extending across the Southern Ocean between 45° S–50° S and 60° S–65° S, and a northern component chiefly represented by samples located to the southwest and to the east of Greenland. In the Pacific, this northern component is restricted to very few sites (off the coasts of northern North America, between ca. 40° N and 55° N, Fig. 6.2), most probably because the Arctic and Subarctic Pacific is very sparsely covered in the ForCenS database (Fig. 6.1). This Domain is characterized by two species (*Neogloboquadrina pachyderma* and *Turborotalita quinqueloba*), which,

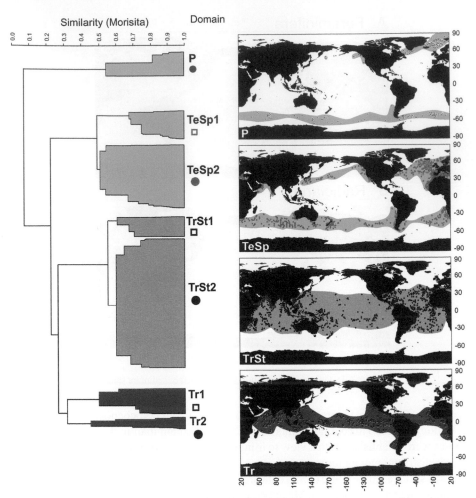

Figure 6.2: Dendrogram of association between 4205 surface sediment foraminiferal samples based on the percentage contributions of 32 species (see Methods for details, cophenetic correlation: 0.805), and geographic distribution of the clusters. Seven outlier samples excluded. Codes for Domain names, P: Polar, TeSp: Temperate/Subpolar, TrSt: Tropical/Subtropical, Tr: Tropical.

on average, account for 85% of its foraminifers. Although both these species are present in > 80% of the samples of this Domain, *N. pachyderma* is by far the most abundant (~ 13 times more than *T. quinqueloba*). *Globigerinita uvula* is very rare both in this Domain (0.01%) and elsewhere, yet in the Polar Domain it is over twice as abundant as in the World Ocean as a whole (Tables 6.1 and 6.3).

The Temperate/Subpolar Domain (TeSp, 1276 samples) covers the middle latitudes of both hemispheres, but its limits vary substantially between longitudes (Fig. 6.2). Its open-ocean samples (TeSp2) are dominated by *Globigerina bulloides* and *Globoconella inflata*, as well as *T. quinqueloba* (in lower proportions, and shared with the Polar Domain). TeSp2 also hosts considerable proportions (~ 30%) of warmer-water foraminifers characteristic of the TrSt Domain (Tables 6.1 and 6.3). *Neogloboquadrina incompta* accounts for larger proportions of the taxocoenoses in the north-eastern Atlantic and in coastal and upwelling areas (Chile-Perú, Namibia, California Current, TeSp1 in Fig. 6.2), with slightly lower SST and, in particular, considerably higher primary production values than TeSp2 (Table 6.1).

The Tropical/Subtropical Domain (TrSt, 2079 samples) is the largest in terms of geographic extension and of samples included. It represents a worldwide belt circumscribed by ~ 35° N to 35° S.

Table 6.1: Foraminiferal biogeographic Domains defined and percentage contribution of their dominant species or species groups (i.e., those whose % abundance is ≥ two times higher in the Domain indicated than in the entire collection, shaded values) (see Table 6.3 and Fig. 6.2). P: Polar; TeSp1: Temperate/Subpolar 1; TeSp2: Temperate/Subpolar 2; TrSt1: Tropical/Subtropical 1; TrSt2: Tropical/Subtropical 2; Tr1: Tropical 1; Tr2: Tropical 2. PP: mean annual primary production (g C m^{-2} year^{-1}); SST: mean annual sea surface temperatura (°C). See Methods for further details.

Domain	P	TeSp1	TeSp2	TrSt1	TrSt2	Tr1	Tr2
Samples included	339	396	880	283	1796	349	162
Mean SST	3.9	14.0	16.5	26.8	25.5	25.7	27.8
Mean PP	477	807	705	1022	421	577	332
Mean diversity	0.616	1.501	1.843	1.943	2.088	1.703	1.497
Mean evenness	0.361	0.615	0.707	0.703	0.726	0.633	0.623
Mean % contribution of dominant species							
P (3 species)	85	5	9	1	0	0	0
TeSp1 (1 species)	8	46	28	8	1	4	0
TeSp2 (3 species)	9	11	50	10	4	4	1
TrSt1 (5 species)	1	0	18	41	15	11	4
TrSt2 (2 species)	0	2	12	15	32	9	5
Tr1 (6 species)	0	3	7	5	5	56	12
Tr2 (5 species)	0	0	2	3	3	15	65

Table 6.2: Radiolarian biogeographic Domains defined and percentage contribution of their dominant species or species groups (i.e., those whose % abundance is ≥ three times higher in the Domain indicated than in the entire collection, shaded values) (see Table 6.4 and Fig. 6.3). Arc: Arctic; Ant: Antarctic; SPo1: Subpolar 1; SPo2: Subpolar 2; SPo3: Subpolar 3; Tra: Transitional; TrSt1: Tropical/Subtropical 1; TrSt2: Tropical/Subtropical 2; TrSt3: Tropical/Subtropical 3. PP: mean annual primary production (g C m^{-2} year^{-1}); SST: mean annual sea surface temperatura (°C). See Methods for further details.

Domain	Arc	Ant	SPo1	SPo2	SPo3	Tra	TrSt1	TrSt2	TrSt3
Samples included	175	119	110	203	73	39	91	304	36
Mean SST	4.7	3.5	14.0	12.3	11.0	14.0	24.5	23.8	26.1
Mean PP	640	221	489	534	527	660	511	489	535
Mean diversity	1.903	1.784	2.203	2.493	2.436	1.685	2.232	2.700	2.167
Mean evenness	0.712	0.599	0.726	0.770	0.749	0.638	0.775	0.757	0.832
Mean % contribution of dominant species									
Arc (11 species)	5.3	0.3	0.8	0.6	0.7	4.2	1.7	0.2	5.0
Ant (6 species)	0.4	8.5	1.1	0.3	3.3	1.7	0.0	0.1	0.0
SPo1 (3 species)	0.6	2.4	8.7	0.7	1.1	0.5	0.0	0.8	0.0
SPo2 (1 species)	0.0	0.0	0.0	0.7	0.0	0.0	0.0	0.1	0.0
SPo3 (5 species)	0.8	0.3	0.0	0.3	3.7	0.2	0.0	0.1	0.0
Tra (6 species)	3.8	2.6	1.2	1.1	1.1	9.9	0.6	0.5	3.0
TrSt1 (29 species)	0.2	0.0	0.5	0.2	0.1	0.5	3.6	0.6	1.0
TrSt2 (1 species)	0.1	0.0	0.1	0.1	0.0	0.0	0.0	0.7	0.0
TrSt3 (4 species)	2.0	0.1	0.4	0.6	0.5	3.5	0.6	0.5	2.9

Two sample groups can be recognized: one chiefly comprised by coastal samples from the northern Indian Ocean and the western Pacific north of the equator (TrSt1), and the other spread across the entire Domain (TrSt2) (Fig. 6.2). The two areas differ little in SST, but the primary production of TrSt1 is over twice as large as that of TsSt2 (Table 6.1). TrSt1 is characterized by the highest proportions of

Table 6.3: Mean percentage contributions of the 20 foraminiferal species characteristic of the Domains illustrated in Fig. 6.2. Greyed cells denote values for species whose mean % in the samples encompassed by the Domain indicated are ≥ two times higher than in the entire collection. See Table 6.1 for Domain names.

Species [Domain]	P	TeSp1	TeSp2	TrSt1	TrSt2	Tr1	Tr2
Neogloboquadrina pachyderma [P]	78.54	5.65	5.39	0.08	0.13	0.16	0.05
Turborotalita quinqueloba [P] [TeSp2]	6.48	2.57	3.31	0.71	0.21	0.09	0.06
Globigerinita uvula [P]	0.01	0.00	0.01	0.00	0.00	0.00	0.00
Neogloboquadrina incompta [TeSp1]	4.57	45.86	11.43	0.35	1.54	3.20	0.28
Globigerina bulloides [TeSp2]	5.87	14.91	27.84	17.58	6.07	4.26	1.15
Globoconella inflata [TeSp2]	2.46	11.58	18.73	0.29	3.71	2.02	0.25
Globigerinita glutinata [TrSt1]	0.79	7.30	6.14	30.42	10.50	4.51	2.60
Globigerina falconensis [TrSt1]	0.09	0.59	3.07	5.90	2.31	0.30	0.09
Globigerinoides tenellus [TrSt1]	0.00	0.02	0.40	3.09	1.35	0.12	0.05
Globoturborotalita rubescens [TrSt1]	0.00	0.14	0.57	1.46	1.14	0.11	0.06
Globorotaloides hexagonus [TrSt1]	0.01	0.05	0.05	0.44	0.23	0.41	0.09
Globigerinoides ruber [TrSt2]	0.00	0.04	0.28	0.00	1.94	0.36	0.12
Candeina nitida [TrSt2]	0.00	0.00	0.00	0.02	0.10	0.03	0.00
Neogloboquadrina dutertrei [Tr1]	0.45	3.55	3.28	5.91	5.76	39.64	11.71
Globorotalia menardii [Tr1]	0.01	0.23	0.75	5.08	3.48	15.93	3.52
Globorotalia tumida [Tr1] [Tr2]	0.01	0.08	0.14	0.73	0.90	4.93	28.25
Globoquadrina conglomerata [Tr1] [Tr2]	0.01	0.00	0.01	0.29	0.36	0.90	1.23
Sphaeroidinella dehiscens [Tr1] [Tr2]	0.00	0.01	0.05	0.09	0.16	0.75	1.84
Beella digitata [Tr1] [Tr2]	0.0	0.1	0.2	0.3	0.4	0.6	0.7
Pulleniatina obliquiloculata [Tr2]	0.04	0.14	0.37	2.75	3.09	5.40	33.74

Globigerinita glutinata, which accounts for > 30% of its foraminifers, accompanied by *Globigerina falconensis, Globigerinoides tenellus, Globoturborotalita rubescens* and *Globorotaloides hexagonus* (~ six to < one percent each). TrSt2 also hosts relatively high values of these foraminifers, but the dominant species is *Globigerinoides ruber*. *Candeina nitida* is also over two times more abundant in TrSt2 than in the entire collection, but its abundances are very low (0.1%) (Table 6.3).

The Tropical Domain (Tr) is entirely superimposed on the Tropical/Subtropical one, but is more restricted latitudinally, spanning from ~ 20° N to 10° S (Fig. 6.2). Foraminiferal data suggest the presence of two different assemblages in this area: Tr1 (349 samples), chiefly distributed in the Indian Ocean, western and eastern Pacific, and eastern Atlantic, and Tr2 (162 samples), mostly in the open-ocean Pacific. Their geographic segregation is marginally consistent, and they share four of their dominant species one of which, however, is over five times more abundant in Tr2 than in Tr1 (*Globorotalia tumida*, Table 6.3). In addition to the above, Tr1 is characterized by very high proportions of *Neogloboquadrina dutertrei* and *Globorotalia menardii*, whereas *Pulleniatina obliquiloculata* is dominant in Tr2 (Table 6.3).

Associations between SST and peak relative abundances of the species covered are generally similar between oceans, but some species show conspicuous differences. For example, in all three oceans (Atlantic, Pacific and Indian), *Orbulina universa* shows a major abundance peak around 11°C–15°C, but secondary peaks occur at 18°C–19°C in the Pacific, 22°C–23°C in the Indian, and 28°C–29°C in the Atlantic. *G. hexagonus*, absent in the Atlantic, peaks at 28°C–29°C in the Indian Ocean, but in the Pacific its highest proportions are at ~ 20°C. In the Pacific, *G. tenellus, G. white* and *G. rubescens* increase monotonically with SST, but in the Atlantic and Indian oceans they drop at temperatures above 23–26°C.

This biogeographic scheme is generally similar to the one proposed by Bé and Tolderlund (1971), as well as those of Hemleben et al. (1989) and Schiebel and Hemleben (2017) (Fig. 6.5B, G, I). However, these published biogeographic divisions are based on partly subjective assumptions, where the gaps are filled in on the basis of ancillary information (SST fields, currents), and therefore depict "tidier" and geographically more consistent areas than those that emerge from objective numerical analyses. Further, they are based on water-column and sedimentary data, where outliers and counterintuitive borders are smoothed out or adjusted. This is not necessarily a drawback, and it probably better captures the large picture, yet some differences emerge. For example, in the present analysis the Transitional Domain is fused with the Subpolar one (as depicted by Schiebel and Hemleben, 2017, for the North Pacific, see below). The fact that much of the World Ocean is characterized by areas where sites belonging to different biogeographic Domains are interspersed (i.e., overlapping Domains) also stands out as a salient feature (see below).

6.3.2 Radiolaria

Geographic partitioning of Radiolarian specific compositions allows identification of nine Domains (Fig. 6.3), all of which differ statistically in their specific composition ($p < 0.01$ for all pairwise contrasts, ANOSIM analysis).

The Arctic Domain (175 samples) is chiefly defined by samples from the Atlantic sector of the Arctic Sea, and a smaller group in the Sea of Okhotsk (Fig. 6.3). Eleven radiolarians characterize this Domain (Table 6.1), with particularly large proportions of *Actinomma leptodermum* and *Cycladophora davisiana* var. *cornutoides* (Table 6.4).

The Antarctic Domain (119 samples) is restricted to the Southern Ocean, and although several Arctic species are present in the Antarctic as well, the numerically dominant radiolarians differ sharply (Table 6.4), with *Antarctissa denticulata-strelkovi* accounting for > 30% of the assemblages in these samples.

The Subpolar Domain (SPo, 386 samples) comprises three groups of samples and species (Fig. 6.3). Spo1 and Spo2 are widely distributed across the World Ocean in both hemispheres, host generally similar assemblages and few characteristic species (Table 6.4). In the Southern Hemisphere, they are fairly well circumscribed to 35° S–55° S, but in the Northern Hemisphere their latitudinal span is wider, especially in the Atlantic Ocean (Fig. 6.3). In contrast, Spo3 is restricted to 73 samples in the Northwestern Pacific, albeit interspersed with SPo1. Five radiolarians are particularly abundant in Spo3 (and scarce elsewhere), especially S*tylochlamydium venustum* and *Siphocampe arachnea* (Table 6.4).

The Transitional Domain is circumscribed to 12 samples in the South Atlantic (~ 25° S–35° S) and 27 samples in the Sea of Japan. Six radiolarians characterize this Domain, the most abundant of which is *Larcopyle buetschlii* (Table 6.4).

As with Foraminifera, the Tropical/Subtropical Domain (TrSt) is geographically the largest, spanning between ~ 40° N and 40° S, and hosts the highest number of characteristic species (Tables 2 and 4). TrSt1 and TrSt2 are interspersed in the eastern Pacific; the Indian and Atlantic oceans are almost exclusively occupied by TrSt1, whereas the western Pacific and coastal areas off Western Australia define TrSt2. These two clusters differ little in their radiolarian assemblages as they share a large number of species and relative abundances are highly correlated (Spearman's R = 0.545, P < 0.001, for the 61 radiolarians listed in Table 6.4). However, while TrSt1 hosts 29 species > three times more abundant in this Domain than in the entire collection (particularly *Didymocyrtis tetrathalamus* and *Pterocorys minythorax*), TrSt2 has only one (*Lophophaena capito*). TrSt3 clusters separately from the previous two, but all its samples are superimposed on TrSt1 and TrSt2 and restricted to the eastern equatorial Pacific (Fig. 6.3). Five radiolarians are particularly abundant in this area, three of which are not characteristic of any other Domain: *Pterocorys zancleus*, *Eucyrtidium erythromystax* and *Carpocanarium papillosum* (Table 6.4).

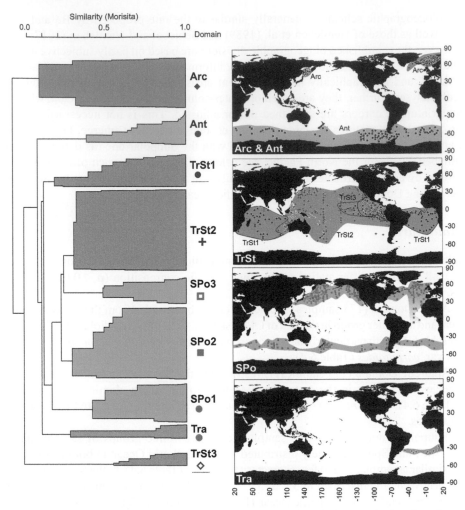

Figure 6.3: Dendrogram of association between 1150 surface sediment radiolarian samples based on the percentage contributions of 139 species (see Methods for details, cophenetic correlation: 0.812), and geographic distribution of the clusters. Codes for Domains names, Arc: Arctic, Ant: SPo: Subpolar, Tra: Transitional, TrSt: Tropical/Subtropical.

In general, peak relative abundances of the species are associated with similar SST intervals in the three major oceans, but proportions of overall taxocoenoses can vary noticeably. For example, *Pseudodictyophimus gracilipes*, *Botryostrobus aquilonaris*, *Phorticium pylonium*, and *Cromyechinus antarctica* are considerably more abundant in the Atlantic than in the Pacific and, especially, the Indian Ocean. *Amphirhopalum ypsilon*, *Acrosphaera murrayana*, and *Botryostrobus auritus-australis* are more abundant in the Pacific and Indian Oceans than in the Atlantic, whereas *Pterocanium praetextum*, *Acanthodesmia vinculata*, *Dictyocoryne truncatum*, *Pterocanium trilobum*, *Pterocorys zancleus*, and *Actinomma medianum* are considerably more abundant in the Indian Ocean than elsewhere.

These subdivisions (Fig. 6.4B) are similar to those proposed by Boltovskoy and Correa (2016) based on the same set of samples, but where the information was processed differently in order to minimize data-related limitations and, in particular, the biasing effects of isothermal submersion (see below). These adjustments of the original data (i.e., elimination of the positive records outside of the range of the MWSST ±1 SD, where SD is the standard deviation of the differences between the MWSST and the actual SSTs of all the records for each species, which were not implemented in this study in order to maximize comparability with the foraminiferal database), as well as the fact

Table 6.4: Mean percentage contributions of the 61 radiolarian species characteristic of the Domains illustrated in Fig. 6.3. Greyed cells denote values for species whose mean % in the samples encompassed by the Domain indicated are ≥ three times higher than in the entire collection. See Table 6.2 for Domain names.

Species [Domain]	Arc	Ant	SPo1	SPo2	SPo3	Tra	TrSt1	TrSt2	TrSt3
Actinomma leptodermum [Arc]	10.97	1.47	1.72	2.36	1.24	4.46	1.96	0.46	0.00
Botryostrobus auritus-australis [Arc] [Tra] [TrSt3]	6.10	0.25	0.48	0.48	0.34	6.99	0.00	0.52	4.98
Carposphaera acanthophora [Arc]	0.73	0.01	0.00	0.14	0.14	0.00	0.00	0.14	0.00
Cromyechinus antarctica [Arc] [Tra]	7.93	0.83	1.33	1.73	0.46	5.28	0.00	0.23	0.00
Cycladophora davisiana var. *cornutoides* [Arc]	14.57	0.31	0.00	0.97	1.31	0.00	0.00	0.07	0.00
Cyrtolagena laguncula [Arc]	7.73	0.04	0.00	0.12	0.04	0.00	0.00	0.03	0.00
Dictyophimus hirundo [Arc] [TrSt1]	1.47	0.33	0.64	0.40	0.62	0.00	1.53	0.27	0.00
Druppatractus irregularis [Arc]	1.58	0.02	0.01	0.17	0.42	0.00	0.00	0.12	0.00
Phormacantha hystrix [Arc] [SPo3]	0.57	0.00	0.00	0.00	1.27	0.19	0.00	0.01	0.00
Rhizoplegma boreale [Arc]	2.85	0.06	0.00	0.02	1.15	0.00	0.00	0.07	0.00
Saccospyris conithorax [Arc]	4.31	0.06	0.00	0.02	1.10	0.00	0.00	0.01	0.00
Antarctissa denticulata-strelkovi [Ant]	0.00	33.44	1.95	1.05	15.59	0.02	0.00	0.09	0.00
Helotholus histricosa [Ant] [Tra]	0.38	12.20	0.96	0.38	0.59	5.07	0.00	0.05	0.00
Lophophaena hispida-cylindrica [Ant]	0.00	1.29	0.00	0.13	0.00	0.00	0.00	0.20	0.00
Saccospyris antarctica [Ant] [SPo3]	0.00	0.98	0.00	0.05	0.57	0.00	0.00	0.01	0.00
Stichopilium bicorne [Ant]	0.00	0.86	0.00	0.05	0.00	0.00	0.00	0.27	0.00
Triceraspyris antarctica [Ant]	0.00	2.16	0.45	0.18	0.00	0.02	0.00	0.02	0.00
Dictyocoryne profunda [SPo1]	0.00	0.08	6.85	1.08	0.05	0.17	0.00	1.14	0.00
Pterocanium korotnevi [SPo1]	0.00	0.07	1.54	0.51	0.81	0.00	0.00	0.02	0.00
Spongotrochus glacialis [SPo1]	1.12	7.02	17.85	0.59	2.34	0.76	0.00	1.10	0.00
Hexacontium laevigatum [SPo2]	0.00	0.04	0.03	0.71	0.00	0.00	0.00	0.15	0.00
Cladoscenium ancoratum [SPo3]	0.00	0.10	0.00	0.04	0.26	0.00	0.00	0.03	0.00
Lithomelissa hystrix [SPo3]	0.58	0.25	0.03	0.23	1.76	0.00	0.00	0.21	0.00
Plectacantha oikiskos [SPo3]	0.34	0.32	0.00	0.08	0.87	0.00	0.00	0.02	0.00
Siphocampe arachnea [SPo3]	1.01	0.09	0.00	0.73	2.93	0.00	0.00	0.04	0.00
Stylochlamydium venustum [SPo3]	1.25	0.27	0.01	0.93	17.94	0.28	0.00	0.51	0.00
Cycladophora davisiana davisiana [Tra]	4.09	1.88	2.57	1.78	3.89	6.26	0.00	0.83	1.06
Larcopyle buetschlii [Tra]	0.55	0.54	0.79	2.04	1.17	35.57	0.60	1.14	0.00
Neosemantis distephanus [Tra]	0.00	0.00	0.00	0.01	0.00	0.45	0.00	0.03	0.00
Acrosphaera spinosa [TrSt1]	0.00	0.01	0.96	0.63	0.08	0.44	4.51	1.87	0.00
Amphirhopalum ypsilon [TrSt1]	0.00	0.00	0.37	0.15	0.00	0.00	2.03	0.27	1.03
Anthocyrtidium ophirense [TrSt1]	0.00	0.00	0.04	0.04	0.00	0.00	4.14	0.35	0.00
Anthocyrtidium zanguebaricum [TrSt1] [TrSt3]	0.00	0.00	0.02	0.02	0.00	0.00	0.90	0.12	1.15
Botryocyrtis scutum [TrSt1]	0.14	0.07	0.00	0.07	0.00	0.00	3.76	0.60	0.00
Carpocanium sp. [TrSt1]	0.00	0.00	0.62	0.21	0.67	0.00	2.82	0.80	0.00
Centrobotrys thermophila [TrSt1]	0.00	0.00	0.00	0.00	0.00	0.00	0.81	0.07	0.00
Cornutella profunda [TrSt1]	0.13	0.08	0.01	0.38	0.40	0.00	1.91	0.20	0.00

Table 6.4 contd. ...

...Table 6.4 contd.

Species [Domain]	Arc	Ant	SPo1	SPo2	SPo3	Tra	TrSt1	TrSt2	TrSt3
Dictyophimus infabricatus [TrSt1]	0.01	0.00	0.00	0.04	0.00	0.00	2.17	0.05	0.00
Didymocyrtis tetrathalamus [TrSt1]	0.00	0.07	2.31	0.43	0.66	0.88	14.59	3.89	0.00
Euchitonia elegans-furcata [TrSt1]	0.02	0.03	1.73	0.35	0.22	0.09	9.68	1.27	0.00
Eucyrtidium acuminatum [TrSt1]	0.04	0.07	0.49	0.68	0.34	0.16	2.56	0.39	0.44
Eucyrtidium hexagonatum [TrSt1]	0.00	0.02	0.14	0.22	0.00	0.00	1.89	0.46	0.50
Lamprocyclas maritalis [TrSt1]	0.02	0.06	0.32	1.28	0.00	0.00	2.83	0.89	1.04
Lamprocyrtis nigriniae [TrSt1]	0.00	0.01	0.04	0.45	0.00	0.00	3.11	0.26	0.94
Lithocampe sp. 1 [TrSt1]	0.00	0.00	0.13	0.00	0.26	0.00	2.18	1.20	0.00
Lithopera bacca [TrSt1]	0.00	0.00	0.02	0.04	0.00	0.00	1.30	0.23	0.00
Lophospyris pentagona pentagona [TrSt1]	0.00	0.02	0.01	0.05	0.00	0.12	1.21	0.36	0.00
Otosphaera polymorpha [TrSt1]	0.00	0.00	0.00	0.00	0.00	0.00	1.82	0.11	0.00
Phormostichoartus corbula [TrSt1]	0.00	0.09	0.03	0.31	0.08	0.00	1.09	0.16	0.00
Pterocanium praetextum [TrSt1]	0.00	0.02	0.17	0.09	0.00	0.00	8.13	0.67	0.00
Pterocanium trilobum [TrSt1]	0.00	0.04	0.26	0.14	0.00	0.00	3.20	0.65	0.00
Pterocorys hertwigii [TrSt1]	0.00	0.00	0.00	0.00	0.00	0.00	0.43	0.11	0.00
Pterocorys minythorax [TrSt1]	0.00	0.00	0.04	0.03	0.00	0.00	13.43	1.00	1.94
Solenosphaera zanguebarica [TrSt1]	0.00	0.00	0.05	0.05	0.00	0.05	2.11	0.65	0.00
Spirocyrtis scalaris [TrSt1]	0.00	0.00	0.00	0.02	0.05	0.00	0.63	0.04	0.00
Spongaster tetras [TrSt1]	0.00	0.04	3.23	0.25	0.00	0.16	4.31	0.80	0.00
Theocorythium trachelium [TrSt1]	0.00	0.05	0.15	0.45	0.22	2.03	5.10	0.62	0.00
Lophophaena capito [TrSt2]	0.15	0.00	0.14	0.11	0.00	0.00	0.00	0.68	0.00
Carpocanarium papillosum [TrSt3]	0.00	0.03	0.01	0.38	0.06	0.00	0.24	0.34	1.88
Eucyrtidium erythromystax [TrSt3]	0.02	0.03	0.32	0.23	0.38	0.00	0.00	0.11	1.90
Pterocorys zancleus [TrSt3]	0.00	0.29	1.31	1.65	1.61	0.02	0.00	1.42	4.67

that Boltovskoy and Correa (2016) also used data from water-column samples, resulted in some differences in the inventories characteristic of the Domains defined. Isothermal submersion is again suggested in the present scheme by the fact that Subpolar assemblages extend far into subtropical and even tropical waters throughout the ocean, and especially in the North Atlantic (Fig. 6.4B).

6.4 Comparison Between Biogeographic Schemes

Differences between the zonations based on Foraminifera and Radiolaria are few but substantial (Fig. 6.4). First, in the Foraminifera the specific composition characteristic of polar waters is identical in the two hemispheres and largely dominated by a single species (*N. pachyderma*, but see Darling et al. 2007), whereas radiolarian assemblages are quite different in Arctic and Antarctic waters (Table 6.4). Second, the Foraminifera show a clearly defined Tropical belt that, although superimposed on the Subtropical one (TrSt in Fig. 6.2), is well characterized by several typical species (Table 6.3). In the Radiolaria, on the other hand, there is no such distinction (Fig. 6.3). Finally, radiolarians suggest the presence of a peculiar assemblage circumscribed to the eastern equatorial Pacific (TrSt3 in Figs. 6.3 and 6.4B), which is absent in Foraminifera (although a few foraminifers tend to depict high relative abundances in TrSt3, like *N. incompta*, *N. dutertrei*, and *G. menardii*).

Comparisons with previous biogeographic divisions based on pelagic organisms starting in the 1960s coincide in the separation of a warmwater from two coldwater (N and S) panoceanic belts

A Foraminifera

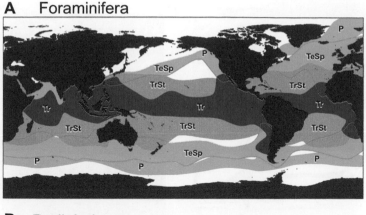

B Radiolaria

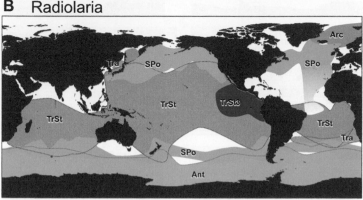

Figure 6.4: Composite image of the biogeographic Domains defined on the basis of Foraminifera and Radiolaria from surface sediment samples. See Figs. 6.2 and 6.3 for Domain names.

(Meisenheimer 1905, Steuer 1933, Bobrinskii et al. 1946). Subsequent studies refined and added detail to these early investigations identifying Subpolar, Transitional and Subtropical Domains (Fager and McGowan 1963, Frost and Fleminger 1968, Beklemishev 1969, 1971, Bé and Tolderlund 1971, Bé 1977, Pierrot-Bults and Spoel 1979, Spoel and Heyman 1983, Backus 1986, McGowan and Walker 1993, Briggs 1995, Schiebel and Hemleben 2017). A common feature of these biogeographic schemes is their general concordance in the position and number of biogeographic units, as well as the fact that they clearly mirror SST fields (Fig. 6.5K). Thus, in most cases pelagic animal assemblages (zooplankton, fishes) define largely similar geographic areas and limits, regardless of the organisms involved. In contrast, attempts at partitioning the World Ocean based on chlorophyll a fields and process-oriented attributes (mixing of the upper water layer, seasonality in chlorophyll a and physical-chemical variables; Longhurst 1998, 2007, Reygondeau et al. 2013), or on combined taxonomic data and oceanographic drivers (ocean gyres, equatorial upwellings, upwelling zones at basin edges, semi-enclosed pelagic basins; Spalding et al. 2012) yielded similar major units, but also a large number of smaller-scale divisions. Longhurst (1998) proposed a system of 12 major units (Biomes), and 51 smaller-scale divisions or Provinces (Fig. 6.5H; subsequently modified to 56 Provinces by Longhurst 2007). Spalding et al. (2012) defined a nested system of four Realms, seven Biomes, and 37 pelagic Provinces.

Ocean-wide species-based biogeographic schemes hardly ever produce such a large number of areas. This might be partly due to the fact that species census data are sparser and more heterogeneous than data on chlorophyll a, nutrients and physical properties, and the circumstance that seasonality is often poorly accounted for in species-based patterns (or not at all, as in the present survey). On the

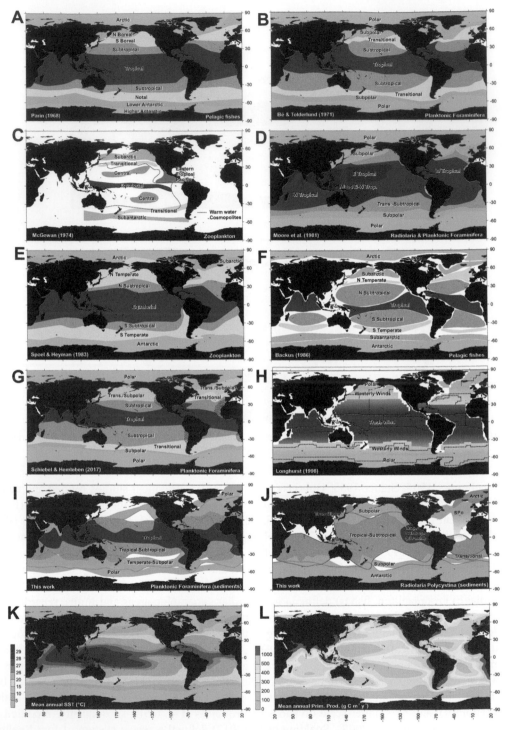

Figure 6.5: Biogeographic zonations of the World Ocean proposed by various authors (A–H), this study (I, J), SST (K, data from Boyer et al. 2013) and primary production (L, data from Ocean Productivity database). C shows the ''core'' regions of presence of 100% of the species characteristic of each of the biogeographic areas outlined. D is based on surface sediment samples of planktonic Foraminifera (Atlantic and Indian Oceans), and Radiolaria Polycystina (Pacific Ocean). Hachured areas in G denote upwelling zones. Reference and materials used for zonations are indicated at the base of each map. Panels A–H and K were reprinted from: Boltovskoy, D. and N. Correa. 2016. Biogeography of Radiolaria Polycystina (Protista) in the World Ocean. Prog. Oceanogr. 149: 82–105, © 2016 Elsevier Inc., with permission from Elsevier.

other hand, this difference suggests that the effects of the biological and physical drivers analyzed by Longhurst (1998, 2007) and Spalding et al. (2012) have minor and/or variable effects on the large time- and space scales of the geographic distribution of open-ocean species assemblages. Thus, the long-standing goal of establishing a single geographic scheme that adequately accounts for the distribution and dynamics of pelagic organisms (Spalding et al. 2012) is probably not a realistic target. Aside from information-related limitations and methodological differences, taxon-specific dissimilarities (Beklemishev et al. 1977), and between-group contrasts in diversity and affinity for different water-types (Boltovskoy 1999, Boltovskoy et al. 1999, Grady et al. 2019) are major impediments. Nevertheless, given the limitations involved, the similarities in the location and limits of the biogeographic zones proposed on the basis of a wide array of methods and organisms is striking, and the coincidences are more salient than the disagreements (Fig. 6.5). Some differences, however, are notorious, such as the segregation of the eastern tropical Pacific from the rest of tropics (Beklemishev 1969, McGowan 1974, Moore et al. 1981), compositional differences (or the lack thereof) of the tropics and the subtropics (McIntyre and Bé 1967, Bé and Tolderlund 1971, McGowan 1974), and differences in the taxonomic composition of the northern and southern Polar and Subpolar zones (Brinton 1962, Bé 1977).

6.5 The Drivers of the Patterns

The geographic distribution of oceanic planktonic organisms responds to multiple drivers, including abiotic (temperature, salinity, vertical stratification, light, seasonality, dissolved inorganic nutrients, oxygen and CO_2, currents and fronts), and biotic (food availability, predation) factors (Beklemishev 1969, Spoel and Heyman 1983, Capriulo 1990, Pierrot Bults and Spoel 1998, Ohtsuka et al. 2015, Marteinsson et al. 2016). With the exception of primary production, our understanding of the effects of food availability and grazing or predation on the global distribution of planktonic heterotrophs is meager and fragmentary. Abiotic factors have been more intensively studied, but a major problem for disentangling their influence (as well as that of primary production) on the geographic patterns of planktonic species is the fact that many of them are spatially correlated. Thus, throughout the World Ocean SST, salinity, and dissolved inorganic nutrients (N, P, Si) are significantly (mostly $p < 0.01$; either positively or negatively) correlated, and only dissolved oxygen (which is rarely limiting) shows an independent pattern (Boltovskoy and Correa 2017). We therefore centered our attention on the two variables that have most frequently been found to be associated with changes in species compositions: primary production and, in particular, SST (Rutherford et al. 1999, Morey et al. 2005, Žarić et al. 2005, Rombouts et al. 2009, Sunagawa et al. 2015, Boltovskoy and Correa 2017, Righetti et al. 2019).

For the Foraminifera, the SST of all the biogeographic partitions illustrated in Fig. 6.2 differ significantly ($p < 0.000$), except for TrSt2 vs. Tr1, whose SST is not significantly different. For primary production, differences are significant ($p < 0.05$) in all cases (Kruskal-Wallis ANOVA and Bonferroni-corrected pairwise Mann-Whitney comparisons).

Radiolarian domains are generally less consistent geographically, which also reflects in the corresponding mean SST values. Of the 36 pairwise contrasts for the nine partitions defined, seven yielded non-significant ($p > 0.05$) differences. In the subpolar and transitional areas, SPo2 does not differ from SPo1 and SPo3, and neither of the SubPolar Domains differ from the Transitional. In the tropics and subtropics, TrSt1 does not differ from TrSt2 and TrSt3. Primary production values are also only partly different between Domains. With the exception of Ant (significantly lower PP than all other Domains), Tra (significant difference with all others but Arc), and Arc (significant difference with all others but Tra and TrSt3), all other pairwise contrasts are not significant (Kruskal-Wallis ANOVA and Bonferroni-corrected pairwise Mann-Whitney comparisons).

It is unlikely that these differences in the association of biogeographic areas with SST and primary production are due to dissimilarities in the ecology of the two protist groups compared, but are most probably an artifact that stems from differences in the amount and quality of information.

For the radiolarians, of particular importance seems to be the fact that these results are based on raw data without any attempts at "correcting" the distributional ranges in order to account for the many processes that distort the sedimentary image of the living assemblages (see above and "Isothermal Submersion" below), as well as methodological constraints associated with the geographic distribution of the samples available. A more detailed analysis of radiolarian biogeography based on sediment and water-column materials, with adjustments to minimize these problems, indicates that the distribution of radiolarian biogeographic Domains is influenced chiefly by SST, and to a lesser extent by nutrients (P, N) and chlorophyll a fields (Boltovskoy and Correa 2016). Thus, while these maps (Fig. 6.4) adequately summarize the patterns in the surface sediments, their representativeness of the living populations is less precise.

6.6 Latitudinal Variations in Species Richness

Most surveys published until the late 1960s (Thorson 1957, Stehli et al. 1969), as well as later reviews (Hillebrand 2004, Righetti et al. 2019) found that species richness decreases more or less evenly from the tropics to the poles. However, in the last decades several investigations based on marine organisms (including planktonic Foraminifera) concluded that species richness peaks in the middle latitudes, dropping at the equator (McGowan and Walker 1993, Rutherford et al. 1999, Rombouts et al. 2009, Tittensor et al. 2010, Yasuhara et al. 2012, Schiebel and Hemleben 2017), and bimodal patterns were also reported for other traits, like copepod body and offspring size and myelination (Brun et al. 2016).

Using a subset of the foraminiferal data compiled by Siccha and Kucera (2017) (the Brown University Foraminiferal Data Base; Prell et al. 1999), Boltovskoy and Correa (2017) compared the latitudinal gradients in species richness of Radiolaria and Foraminifera concluding that the latter, indeed, show a trough at the equator, whereas radiolarian species numbers are highest at 0°. A reassessment of this trend using the much larger database of Siccha and Kucera (2017) confirms this contrast. Foraminiferal species richness shows a plateau around the equator, whereas radiolarian assemblages increase in richness more or less monotonically from the poles to the equator (Fig. 6.6A and B).

Although this comparison has some caveats, including the fact that sampling coverage is dissimilar for the two groups (see Boltovskoy and Correa 2017), and taxonomic uncertainties are significantly higher for radiolarians, it is likely that the contrast is not a methodological or information-related artifact. Peak species richness values offset from the equator can be explained by several mechanisms. Because sedimentary materials integrate the output of a three-dimensional space (latitude, longitude and depth) onto a two-dimensional plane (the ocean floor), the remains of organisms that inhabit different depth layers are collapsed on the bottom. The middle latitudes are affected the most by seasonal changes, whereby the same areas are occupied by different assemblages throughout the year, whose integration in the sedimentary record (and in time- and depth-averaging water-column collections) reflects long-term trends, concealing their dynamic shifts and the circumstance that many of the species involved do not coexist there in time (Angel 1993, Righetti et al. 2019). Further, isothermal submersion of coldwater species towards the tropics ends up contributing sizable numbers of polar and subpolar remains to areas where these species are only found at depth, and are likely mostly represented by terminal, unviable populations (Boltovskoy and Correa 2017) (Fig. 6.7C). Interestingly, as opposed to several zooplanktonic taxa, which show an equatorial trough in species richness, marine phytoplankton depicts highest species richness values at the equator (Barton et al. 2010, Righetti et al. 2019; Fig. 6.6C) (see also Vallina et al. 2014, Rodríguez-Ramos et al. 2015). Because of their dependence on light, isothermal submersion is not a viable survival alternative for photosynthetic organisms, which supports the assumption that their diversity-latitude pattern is not affected by this mechanism as those of many zooplankton. Further, Righetti et al. (2019) also found that phytoplankton species seasonal turnover rates are highest around 40° N and 40° S, thus implying that time-averaged (annual or multiannual) inventories can be inflated in that latitudinal belt.

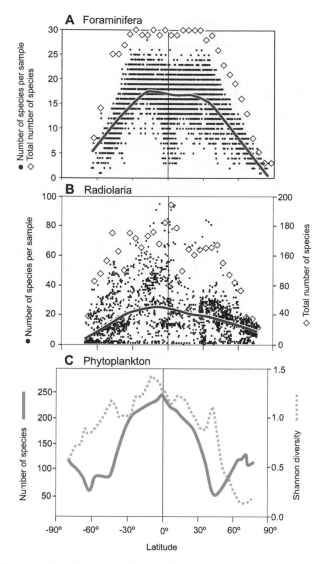

Figure 6.6: A and B. Numbers of species per sample as a function of latitude (dots) and total species recorded in 5° latitude bands worldwide (large diamonds, right-hand scale in B). Totals for Foraminifera: 4205 surface sediment samples, 36 species; for Radiolaria: 1973 surface sediment samples, 239 species. Lines denote LOESS regressions with polynomial fitting (smoothing factor = 0.5). C. Numbers of phytoplankton species (total: 1298 species of seven phytoplanktonic phyla; adapted from Righetti et al. 2019) and phytoplankton diversity (adapted from Barton et al. 2010) as a function of latitude worldwide.

Another important factor is the Mid-Domain Effect, a null model initially proposed by Colwell and Hurtt (1994), and later reformulated by Brayard et al. (2005). The Mid-Domain Effect demonstrates that the random distribution of species in a space bounded by the continents, the poles, and the equatorial belt produces highest species richness values at the center of the domain (i.e., in the middle latitudes), where range overlaps are greatest (Beaugrand et al. 2013, Letten et al. 2013).

The reasons why radiolarians and foraminifers show different latitudinal species richness gradients are still unclear. Historical and evolutionary explanations have been proposed (Powell and Glazier 2017), but the latitude-diversity patterns derived for Radiolaria in that survey have widely overlapping confidence intervals and do not agree with current information for this group. Aside from methodological and information-related confounding factors, for sedimentary materials differences

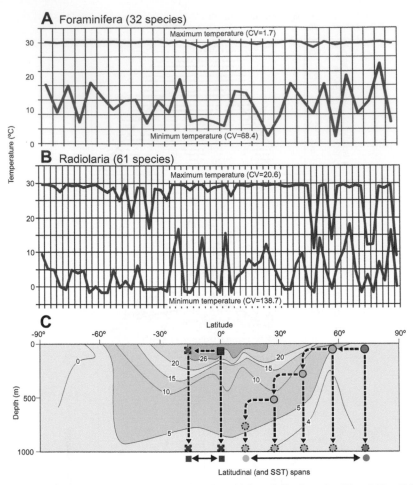

Figure 6.7: Maximum and minimum SST values of the sites where 32 foraminiferal species (A) and 61 radiolarian species (B) were recorded. Data used are restricted to samples where the relative abundance of the species are higher than their overall mean abundances in the entire respective collections, and limited to species present in > 38 samples (Foraminifera), and > 100 samples (Radiolaria). Total (surface sediment) samples covered, Foraminifera: 4205, Radiolaria: 1303. (C) Schematic illustration of the isothermal submersion effect, a mechanism through which the fossilizable remains of coldwater species extend their latitudinal distribution equatorwards, superimposed on the vertical temperature profile of the Pacific Ocean at around 160° W (in °C, simplified from Reid et al. 1978). Squares: warmwater species; circles: coldwater species: darker fill: home range; lighter fill: expatriation range; solid outline: living individuals; dashed outline: dead individuals. CV: Coefficient of Variation, SST: Sea Surface Temperature.

in the resistance to dissolution of biogenic silica (Radiolaria) and calcium carbonate (Foraminifera) as a function of water depth and temperature may play a key role. Inter-specific differences in preservation potential in association with water temperature preferences are dissimilar as well. While in Foraminifera, coldwater species are generally more solution-resistant than the warmwater ones (Vincent and Berger 1981), no such relationship seems to exist in Radiolaria (Johnson 1974, De Wever et al. 2001). This is of particular importance because the species responsible for the bimodal diversity peaks in Foraminifera are chiefly cold and temperate water forms (Boltovskoy and Correa 2017). The specific diversity of the two groups is sharply dissimilar, which may suggest that the diversity of the Foraminifera reaches a saturation level at ~ 30° N and 30° S (Fig. 6.6A), whereas radiolarians, which are much more specious, have no such limitation. Interestingly, global species richness has been identified as one of the factors that significantly affect the slope of the species richness vs. latitude gradient worldwide (Hillebrand 2004). Only two foraminifers have MWSST above 27°C, as opposed to 29 radiolarians. Also, the proportions of their overall inventories restricted to- or characteristic of different water-mass types are different. For example, only 6% of the foraminiferal species have MWSST values below 15°C, as opposed to 40% for the radiolarians.

6.7 Isothermal Submersion

Isothermal submersion is the mechanism by which the expatriated populations of coldwater and temperate species that inhabit shallow layers at high latitudes move equatorwards and sink deeper in the water-column where they can survive longer (and displace farther) (Briggs 1995, Alvariño 1964, Kling 1976, Wiebe and Boyd 1978). Thus, these expatriates end up dying and, in the case of fossilizable organisms, feeding the bottom deposits of lower-latitude areas than those occupied by their living populations (Boltovskoy 1988, 2017, Boltovskoy and Correa 2016, 2017).

In the two databases analyzed, the biasing effects of isothermal submersion are supported by the following observations. When tallying the maximum and minimum SST of the sites where the species analyzed were recorded, minimum temperatures vary widely, whereas maximum temperatures are much more homogeneous and mostly coincident with the highest SST oceanwide. For the Foraminifera, the variability in the minimum SST is around 40 times higher than the maximum SST, while for the Radiolaria it is ~ seven times higher (Fig. 6.7A, B). The fact that coldwater species are chiefly responsible for this effect is supported by a comparison of the temperature span vs. the MWSST of the species examined. For both groups there is a significant trend for wider overall thermal spans with lower species-specific MWSST (i.e., the SST where the species peaks in abundance) (Fig. 6.8). Warmwater species have smaller thermal spans than coldwater species because their expatriation toward colder-water areas leads to death and sedimentation in close vicinity of their living range, whereas coldwater species can survive longer and displace farther toward the tropics taking advantage of the thermal stratification (Fig. 6.7C). Comparison of the thermal ranges of 25 foraminiferal species on the basis of water-column versus sedimentary materials (see Fig. 17 in Boltovskoy and Correa 2016) confirms this conclusion. The higher presence of coldwater species at low latitude sediments than that of warmwater species at high latitudes is particularly noticeably when water-column and sedimentary materials are compared, especially in the bands around 30° N–40° N and 30° S–40° S, which is where most planktonic species turnovers typically occur worldwide (see Fig. 12 in Boltovskoy and Correa 2016).

This trend is very consistent for all the Foraminifera analyzed (Fig. 6.8A), and for ~ 90% of the Radiolaria, yet in the latter there are several noticeable outliers (Fig. 6.8B). Of these, four are virtually restricted to northern polar and subpolar waters (*Artostrobus jorgenseni, Ceratospyris borealis, Lithocampe platycephala* and *Pseudodictyophimus gracilipes*), and two to the Southern Ocean (*A. denticulata-strelkovi* and *Triceraspyris antarctica*) but, despite their clear association with cold waters (MWSST ~ 3°C–7°C), their occurrence in warmwater areas is comparatively low, which may suggest that their capabilities of adjusting their preferred water temperature by sinking are physiologically limited.

The trends discussed are generally robust, and they do not seem to be affected by the depths at which these species live. Several foraminifers are known to inhabit intermediate waters (*G. bulloides, Globorotalia truncatulinoides, Globoquadrina conglomerata, Globorotalia crassaformis, Sphaeroidinella dehiscens, G. tumida, C. nitida*; Berger 1969, Bé 1977, Schiebel and Hemleben 2017), yet they do not seem to deviate from the overall trend. For the radiolarians the information available is scarcer, but most of those included in the analysis are likely surface or subsurface dwelling, at least in warmwater areas (abundance maxima are generally deeper in colder waters, Boltovskoy 2017). *Pseudodictyophimus gracilipes, Actinomma leptodermum, Antarctissa denticulata-strelkovi*, and *Ceratospyris borealis*, on the other hand, are usually found in deeper waters (Kling and Boltovskoy 1995, Boltovskoy et al. 2010, Boltovskoy and Correa 2017), which may explain the fact that they do not follow the overall pattern (Fig. 6.8B).

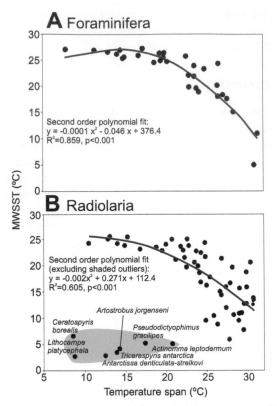

Figure 6.8: SST span of 32 foraminiferal (**A**) and 61 radiolarian (**B**) species (see caption to Fig. 6.7 for selection criteria) in the surface sediments as a function of their preferred thermal regimes, as indicated by their corresponding MWSST (Mean Weighed Sea Surface Temperature, see text for explanation). Graphs and regression lines are based on raw data, regression equations, R and p values are based on Box-Cox transformed (normalized) data. In panel B, the regression excludes the outliers (shaded).

6.8 Caveats, Concluding Remarks and Future Challenges

The two databases used for this study share the condition of having been critically assembled from multiple primary sources, but the fact that they deal with different organisms imposes several significant differences. Planktonic foraminifers are much less diverse than radiolarians, the volume of information available for them, including their geographic distribution, is considerably higher, and their species-level taxonomy has historically been much more stable. Further, foraminiferal distribution surveys routinely cover all or a very large proportion of the species described, whereas those on Radiolaria very often limit their scope to the 20–50 most common morphotypes only. A major problem with such data is associated with the absence of species because null records may stem from two different circumstances: the species was looked for but not found, or the species was present but was excluded from the counting categories used. Restricting the species used in the analyses to the ones most often identified in the literature (as done in this survey) mitigates this problem, but obviously is a compromise, rather than a solution.

Another important issue is that, because the sedimentary record is skewed by a number of processes, sediment-based records of shell-bearing protists are not the best source for interpreting the distribution patterns of their (mostly upper-layer) living populations. Fragmentation and dissolution of the biogenic remains, reworking of sediments, winnowing and lateral advection of shells, seasonally and interannually variable output rates, and the integration of vertically heterogeneous assemblages are among the most important mechanisms that distort the sedimentary record of the planktonic patterns. On the other hand, utilization of bottom deposits also has advantages: thanatocoenoses require smaller sample sizes, materials are more readily available, and they eliminate the small time-scale variability which obscures long-term trends (Boltovskoy 1994). Additionally, a major advantage is the availability of data, which for these protists is considerably higher for sedimentary than for water-column materials.

Thus, although in general terms the schemes shown in Fig. 6.4 do reflect the biogeography of the living populations, some distortion is likely, especially in the equatorwards reaches of the coldwater domains.

On the other hand, for paleoenvironmental reconstructions use of sediment-based biogeographic data does not present a problem. The aim of microfossil-based palaeoceanographic investigations is to identify surface sedimentary assemblages tightly coupled with (usually upper layer) environmental variables (e.g., temperature, productivity), and then use this relationship to investigate changes in these same parameters back in time, as derived from changing microfossil compositions in downcore samples (Guiot and de Vernal 2007). Thus, as long as the relationship between recent sedimentary assemblage composition and the variable of interest is strong, its application to downcore data is in principle unrestrained by considerations of the changes involved in the passage from the plankton to the sediments, except when no-analog conditions are involved (Kucera 2007).

The biogeographic partitions illustrated in this study (Figs. 6.4 and 6.5) are obviously dependent upon the methodological approaches used, so one should not expect to arrive at exactly the same results if a different set of instruments or data were employed. However, the fact that the zonations are in line with those obtained for these and other groups based on wide variety of sources of information and methods (Fig. 6.5) suggests that potential differences would be restricted to details, rather than affect the large picture. On the other hand, our study is based on morphospecies only, many of which conceivably include several species or genotypes with significant ecologic and distributional differences, both in Foraminifera (de Vargas et al. 1999, Kucera and Darling 2002, Darling et al. 2007, Darling and Wade 2008, Aurahs et al. 2009, Seears et al. 2012, Ujiie et al. 2012), and in Radiolaria (Ishitani et al. 2012, 2014, Biard et al. 2015, Darling et al. 2017). Although the opposite situation has also been described (i.e., morphologically dissimilar organisms described as new species were found to be genetically indistinguishable: André et al. 2013, Dolan 2016), it is likely that the number of morphospecies underestimates the true genetic diversity of both these groups (Appeltans et al. 2012). Thus, it is conceivable that further refinements in the taxonomy of Foraminifera, and especially Radiolaria, will modify our current understanding of their worldwide distribution patterns, as already shown in some surveys (de Vargas et al. 1999, Darling et al. 2017).

Acknowledgments

This work was supported by grant PICT 2015 2598, from the Agencia Nacional de Promoción Científica y Tecnológica (Argentina).

References

Adl, S.M., D. Bass, C.E. Lane, J. Lukes, C.L. Schoch, A. Smirnov et al. 2018. Revisions to the classification, nomenclature, and diversity of eukaryotes. J. Eukaryot. Microbiol. 66: 4–119.
Alvariño, A. 1964. Bathymetric distribution of chaetognaths. Pac. Sci. 18: 64–82.
Anderson, O.R. 1983. Radiolaria. Springer, New York (USA).
André, A., A. Weiner, F. Quillévéré, R. Aurahs, R. Morard, C.J. Douady et al. 2013. The cryptic and the apparent reversed: lack of genetic differentiation within the morphologically diverse plexus of the planktonic foraminifer *Globigerinoides sacculifer*. Paleobiology 39: 21–39.
Angel, M.V. 1993. Biodiversity of the pelagic ocean. Conserv. Biol. 7: 760–772.
Appeltans, W., S.T. Ahyong, G. Anderson, M.V. Angel, T. Artois, N. Bailly et al. 2012. The magnitude of global marine species diversity. Curr. Biol. 22: 2189–2202.

Aurahs, R., G.W. Grimm, V. Hemleben, C. Hemleben and M. Kucera. 2009. Geographical distribution of cryptic genetic types in the planktonic foraminifer *Globigerinoides ruber*. Mol. Ecol. 18: 1692–706.

Backus, R.H. 1986. Biogeographic boundaries in the open ocean. pp. 9–13. *In*: A.C. Pierrot-Bults, S. van der Spoel, B.J. Zahuranec and R.K. Johnson [eds.]. Pelagic Biogeography. UNESCO Press, Paris (France).

Barton, A.D., S. Dutkiewicz, G. Flierl, J. Bragg and M.J. Follows. 2010. Patterns of diversity in marine phytoplankton. Science 327: 1509–11.

Bé, A.W.H. and D.S. Tolderlund. 1971. Distribution and ecology of living planktonic Foraminifera in surface waters of the Atlantic and Indian Oceans. pp. 105–149. *In*: B.M. Funnel and W.R. Riedel [eds.]. The Micropaleontology of Oceans. Cambridge University Press, Cambridge (UK).

Bé, A.W.H. 1977. An ecological, zoogeographic and taxonomic review of recent planktonic Foraminifera. pp. 1–100. *In*: A.T.S. Ramsay [ed.]. Oceanic Micropaleontology. Academic Press, London (UK).

Beaugrand, G., I. Rombouts and R.R. Kirby. 2013. Towards an understanding of the pattern of biodiversity in the oceans. Glob. Ecol. Biogeogr. 22: 440–449.

Beklemishev, C.W. 1969. Ekologiya i biogeografiya pelagiali [Ecology and biogeography of the open ocean]. Nauka, Moscow (Russia).

Beklemishev, C.W. 1971. Distribution of plankton as related to micropaleontology. pp. 75–87. *In*: B.M. Funnel and W.R. Riedel [eds.]. The Micropaleontology of Oceans. Cambridge University Press, Cambridge (UK).

Beklemishev, C.W., N.V. Parin and G.I. Semina. 1977. Biogeografiya okeana. Pelagial [Biogeography of the ocean. Pelagial]. pp. 219–261. *In*: M.E. Vinogradov [ed.]. Okeanologiya. Biologiya Okeana. 1. Biologicheskaya struktura okeana [Oceanology. Biology of the ocean. 1. Biological structure of the ocean]. Nauka, Moscow (Russia).

Berger, W.H. 1969. Ecologic patterns of living planktonic Foraminifera. Deep-Sea Res. 16: 1–24.

Berger, W.H. 1974. Deep-sea sedimentation. pp. 213–241. *In*: C.A. Burk and C.I. Drake [eds.]. The Geology of Continental Margins. Springer, New York (USA).

Biard, T., L. Pillet, J. Decelle, C. Poirier, N. Suzuki and F. Not. 2015. Towards an integrative morpho-molecular classification of the Collodaria (Polycystinea, Radiolaria). Protist 166: 374–388.

Bobrinskii, N.A., N.A. Zenkevitch and Y.A. Birstein. 1946. Geografiya zhivotnykh [The geography of animals]. Sovetskaya Nauka, Moscow (Russia).

Boltovskoy, D. and W.R. Riedel. 1987. Polycystine Radiolaria of the California current region: seasonal and geographic patterns. Mar. Micropaleontol. 12: 65–104.

Boltovskoy, D. 1988. Equatorward sedimentary shadows of near-surface oceanographic patterns. Speculations Sci. Technol. 11: 219–232.

Boltovskoy, D. 1994. The sedimentary record of pelagic biogeography. Prog. Oceanogr. 34: 135–160.

Boltovskoy, D. [ed.]. 1999. South Atlantic Zooplankton. Backhuys Publishers, Leiden (The Netherlands).

Boltovskoy, D., M.J. Gibbons, L. Hutchings and D. Binet. 1999. General biological features of the South Atlantic. pp. 1–42. *In*: D. Boltovskoy [ed.]. South Atlantic Zooplankton. Backhuys Publishers, Leiden (The Netherlands).

Boltovskoy, D., S.A. Kling, K. Takahashi and K. Bjørklund. 2010. World atlas of distribution of Recent Polycystina (Radiolaria). Palaeontologia Electronica 13: 1–229.

Boltovskoy, D. and N. Correa. 2016. Biogeography of Radiolaria Polycystina (Protista) in the World Ocean. Prog. Oceanogr. 149: 82–105.

Boltovskoy, D. 2017. Vertical distribution patterns of Radiolaria Polycystina (Protista) in the World Ocean: living ranges, isothermal submersion, and settling shells. J. Plankton Res. 39: 330–349.

Boltovskoy, D. and N. Correa. 2017. Planktonic equatorial diversity troughs: fact or artifact? Latitudinal diversity gradients in Radiolaria. Ecology 98: 112–124.

Boltovskoy, D., O.R. Anderson and N. Correa. 2017. Radiolaria and Phaeodaria. pp. 731–763. *In*: J.M. Archibald, A.G.B. Simpson and C.H. Slamovits [eds.]. Handbook of the Protists. Springer, Cham (Switzerland).

Boyer, T.P., J.I. Antonov, O.K. Baranova, C. Coleman, H.E. Garcia, A. Grodsky et al. 2013. World Ocean Database 2013. NOAA, Silver Spring (USA).

Brayard, A., G. Escarguel and H. Bucher. 2005. Latitudinal gradient of taxonomic richness: combined outcome of temperature and geographic mid-domains effects? J. Zool. Syst. Evol. Res. 43: 178–188.

Briggs, J.C. 1995. Global Biogeography. Elsevier, Amsterdam (The Netherlands).

Brinton, E. 1962. The distribution of Pacific euphausiids. Bull. Scripps Inst. Oceanogr., Univ. Calif. 8: 51–270.

Brun, P., M.R. Payne and T. Kiørboe. 2016. Trait biogeography of marine copepods—an analysis across scales. Ecol. Lett. 19: 1403–1413.

Capriulo, G.M. [ed.]. 1990. Ecology of Marine Protozoa. Oxford University Press, Oxford (UK).

Clarke, I.R. and R.M. Warwick. 2001. Change in Marine Communities: An Approach to Statistical Analysis and Interpretation. Primer-E Ltd., Plymouth (UK).

Clarke, K.R. 1993. Non-parametric multivariate analysis of changes in community structure. Aust. J. Ecol. 18: 117–143.

Colwell, R.K. and G.C. Hurtt. 1994. Nonbiological gradients in species richness and a spurious Rapoport effect. Amer. Natur. 144: 570–595.

Darling, K.F., M. Kucera and C.M. Wade. 2007. Global molecular phylogeography reveals persistent Arctic circumpolar isolation in a marine planktonic protist. Proc. Natl. Acad. Sci. U.S.A. 104: 5002–7.

Darling, K.F. and C.M. Wade. 2008. The genetic diversity of planktic foraminifera and the global distribution of ribosomal RNA genotypes. Mar. Micropaleontol. 67: 216–238.

Darling, K.F., C.M. Wade, M. Siccha, G. Trommer, H. Schulz, S. Abdolalipour et al. 2017. Genetic diversity and ecology of the planktonic foraminifers *Globigerina bulloides*, *Turborotalita quinqueloba* and *Neogloboquadrina pachyderma* off the Oman margin during the late SW Monsoon. Mar. Micropaleontol. 137: 64–77.

de Vargas, C., R. Norris, L. Zaninetti, S.W. Gibb and J. Pawlowski. 1999. Molecular evidence of cryptic speciation in planktonic foraminifers and their relation to oceanic provinces. Proc. Natl. Acad. Sci. U.S.A. 96: 2864–2868.

de Vargas, C., A.G. Sáez, L.K. Medlin and H.R. Thierstein. 2004. Super-Species in the calcareous plankton. pp. 251–298. *In*: H.R. Thierstein and J.R. Young [eds.]. Coccolithophores: From Molecular Proccesses to Global Impact. Springer, Berlin (Germany).

de Vargas, C., S. Audic, N. Henry, J. Decelle, F. Mahé, R. Logares et al. 2015. Eukaryotic plankton diversity in the sunlit ocean. Science 348: 1–11.

De Wever, P., P. Dumitrica, J. Caulet, C. Nigrini and M. Caridroit. 2001. Radiolarians in the Sedimentary Record. Gordon and Breach, Amsterdam (The Netherlands).

Dolan, J.R. 2016. Planktonic protists: little bugs pose big problems for biodiversity assessments. J. Plankton Res. 38: 1044–1051.

Fager, E.W. and J.A. McGowan. 1963. Zooplankton species groups in the North Pacific. Science 140: 453–460.

Frost, B. and A. Fleminger. 1968. A revision of the genus *Clausocalanus* (Copepoda: Calanoida) with remarks on distribution patterns in diagnostic characters. Bull. Scripps Inst. Oceanogr., Univ. Calif. 12: 1–235.

Grady, J.M., B.S. Maitner, A.S. Winter, K. Kaschner, D.P. Tittensor, S. Record et al. 2019. Metabolic asymmetry and the global diversity of marine predators. Science 363: eaat4220.

Guiot, J. and A. de Vernal. 2007. Transfer functions: methods for quantitative paleoceanography based on microfossils. pp. 523–563. *In*: C. Hillaire-Marcel and A. de Vernal [eds.]. Proxies in Late Cenozoic paleoceanography. Elsevier, Amsterdam (The Netherlands).

Hammer, Ø., D.A.T. Harper and P.D. Ryan. 2001. PAST: Paleontological statistics software package for education and data analysis. Palaeontologia Electronica 4: 1–9.

Hartwell, S.I. and L.W. Claflin. 2005. Cluster analysis of contaminated sediment data: Nodal analysis. Environ. Toxicol. Chem. 24: 1816–1834.

Hemleben, C., M. Spindler and O.R. Anderson. 1989. Modern Planktonic Foraminifera. Springer, New York (USA).

Hillebrand, H. 2004. On the generality of the latitudinal diversity gradient. Amer. Natur. 163: 191–211.

Ishitani, Y., Y. Ujiié, C. de Vargas, F. Not and K. Takahashi. 2012. Two distinct lineages in the radiolarian Order Spumellaria having different ecological preferences. Deep-Sea Res. 61-64: 172–178.

Ishitani, Y., Y. Ujiié and K. Takishita. 2014. Uncovering sibling species in Radiolaria: evidence for ecological partitioning in a marine planktonic protist. Mol. Phylogen. Evol. 78: 215–222.

Johnson, T.C. 1974. The dissolution of siliceous microfossils in surface sediments of the eastern tropical Pacific. Deep-Sea Res. 21: 851–864.

Kling, S.A. 1976. Relation of radiolarian distribution to subsurface hydrography in the North Pacific. Deep-Sea Res. 23: 1043–1058.

Kling, S.A. and D. Boltovskoy. 1995. Radiolarian vertical distribution patterns across the southern California Current. Deep-Sea Res. 42: 191–231.

Kucera, M. and K.F. Darling. 2002. Cryptic species of planktonic foraminifera: their effect on palaeoceanographic reconstructions. Phil. Trans. R. Soc. Lond. A 360: 695–718.

Kucera, M. 2007. Planktonic Foraminifera as tracers of past oceanic environments. pp. 213–262. *In*: C. Hillaire-Marcel and A. de Vernal [eds.]. Proxies in Late Cenozoic paleoceanography. Elsevier, Amsterdam.

Lazarus, D.B., N. Suzuki, J.-P. Caulet, C. Nigrini, I. Goll, R. Goll et al. 2015. An evaluated list of Cenozoic-Recent radiolarian species names (Polycystinea), based on those used in the DSDP, ODP and IODP deep-sea drilling programs. Zootaxa 3999: 301–333.

Letten, A.D., S.K. Lyons, A.T. Moles and R. Pearson. 2013. The mid-domain effect: it's not just about space. J. Biogeogr. 40: 2017–2019.

Lisitzin, A.P. 1971a. Distribution of siliceous microfossils in suspension and in bottom sediments. pp. 173–195. *In*: B.M. Funnell and W.R. Riedel [eds.]. The Micropaleontology of Oceans. Cambridge University Press, Cambridge (UK).

Lisitzin, A.P. 1971b. Distribution of carbonate microfossils in suspension and in bottom sediments. pp. 197–218. *In*: B.M. Funnell and W.R. Riedel [eds.]. The Micropaleontology of Oceans. Cambridge University Press, Cambridge (UK).

Longhurst, A. 1998. Ecological Geography of the Sea. Academic Press, San Diego (USA).

Longhurst, A. 2007. Ecological Geography of the Sea. Academic Press, Burlington (USA).

Marteinsson, V.P., R. Groben, E. Reynisson and P. Vannier. 2016. Biogeography of marine microorganisms. pp. 187–207. *In*: L.J. Stal and M.S. Cretoiu [eds.]. The Marine Microbiome. Springer, Cham (Switzerland).

McGowan, J.A. 1974. The nature of oceanic ecosystems. pp. 9–28. *In*: C.B. Miller [ed.]. The Biology of the Oceanic Pacific. Oregon State University Press, Corvallis (USA).

McGowan, J.A. and P.W. Walker. 1993. Pelagic diversity patterns. pp. 203–214. *In*: R.E. Ricklefs and D. Schluter [eds.]. Species Diversity in Ecological Communities. Historical and Geographic Perspectives. University of Chicago Press, Chicago (USA).

McIntyre, A. and A.W.H. Bé. 1967. Modern Coccolithophoridae of the Atlantic Ocean-I. Placoliths and Cyrtholiths. Deep-Sea Res. 14: 561–597.

Meisenheimer, J. 1905. Die tiergeographischen Regionen des Pelagials, auf Grund der Verbreitung der Pteropoden. Zool. Anz. 29: 155–163.

Moore, T.C., W.H. Hutson, N. Kipp, J.D. Hays, W. Prell, P. Thompson et al. 1981. The biological record of the ice-age ocean. Palaeogeogr. Palaeoclimatol. Palaeoecol. 35: 357–370.

Morey, A.E., A.C. Mix and N.G. Pisias. 2005. Planktonic foraminiferal assemblages preserved in surface sediments correspond to multiple environment variables. Quaternary Science Reviews 24: 925–950.

Morisita, M. 1959. Measuring of interspecific association and similarity between communities. Memories of the Faculty of Science, Kyushu University, Series E (Biology) 3: 65–80.

Ohtsuka, S., T. Suzaki, T. Horiguchi, N. Suzuki and F. Not [eds.]. 2015. Marine Protists. Diversity and Dynamics. Springer, Tokyo (Japan).

Parin, N.V. 1968. Ichthyofauna okeanskoi epipelagiali [Ichthyofauna of the oceanic epipelagal]. Nauka, Moscow (Russia).

Pielou, E.C. 1975. Ecological Diversity. Wiley, New York (USA).

Pierrot-Bults, A.C. and S. Spoel [eds.]. 1979. Zoogeography and Diversity of Plankton. Bunge, Utrecht (The Netherlands).

Pierrot Bults, A.C. and S.v.d. Spoel [eds.]. 1998. Pelagic Biogeography ICoPB II. Proceedings of the 2nd International Conference. Final report of SCOR/IOC Working Group 93 "Pelagic Biogeography". UNESCO Press, Paris (France).

Powell, M.G. and D.S. Glazier. 2017. Asymmetric geographic range expansion explains the latitudinal diversity gradients of four major taxa of marine plankton. Paleobiology 43: 196–208.

Prell, W., A. Martin, J. Cullen and M. Trend. 1999. The Brown University Foraminiferal Data Base. IGBP PAGES/World Data Center-A for Paleoclimatology. Data Contribution Series # 1999-027. NOAA/NGDC Paleoclimatology Program, Boulder (USA).

Reid, J.L., E. Brinton, A. Fleminger, E.L. Venrick and J.A. McGowan. 1978. Ocean circulation and marine life. pp. 65–130. *In*: H. Charnock and G. Deacon [eds.]. Advances in Oceanography, Plenum Press, New York (USA).

Reygondeau, G., A. Longhurst, E. Martinez, G. Beaugrand, D. Antoine and O. Maury. 2013. Dynamic biogeochemical provinces in the global ocean. Global Biogeochem. Cycles 27: 1046–1058.

Righetti, D., M. Vogt, N. Gruber, A. Psomas and N.E. Zimmermann. 2019. Global pattern of phytoplankton diversity driven by temperature and environmental variability. Science Advances 5: eaau6523.

Rodríguez-Ramos, T., E. Marañón and P. Cermeño. 2015. Marine nano- and microphytoplankton diversity: redrawing global patterns from sampling-standardized data. Glob. Ecol. Biogeogr. 24: 527–538.

Rombouts, I., G. Beaugrand, F. Ibanez, S. Gasparini, S. Chiba and L. Legendre. 2009. Global latitudinal variations in marine copepod diversity and environmental factors. Proceedings of the Royal Society, Biological Sciences 276: 3053–62.

Rutherford, S., S. D'Hondt and W. Prell. 1999. Environmental controls on the geographic distribution of zooplankton diversity. Nature 400: 749–753.

Schiebel, R. and C. Hemleben. 2017. Planktic Foraminifers in the Modern Ocean. Springer, Berlin (Germany).

Seears, H.A., K.F. Darling and C.M. Wade. 2012. Ecological partitioning and diversity in tropical planktonic foraminifera. BMC Evol. Biol. 12: 54.

Shannon, C.E. and W. Weaver. 1963. The Mathematical Theory of Communication. University of Illinois Press, Urbana (US).

Siccha, M. and M. Kucera. 2017. ForCenS, a curated database of planktonic foraminifera census counts in marine surface sediment samples. Scientific Data 4: 170109.

Sierra, R., M.V. Matz, G. Aglyamova, L. Pillet, J. Decelle, F. Not et al. 2013. Deep relationships of Rhizaria revealed by phylogenomics: a farewell to Haeckel's Radiolaria. Mol. Phylogen. Evol. 67: 53–9.

Spalding, M.D., V.N. Agostini, J. Rice and S.M. Grant. 2012. Pelagic provinces of the world: A biogeographic classification of the world's surface pelagic waters. Ocean Coast. Manage. 60: 19–30.

Spoel, S. and R.P. Heyman. 1983. A comparative atlas of zooplankton. Biological Patterns in the Oceans. Springer, Berlin (Germany).

Stehli, F.G., R.G. Douglas and N.D. Newell. 1969. Generation and maintenance of gradients of taxonomic diversity. Science 164: 947–949.

Steuer, A. 1933. Zur planmassigen Erforschung der geographischen Verbreitung des Haliplanktons, besonders der Copepoden. Zoogeographica 1: 269–302.

Sunagawa, S., L.P. Coelho, S. Chaffron, J.R. Kultima, K. Labadie, G. Salazar et al. 2015. Structure and function of the global ocean microbiome. Science 348: 207–310.

Thorson, G. 1957. Bottom communities. pp. 461–534. *In*: J.W. Hedgpeth [ed.]. Treatise on Marine Ecology and Palaeoecology. Memoirs of the Geological Society of America, Washington DC (USA).

Tittensor, D.P., C. Mora, W. Jetz, H.K. Lotze, D. Ricard, E.V. Berghe et al. 2010. Global patterns and predictors of marine biodiversity across taxa. Nature 466: 1098–101.

Ujiie, Y., T. Asami, T. de Garidel-Thoron, H. Liu, Y. Ishitani and C. de Vargas. 2012. Longitudinal differentiation among pelagic populations in a planktic foraminifer. Ecol. Evol. 2: 1725–37.

Vallina, S.M., M.J. Follows, S. Dutkiewicz, J.M. Montoya, P. Cermeno and M. Loreau. 2014. Global relationship between phytoplankton diversity and productivity in the ocean. Nat. Commun. 5: 4299.

Vincent, E. and W.H. Berger. 1981. Planktonic Foraminifera and their use in paleoceanography. pp. 1025–1119. *In*: C. Emiliani [ed.]. The Oceanic Lithosphere. The Sea. Wiley, Hoboken (USA).

Wiebe, P.H. and S.H. Boyd. 1978. Limits of *Nematoscelis megalops* in the Northwestern Atlantic in relation to Gulf Stream cold-core rings. Part I. Horizontal and vertical distributions. J. Mar. Res. 36: 119–142.

Yasuhara, M., G. Hunt, H.J. Dowsett, M.M. Robinson and D.K. Stoll. 2012. Latitudinal species diversity gradient of marine zooplankton for the last three million years. Ecol. Lett. 15: 1174–1179.

Žarić, S., B. Donner, G. Fischer, S. Mulitza and G. Wefer. 2005. Sensitivity of planktic foraminifera to sea surface temperature and export production as derived from sediment trap data. Mar. Micropaleontol. 55: 75–105.

CHAPTER *7*

Metazooplankton Dynamics in Coastal Upwelling Systems

Rita F.T. Pires[1,2,]* and *Antonina dos Santos*[1,2]

7.1 Introduction

Upwelling is a worldwide wind-driven oceanographic process associated with the advection of deeper and nutrient-rich waters into the upper euphotic layer (Pond and Pickard 1983, Chavez and Messié 2009), promoting phytoplankton growth and biological productivity. Upwelling dynamics and associated oceanographic features drive important exchanges between oceanic and coastal regions (Simpson and Sharples 2012). Nutrient-rich upwelled waters allow high levels of primary and secondary production, manifested by high biomass of both phytoplankton and zooplankton (Kampf and Chapman 2016). Consequently, upwelling regions support significant pelagic fish species, explored by economically important fisheries (Cury et al. 2000), and concentrate around 90% of worldwide coastal fisheries (Pond and Pickard 1983). Indeed, coastal upwelling systems, located along western continental margins, are the most studied upwelling areas. However, upwelling is also associated with open ocean systems, such as the equatorial upwelling along the equatorial Pacific Ocean (Wyrtki 1981), and localized upwelling areas associated with island-mass effects (Shiozaki et al. 2014), where islands, seamounts and other prominent submarine topographic features, cause a deflection of currents, uplifting deep-water masses into ocean surface layers.

This chapter aims to revise the incompletely understood interactions within zooplanktonic assemblages in the most relevant world coastal upwelling systems. The chapter starts with a brief overview of the wind-driven upwelling process, including Ekman transport and the Coriolis effect. Then, a physical-ecological characterization of upwelling dynamics is discussed for the four most productive and persistent upwelling systems of the world, the California and Humboldt Current systems (Pacific), and the Iberian, Canary and Benguela Current systems (Atlantic). Upwelling regimes will be examined in terms of the coastal forcing behind its occurrence, oceanographic patterns generated, as well as their intensity and frequency. The physical, chemical and biological variability introduced in these systems by the transition between upwelling and relaxation events will also be discussed.

[1] Instituto Português do Mar e da Atmosfera (IPMA), Av. Doutor Alfredo Magalhães Ramalho, 6, 1495-165 Algés, Portugal.
[2] Interdisciplinary Centre of Marine and Environmental Research (CIIMAR), Portugal.
  Email: antonina@ipma.pt
* Corresponding author: rita.pires@ipma.pt

The mesoscale features usually associated with coastal upwelling systems, including fronts, eddies and filaments, known to influence the distribution and dispersal of planktonic organisms (Pineda et al. 2007), will also be covered throughout the chapter.

The oceanographic setting in upwelling regions is particularly interesting and, despite comprising highly studied regions, several questions on the biophysical interactions in these systems have yet to be fully understood, namely the impact of the highly variable spatial and temporal variability scales on planktonic organisms. Despite the potential for high offshore transport of these organisms, upwelling-related oceanographic features have been suggested as promoters of retention (dos Santos et al. 2008, Morgan et al. 2018). This chapter specifically explores the processes that influence zooplankton dynamics in coastal wind-driven upwelling systems, at multiple temporal and spatial scales. The influence of oceanic conditions on zooplankton distribution, diversity, productivity and behaviour, either promoting retention or dispersal, will be addressed. Studies exploring vertical migrations, and their impacts on zooplankton transport direction and dispersal under upwelling conditions, are also summarized. As biologically dynamic areas, upwelling systems support a rich and productive, though relatively simple, trophic web, characterized by a wasp-waist structure (Kampf and Chapman 2016). The tropic interactions involving zooplankton in coastal upwelling systems are additionally addressed.

Coastal regions are expected to be highly affected by predicted climate changes (Bakun et al. 2015), but their impact on upwelling regions is still uncertain. Thus, this chapter also outlines current knowledge on the expected impacts of climatic changes on upwelling systems of the world. The chapter concludes with the analysis of the impacts of model-predicted climate change scenarios on upwelling intensity and patterns (e.g., seasonality), and how these could lead to alterations in zooplankton abundance and a mismatch between plankton functional groups and production cycles.

7.2 Wind-driven Coastal Upwelling

7.2.1 Fundamental Concepts and Processes

Continental shelf regions are highly hydrodynamic and turbulent systems that, particularly during spring and summer, due to equatorward winds, tend to be influenced by coastal upwelling (Pond and Pickard 1983, Chavez and Messié 2009). Coastal upwelling is a divergence induced by the contrasting low/high pressure gradients between warm land areas and cold ocean surfaces (Simpson and Sharples 2012), which result in alongshore geostrophic winds. The persistence of alongshore winds promotes wind-driven coastal upwelling, that leads to changes in the Ekman layer, driven by the Coriolis force and the Earth's rotation (Chavez and Messié 2009). These conditions are observed for as long as the wind forcing is maintained, and consist on the seaward transport of surface waters that are then replaced by colder, denser and nutrient-rich waters, originating from deeper intermediate water layers (100–300 m depth) (Barth 1989, Simpson and Sharples 2012). Ekman transport is directed, at right-angles, to the right or left of the dominant winds in the northern or southern hemisphere, respectively, and current velocities decrease with depth, being highest at surface (Pond and Pickard 1983).

Local, seasonal and large-scale variability in hydrodynamic conditions can influence the frequency and intensity of upwelling, impacting nutrient concentration, plankton productivity, and the survival and development of pelagic communities (Morgan et al. 2012). Upwelling events can last from days to weeks, although highly persistent events occur in the major upwelling regions (Hutchings et al. 1995). Upwelling events are interrupted during periods of weaker winds, known as relaxation periods or downwelling, especially during winter, which promote poleward flows (Simpson and Sharples 2012). This short-term intercalated upwelling relaxation periods are, in fact, one of the factors contributing to the high productivity of these regions. Highest biological production usually occurs after the onset of weaker winds, allowing the growth of phytoplankton and zooplankton after the stabilization of the upper water layers (Kampf and Chapman 2016).

Coastal upwelling systems are usually associated with mesoscale features, including jets, fronts, eddies and filaments, whose location is dependent on local bathymetry and topography (Chavez and

Messié 2009). These features are important for the supply of nutrients and organic matter throughout continental shelves and off the continental margin, extending the influence of coastal upwelling into oceanic domains (Álvarez-Salgado et al. 2001, Hernández-León et al. 2007). Fronts are regions of high hydrodynamic variability at the offshore upwelling boundary, characterized by strong currents, high lateral and vertical water mixing, and abrupt changes in physical and chemical water properties (Barth 1989). Upwelling coastal jets are unstable structures influenced by the cross-shore wind component, which results in the formation of mesoscale eddies, usually associated with the off-shelf propagation of cold and chlorophyll-rich filaments of surface water (Álvarez-Salgado et al. 2001). The area of influence of upwelling filaments is known as the coastal transition zone but can extend for hundreds of kilometres from the coast, potentially influencing conditions in the open ocean (Haynes et al. 1993).

Geographic coastal features, such as headlands and capes, disrupt the upwelling jet flow, altering the wind and coastal forcing (Haynes et al. 1993, Chavez and Messié 2009). These structures favour the generation of upwelling shadows, which are usually sheltered from the main upwelling areas and exposed to weaker currents, being located downstream of the coastal jet flow (Graham et al. 1992). Plankton retention is promoted in these sheltered areas due to current recirculation (Pires et al. 2013, Morgan et al. 2018), providing good environmental conditions for reproduction and feeding of many species.

Intense water column mixing and high nutrient concentrations in the euphotic zone, characteristic of coastal upwelling systems, are linked with enhanced phytoplankton primary productivity, also resulting in high zooplankton abundance (Bakun et al. 2010; see Fig. 7.1 for an illustrative example). The highly productive waters of upwelling regions support fish assemblages often dominated by small planktivorous species (e.g., sardines and anchovies), that use these areas as spawning and nursery grounds (Cury et al. 2000). These pelagic fish species are important prey for higher trophic levels

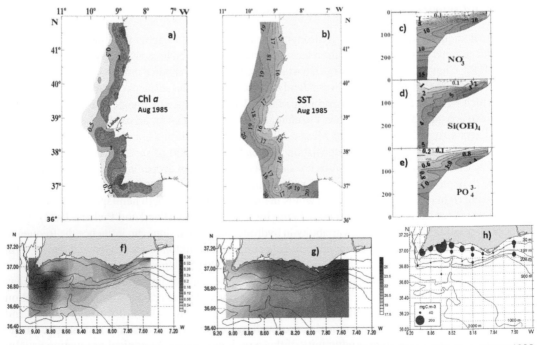

Figure 7.1: Distribution of physical and biological variables, during different summer upwelling periods (a–e, August 1985; f–h, August 2010), in the Portuguese coast (Iberian upwelling system). (a, f) chlorophyll-*a* concentration (µg L^{-1}); (b, g) sea surface temperature (°C); (c, d, e) vertical distribution of nutrient concentrations (µM), including nitrate (NO3$^-$), silicate (Si(OH)$_4$)) and phosphate (PO$_4$$^{3-}$), respectively, for a northern cross-shore section (~ 41° N, see panels a-b); and (c, d, e) zooplankton biomass (mg C m^{-3}). Sources: a–e, adapted from Moita 2001; f–g, courtesy of the MedEx project.

(see Section 7.2.4), namely seabirds and marine mammals, and are exploited worldwide by several highly valuable commercially fisheries (Kampf and Chapman 2016). Moreover, upwelling events also play an important role due to the horizontal exportation of phytoplankton biomass into surface ocean layers, and the vertical exportation of derived particulate organic material into deeper ocean waters (Simpson and Sharples 2012).

7.2.2 Characterization of the Physical and Chemical Conditions in Upwelling Areas

Upwelling areas and fronts exhibit sharp spatial gradients in chemical and physical water properties, such as density, temperature, pH, and nutrient concentrations (Barth 1989, Chavez and Messié 2009). Under upwelling dynamics, the advection of deeper, colder and denser water masses into surface promotes changes in mixed layer depth and a high exchange between near- and offshore waters, impacting the productivity of both systems (Hutchings et al. 1995). This is particularly important, given that the mixed layer corresponds to highly turbulent water strata, with high biological activity, where nutrients are quickly used and transformed (Peterson 1998, Kampf and Chapman 2016).

Satellite-retrieved sea surface temperature (SST) images represent a straightforward approach to identify the occurrence of coastal upwelling events (Fig. 7.1), and the presence of colder waters close to the coast is typically considered an upwelling signature (Chavez and Messié 2009). As for SST, ocean colour remote sensing of chlorophyll *a* concentration, a proxy for phytoplankton biomass, can be also used to identify upwelling, since higher values are indicators of increased productivity (Figs. 7.1 and 7.2). However, during less intense and short upwelling events, the stimulatory effects on primary production may be restricted to the sub-surface waters (Jahnke 2010), and can go unnoticed using remote sensing.

Nutrient enrichment is one of the most biologically relevant consequences of upwelling, although some delay may occur between the beginning of an upwelling event and the increased nutrient uptake by phytoplankton (Hutchings et al. 1995). High concentrations of dissolved inorganic macro- (nitrate, silica, phosphorus) and micronutrients (e.g., iron), as well as high availability of dissolved organic carbon, are associated with upwelling events (Fig. 7.1). Nutrients are assimilated by phytoplankton, being the main drivers of its abundance, photosynthesis, primary production and oxygen production in these regions (Hutchings et al. 1995, Chavez and Messié 2009). Nitrogen is an essential component of the ocean's biogeochemical cycle, and frequently is a limiting nutrient for marine phytoplankton. Moreover, in the case of diatoms, main phytoplanktonic components in upwelling systems (see Sections 7.2.3 and 7.2.4), silica is a strict requirement for the formation of their frustules. Yet, micronutrients, such as iron, can also limit phytoplankton growth, particularly in the upwelling areas of the California and Humboldt (Chavez and Messié 2009).

Increased phytoplankton production in upwelling areas is grazed by zooplankton, thereby enhancing global biological production and respiration (oxygen consumption). Oxygen depletion in upwelling areas can be enhanced by the quick offshore exportation of phytoplankton, that will thus be less consumed and more available to be vertically exported into the deep ocean (Bakun 2017). Therefore, due to intensified biological activity and sinking and decomposition of organic matter, oxygen limitation is common in upwelling areas (Criales-Hernández et al. 2008). These processes, together with the low concentration of oxygen in upwelled waters, are responsible for the Oxygen Minimum Zones (OMZs), where oxygen saturation in seawater is at its lowest, typically located in the mesopelagic domain of most coastal upwelling systems. The OMZs are generally considered as the minimum oxygen concentration that plankton can support, affecting their vertical distribution and survival (Criales-Hernández et al. 2008). OMZ is a common feature in the Humboldt Current system, namely off northern Chile, where this oxygen-depleted layer is very shallow during spring and summer (Chavez and Messié 2009). In the California Current system, OMZ is observed in deeper shelf waters, during summer (Simpson and Sharples 2012). Off southwestern Africa, specific

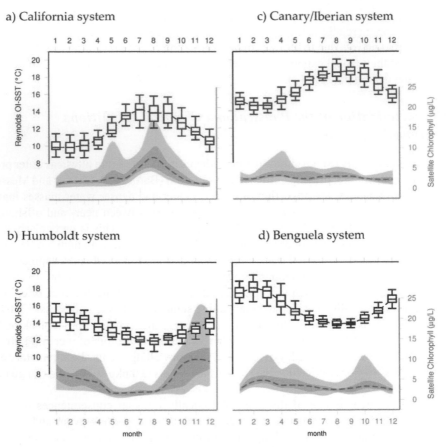

Figure 7.2: Average monthly Sea Surface Temperature (SST, °C; upper line, with box and whisker) and satellite-retrieved chlorophyll concentration (μg L⁻¹; grey shadow area) for different coastal upwelling systems, obtained and adapted from IGMETS/WGZE accessed through www.st.nmfs.noaa.gov/copepod/time-series/. (a) California (Newport Line NH-5, Newport-Oregon site; Associated investigators: Fisher and others); (b) Humboldt (Concepción Station 18, Chilean coast site; Associated investigator: Escribano); (c) Canary/Iberia (Cascais bay site; Associated investigators: Dos Santos and others); and (d) Benguela (Walvis Bay 23S off-shore, northern Benguela Current site; Associated investigators: Kreiner and others).

Further information on the processing of these data sets is available in O'Brien and Oakes 2021 (this book, Chapter 9).

upwelling conditions promote occasional and large coastal emissions of toxic hydrogen sulphide from the bottom sediments into the water column, with important consequences for local ecosystems (Weeks et al. 2004).

Upwelling events also impact the global ocean carbon cycle, given the direct relationship between high nitrate concentrations and increased primary production (Hutchings et al. 1995). These conditions can lead to the increased absorption of atmospheric carbon dioxide into the surface waters, and to higher exportation of particulate organic carbon into deeper layers of the water column, as reported for the California system (Chavez and Messié 2009), considering that upwelling can transport the fixed carbon before remineralisation. Despite the uncertainty on their relevance as sinks or sources of carbon dioxide, upwelling regions play an important role in carbon exportation and recycling (Jahnke 2010), with filaments and fronts contributing to the aggregation of particulate organic carbon. For example, the extension of upwelling filaments off the Iberian and Canary systems supplies dissolved organic carbon into the oligotrophic north-eastern subtropical Atlantic gyre (Haynes et al. 1993, Álvarez-Salgado et al. 2001). Physically, carbon is cycled within oceans between various layers by upwelling and downwelling, as it is consumed by plankton in these productive regions (Hutchings et al. 1995, Jahnke 2010).

7.2.3 Main Wind-driven Coastal Upwelling Regions Worldwide

The main coastal upwelling areas are concentrated in western continental shelves and comprise major biogeographic transition areas: the California Current and the Humboldt Current systems, in the Pacific Ocean; and the Iberian/Canary Current and the Benguela Current systems, in the Atlantic Ocean (Chavez and Messié 2009). Several similarities can be found between these areas in terms of hydrodynamics and ecosystem structure, such as the common occurrence of equatorward winds and poleward currents, as well as the narrow shelves intersected by submarine canyons. However, each system has its own particularities that will be highlighted below.

In some areas, coastal upwelling events are year-round persistent and are closely related with the long-term changes in wind regimes of subtropical gyres (Haynes et al. 1993), while in other areas upwelling events are seasonal (Kampf and Chapman 2016). Moreover, in each system, upwelling presents interannual variability. The eastern Pacific upwelling systems are highly influenced by the El Niño-Southern Oscillation (ENSO) shifts (Collins et al. 2010), with consequences on the entire food web. Shifts are mostly driven by changes in pressure gradients, with consequent large-scale alterations of the wind regimes (Simpson and Sharples 2012). El Niño events occur when winds are weaker, resulting in lower upwelling intensity (Simpson and Sharples 2012). On the other hand, La Niña periods (Ayón et al. 2008) correspond to the strengthening of trade winds, resulting in lower water temperatures than usual in the South American coast (Ayón et al. 2008). These events alter phytoplankton primary production, also affecting the diversity, abundance and distribution of zooplanktonic species in the region (Carrasco and Santander 1987). In the case of north-eastern Atlantic upwelling systems, another large-scale climate index, the North Atlantic Oscillation (NAO), affects the Iberian and Canary systems (Haynes et al. 1993). NAO variability is caused by changes in the position and intensity of pressure centres, which determine a positive (winter) or negative (summer) NAO phase (Hurrell et al. 2003). Since NAO drives long-term changes in the wind patterns, it also influences upwelling intensity (Santos et al. 2005).

Interannual and short-term variability, as well as the high dynamics of upwelling systems, contribute to the low diversity of planktonic communities, characterized by high abundance of a few species. Upwelling systems are usually characterized by high abundance of large phytoplanktonic cells (e.g., diatoms) (see Section 7.2.3). Crustaceans are also highly abundant, mainly small copepod species and euphausiids, with *Calanus* being a common genus in all upwelling systems (Longhurst 1967, Peterson 1998). Gelatinous plankton is also an important component of upwelling systems, often associated with mixed waters, and may be related with the high resistance of polyps to hypoxic conditions (Miller and Graham 2012). According to Brodeur et al. (2011), in upwelling systems, gelatinous plankton mainly impact lower trophic levels, and a relatively small fraction of their production is transferred to higher levels of the food web, when compared to forage fishes.

The California Current System

In the regions off California and Oregon (north-eastern Pacific), upwelling is persistent from March to October, despite the year-round occurrence of upwelling-favourable winds (Hickey 1998). Winds are more intermittent towards the Oregon shelf at north, where downwelling periods are frequent. On the other hand, upwelling is more persistent at northern and central California than at south, and downwelling periods are less frequent (Hickey and Banas 2008). The California Current flows southward near the shelf slope, following the north-south orientation of the coastline, which is disrupted by several headlands (Hickey 1998). The upwelling influence can extend up to 100 km from shore (Kampf and Chapman 2016), commonly generating upwelling filaments in the southern area of the system, linked to the instabilities in the flow of the California Current driven by persistent eddies (Hickey and Banas 2008). The system is also influenced by ENSO events.

In terms of planktonic communities, the California Current system has been studied since 1949 by the California Cooperative Oceanic Fisheries Investigation (CalCOFI) programme. Phytoplanktonic assemblages were examined by Wilkerson et al. (2000) for the southern part of the system, showing the dominance of diatoms and other large phytoplankton cells. In the same area, Longhurst (1967) studied the zooplankton communities during upwelling and non-upwelling conditions. Under upwelling conditions, the region is characterized by low zooplankton diversity and high dominance of a few herbivorous crustacean species, such as the copepods *Calanus pacificus*, *Metridia pacifica* and *Acartia tonsa* (Longhurst 1967). On the other hand, non-upwelling conditions present high abundance of microzooplankton, such as radiolarians, dinoflagellates and tintinnids (Longhurst 1967). For the Oregon region, Hubbard and Pearcy (1971) registered high abundances of the tunicates *Salpa fusiformis* and *Soestia zonaria* during upwelling periods. Zooplanktonic assemblages of the southern California present cross-shore differences, with smaller patches nearshore during spring and summer (Star and Mullin 1981). Smith et al. (1986) compared weaker, moderate and stronger upwelling conditions, concluding that species diversity and abundance decreased with enhanced upwelling intensity. Off California, most copepods are highly abundant during the spring upwelling period (Fig. 7.3a). Off Oregon, different distributions can be observed between the offshore region

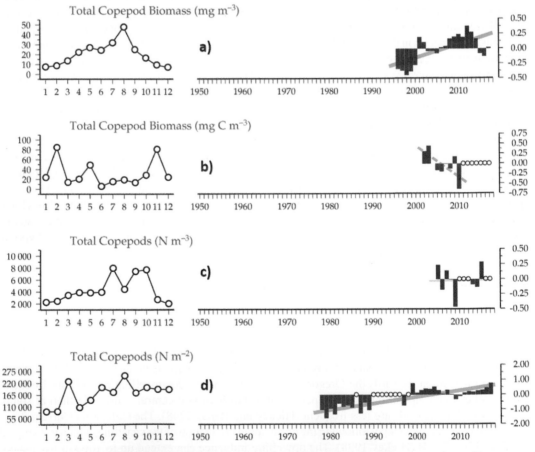

Figure 7.3: Average monthly and yearly total abundance or biomass of copepods for different coastal upwelling systems, obtained and adapted from IGMETS/WGZE accessed through www.st.nmfs.noaa.gov /copepod/time-series/. (a) California (Newport Line NH-5, Newport-Oregon site; Associated investigators: Fisher and others), (b) Humboldt (Concepción Station 18, Chilean coast site; Associated investigator: Escribano), (c) Canary/Iberia (Cascais bay site; Associated investigators: Dos Santos and others); and (d) Benguela (Walvis Bay 23S off-shore, northern Benguela Current site; Associated investigator: Kreiner et al.) Further information on the processing of these data sets is available in O'Brien and Oakes 2021 (this book, Chapter 9).

and the mid-shelf, where the dominant species are *Calanus marshallae, Pseudocalanus mimus, Centropages abdominalis*, and *Acartia longiremis* (Morgan et al. 2003). Roemmich and McGowan (1995) examined 43 years of data on macrozooplankton biomass, and reported a decreasing tendency related with the increase in SST. High abundances of gelatinous organisms were also registered in the region of the northern California Current, constituting the majority of zooplanktonic organisms (Brodeur et al. 2011).

The Humboldt Current System

In the Humboldt Current system, off Chile and Peru in the south-eastern Pacific Ocean, the coastal circulation is highly influenced by the Peru Current flowing equatorward and the Peru-Chile counter-current flowing southward (Penven et al. 2005). This system is characterized by upwelling-favourable southwestern winds that persist year-round in the Peru and northern Chile shelves, and seasonally in the southern and central Chile, being especially intense during spring and summer (Escribano et al. 2007). The local circulation is also associated with the south-eastern Pacific trade winds, responsible for seasonal and inter-annual shifts in the climatic conditions of the region linked to ENSO periods, and stronger upwelling conditions are detected during La Niña periods (Ayón et al. 2008). The OMZ is especially evident in the upwelling areas of the eastern Pacific, linked with the equatorial subsurface waters (Escribano et al. 2000), influencing the distribution of planktonic assemblages, which can avoid it, reside in it or even occupy it occasionally (Escribano et al. 2009). When wind reversals generate upwelling, the hypoxic water can be transported to nearshore locations, causing disturbances in the zooplankton and communities of higher trophic levels. Upwelling filaments in the region are considered shallow, intermittent and can extend up to several hundred kilometres offshore (Sobarzo and Figueroa 2001).

The planktonic assemblages associated with the Humboldt Current system show distinct characteristics, whether we consider the upwelling region or the subtropical waters to west, but also between the northern and southern branches of the system (Ayón et al. 2008). As in the California Current system, phytoplankton assemblages are dominated by large cells, as diatoms, microflagellates, and the cilliate *Mesodinium rubrum* as the upwelling events progress (Peterson et al. 1988). Higher zooplankton abundances have been reported in offshore areas in the north (Ayón et al. 2008). Off Peru, in the Pisco to San Juan region, zooplankton assemblages of the Humboldt system are primarily composed by copepods, together with euphausiids (Escribano and Hidalgo 2000, Escribano et al. 2000). Carrasco and Santander (1987) attributed the nearshore presence of carnivorous copepods to the influence of subtropical warmer waters. Distinct cross-shore groups, closely linked to the upwelling front, are detected in the spatial distribution of the communities (Santander 1981, Escribano et al. 2000), with *Acartia tonsa* and *Centropages brachiatus* dominating over the shelf, and siphonophores, bivalve larvae, Foraminifera and Radiolaria being more abundant over the shelf slope. In oceanic areas, other copepods, such as *Mecynocera clausi, Pleuromamma gracilis, Oithona plumifera* and *Corycaeus* sp., are dominant. However, off northern Chile, *Paracalanus parvus* and *Calanus chilensis* are the main species (Escribano and Hidalgo 2000, Escribano et al. 2009). Most copepods in this system show a low amplitude seasonal cycle (Fig. 7.3b).

As mentioned above, the OMZ is an important feature off the Humboldt Current system, occurring below 60 m depth (Escribano et al. 2009). Several copepod species, such as *C. chilensis* and *C. brachiatus*, are limited to the upper water layers and may not perform diel vertical migrations, due to the OMZ (Escribano and Hidalgo 2000, Escribano et al. 2009). However, *Eucalanus inermis* is an abundant species in upwelling areas that can adapt to the OMZ (Hidalgo et al. 2005), as well as *Euphausia mucronata* that can temporarily occupy this layer (Escribano et al. 2000, 2009).

Gelatinous zooplankton is abundant in the upper layers of the water column in the Humboldt upwelling system (Pagès et al. 2001). These organisms can be abundant in hypoxic layers, during years of persistent downwelling, due to the deeper thermocline and hypoxia (Pagès et al. 2001).

The Canary and Iberian Systems

In the case of the Canary Current system, the upwelling intensity is variable throughout the year, lasting from summer (April-May) to autumn (October) (Haynes et al. 1993, Álvarez-Salgado et al. 2007). The system is closely associated with the North Atlantic Subtropical gyre and the Portugal Current system, in the Iberian region, only disrupted at the Strait of Gibraltar, where the outflow of Mediterranean waters is of noteworthy influence on local hydrodynamics (Peliz et al. 2013). Moreover, the entire system is influenced by NAO and the Atlantic Multidecadal Oscillation, mainly during fall and winter (Haynes et al. 1993, Hurrell et al. 2003). The Canary Islands are a distinct feature not found in other upwelling systems, acting as obstacles to the circulation of the Canary Current and influencing local hydrodynamic variability (Hernández-León et al. 2007). In this region, the upwelling is more intense in areas under the influence of capes (Relvas et al. 2007). Throughout the year, offshore waters are usually highly stratified. Despite the off-shore extension of upwelling filaments for hundreds of kilometres from the coast, the impact is mainly restricted to the upper 100 m (Hernández-León et al. 2007, Relvas et al. 2007).

In the case of the phytoplankton assemblages, picoplanktonic cyanobacteria and small flagellates are predominant in oceanic areas through the year, but during the upwelling period, abundant large diatoms dominate phytoplankton (Hernández-León et al. 2007). Zooplankton in the Canary and Iberian system show differences in species composition both spatially, between nearshore and offshore (Domínguez et al. 2017), and temporally, particularly during the upwelling season (Thiriot 1978, Hernández-León et al. 2007). Copepods are the main component of the zooplanktonic assemblages (Domínguez et al. 2017), with small size calanoids and cyclopoids dominating the nearshore upwelling waters, and larger individuals dominating offshore regions (Postel 1990). *Calanoides carinatus* is a common species at shallower waters under upwelling conditions, presenting a seasonal migration pattern strongly related with this process (Hernández-León et al. 2007). *Calanus helgolandicus, Oithona plumifera, Clausocalanus arcuicornis* and *Temora stylifera* are common species throughout the year, and *Acartia negligens, Paracalanus parvus parvus, Ctenocalanus vanus, Metridia lucens lucens, Acartia danae* and *Oncaea curta* are detected during the upwelling season (Hernández-León et al. 2007). Large *Temora stylifera* copepods are observed nearshore, decreasing in size towards offshore regions, and apparently associated with upwelling and precipitation conditions (Domínguez et al. 2017). Euphausiids are represented by *Nyctiphanes couchii*, highly abundant in the Morocco region, and *Nematoscelis megalops* and *Euphausia krohnii*, common in the Mauritanian shelf (Thiriot 1978, Hernández-León et al. 2007). Gelatinous plankton is represented by Siphonophora, mainly *Muggiaea atlantica, Abylopsis tetragona, Chelophyes appendiculata* and *Agalma* sp., which are usually detected in oceanic waters off the shelf slope (Thiriot 1978).

Off the Iberian shelf, zooplankton biomass does not show a clear seasonal pattern. However, the highest values occur during early spring or summer/autumn, and the lowest in winter. Moreover, interannual variation in zooplankton biomass can be also observed (O'Brien et al. 2013) (see Fig. 7.3c).

The Benguela Current System

The upwelling system associated with the Benguela Current (south-eastern Atlantic) extends from the Namibian shelf to Cape Agulhas in South Africa (Fennel 1999). At north, the system is influenced by the warm Angola Current (Shannon et al. 1992), while at south it is driven by the confluence of the warm Agulhas Current and the cold south Atlantic flow of the Benguela Current (Hutchings et al. 1986, Bode et al. 2014). The Benguela current flows northwards, along the shelf-edge of the western African coast (Fennel 1999). Persistent coastal upwelling is observed during summer (September to March) in the southern region, under south-easterly winds (Andrews and Hutchings 1980, Fennel 1999), with its influence extending up to 500–750 km from the coast (Lutjeharms and Stockton 1987). Intermittent and seasonal upwelling are also observed east of Cape Agulhas, and in the north domain of this system (Gibbons and Hutchings 1996). The large environmental variability caused by weaker ENSO-like events, known as the Benguela Niño (Shannon et al. 1992), together with the high fishing effort over the region, that led to periods of over-exploitation of several species (Kampf and Chapman 2016), induces notorious inter-annual and decadal fluctuations in larval survival and fish recruitment (Parada et al. 2003).

Phytoplankton assemblages are more abundant nearshore (Olivar and Barangé 1990) and, under upwelling conditions, are dominated by chain-forming diatoms species. Under these conditions, microzooplankton, including Foraminifera, microflagellates and ciliates, are also important (Verheye 1991, Gibbons and Hutchings 1996). Zooplankton assemblages are dominated by a few cold, temperate and neritic species (Gibbons and Hutchings 1996), and show similar composition at the north and south areas of the system (Pagès et al. 1991). Cross-shore differences in abundance and composition have been reported for the southern area, during upwelling (Olivar and Barangé 1990, Pagès and Gili 1992). In upwelling conditions, the nearshore region west off South Africa is characterized by low diversity (Pagès and Gili 1992), while at east, the Agulhas Current harbours high diversity due to the mixture of species from the Indian Ocean and lower latitudes. Zooplankton, more abundant in the nearshore region of the south of the system (Andrews and Hutchings 1980, Hutchings et al. 1986), are dominated by crustaceans and cold-water species (Hutchings et al. 1986), mainly copepods and euphausiids (Verheye 1991, Bode et al. 2014). For copepods, long-term shifts (Fig. 7.3d) in the species composition have been reported (Verheye et al. 1998). Notwithstanding, calanoid and cyclopoid are the dominant groups (Verheye 1991, Bode et al. 2014), with smaller species being more common in the mid-shelf and larger species nearshore. *Calanoides carinatus* and *Centropages brachiatus* are commonly observed at nearshore, under upwelling conditions (Verheye 1991, Hutchings et al. 1986, Gibbons and Hutchings 1996). However, high dominance of *Nannocalanus minor* has been related with periods of weaker upwelling, and lower abundances of *C. carinatus* and *M. lucens* (Bode et al. 2014). Euphausiids are represented by *Nyctiphanes capensis* and *Euphausia lucens,* dominant in the north and south domains of the system, respectively (Olivar and Barangé 1990). Upwelling conditions are also linked with high abundance and diversity of gelatinous organisms, decreasing during periods of low upwelling intensity (Pagès and Gili 1992). These organisms are distributed over the slope, where the circulation is more persistent (Shelton and Hutchings 1990). Siphonophora (*Muggiaea atlantica*) and hydromedusae (e.g., *Leuckartiara octona*) are common nearshore (Pagès and Gili 1992, Gibbons and Hutchings 1996), while *Liriope tetraphylla* is abundant over oceanic areas (Pagès and Gili 1992).

7.2.4 *Trophic Interactions in Upwelling Systems*

Upwelling regions support a highly productive food web and, given the nutrient-rich waters uplifted to the surface, also contribute to the recycling of nutrients that would otherwise be unavailable for planktonic organisms (Chavez and Messié 2009, Bakun et al. 2010). The food web structure in upwelling systems is simple, including phytoplankton and zooplankton, as well as forage and predatory fish.

Upwelling results, firstly, in the development of phytoplankton, particularly diatoms (Hutchings et al. 1995, Wilkerson et al. 2000), that benefit from the nutrient enrichment of surface layers, supporting zooplanktonic organisms (Bakun et al. 2015). Shifts in species composition at low trophic levels will thus impact the entire food web. Strong upwelling events intensify vertical mixing, leading to a decrease in light availability in the mixed layer, which may result in lower phytoplankton abundance and primary productivity (Chavez and Messié 2009). During upwelling relaxation periods, water stratification increases, diatom relative abundance decreases, and the primary productivity supporting fish species will be lower, given the dominance of small phytoplankton species, associated with long retentive food webs, supported by regenerated nutrients (Kampf and Chapman 2016).

Zooplanktonic organisms have a major role in the food web, functioning as the link between phytoplankton and the upper trophic levels. A close relationship between high zooplankton abundances and upwelling periods was reported for the major upwelling systems (for Humboldt system, Escribano et al. 2007). Long-term variability of copepod abundance (Fig. 7.3), as the main component of metazooplankton, can be associated with most of the secondary production in upwelling systems (Simpson and Sharples 2012). Considering their relatively longer generation times (Peterson 1998, Hernández-León et al. 2007), zooplanktonic organisms can be coupled with phytoplanktonic assemblages, for extended periods.

The food web of upwelling systems follows a wasp-waist structure (Kampf and Chapman 2016), since intermediate trophic levels are dominated by large populations of one or a few species of planktivorous pelagic fish, namely Clupeidae, that exert strong control on both their prey and predator populations (Cury et al. 2000). In fact, this is a shared aspect between the food webs of the major upwelling regions. Anchovy (*Engraulis encrasicolus, Engraulis capensis*) and sardine (*Sardina pilchardus, Sardinops sagax* and *Sardinella* spp.) species are the main pelagic fishes for all four major coastal upwelling systems (Silva 2003, Miller et al. 2006, Demer and Zwolinski 2014). These forage fishes are highly exploited in commercial fisheries. Other fish species, such as hake and tuna, use upwelling systems as feeding areas, despite spending a part of their life cycles in offshore waters. Species positioned at higher trophic levels, such as seabirds and whales, also use upwelling areas (Kampf and Chapman 2016).

Fisheries are highly susceptible to shifts in primary production, and a disrupting factor in the trophic structure of upwelling systems, as in the case of the Humboldt region (Kampf and Chapman 2016). For the past decades, the abundance of Clupeids has registered high interannual variability worldwide due both to overfishing and environmental conditions (Stenevik et al. 2003, Chavez and Messié 2009). For example, with the over-exploitation of the stock of Namibian sardine, in the Benguela system, the abundance of gelatinous organisms increased (Lynam et al. 2006). Indeed, gelatinous organisms can be important components of the food web in upwelling systems, such as the California system, where they are highly abundant and considered to limit the productivity available for higher trophic levels (Brodeur et al. 2011). However, recent studies have suggested that jellyfish are key preys, not only for turtles and sunfish, but also for other top predators, including penguins, albatrosses and tuna (Hays et al. 2018).

Under less intense upwelling events, limited nutrient availability is associated with lower productivity, while stronger events imply large turbulence and seaward transport of fish larvae, dictating fish production (Chavez and Messié 2009). Shifts in the duration of the upwelling events may result in a mismatch between functional groups of prey and predators. In the case of the Benguela Current system, the long-term increase of zooplankton abundance, in association with higher upwelling intensity detected in the west coast off South Africa between the 50s and 90s of the last century (Verheye et al. 1998), resulted in a cascading trophic effect. This trend was combined with shifts on the dominant crustacean species, and related with the dominant pelagic fish species: calanoid copepods, euphausiids and amphipods dominated in years of high abundance of *Sardinops sagax*, while during years of dominance of *Engraulis capensis* cyclopoid copepods were the most abundant zooplankton (Verheye 2000). Since 1995, the decrease in the abundance of large copepods in the southern part of this system was accompanied by higher abundances of anchovy (Hutchings et al. 2006, Verheye et al. 1998). In the north, a long-term increase has been observed since 1983 (Hutchings et al. 2006). A decrease in zooplankton biomass was also observed for the California (Roemmich and McGowan 1995) and Humboldt (Alheit and Bernal 1993) systems, with consequences for the abundance of organisms in higher trophic levels.

7.3 Zooplankton Dynamics and Behaviour in Upwelling Systems

Upwelling regions are widely studied, but still incompletely understood, particularly in what concerns the main biophysical interactions at play. Considering that upwelling areas support important fisheries worldwide, no correct assessment of fish populations can be obtained without knowledge on planktonic communities, which comprise the main food source of many harvested species that strongly depend on the intensity of the upwelling events. Hence, zooplankton dynamics, as well as underlying factors driving variability, and the impacts of temporal and spatial changes induced by upwelling, are important and insufficiently understood questions for these regions.

The transport of eggs and early life stages is an important factor determining the recruitment of planktonic organisms in the marine environment (Pineda et al. 2007), and is affected by the high temporal variability of the Ekman transport, both at seasonal and inter-annual levels (Stenevik et al. 2003). Consequently, the distribution of planktonic organisms can be highly affected by the hydrodynamic fluctuations in the frequency and intensity of upwelling events (Rivera et al. 2019), but also by nutrient limitations and other physical, chemical and biological variables (Kampf and Chapman 2016). Bakun (1996) refers the processes of enrichment, concentration and retention as highly important for the recruitment of marine larvae in upwelling systems. Moreover, mortality rates for early life stages of zooplanktonic organisms are naturally high, and result in variable survival, which is also closely related with local environmental conditions (Bakun 1996). In areas under the influence of river plumes, upwelling may also contribute to the enrichment of offshore areas through the exportation of these nutrient-rich waters, promoting high secondary productivity, based on increased copepod egg production, supporting the growth and high physiological condition of fish larvae (Chícharo et al. 2003).

Strong upwelling events imply enhanced offshore transport of surface waters and, consequently, phytoplankton and zooplankton, towards frontal regions and associated eddies (Barth 1989), regions where the prey availability usually is also higher. Therefore, planktonic organisms may depict aggregated distributions over upwelling areas (Peterson 1998, Escribano et al. 2002, Hutchings et al. 2006), particularly those taxa able to adopt and maintain specific vertical positions. Accordingly, higher zooplankton abundances may reflect the persistence of upwelling-favourable winds (Verheye 2000).

In upwelling areas, coastline features such as headlands may also function as retention areas for planktonic organisms, given that they can alter the circulation and wind forcing patterns by promoting recirculation currents and the generation of upwelling shadows (Graham et al. 1992). Other oceanographic features, such as eddies, associated with the offshore larval transport promoted by upwelling filaments, can also result in plankton retention (Peterson 1998, Rivera et al. 2019). In upwelling areas, retentive processes will determine which species compose the planktonic assemblages during upwelling events (Peterson 1998).

Extreme upwelling regimes, that induce high turbulence in coastal regions can, however, break up plankton aggregates into smaller components (Kampf and Chapman 2016). Early views of the influence of upwelling on planktonic organisms considered that upwelling events enhanced offshore advection of larvae, and relaxation periods would favour their return to coastal grounds (Menge et al. 2004). This would suggest lower survival for larvae transported farther from the nearshore area. Indeed, events of upwelling relaxation impact the distribution and abundance of crustacean larvae (Morgan et al. 2012). However, recent studies report the retention of invertebrate larvae in coastal regions despite upwelling dynamics (Santos et al. 2004, Pires et al. 2013, Morgan et al. 2018), and low dispersal from the adult grounds. Then, the position of larvae, at vertical and horizontal levels, are determinant for the direction of their transport, retention or offshore exportation pathways of dispersal (Stenevik et al. 2003, Pires et al. 2013).

For invertebrate larvae, exposed to eventual seaward advection under upwelling conditions (Morgan et al. 2009), the ability to perform diel vertical migrations will impact their distribution (Dos Santos et al. 2007, 2008). Accordingly, invertebrate larvae are known to occupy shallower water layers during the night and descend to deeper waters during the day, being exposed to lateral displacement by upwelling currents (Dos Santos et al. 2008). Different strategies have been reported for the vertical position adopted by the larval stages to cope with offshore transport in upwelling regions. For instance, differences in the vertical distribution of distinct ontogenetic stages have been observed in these systems, with different larval stages occupying distinct ocean layers (Bartilotti et al.

2014). Furthermore, larvae actively promote onshore transport either by descending to deeper water layers (Pires et al. 2013) or exploring neuston layers, both mechanisms being enhanced by internal waves (Morgan et al. 2009) or sea breezes (Queiroga et al. 2007).

Upwelling relaxation periods, corresponding to a shift on the dominant currents and winds, can also drive changes in larval behaviour and transport, impacting larval recruitment (Morgan et al. 2012). For larvae accumulated offshore at upwelling fronts, relaxation periods can lead to onshore transport, particularly if a shallower distribution is maintained. For upwelling regions where relaxation periods are less frequent, such as the northern and central California system, the net seaward transport can be higher and impact larval recruitment (Navarrete et al. 2005). Thus, the nearshore retention of larvae tends to be higher in regions presenting less persistent upwelling events, and where the upwelling fronts occupy narrow nearshore bands. In fact, as a consequence of alongshore circulation in coastal areas, influenced by upwelling conditions, plankton tends to aggregate in bands parallel to the coastline (Bartilotti et al. 2014). The season of the year, time of the day, depth and distance to the coast of the spawning process are also important for larval transport, retention and survival (Stenevik et al. 2003, Dos Santos and Peliz 2005, Miller et al. 2006, Pochelon et al. 2017).

7.4 Effects of Climate Change Scenarios on Upwelling and Zooplankton

Coastal regions are expected to be the most affected marine domains under climate change scenarios (Harley et al. 2006, Bakun et al. 2015). Despite their resilience to natural climate change, coastal upwelling regions are highly exposed to anthropogenic stressors (Bakun 1990, Collins et al. 2010). Hence, for these ecosystems, it is important to understand how upwelling frequency, intensity and extension will respond to these environmental changes, but also to explore what major shifts may be expected on ecological processes and fishery resources. Modelling approaches using ocean simulations, forced with atmospheric data, have been used to explore the impact of anticipated environmental changes on the hydrodynamics in upwelling regions (Penven et al. 2005). The impact of future climate changes on planktonic organisms is still far from completely explored. Nevertheless, early stages of planktonic organisms are known to respond quickly to variations in ocean dynamics (Richardson 2008). Additionally, modelling studies in upwelling systems have also addressed biological and physical processes conjointly, particularly to understand plankton dispersal (Stenevik et al. 2003, Miller et al. 2006, Pires et al. 2013), which can help the development of more accurate simulations for models of climate changes.

A long-term intensification of upwelling-favourable winds during the last decades has been reported for most of the major upwelling areas (Bakun 1990, Shannon et al. 1992, Sydeman et al. 2014), except the Iberian system where this trend is not clear (Barton et al. 2013). Under global warming, studies suggest that upwelling intensification is expected to continue due to changes in the location, extension and intensity of pressure gradients between continental and oceanic areas (Bakun 1990, Rykaczewski et al. 2015). This scenario would follow the continued rising of atmospheric greenhouse gases. In this case, upwelling areas may undergo more turbulence and an increase in nutrient concentrations and productivity, resulting in more acidified and/or oxygen-depleted water masses (Rykaczewski and Dunne 2010, Gruber et al. 2012).

Under these conditions, pelagic organisms, although benefiting from higher nutrient inputs in surface waters, would also be negatively impacted due to faster offshore transport (Rykaczewski and Dunne 2010, Sydeman et al. 2014). Changes in the trophic structure and habitat will be also promoted by intensified upwelling, which may imply higher phytoplankton and zooplankton abundances (Shannon et al. 1992), although many uncertainties still surround this scenario (Rykaczewski and Dunne 2010). Thus, the effects of climate change will eventually manifest mainly on the lower trophic levels, and indirectly on higher trophic levels (Bakun et al. 2015). Moreover, organisms inhabiting benthic and intermediate layers may suffer higher mortality rates or be forced to move into different regions, given their lower tolerance to oxygen-depleted waters (Rykaczewski and Dunne 2010). Major circulation shifts caused by climate changes can also impact the transport and vertical distribution of planktonic organisms, resulting in expansions, retractions or redistributions of their habitats (Parada et al. 2003, Harley et al. 2006). Hence, species may undergo shifts on their behaviour, reproduction and growth, affecting the composition and abundance, at the community level (Bakun et al. 2015). An increasing trend in jellyfish abundance has been reported for several upwelling regions of the world (Condon et al. 2013). Being more resilient to climatic changes, increases in the abundance of gelatinous species may also be linked with anthropogenic stressors, including overfishing, eutrophication, and ocean deoxygenation (Brodeur et al. 2011). However, studies show that there is no strong evidence (Condon et al. 2013) to support this view of a jellification of the coastal upwelling areas, given the singularities of the life cycle of these species and the lack of long-term time series on jellyfish around the world. Increases in phytoplankton, specifically dinoflagellates, have also been reported for upwelling systems (Ryan et al. 2009). However, the upwelling enhancement under predicted climate change scenarios is still highly uncertain, and opposite trends have also been reported, namely lower upwelling intensity and lower effects on water stratification (Auad et al. 2006).

Increased and highly variable seawater temperatures are expected in future scenarios of climatic changes for upwelling regions, due to changes in wind patterns (Bakun et al. 2015). In upwelling regions, larger variability in seawater temperature will eventually lead to higher stratification of the water column, lower water mixing and changes in the circulation patterns that may be detrimental to zooplanktonic organisms (Roemmich and McGowan 1995). High temperatures induce higher developmental and metabolic rates of eggs and larvae, resulting in smaller organisms (Miller et al. 2006). Furthermore, ocean surface warming and enhanced vertical stratification will probably result in lower availability of new nutrients in the euphotic zone (Roemmich and McGowan 1995), and lower primary productivity in upwelling systems, although ecosystem responses may be distinct at local or global scales (Rykaczewski and Dunne 2010).

The carbon cycle is also expected to suffer major alterations in the future, following the continued increase in atmospheric carbon dioxide (Rykaczewski and Dunne 2010, Bakun et al. 2015). Higher levels of carbon dioxide imply an increased dissolution in oceanic waters and, consequently, acidified waters. Ocean acidification is particularly adverse on planktonic organisms that form calcium carbonate exoskeletons or shells (Gruber et al. 2012), but also on organisms at higher trophic levels (see Campoy et al. 2021, this book, Chapter 4), due to decreased prey availability (Bakun et al. 2015). For example, negative impacts of ocean acidification on the behaviour and development of early fish larvae, namely on otolith formation, but also on eggs, hatching and survival may occur (Frommel et al. 2014, Stiasny et al. 2016). However, available research is still unclear, with some species of fish larvae showing some resilience to more acidic conditions (Frommel et al. 2013). Nevertheless, the higher input of carbon dioxide can promote the enhancement of primary production by itself (Hutchings et al. 1995).

Climate change is expected to reinforce oxygen limitation in upwelling regions, a tendency already being observed in several areas (Chan et al. 2008). In this case, higher water stratification can occur, as well as reduced oxygenation of deeper layers (Rykaczewski and Dunne 2010). This scenario is linked with increased nutrient uptake by phytoplankton and respiration rates, under heightened upwelling events, and can be especially adverse for benthic organisms (Chan et al. 2008).

In summary, the variability expected in scenarios of climate change can strongly alter the composition, growth, reproduction and mortality rates of planktonic organisms, ultimately altering the nutritional quality at the base of the food web, and leading to extensive consequences for higher trophic levels and the entire ecosystem functioning in upwelling areas.

7.5 Conclusions and Future Work

Despite being located in distinct regions of the world, all major upwelling regions present several similarities. The food web structure is relatively simple, with zooplanktonic organisms representing an important trophic link and supporting highly productive ecosystems. Zooplanktonic assemblages are characterized by low diversity and high abundance of crustaceans, especially calanoid copepods, which support high biomass at upper trophic levels. Despite the potential offshore transport promoted by upwelling, zooplanktonic species, especially those performing diel vertical migrations, are able to limit their cross-shore dispersion, being retained in alongshore bands over the continental shelf. The inherent variability in upwelling intensity, related with the dominant wind patterns and other environmental conditions, leads to changes in phytoplankton primary production, consequently affecting the zooplanktonic organisms and fishery resources.

The high productivity of upwelling systems, but also its high vulnerability and the uncertainty surrounding the impacts of future climate change on upwelling intensity, makes the assessment of present conditions in upwelling regions of exceptional importance. Understanding the variability in coastal upwelling processes may provide the scientific community a means to detect major shifts in these systems and improve the simulations on future scenarios. In order to establish accurate models addressing the effects of environmental changes to which these ecosystems are and will be exposed to, there is a need to implement sustained long-term monitoring, not only of the hydrodynamic conditions, but also of the planktonic assemblages. Considering that ocean productivity is heavily linked with hydrodynamic forcing, high-resolution time-series of planktonic communities are required for the examination of the factors driving species and community patterns in upwelling systems. These steps are of extreme relevance for accurate estimates of fishery productivities and the management of marine resources, but also for the evaluation of the environmental quality of ocean waters. Moreover, at a fundamental level, we need to look at these processes at local scales, examining short-term variability driving the systems and, from the analysis of specific events, seek to resolve the global trends. Key and vulnerable species must be identified, based on the analysis of trophic and ecological interactions.

Considering the coastal upwelling regions, and despite being some of the most studied regions worldwide, there is still limited local *in situ* knowledge on metazooplankton assemblages. This knowledge is essential for improving simulations of climate change and evaluating the responses of specific regional ecosystems. One of the topics that must be clarified for upwelling regions is the impact of changes in the source of upwelled water masses on biological productivity. Moreover, the comparison of zooplanktonic assemblages and how they change in respect with upwelling intensity and water stratification, as well as the evaluation of the influence of internal waves on the transport pathways, are required, especially under a climate change scenario. Further, it is essential to be able to detect shifts in circulation intensity and patterns, identifying specific regions exposed to higher risks of eutrophication, hypoxia and acidification. Finally, and taking into account the 17 United Nations sustainable development goals, to be achieved by 2030, especially the 14th (Life Below water), and the fact that coastal upwelling areas support most of the world's fisheries, research efforts should be dedicated to these vulnerable ecosystems of the world's oceans.

Acknowledgements

PLANTROF (Programa Operacional Mar 2020 Portaria n° 118/2016), HiperSea (Portugal 2020, Aviso 03/SI/2017, Projecto 33889), EMODnet (EASME/EMFF/2016/1.3.1.2-Lot5/SI2.750022-Biology), MedEx (MARIN-ERA/MAR/0002/2008) and JellyFisheries (PTDC/MAR-BIO/0440/2014) projects. RFTP was supported by FCT (Fundação para a Ciência e a Tecnologia) through a PhD scholarship (SFRH/BD/139269/2018). We are grateful to Todd O'Brien for the preparation of Figs. 7.2 and 7.3 and for the Copepod Project (https://www.st.nmfs.noaa.gov/copepod/) developed under the ICES Working Group on Zooplankton Ecology (WGZE), and to the book editors for the help throughout the process.

References

Alheit, J. and P. Bernal. 1993. Effects of physical and biological changes on the biomass yield of the Humboldt Current ecosystem. pp. 53–68. *In*: K. Sherman, L.M. Alexander and B.D. Gold [eds.]. Large Marine Ecosystems: Stress Mitigation, and Sustainability. American Association for the Advancement of Science, Washington, USA.

Álvarez-Salgado, X.A., M.D. Doval, A.V. Borges, I. Joint, M. Frankignoulle, E.M.S. Woodward et al. 2001. Off-shelf fluxes of labile materials by an upwelling filament in the NW Iberian upwelling system. Progr. Oceanogr. 51: 321–337.

Andrews, W.R.H. and L. Hutchings. 1980. Upwelling in the southern Benguela Current. Progr. Oceanogr. 9(1): 1–81.

Auad, G., A. Miller and E. Di Lorenzo. 2006. Long-term forecast of oceanic conditions off California and their biological implications. J. Geophys. Res. 111: C09008.

Ayón, P., M.I. Criales-Hernández, R. Schwamborn and H. Hirche. 2008. Zooplankton research off Peru: A review. Progr. Oceanogr. 79: 238–255.

Bakun, A. 1990. Climate change and intensification of coastal ocean upwelling. Science 247(4939): 198–201.

Bakun, A. 1996. Patterns in the ocean: Ocean processes and marine population dynamics. California Sea Grant in cooperation with Centro de Investigaciones Biologicas del Noroeste, La Paz, Mexico.

Bakun, A., D.B. Field, A. Redondo-Rodríguez and S.J. Weeks. 2010. Greenhouse gas, upwelling-favorable winds, and the future of coastal ocean upwelling ecosystems. Glob. Change Biol. 16: 1213–1228.

Bakun, A., B.A. Black, S.J. Bograd, M. García-Reyes, A.J. Miller, R.R. Rykaczewski et al. 2015. Anticipated effects of climate change on coastal upwelling ecosystems. Curr. Clim. Change Rep. 1(2): 85–93.

Bakun, A. 2017. Climate change and ocean deoxygenation within intensified surface-driven upwelling circulations. Phil. Trans. R. Soc. A 375: 20160327.

Barth, J.A. 1989. Stability of a coastal upwelling front 1. Model development & a stability theorem. J. Geophys. Res. 94: 10844–10856.

Bartilotti, C., A. dos Santos, M. Castro, Á. Peliz and A.M.P. Santos. 2014. Decapod larval retention within distributional bands in a coastal upwelling ecosystem. Mar. Ecol. Prog. Ser. 507: 233–247.

Barton, E.D., D.B. Field and C. Roy. 2013. Canary Current upwelling: More or less? Progr. Oceanogr. 116: 167–178.

Bode, M., A. Kreiner, A.K. van der Plas, D.C. Louw, R. Horaeb, H. Auel et al. 2014. Spatio-temporal variability of copepod abundance along the 20° S monitoring transect in the northern Benguela upwelling system from 2005 to 2011. PloS One 9(5): e97738.

Brodeur, R.D., J.J. Ruzicka and J.H. Steele. 2011. Investigations alternate trophic pathways through gelatinous zooplankton and planktivorous fishes in an upwelling ecosystem using end-to-end models. pp. 57–63. *In*: K. Omori, X. Guo, N. Yoshie, N. Fujii, I.C. Handoh, A. Isobe et al. [eds.]. Interdisciplinary Studies on Environmental Chemistry–Marine Environmental Modelling and Analysis. Tokyo, Japan.

Campoy, A.N., J. Cruz, J. Barcelos e Ramos, F. Viveiros, P. Range and M.A. Teodósio. 2021. Ocean acidification impacts on zooplankton. pp. 64–82. *In*: M.A. Teodósio and A.B. Barbosa [eds.]. Zooplankton Ecology. CRC Press.

Carrasco, S. and H. Santander. 1987. The El Niño event and its influence on the zooplankton off Peru. J. Geophys. Res. 92(C13): 14,405–14,410.

Chan, F., J.A. BArth, J. Lubchenco, A. Kirincich, H. Weeks, W.T. Peterson et al. 2008. Emergence of anoxia in the California current large marine ecosystem. Science 319: 920.

Chavez, F.P. and M. Messié. 2009. A comparison of Eastern boundary upwelling ecosystems. Progr. Oceanogr. 83: 80–96.

Chícharo, M.A., E. Esteves, A.M.P. Santos, A. dos Santos, Á. Peliz and P. Ré. 2003. Are sardine larvae caught off northern Portugal in winter starving? An approach examining nutritional conditions. Mar. Ecol. Prog. Ser. 257: 303–309.

Collins, M., S. An, W. Cai, A. Ganachaud, E. Guilyardi, F. Jin et al. 2010. The impact of global warming on the tropical Pacific Ocean and El Niño. Nat. Geosci. 3: 391–397.

Condon, R.H., C.M. Duarte, K.A. Pitt, K.L. Robinson, C.H. Lucas, K.R. Sutherland et al. 2013. Recurrent jellyfish blooms are a consequence of global oscillations. P. Natl. Acad. Sci. USA. 110(3): 1000–1005.

Criales-Hernández, M.I., R. Schwamborn, M. Graco, P. Ayón, H.J. Hirche and M. Wolff. 2008. Zooplankton vertical distribution and migration off Central Peru in relation to the oxygen minimum layer. Helgol. Mar. Res. 62(Suppl. 1): S85–S100.

Cury, P., A. Bakun, R.J.M. Crawford, A. Jarre, R. A. Quiñones, L.J. Shannon et al. 2000. Small pelagics in upwelling systems: patterns of interaction and structural changes in "wasp-waist" ecosystems. ICES J. Mar. Sci. 57: 603–618.

Demer, D.A. and J.P. Zwolisnki. 2014. Corroboration and refinement of a method for differentiating landings from two stocks of Pacific sardine (*Sardinops sagax*) in the California Current. ICES J. Mar. Sci. 71(2): 328–335.

Domínguez, R., S. Garrido, A.M.P. Santos and A. dos Santos. 2017. Spatial patterns of mesozooplankton communities in the Northwestern Iberian shelf during autumn shaped by key environmental factors. Est. Coast. Shelf Sci. 198: 257–268.

Dos Santos, A. and Á. Peliz. 2005. The occurrence of Norway lobster (*Nephrops norvegicus*) larvae off the Portuguese coast. J. Mar. Biol. Ass. U.K. 85: 937–941.

Dos Santos, A., A.M.P. Santos and D.V.P. Conway. 2007. Horizontal and vertical distribution of cirripede cyprid larvae in an upwelling system off the Portuguese coast. Mar. Ecol. Prog. Ser. 329: 145–155.

Dos Santos, A., A.M.P. Santos, D.V.P. Conway, C. Bartilotti, P. Lourenço and H. Queiroga. 2008. Diel vertical migration of decapod larvae in the Portuguese coastal upwelling ecosystem: implications for offshore transport. Mar. Ecol. Prog. Ser. 359: 171–183.

Escribano, R. and P. Hidalgo. 2000. Spatial distribution of copepods in the north of the Humboldt Current region off Chile during coastal upwelling. J. Mar. Biol. Assoc. U.K. 80: 283–290.

Escribano, R., V.H. Marin and C. Irribarren. 2000. Distribution of *Euphausia mucronata* at the upwelling area of Peninsula Mejillones, northern Chile: the influence of the oxygen minimum layer. Sci. Mar. 64(1): 69–77.

Escribano, R., V. Marin, P. Hidalgo and G. Olivares. 2002. Physical–biological interactions in the nearshore zone of the northern Humboldt Current ecosystem. pp. 145–175. *In*: J.C. Castilla and J.L. Largier [eds.]. The Oceanography and Ecology of the Nearshore and Bays in Chile. Ediciones Universidad Católica de Chile.

Escribano, R., P. Hidalgo, H. González, R. Giesecke, R. Riquelme-Bugueño and K. Manríquez. 2007. Seasonal and inter-annual variation of mesozooplankton in the coastal upwelling zone off central-southern Chile. Progr. Oceanogr. 75: 470–485.

Escribano, R., P. Hidalgo and C. Krautz. 2009. Zooplankton associated with the oxygen minimum zone system in the northern upwelling region of Chile during March 2000. Deep-Sea Res. PT II. 56: 1083–1094.

Fennel, W. 1999. Theory of the Benguela upwelling system. J. Phys. Oceanogr. 29(2): 177–190.

Frommel, A.Y., A. Schubert, U. Piatkowski and C. Clemmesen. 2013. Egg and early larval stages of Baltic cod, *Gadus morhua*, are robust to high levels of ocean acidification. Mar. Biol. 160: 1825–1834.

Frommel, A.Y., R. Maneja, D. Lowe, C.K. Pascoe, A.J. Geffen, A. Folkvord et al. 2014. Organ damage in Atlantic herring larvae as a result of ocean acidification. Ecol. Appl. 24(5): 1131–1143.

Gibbons, M.J. and L. Hutchings. 1996. Zooplankton diversity and community structure around southern Africa, with special attention to the Benguela upwelling system. S. Afr. J. Mar. Sci. 92: 63–76.

Graham, W.M., J.G. Field and D.C. Potts. 1992. Persistent "upwelling shadows" and their influence on zooplankton distributions. Mar. Biol. 114: 561–570.

Gruber N., C. Hauri, Z. Lachkar, D. Loher, T. Frölicher and G.-K. Plattner. 2012. Rapid progression of ocean acidification in the California Current system. Science 337: 220.

Harley, C.D.G., A.R. Hughes, K.M. Hultgren, B.G. Miner, C.J.B. Sorte, C.S. Thomber et al. 2006. The impacts of climate change in coastal marine systems. Ecol. Lett. 9: 228–241.

Haynes, R., E.D. Barton and I. Pilling. 1993. Development, persistence, and variability of upwelling filaments off the Atlantic coast of the Iberian Peninsula. J. Geophys. Res. 98: 22681–22692.

Hays, G.C., T.K. Doyle and J.D.R. Houghton. 2018. A paradigm shift in the trophic importance of jellyfish? Trends Ecol. Evol. 33(11): 874–884.

Hernández-León, S., M. Goméz and J. Arístegui. 2007. Mesozooplankton in the Canary Current system: The coastal-ocean transition zone. Progr. Oceanogr. 74: 397–421.

Hickey, B.M. 1998. Coastal Oceanography of Western North America from the tip of Baja California to Vancouver Island. pp. 345–393. *In*: K.H. Brink and A.R. Robinson [eds.]. The Sea. Wiley and Sons, Inc., New York, USA.

Hickey, B.M. and N.S. Banas. 2008. Why is the northern end of the California Current system so productive? Oceanography 21(4): 90–107.

Hidalgo, P., R. Escribano and C.E. Morales. 2005. Annual life cycle of the copepod *Eucalanus inermis* at a coastal upwelling site off Mejillones (23° S), northern Chile. Mar. Biol. 146: 995–1003.

Hubbard, L.T. and W.G. Pearcy. 1971. Geographic distribution and relative abundance of Salpidae off the Oregon coast. J. Fish. Res. Board Can. 28(12): 1831–1836.

Hurrell, J.W., Y. Kushnir, G. Ottersen and M. Visbeck. 2003. An overview of the North Atlantic Oscillation: Climatic Significance and Environmental Impact. Geophysical Monograph 134. American Geophysical Union, Washington, USA.

Hutchings, L., D.A. Armstrong and B.A. Mitchell-Innes. 1986. The frontal zone in the Southern Benguela Current. pp. 67–94. *In*: J.C.J. Nihoul [ed.]. Marine Interfaces Ecohydrodynamics, Elsevier Oceanography Series 42. Elsevier, Amsterdam, Netherlands.

Hutchings, L., G.C. Pitcher, T.A. Probyn and G.W. Bailey. 1995. The chemical and biological consequences of coastal upwelling. pp. 64–81. *In*: C.P. Summerhayes, K.C. Emeis, M.V. Angel, R.L. Smith and B. Zeitzschel [eds.]. Upwelling in the Ocean: Modern Processes and Ancient Records. Wiley & Sons, New York, USA.

Hutchings, L., H.M. Verheye, J.A. Huggett, H. Demarcq, R. Cloete, R.G. Barlow et al. 2006. Variability of plankton with reference to fish variability in the Benguela Current Large Marine Ecosystem—An overview. pp. 91–124. *In*: V. Shannon, G. Hempel, P. Malanotte-Rizzoli, C. Moloney and J. Woods [eds.]. Benguela: Predicting a Large Marine Ecosystem. Large Marine Ecosystems 14. Elsevier, Amsterdam, Netherlands.

Jahnke, R.A. 2010. Global synthesis. pp. 597–615. *In*: K.K. Liu, L. Atkinson, R. Quiñones and L. Talaue-McManus [eds.]. Carbon and Nutrient Fluxes in Continental Margins: A Global Synthesis. Springer, New York, USA.

Kampf, J. and P. Chapman. 2016. Upwelling Systems of the World: A Scientific Journey to the Most Productive Marine Ecosystems. Springer International Publishing, Switzerland.

Longhurst, A.R. 1967. Diversity and trophic structure of zooplankton communities in the California Current. Deep-Sea Res. 14: 393–408.

Lutjeharms, J.R.E. and P.L. Stockton. 1987. Kinematics of the upwelling front off southern Africa. S. Afr. J. Mar. Sci. 5(1): 35–49.

Lynam, C.P., M.J. Gibbons, B.E. Axelsen, C.A.J. Sparks, J. Coetzee, B.G. Heywood et al. 2006. Jellyfish overtake fish in a heavily fished ecosystem. Curr. Biol. 16(13): R492.

Menge, B.A., C. Blanchette, P. Raimondi, T. Freidenburg, S. Gaines, J. Lubchenco et al. 2004. Species interaction strength: Testing model predictions along an upwelling gradient. Ecol. Monogr. 74(4): 663–684.

Miller, D.C.M., C.L. Moloney, C.D. van der Lingen, C. Lett, C. Mullon and J.G. Field. 2006. Modelling the effects of physical–biological interactions and spatial variability in spawning and nursery areas on transport and retention of sardine *Sardinops sagax* eggs and larvae in the southern Benguela ecosystem. J. Mar. Syst. 61: 212–229.

Miller, M.-E.C. and W.M. Graham. 2012. Environmental evidence that seasonal hypoxia enhances survival and success of jellyfish polyps in the northern Gulf of Mexico. J. Exp. Mar. Biol. 432-433: 113–120.

Moita, M.T. 2001. Estrutura, variabilidade e dinâmica do fitoplâncton na costa de Portugal Continental. Ph.D. Thesis, Faculdade de Ciências da Universidade de Lisboa, Lisbon, 272 pp.

Morgan, C.A., W.T. Peterson and R.L. Emmett. 2003. Onshore-offshore variations in copepod community structure off the Oregon coast during the summer upwelling season. Mar. Ecol. Prog. Ser. 249: 223–236.

Morgan, S.G., J.L. Fisher, S.H. Miller, S.T. McAfee and J.L. Largier. 2009. Nearshore larval retention in a region of strong upwelling and recruitment limitation. Ecology 90(12): 3489–3502.

Morgan, S.G., J.L. Fisher, S.T. McAfee, J.L. Largier and C.M. Halle. 2012. Limited recruitment during relaxation events: Larval advection and behavior in an upwelling system. Limnol. Oceanogr. 57(2): 457–470.

Morgan, S.G., S.H. Miller, M.J. Robart and J. Largier. 2018. Nearshore larval retention and cross-shelf migration of benthic crustaceans at an upwelling center. Front. Mar. Sci. 5: 161.

Navarrete, S.A., E.A. Wieters, B.R. Broitman and J.C. Castilla. 2005. Scales of benthic-pelagic coupling and the intensity of species interactions: From recruitment limitation to top-down control. P. Natl. Acad. Sci. USA 102(50): 18046–18051.

O'Brien, T.D., P.H. Wiebe and T. Falkenhaug. 2013. ICES Zooplankton Status Report 2010/2011. ICES Cooperative Research Report No. 318, pp. 208.

O'Brien, T.D. and S. Oakes. 2021. Visualizing and exploring zooplankton spatio-temporal variability. pp. 192–224. *In*: M.A. Teodósio and A.B. Barbosa [eds.]. Zooplankton Ecology. CRC Press.

Olivar, M.P. and M. Barangé. 1990. Zooplankton of the northern Benguela region in a quiescent upwelling period. J. Plank. Res. 12(5): 1023–1044.

Pagès, F., H.M. Verheye, J.-M. Gili and J. Flos. 1991. Short-term effects of coastal upwelling and wind reversals on epiplanktonic cnidarians in the southern Benguela ecosystem. S. Afr. J. Mar. Sci. 10(1): 203–211.

Pagès, F. and J.-M. Gili. 1992. Influence of Agulhas waters on the populations structure of planktonic Cnidarians in the southern Benguela region. Sci. Mar. 56(2-3): 109–123.

Pagès, F., H.E. González, M. Ramón, M. Sobarzo and J.-M. Gili. 2001. Gelatinous zooplankton assemblages associated with water masses in the Humboldt Current system, and potential predatory impact by *Bassia bassensis* (Siphonophora: Calycophorae). Mar. Ecol. Progr. Ser. 210: 13–24.

Parada, C., C.D. van der Lingen, C. Mullon and P. Penven. 2003. Modelling the effect of buoyancy on the transport of anchovy (*Engraulis capensis*) eggs from spawning to nursery grounds in the southern Benguela: an IBM approach. Fish. Oceanogr. 12(3): 170–184.

Peliz, Á., D. Boutov, R.M. Cardoso, J. Delgado and P.M.M. Soares. 2013. The Gulf of Cadiz-Alboran Sea sub-basin: Model setup, exchange and seasonal variability. Ocean. Model. 61: 49–67.

Penven, P., V. Echevin, J. Pasapera, F. Colas and J. Tam. 2005. Average circulation, seasonal cycle, and mesoscale dynamics of the Peru Current system: A modelling approach. J. Geophys. Res. 110: C10021.

Peterson, W.T., D.F. Arcos, G.B. McManus, H. Dam, D. Bellantoni, T. Johnson et al. 1988. The nearshore zone during coastal upwelling: Daily variability and coupling between primary and secondary production off Chile. Prog. Oceanogr. 20: 1–40.

Peterson, W. 1998. Life cycle of copepods in coastal upwelling zones. J. Mar. Syst. 15: 313–326.

Pineda, J., J.A. Hare and S. Sponaugle. 2007. Larval transport and dispersal in the coastal ocean and consequences for population connectivity. Oceanography 20(3): 22–39.

Pires, R.F.T., M. Pan, A.M.P. Santos, Á. Peliz, D. Boutov and A. Dos Santos. 2013. Modelling the variation in larval dispersal of estuarine and coastal ghost shrimp: *Upogebia* congeners in the Gulf of Cadiz. Mar. Ecol. Prog. Ser. 492: 153−168.

Pochelon, P.N., R.F.T. Pires, J. Dubert, R. Nolasco, A.M.P. Santos, H. Queiroga et al. 2017. Decapod larvae distribution and species composition off the southern Portuguese coast. Cont. Shelf. Res. 151: 53–61.

Pond, S. and G.L. Pickard. 1983. Introductory Dynamical Oceanography. Butterworth-Heinemann Ltd, Oxford.

Postel, L. 1990. The meso-zooplankton response to coastal upwelling off West Africa with particular regard to biomass. Meereswis-senschaftliche Berichte, Warnemünde. 1: 127.

Queiroga, H., T. Cruz, A. dos Santos, J. Dubert, J.I. González-Gordilo, J. Paula et al. 2007. Oceanographic and behavioural processes affecting invertebrate larval dispersal and supply in the western Iberia upwelling ecosystem. Progr. Oceanogr. 74: 174–191.

Relvas, P., E.D. Barton, J. Dubert, P.B. Oliveira, Á. Peliz, J.C.B. da Silva et al. 2007. Physical oceanography of the western Iberia ecosystem: Latest views and challenges. Progr. Oceanogr. 74: 149–173.

Richardson, A.J. 2008. In hot water: zooplankton and climate change. ICES J. Mar. Sci. 65: 279–295.

Roemmich, D. and J. McGowan. 1995. Climate warming and the decline of zooplankton in the California Current. Sci. 267: 1324–1328.

Rivera, M.J., G. Guzmán and S. Palma. 2019. Cross-shelf distribution of decapod larvae in a coastal upwelling zone of northern Chile. Cont. Shelf Res. 181: 50–71.

Ryan, J.P., A.M. Fischer, R.M. Kudela, J.F.R. Gower, S.A. King, R. Marin III et al. 2009. Influences of upwelling and downwelling winds on red tide bloom dynamics in Monterey Bay, California. Cont. Shelf. Res. 29: 785–795.

Rykaczewski, R.R. and J.P. Dunne. 2010. Enhanced nutrient supply to the California Current ecosystem with global warming and increased stratification in na earth system model. Geophys. Res. Lett. 37: L21606.

Rykaczewski, R.R., J.P. Dunne, W.J. Sydeman, M. García-Reyes, B.A. Black and S.J. Bograd. 2015. Poleward displacement of coastal upwelling-favourable winds in the ocean's eastern boundary currents through the 21st century. Geophys. Res. Lett. 42: 6424–6431.

Santander, B.H. 1981. The zooplankton in an upwelling area off Peru. Coast. Est. S. 1: 411–416.

Santos, A.M.P., Á. Peliz, J. Dubert, P.B. Oliveira, M.M. Angélico and P. Ré. 2004. Impact of a winter upwelling event on the distribution and transport of sardine (*Sardina pilchardus*) eggs and larvae off western Iberia: a retention mechanism. Cont. Shelf Res. 24: 149–165.

Santos, A.M.P., A.S. Kazmin and Á. Peliz. 2005. Decadal changes in the Canary upwelling system as revealed by satellite observations: Their impact on productivity. J. Mar. Res. 63: 359–379.

Shannon, L.V., R.J. Crawford, D.E. Pollock, L. Hutchings, A.J. Boyd, J. Taunton-Clark et al. 1992. The 1980s—A decade of change in the Benguela ecosystem. S. Afr. J. Mar. Sci. 12(1): 271–296.

Shelton, P.A. and L. Hutchings. 1990. Ocean stability and anchovy spawning in the southern Benguela region. Fish. Bull. U.S. 88: 323–338.

Shiozaki, T., T. Kodama and K. Furuya. 2014. Large-scale impact of the island mass effect through nitrogen fixation in the western South Pacific Ocean. Geophys. Res. Lett. 41: 2907–2913.

Silva, A. 2003. Morphometric variation among sardine (*Sardina pilchardus*) populations from the northeastern Atlantic and the western Mediterranean. ICES J. Mar. Sci. 60: 1352–1360.

Simpson, J.H. and J. Sharples. 2012. Introduction to the Physical and Biological Oceanography of Shelf Seas. Cambridge University Press, Cambridge.

Smith, S.L., B.H. Jones, L.P. Atkinson and K.H. Brink. 1986. pp. 195–213. *In*: J.C.J. Nihoul [ed.]. Marine Interfaces Ecohydrodynamics, Elsevier Oceanography Series 42. Elsevier, Amsterdam, Netherlands.

Sobarzo, M. and D. Figueroa. 2001. The physical structure of a cold filament in a Chilean upwelling zone Península Mejillones, Chile, 23° S. Deep-Sea Res. Pt I. 48: 2699–2726.

Star, J.L. and M.M. Mullin. 1981. Zooplanktonic assemblages in three areas of the North Pacific as revealed by continuous horizontal transects. Deep-Sea Res. 28A(11): 1303–1322.

Stenevik, E.K., M. Skogen, S. Sundby and D. Boyer. 2003. The effect of vertical and horizontal distribution on retention of sardine (*Sardinops sagax*) larvae in the Northern Benguela—observations and modelling. Fish. Oceanogr. 12(3): 185–200.

Stiasny, M.H., F.H. Mittermayer, M. Sswat, R. Voss, F. Jutfelt, M. Chierici et al. 2016. Ocean acidification effects on Atlantic Cod larval survival and recruitment to the fished population. PLoS ONE 11(8): e0155448.

Sydeman, W.J., M. García-Reyes, D.S. Schoeman, R.R. Rykaczewski, S.A. Thompson, B.A. Black et al. 2014. Climate change and wind intensification in coastal upwelling ecosystems. Science 345(6192): 77–80.

Thiriot, A. 1978. Zooplankton communities in the west African upwelling area. pp. 32–61. *In*: R. Boje and M. Tomczak [eds.]. Upwelling Ecosystems. Springer-Verlag, Berlin, Germany.

Verheye, H.M. 1991. Short-term variability during an anchor station study in the southern Benguela upwelling system: Abundance, distribution and estimated production of mesozooplankton with special reference to *Calanoides carinatus* (Krøyer, 1849). Progr. Oceanogr. 28: 91–119.

Verheye, H.M., A.J. Richardson, L. Hutchings, G. Marska and D. Gianakouras. 1998. Long-term trends in the abundance and community structure of coastal zooplankton in the Southern Benguela system, 1951–1996. S. Afr. J. Mar. Sci. 19: 317–332.

Verheye, H.M. 2000. Decadal-scale trends across several marine trophic levels in the Southern Benguela upwelling system off South Africa. Ambio 29(1): 30–34.

Weeks, S.J., B. Currie, A. Bakun and K.R. Peard. 2004. Hydrogen sulphide eruptions in the Atlantic Ocean off southern Africa: implications of a new view based on SeaWiFS satellite imagery. Deep-Sea Res. 51: 153–172.

Wilkerson, F.P., R.C. Dugdale, R.M. Kudela and F.P. Chavez. 2000. Biomass and productivity in Monterey Bay, California: contribution of the large phytoplankton. Deep-Sea Res. Pt II 47: 1003–1022.

Wyrtki, K. 1981. An estimate of Equatorial upwelling in the Pacific. J. Phys. Oceanogr. 11: 1205–1214.

CHAPTER 8

Interactions in Plankton Food Webs
Seasonal Succession and Phenology of Baltic Sea Zooplankton

Monika Winder[1],* and *Øystein Varpe*[2]

8.1 Introduction

In the world's largest ecosystems, the oceans, marine plankton form complex communities and interact in many different ways (Lima-Mendez et al. 2015). Food web interactions determine the transfer of energy from primary producers to higher trophic levels, cycling of carbon and energy within the pelagic system, and export of pelagic production to the seafloor, and thus global biogeochemical processes (Falkowski et al. 1998, Steinberg and Landry 2017). Zooplankton perform an important ecosystem service as the main group of grazers on the pelagic primary production and as the main prey of fish, and therefore link lower and higher trophic levels of aquatic food webs. In some systems, the zooplankton level also comprises the largest standing biomass across groups, larger than both the primary producers lower in the food web, and fish and other predator levels higher in the web (Gasol et al. 1997). Zooplankton, particularly copepods, are the major food source for commercially important fish in marine systems, and by grazing on phytoplankton, zooplankton also control phytoplankton and their blooms.

Plankton communities are formed by a large diversity of taxa from viruses, heterotrophic prokaryotes and other unicellular organisms, including flagellates, ciliates to multicellular organisms such as rotifers, diverse crustaceans from copepods to shrimps, chaetognaths and jellyfish. Planktonic organisms range from around 0.02 μm to 200 cm (Sieburth et al. 1978, Steinberg and Landry 2017) (Fig. 8.1), varying with more than six orders of magnitude. The majority of biological activity (> 90%) in aquatic food webs takes place in microorganisms, which includes organisms feeding on at least four different trophic levels (Calbet and Landry 2004). For example, heterotrophic nanoflagellates eat picophytoplankton, that in turn are preyed upon by ciliates (microzooplankton) and further

[1] Department of Ecology, Environment and Plant Sciences, Stockholm University, 106 91 Stockholm, Sweden.

[2] Department of Biological Sciences, University of Bergen, 5020 Bergen, Norway & Norwegian Institute for Nature Research, 5006 Bergen, Norway.
 Email: oystein.varpe@uib.no

* Corresponding author: monika.winder@su.se

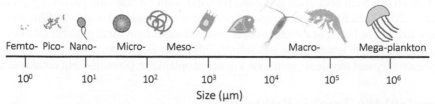

Figure 8.1: Size spectra of plankton illustrated using dominant species of the Baltic Sea. Size range from femto- (0.02–0.2 µm; e.g., viruses), pico- (0.2–2 µm; e.g., heterotrophic prokaryotes, picophytoplankton), nano- (2–20 µm; e.g., phytoplankton), micro- (20–200 µm; e.g., flagellates, ciliates), meso- (0.2–20 mm, e.g., cladocerans, copepods), macro- (2–20 cm; e.g., shrimps) to megaplankton (20–200 cm; e.g., jellyfish) (Sieburth et al. 1978).

consumed by copepods (mesozooplankton), and thus connect the microbial loop with the classical grazing food web (phytoplankton-zooplankton-fish) (Sommer et al. 2002, Basedow et al. 2016). In the interaction network, zooplankton have a key role by concentrating and channelling carbon and essential biochemicals from primary producers to upper trophic levels, such as fish (Varpe et al. 2005, Winder et al. 2017b).

While the seasonal succession of phytoplankton groups and their traits and driving mechanisms are much studied and well described (Sommer et al. 2012, 2017, Weithoff and Beisner 2019), the succession of zooplankton species and their traits is less well synthesized, which is probably a result of more complex behaviours, longer generation times and life cycles in zooplankton (Romagnan et al. 2015). For phytoplankton, increasing abundances are initiated by increasing sunlight and shoaling of the mixed-layer depth, referred to as the critical depth hypothesis (Sverdrup 1953), after intense vertical mixing that redistributes nutrients throughout the water column, although this has been challenged by the critical turbulence and disturbance-recovery hypotheses (Behrenfeld and Boss 2014). The seasonal development of zooplankton follows the spring phytoplankton bloom, with a succession from grazing to predatory organisms (Sommer et al. 2012). Autumn water-column mixing and reduced day length redistribute cells in the water column and terminate the annual succession. Phyto- and zooplankton seasonal succession and species replacement are affected by physical factors, nutrient or food limitation and biotic interactions, such as food quality, parasitism, or fish predation (Sommer et al. 2012, Behrenfeld and Boss 2014, Romagnan et al. 2015).

Understanding the seasonal succession of zooplankton, as well as their responses to changing environmental conditions, requires detailed knowledge of their population dynamics through the annual cycle as well as their interactions with food, competitors and predators. This is particularly so in temperate and polar systems where seasonal cycles of pelagic production are strong.

Zooplankton diversity is large also with respect to life history traits, morphologies, behaviours and feeding modes. Studying the evolutionary drivers behind this diversity, and the underlying trade-offs that have shaped them, is important to understand how different species perform in a seasonal environment, how they interact, and during which periods they are present and abundant. In seasonal environments, the timing of life history events within the annual cycle and their timing with regard to ecological processes require particular attention. This study is called phenology, a word also increasingly used for the actual timing, such as breeding phenology or diapause phenology.

In this chapter, the seasonal aspects of interactions between primary consumers and zooplankton are highlighted and the resulting annual cycles of zooplankton dynamics described. The foci are intra-annual population dynamics as well as the degree that these patterns repeats themselves between years. Particular attention is given to the treatment of temporal interactions in planktonic food webs, the role of phenology and the accompanying life history adaptations to seasonality. Observed dynamics are in turn related to life history traits, likely ecological drivers and environmental constraints. The Baltic Sea is used as a case study. The Baltic Sea is a strongly seasonal ecosystem and one of the best studied brackish water ecosystems with extensive multi-year monitoring programmes, including several stations with sampling designs that have high sampling frequency (Griffiths et al. 2017,

Reusch et al. 2018). Such resolution is needed for solid work on plankton phenology (Ji et al. 2010, Mackas et al. 2012) and for comparing model predictions on seasonal timing and life histories with data (Varpe 2012). The Baltic Sea stretches over a considerable latitudinal gradient (about 13 degrees) with strong gradients in temperature and salinity, allowing rich opportunities for illustrating how varying seasonality and physical conditions impact zooplankton (Snoeijs-Leijonmalm et al. 2017).

8.2 Zooplankton Trophic Pathways

Plankton community structure determines energy availability for higher trophic levels (Stibor et al. 2004), global biogeochemical cycles (Litchman et al. 2015) and remineralization of macronutrients (Calbet and Landry 2004). The aquatic food web, the pelagic one in particular, is often size structured where small organisms are eaten by larger ones (Barnes et al. 2010) (Fig. 8.2). The size ratio between primary consumers and phytoplankton is typically assumed to be 10.1 (Hansen et al. 1994); however, prey size vary among different predator groups, and the ratio is smaller within microbial organisms and higher for mesozooplankton species. Consequently, several trophic levels separate small phytoplankton cells from the larger zooplankton organisms (Stibor et al. 2004). There are, however, exceptions to the size-structured food web where predators are smaller than prey, such as pallium feeding (Jacobson and Anderson 2008) or peduncle-feeding heterotrophic dinoflagellates (Ok et al. 2017). Or prey may be orders of magnitude smaller than themselves within mesozooplankton, such as mucous-mesh grazers (e.g., appendicularians, pelagic tunicates) (Conley et al. 2018).

Within the microbial loop, bacteria are being grazed by heterotrophic nanoflagellates (HNF), which together with small-sized phytoplankton cells are being grazed upon by microzooplankton, which are then prey for mesozooplankton. Lager-sized phytoplankton, however, forms a more direct route, the classical algae-zooplankton-fish link (Fig. 8.2). Part of this production is recycled within the

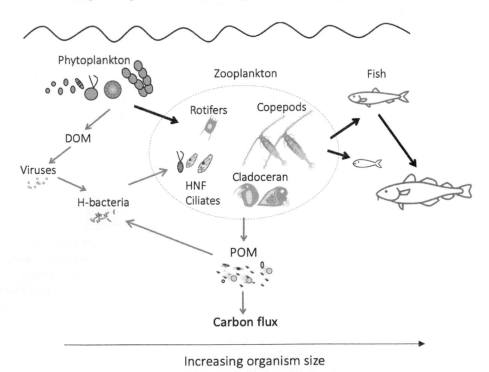

Figure 8.2: A basic model of the pelagic food web with key players, illustrated using dominant taxonomic groups of the Baltic Sea. Trophic (black), detrital (grey) pathways and sedimentation to the seafloor are shown. Organism size increases from left to right. Trophic interactions within the zooplankton assemblage (dashed line) are not shown. DOM = dissolved organic material, POM = particulate organic material, H-bacteria = heterotrophic prokaryotes, HNF = heterotrophic nanoflagellates.

detrital food web through non-living organic matter that is leached out from all organisms (dissolved organic material, DOM) or derived from plant and animal tissue, such as discarded appendicularian houses and crustacean exoskeletons after moulting, or faeces and classified as particulate organic matter (POM) (Fig. 8.2). This non-living organic matter is returned to the food web through its incorporation into bacterial biomass that in turn is grazed by protists or deposited to the seafloor. In addition, mixotrophic organisms that are capable of both photosynthesis and phagotrophy, or the combination of both primary and secondary production in the same organisms, are common within protists, consequently altering carbon and energy flows in food webs (Flynn et al. 2019). Thus, plankton is characterized by a multitude of trophic and non-trophic interactions at the microbial scale (Lima-Mendez et al. 2015, Basedow et al. 2016).

Size-structured trophic interactions imply that the size structure of the phytoplankton community is an important factor determining carbon flow as it affects the relative importance of the microbial and classical food web and consequently food web length and transfer efficiency of zooplankton to upper trophic levels. The size structure of the phytoplankton community is in turn controlled by the supply of growth-limiting dissolved inorganic nutrients, such as nitrogen or phosphorus (Sommer et al. 2002). Under oligotrophic condition, small cells of primary producers have a relative advantage due to their larger surface to volume ratio that facilitates nutrient uptake, compared to larger cells (Falkowski and Oliver 2007). Under eutrophic conditions, the selective pressure for small size is reduced and larger-sized phytoplankton or colony forming species dominate (Sommer et al. 2002). In oligotrophic and eutrophic systems with poorly edible phytoplankton blooms, up to 75% of the daily primary production is transferred within the microbial loop (Landy and Calbet 2004). This implies that the majority of production occurs at the lower end of the size spectrum.

Aquatic food webs are characterized by low transfer efficiency of matter and energy across trophic levels, which is in the range of about 10–20% because of respiration, excretion, egestion or sloppy feeding between trophic levels (Sommer et al. 2002). Given a predator prey size ratio of 10, for consumers in the 1 mm size range, such as copepods occupying trophic level three, less than 1% primary production remains available. Assuming the rules of ecological efficiency, an increase in transfer efficiency is expected with an increasing contribution of larger phytoplankton cell sizes and thus in more productive regions compared to nutrient poor regions. However, other factors than size influence carbon transfer efficiency within the plankton food web, such as food quality, mixotrophy or patchiness. Food quality, including essential macromolecules for consumers that are produced by phytoplankton such as polyunsaturated fatty acids or sterols, varies greatly among phytoplankton organisms and major taxonomic groups (Galloway and Winder 2015). Phytoplankton consisting of species high in essential compounds increase transfer to upper trophic levels (Burian et al. 2019). Common among microzooplankton are mixotrophic plankton that can simultaneously exploit inorganic resources and living prey (Stoecker et al. 2009, Flynn et al. 2019), which enhance trophic transfer (Ward and Follows 2016). In addition, micro-scale variability in plankton distribution (patchiness) and predator-prey overlap enhances trophic transfer in oligotrophic oceans (Priyadarshi et al. 2019).

There are also alternative pathways from primary producers to higher trophic levels. For example, some mesozooplankton, like the cladoceran *Bosmina* are able to feed directly on heterotrophic prokaryotes (bacteria and archaea), reducing the number of trophic steps and energy loss within the microbial loop (Sommer et al. 2002). Filamentous and toxic cyanobacteria are less edible and digestible for zooplankton (Sommer 1989) and the energy (carbon) is reaching higher trophic levels via the microbial food web through cell lysis and excretion, which reduces energy transfer to higher trophic levels by including additional trophic levels.

Nutrient availability and consequently phytoplankton cell size and food web structure are strongly affected by climate change, which affects vertical mixing of the water column (Falkowski and Oliver 2007, Winder and Sommer 2012). Increasing temperature is strengthening vertical stratification and decreasing nutrient supply to the photic zone due to reduced mixing, which is favouring small-sized phytoplankton in oligotrophic regions (Winder et al. 2009a, Winder and Sommer 2012) and is expected to enhance carbon cycling within the microbial loop. In systems with high dissolved phosphorus availability and nitrogen limitation, such as the Baltic Sea, warming favours filamentous cyanobacteria that are able to fix atmospheric nitrogen (diazotrophs, e.g., *Nodularia*) (Paerl and Huisman 2008). Filamentous cyanobacteria are assumed to be less edible for zooplankton organisms and carbon then enters the food web via bacterial decomposition and the microbial loop (Loick-Wilde et al. 2019).

8.3 Zooplankton Life Histories and Adaptations to Seasonality

As for any organism, zooplankton species cannot maximize all traits that contribute to fitness. Instead, there are trade-offs, for instance between offspring size and numbers, adult body size and survival, and between growth and reproduction (Stearns 1992). A large range of solutions to these trade-offs have evolved, giving rise to biodiversity. Allan (1976) illustrated this variation in his analysis of growth rates and life history traits in zooplankton. He focused on the major groups of rotifers, cladocerans, and copepods. These three groups are also among the most central zooplankton groups in the Baltic Sea. In Allan's analysis, rotifers have the highest potential for population growth among the three groups, achieved particularly through short development times from egg to first reproduction. Adult body size is also smallest in the rotifers. Cladocerans are similar to the rotifers with respect to several life history traits, but longer developmental times lead to somewhat lower maximum intrinsic rate of increase, despite higher fecundity and longer lifespan in the cladocerans. Both rotifers and cladocerans have parthenogenetic reproduction when conditions are favourable. Copepods have considerably lower growth potential because of longer developmental times combined with larger adult body size. Copepods have few or only one generation per year and may even use multiple years to reach maturity and reproduction. Also, copepods always reproduce sexually, whereas sexual reproduction in rotifers and cladocerans primarily takes place when resting eggs are produced at the onset of unfavourable conditions.

Allan (1976) stressed the link between life history traits and the seasonal fluctuations of zooplankton, a focus also adopted in this chapter. This section gives a brief overview of life history traits that are important for understanding zooplankton adaptations to seasonality and for interpreting phenology and population dynamics in seasonal environments. In doing so, annual routines are referred to as an organism's regular schedule of activities or behaviours over the annual cycle (McNamara and Houston 2008). The concept of annual routines helps clarify how different activities over the year are linked and how changes in one activity usually lead to changes in others through temporal trade-offs. From an evolutionary perspective, it can be asked what the optimal annual routine would be given the environment (McNamara and Houston 2008, Varpe 2012) and how it may change in response to environmental change, as well as the consequences of no response to expected change (Feró et al. 2008). The ability of zooplankton to respond to changed seasonal timing of food availability vary depending on life history strategy (Winder and Schindler 2004b).

Schematically speaking, it can be useful to group zooplankton adaptations to seasonality as related to two parts of the year: the unproductive part when food is low (often winter like conditions), or productive part when food is high (Varpe 2017). The distinct seasonality of primary production observed in higher-latitude environments (Winder and Cloern 2010, Ji et al. 2013) exemplifies such productivity regimes, and many life history adaptations of zooplankton have evolved in response. The food availability for zooplankton in the Baltic Sea is also highly seasonal (Hjerne et al. 2019).

The unproductive period is usually spent in one of two main forms: a seed like stage, usually referred to as resting eggs, or as a well-developed juvenile or an adult stage. Resting eggs, and the accompanying embryonic dormancy, is common across many taxa including rotifers, copepods and cladocerans (Marcus 1996, Holm et al. 2018). Resting eggs usually sink to the bottom and are in the sediments until hatching, typically prior to the next productive season. However, resting eggs can live long and do not necessarily hatch before after several years (De Meester 1993). Hence, they form the analogy to a seed-bank, which is viewed as a bet-hedging strategy. Resting eggs can for instance remain viable in anoxic sediments and reoxygenation of anoxic sediments activates a large pool of buried zooplankton eggs (Broman et al. 2015). Furthermore, resting eggs have not only evolved as an adaptation to low food availability. Harsh conditions, such as warm or cold temperatures have also been highlighted, both as proximate and ultimate driver (Holm et al. 2018). Furthermore, some species may enter dormancy to avoid periods of high predation risk, such as the summer diapause in some daphnids (Pijanowska and Stolpe 1996). Diapause during summer may also be selected because of warm waters (Chinnery and Williams 2003) or a food bottleneck caused by competition from other species (Santer and Lampert 1995). There are different forms of resting eggs and dormancy. These can be viewed as a continuum from quiescent (facultative) to diapause (obligatory), terms adopted from the literature on insect diapause (Danks 2002). Quiescent eggs are arresting development in direct response to unfavourable conditions but are also resuming development once conditions are favourable again. Diapausing resting eggs on the other hand are produced in order to remain in arrested development for a period of time and may require other cues to get activated—such as photoperiod. Diapausing resting eggs are common in cladocerans and rotifers and require a cooling period before development can resume (Viitasalo and Katajisto 1994). Copepods produce either quiescent eggs that are prevented from hatching by environmental conditions or true diapausing eggs that require a resting period before hatching can proceed (Viitasalo and Katajisto 1994).

The alternative to resting eggs is to spend the unproductive season in a near mature or even adult stage. In many of the world's oceans, such overwintering is found in calanoid copepods (Conover 1988, Atkinson 1998), then often combined with a seasonal migration to great depth. Predator avoidance is regarded a key benefit of the migration, along with metabolic benefits, including metabolic dormancy, and possibly benefits regarding where water currents keep or bring the individual (Kaartvedt 2000, Irigoien 2004, Varpe 2012). The seasonal migration to deeper waters is accompanied by considerable energy storage to fuel metabolic needs during the unproductive winter (Record et al. 2018). Some species may even carry reserves to fuel reproduction after or towards the end of the overwintering period, hence they are capital breeders (Varpe et al. 2009). Even larger and longer-lived zooplankton, such as krill, display several of the same adaptations, notably large body size, wintering in mature stages, and energy storage (Hagen 1999).

The productive season is either started as a well-developed stage that can reproduce early, or in the case of the resting egg solution, as individuals that need time to develop and grow before maturity and the first reproductive spell. Depending on season length and life history, there may be many generations per season, one generation as in the case of an annual life cycle, or it may take more than a year to complete a generation. Many zooplankton species are small and with relatively large growth potential, and able to have multiple generations per year, such as the smaller microzooplankton. Their potential for rapid population growth could lead to top-down control of their food source, the phytoplankton (Landry and Calbet 2004, Boyce et al. 2015). Other forms have one generation per year, but with potential for two generations if conditions change, such as reported for a calanoid copepod (*Leptodiaptomus ashlandi*) in response to climate warming induced environmental changes in Lake Washington (Winder et al. 2009b). Zooplankton with slower growth, larger body size and a multi-year life span can, for instance, be found in colder high-latitude environments (Conover 1988). Some longer lived species are also iteroparous, in the sense that the same individuals may reproduce in consecutive years (Varpe and Ejsmond 2018). Zooplankton species with adult forms living for more than one year are, however, relatively rare.

In addition to seasonality in food availability, temperature clearly impacts growth and thereby annual routines and life histories. Growth and development are faster when water temperatures are higher, impacting key life history traits such as time from egg to first reproduction. Temperature variability also leads to other patterns of life history diversity, such as the intraspecific patterns of larger body size at colder temperatures because of unequal responses of growth and development rates to temperature (Forster and Hirst 2012).

Seasonal food availability often leads to time constraints on development (Sainmont et al. 2014). Interestingly, plastic responses can be expected in response to where in the season an individual is. The light environment may be one source of information organisms can use as cue to base such plasticity on (Johansson et al. 2001). The plasticity can include behavioural responses, such as increased feeding intensity towards the end of a season, but also life history responses such as reduced body size. Similar responses may arise in response to varying predation risk (Bjærke et al. 2014), hence it is also important to understand how risk may vary over the year and through the productive period (Varpe and Fiksen 2010). There are multifaceted predictions on the interactions between environmental conditions (food availability and risk) and life history traits such as body size, energy storage and voltinism (Ejsmond et al. 2018). For instance, as a consequence of fitting more generations into one feeding season, the prediction is that body size will decline and that the potential for energy storage thereby is reduced (Ejsmond et al. 2018).

Some species are also zooplankton only for a relatively short period of their life, and benthic for the rest of their life. This group is called meroplankton. They are planktonic as young, and during a relatively brief time-window within the productive season, before they settle at the seabed. Bivalves, barnacles, snails, crabs or eggs and larval stages of nektonic organisms (e.g., fishes, shrimps) are examples of groups where many have a planktonic stage, with dispersal as the main adaptive value. The benthic form may grow larger and live longer, with indeterminate growth and thereby growth and reproduction co-occurring through large parts of life (Heino and Kaitala 1999).

8.4 The Baltic Sea—A Place where Freshwater and Marine Zooplankton Meet

8.4.1 General Description of the Baltic Sea

The Baltic Sea is a semi-enclosed postglacial sea with a surface of 415,000 km^2 stretching over large latitudinal (53° N to 66° N) and ecological gradients (Snoeijs-Leijonmalm et al. 2017) (Fig. 8.3). The Baltic Sea was gradually formed after the retreat of the ice during the last glaciation and is a young sea, some 14,000 to 10,000 years old. It consists of different basins that vary in temperature, salinity and food web structure, such as primary production or terrestrial carbon input (HELCOM 2007). Besides a strong temperature gradient, the Baltic Sea is also characterized by a strong salinity gradient from near freshwater (salinity of 2) in the innermost parts to marine water (salinity of 30) at the entrance to the North Sea and a permanent halocline that separates the surface water from the more saline bottom water at about 70 m depth (Carstensen et al. 2014). The Baltic Sea is a shallow sea with an average water depth of 58 m and maximum depth of 459 m (Fig. 8.3), and a water residence time above 30 years (HELCOM 2007).

The Baltic Sea has high productivity with intensive fisheries that contributes to 1.2% at a global scale despite its small area (0.11% of the total ocean). Primary productivity is changing along the latitudinal gradient with about 10 times higher primary production in the southern parts compared to the northern parts, primarily due to higher terrestrial input of dissolved organic carbon and reduced underwater light in the north, while heterotrophic prokaryote production varies less along this gradient (Andersson et al. 2017). This results in over 50% bacterial production at the base of the food web in northern basins due to high terrestrial carbon inflow and low phosphorus availability (Sandberg et al. 2004). In addition, there is a strong gradient in eutrophication and reduced phosphorus levels towards the Bothnian Bay, reducing primary production in the north (HELCOM 2009a).

The Baltic Sea is an area that experienced warming over the last century with 1.5°C between 1871 and 2011 during the spring season (The BACC Teach 2015), which is high compared to other marine areas (Reusch et al. 2018). This effect of global warming has led to many related alterations of the physical environments, such as strengthening of vertical water stratification, expansion of anoxia and sea ice decline (Carstensen et al. 2014, The BACC Teach 2015, Liblik and Lips 2019). A

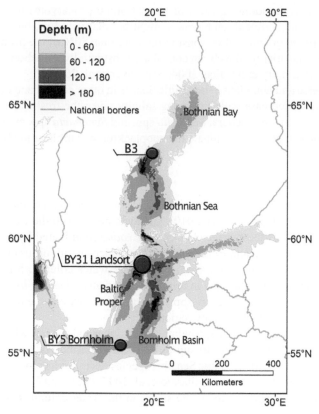

Figure 8.3: Bathymetry map of the Baltic Sea with location of zooplankton sampling stations and major basins. The Baltic proper BY31 Landsort Deep station (large filled circle) is the main focus of this study. The southern Baltic proper station BY5 Bornholm basin and the northern station in the coastal Bothnian Sea B3 are used for spatial comparisons (small circles).

multitude of anthropogenic stressors, including eutrophication, harvesting, and pollutants, have also led to rapid alterations of the system. Environmental conditions also affect zooplankton dynamics, with temperature, food availability and predation pressure being important drivers, which affect abundances of key zooplankton species (Möllmann 2000).

8.4.2 Zooplankton Species Composition of the Baltic Sea

The taxonomic diversity of zooplankton in the Baltic Sea is relatively low due to the young age of the sea and because only few species are endemic to brackish conditions in general, which also holds for the Baltic Sea (HELCOM 2009b, Ojaveer et al. 2010). In this brackish water environment, both limnic and marine species meet their physiological limits. In addition, the strong gradient in salinity and temperature forms physical, physiological or resource-related barriers for crustacean zooplankton to spread (Viitasalo et al. 1990, Vuorinen 1998). Low diversity, however, suggests that species redundancy is in general low and that new introduced species can have big impacts on the ecosystem.

Nano- and Microzooplankton

Phagotrophic protists, including heterotrophic nanoflagellates (HNF; 2–20 μm), dinoflagellates and ciliates ranging in size from about 1 μm to greater than 100 μm are the most numerous and species rich group of zooplankton, both in the Baltic Sea and more generally (Ojaveer et al. 2010).

Phagotrophic protists are important grazers of heterotrophic prokaryotes and picophytoplankton. Their contribution to total zooplankton biomass might be relatively low, but phagotrophic protists contribute substantially to grazing, often consuming more than 50% of the primary production, and carbon cycling within the microbial web can contribute substantially to carbon and nutrient turnover (Sherr and Sherr 2002, Landry and Calbet 2004, Schmoker et al. 2013).

Rotifers are diverse and abundant in the Baltic Sea, as in other coastal and estuarine systems and diversity and abundance decrease with increasing salinity given the freshwater origin of this group (Ojaveer et al. 2010). Two rotifer genera, *Keratella* spp. and *Synchaeta* spp. are most abundant in the central Baltic Sea and can at times dominate the zooplankton assemblage in abundance.

Mesozooplankton

The dominant mesozooplankton groups by biomass in the Baltic Sea are cladocerans (Cladocera) and copepods (Copepoda) (Ojaveer et al. 2010). Cladocerans include the fresh and brackish-water *Bosmina coregoni* as well as the marine cladoceran *Evadne* (dominated by *E. nordmanni* and the less abundant *E. anonyx* species) and *Podon* spp. The introduced carnivorous cladoceran *Cercopagis pengoi* (Ojaveer et al. 2010) appears sporadically in the water column. Small to medium sized (0.6–1.5 mm) copepod species dominate in terms of abundance: *Temora longicornis*, *Eurytemora affinis*, *Acartia* spp. and *Pseudocalanus* spp., and to lesser extent *Centropages hamatus* and *Limnocalanus macrurus* in the low salinity region.

The calanoid copepod *Temora* is a euryhaline and eurythermal species with a wide geographical distribution ranging from sub-tropical to sup-polar coastal marine waters (Continuous Plankton Recorder Survey Team 2004). In the Baltic Sea, this species occurs at its physiological salinity limits and occurs primarily in offshore waters (Viitasalo et al. 1995, Ojaveer 1998). *Temora* is a broadcast spawner and overwinters as active copepodite stages in the water column with no signs of resting stages.

Eurytemora affinis is a euryhaline zooplankton species with a wide distribution in the Northern hemisphere and a dominant species in coastal and estuarine systems, commonly inhabiting brackish waters (Winkler et al. 2011). This species occurs across a wide range of salinities, from freshwater to marine systems (Viitasalo et al. 1994, Lee 2016). This is an egg-carrying species and females carry eggs in a sac until hatching. This copepod is thought to produce diapausing eggs in autumn in the Baltic Sea that overwinter in the sediment (Katajisto et al. 1998). Egg clutch size of this species seems to be unaffected by salinity and temperature, while hatching success is reduced at lower salinity (Karlsson et al. 2018). *E. affinis* is an important grazer and central prey for fish in the Baltic Sea (Diekmann et al. 2012).

Three species of *Acartia* occur in the Baltic Sea, *A. bifolosa*, *A. longiremis* and *A. tonsa*. This copepod is a broadcast spawner and eggs of *Acartia* are found in the sediment, suggesting that a large proportion of eggs spawned by females reach the bottom prior to hatching (Katajisto 2003). Egg production with maximum values of 12 eggs female^{-1} day^{-1} is in general low in the Baltic Sea (Koski et al. 1999). It is thought that *A. bifolosa* produce resting eggs in the Baltic Sea in the form of quiescence, with no obligatory diapausing phase. Egg hatching is thought to occur throughout the year in the Baltic Sea and the hatching success is dependent on bottom temperature, which also affects the development rate of the eggs (Katajisto 2003).

The larger-sized copepod *Pseudocalanus* spp. (mainly *P. acuspes*) is regarded as a glacial arctic relict in the Baltic Sea and is most abundant in deeper water layers below the halocline (Renz et al. 2007). Lower abundances of this species are typically observed in years with low salinities (Möllmann 2000). Egg production of *Pseudocalanus* is highest in April reaching up to 3.6 eggs female^{-1} day^{-1} and is strongly related to food availability during the spring bloom (Koski et al. 1998, Renz et al. 2007). This copepod is an important prey for larval and adult planktivorous fish such as sprat and herring (Möllmann et al. 2003).

Other, less abundant copepod species include the marine larger-sized (ca 1.4 mm) copepod *Centropages hamatus*, which has a coastal distribution pattern (Durbin and Kane 2007) and reaches relatively low abundances in the Baltic Sea, but can at times be an important prey for planktivorous fish (Saage et al. 2009). The copepod *Limnocalanus macrurus* dominates in the Bothnian Bay and Bothnian Sea and with a carnivorous diet in the later copepodite stages (Dahlgren et al. 2012). As a cold-stenothermic species, it prefers temperatures below 11°C and occurs mainly below the thermocline (Hutchinson 1967). Similar to many high-latitude copepods, this species stores lipids in the form of wax esters (Vanderploeg 1998), which are used for metabolism as well as reproduction in winter to early spring (Dahlgren et al. 2012). Stored wax esters give *Limnocalanus* the possibility to survive starvation periods and reproduce at low food levels (Hirche et al. 2003).

Appendicularians, like *Oikopleura dioica* and *Fritillaria borealis*, appear occasionally in the zooplankton assemblage in the central Baltic Sea, but do not form extensive blooms as in the oceans (Ojaveer et al. 2010, Andersson et al. 2017).

Macroplankton and Megazooplankton

Macrozooplankton is mainly represented by Cnidaria (jellyfish) with the most dominant species being the scyphozoan *Aurelia aurita* that occurs throughout the Baltic Sea and the lion's mane *Cyanea* spp. that are restricted to the more saline waters (Andersson et al. 2017). Comb jellies, Ctenophora, are also present in the Baltic Sea, including the non-indigenous *Mnemiopsis leidyi*. Additional macrozooplankton species are mysids that reside close to the seafloor during the day and ascend in the water column during night (Rudstam et al. 1989). Other macrozooplankton forms common in true marine systems, including chaetognathes, krill and shrimps, are absent from the central Baltic Sea (Ojaveer et al. 2010), and appear occasionally at the entrance of the Baltic Sea after penetration of more saline Atlantic oceanic water bodies (Telesh et al. 2008). Given the absence of these groups, it can be expected that top-down control on mesozooplankton is mainly dominated by fish predation rather than invertebrates. Selection pressures to avoid visually searching predators, such as through diel vertical migration, transparency or small body size should therefore be particularly strong in the Baltic Sea.

Meroplankton

Meroplanktonic groups, including eggs and larvae of benthic and nektonic organisms, are occasionally important components of the zooplankton assemblage. These include mainly planktonic larvae of bivalves (*Macoma balthica, Cerastoderma glaucum, Mya arenaria, Mytilus trossulus*), gastropods, polychaetes and the bay barnacle *Amphibalanus improvises* (Andersson et al. 2017). As such, meroplankton is not a separate size class but spans across the micro-, meso- and macrozooplankton range.

8.5 Seasonal Monitoring, Methods and Environmental Conditions

The responses of plankton species and communities to environmental conditions set the framework for food web interactions that ultimately determine the transfer of energy and nutrients to higher trophic levels. The temporal dynamics of plankton communities are integral to understanding the function of planktonic food webs. The seasonal development and its environmental factors control plankton community dynamics and composition, which is often an annually repeated process depending on biotic community interactions (predation, herbivory, or competition), reproduction, resource availability and top-down control by predation (Winder and Cloern 2010, Sommer et al. 2012). Continuous monitoring programmes over an extended period of the year are needed to understand the complex seasonal patterns of environmental conditions and plankton communities, including interannual variability and how variables respond to changes in the environment (Ji et al. 2010, Cloern et al. 2016).

Seasonal dynamics in the Baltic Sea are here illustrated using a pelagic monitoring dataset with high temporal resolution from monthly sampling during the winter (November–February), weekly during the spring bloom (March–April) and bi-weekly during the remaining season over a decadal time period (12 years; 2007–2018, for phytoplankton the time period 2007–2011 is used) at an offshore monitoring station in the northern Baltic Proper, the central part of the Baltic Sea (station BY31) with a depth of 495 m (Fig. 8.3). For phytoplankton, integrated water samples were taken with a sampling hose from 0 to 20 m depth and preserved with acid Lugol's solution (Hjerne et al. 2019). Phytoplankton > 2 μm were counted after sedimentation using an inverted microscope. Zooplankton samples were collected using a 90 μm-WP2 net with a closing system from the upper 0–30 m and 30–60 m water depth strata. For zooplankton species, data are shown from the upper 0–30 m where most of the species are most abundant, except for *Pseudocalanus* which is most abundant in the 30–60 m water column. Abiotic and chemical variables are measured at a 5 m depth interval from 0–30 m followed by a 10 m interval until 100 m and 25 m interval until 150 m (Hjerne et al. 2019). Seasonal dynamics are further compared to a more southern (southern Baltic proper, station BY5) and a more northern monitoring station (Bothnian Sea, B3) (Fig. 8.3) with a monthly sampling interval during the ice-free period and same sampling procedure. Sampling and counting of phytoplankton and zooplankton is described elsewhere (Telesh et al. 2008, HELCOM 2014). Data are available from the Swedish Meteorological and Hydrological Institute (SMHI) (http.//sharkdata.se/; https.//sharkweb.smhi.se/).

For plankton, seasonal variability in population dynamics and phenology are characterized by changes in abundances, timing of the seasonal peak and season duration (Ji et al. 2010). For the zooplankton assemblage in the Baltic Sea, three phenological indices for zooplankton were identified. Average summer abundance was defined between May and September and calculated for each year. The timing of the seasonal peak was defined as the centre of gravity of abundance, as applied by Edwards and Richardson (2004) and season duration as the number of days between the 25th and 75th percentile of the seasonal year-specific cumulative abundance (Ji et al. 2010). Data were daily linearly interpolated between observation days for calculating the indices, although it is acknowledged that this may not be the best strategy for weekly to monthly sampling frequencies.

8.6 Seasonal Dynamics of Environmental Conditions and Phytoplankton

In the Baltic Sea, physical and chemical conditions vary greatly over the course of the year. The winter period from November to February is typically characterized by vertical water-column mixing, low water temperature, dropping below 4°C in the upper water column, and reduced light availability, as illustrated for the central sampling station BY31 (Fig. 8.4). Dissolved inorganic nutrients are typically well mixed over the entire water column, and winter mixing redistributes nitrogen (nitrate) from the deep water to the surface, reaching highest seasonal values between January and March (Fig. 8.4b). The northern Baltic Proper has experienced few ice winters over the last decade usually with fewer than 10 ice days, whereas several years in the 80s had close to 100 ice days (Hjerne et al. 2019).

Phytoplankton increases rapidly in March–April as a response to increasing light availability, water temperature (Fig. 8.4b) and water column stratification, resulting in marked spring bloom peaks that typically occur between mid to late April and are dominated by diatoms and dinoflagellates (Hjerne et al. 2019) (Fig. 8.5). As a consequence, dissolved nutrient concentrations (nitrate, phosphate) and Secchi depth are decreasing, while oxygen concentrations are at its highest. Salinity decreases slightly in March due to increased river inflow (Fig. 8.4b).

The summer period is characterized by steep vertical temperature and oxygen gradients, and a stratified water column with average temperature of about 15°C in the upper 20 m during the

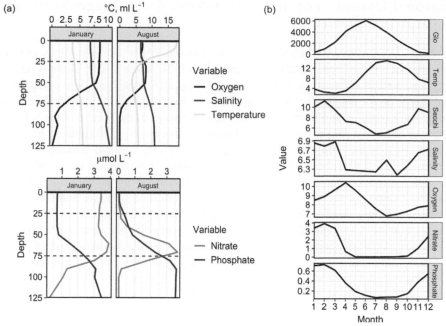

Figure 8.4: Physical and chemical conditions in the Baltic Sea station BY31. Shown are multi-year monthly averages of (a) vertical profiles of temperature (°C), salinity, oxygen (ml L^{-1}), nitrate (nitrite + nitrate) (μmol L^{-1}) and phosphate (μmol L^{-1}) during January and August, representing the mixing winter and stratified summer period, respectively, and (b) seasonal patterns of global irradiance (Glo, W m^{-2}) as a proxy for PAR, temperature (°C, Temp), Secchi depth (m), salinity, oxygen (ml L^{-1}), nitrate and phosphate (μmol L^{-1}) from the upper 20 m water strata. The two horizontal dashed lines in (a) represent approximately the upper and lower thermocline depth, respectively. Data are from 2007–2018 (Glo from 2007–2011, see Hjerne et al. 2019 for station description) available from the Swedish Meteorological and Hydrological Institute (SMHI) (https.//sharkweb.smhi.se/).

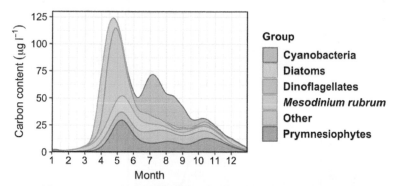

Figure 8.5: Average multi-year seasonal biomass dynamics of major phytoplankton taxonomic groups at the Baltic Sea station BY31. Phytoplankton data include the period 2007–2011 (available at https.//sharkweb.smhi.se/). Data are daily interpolated between observation days and smoothed using a kernel density estimate to generate the graph. The temporal resolution of the underlying sampling is monthly during winter (Nov–Feb), weekly during the spring bloom (Mar–Apr) and bi-weekly during the remaining season.

summer months from June/July to August (Fig. 8.4a). The euphotic zone reaches to about 10–20 m with the thermocline depth at about 25 m. Concentrations of nitrate and phosphate are low in the upper water column and increase with depth below the thermocline (Fig. 8.4). During the summer months, the phytoplankton assemblage is diverse but includes prominent cyanobacteria blooms, with *Aphanizomenon flos-aquae* and *Nodularia spumigena* as abundant species (Fig. 8.5).

8.7 Seasonal Dynamics of the Zooplankton Assemblage: Multi-year Average Perspective

Seasonal fluctuations in zooplankton abundance are a result of how annual routines and life history strategies interact with environmental conditions (Allan 1976, Varpe 2012). There is large diversity in strategies as well as in environmental conditions and preferences, leading to complex dynamics and interactions at the community level. Smaller and fast reproducing zooplankton often show higher temporal variability and sharp peaks, and rapid increases at the order of days, while temporal fluctuations of larger and slower reproducing species are more dampened (Klais et al. 2016).

In the Baltic Sea, the zooplankton assemblage shows strong seasonal fluctuations with distinct peak periods for the different taxonomic groups (Fig. 8.6). This seasonal cycle indicates a set of environmental conditions controlling population abundance, driven by temperature, food availability and predation. Seasonal plankton food web interactions are described below, and the annual cycle is used as a main structure to revisit life history strategies and trophic interactions. Four characteristic periods are identified that divide the annual cycle into: overwintering; stratification onset and the spring phytoplankton bloom period; summer stratification period and the dominance of cyanobacteria bloom; termination of stratification, decline in plankton productivity and onset of autumn mixing.

8.7.1 Overwintering

Zooplankton densities in the water column are typically low during the winter period, from November to February (Fig. 8.6), consisting mainly of nauplii and immature copepod stages, while adult stages

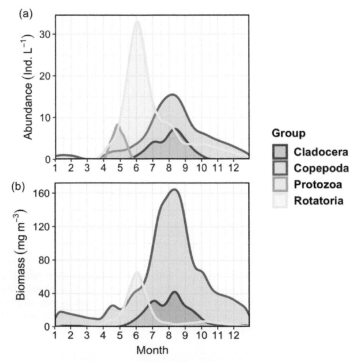

Figure 8.6: Average multi-year seasonal biomass dynamics of the major zooplankton taxonomic groups protozoans, rotifers, cladocerans and copepods at the Baltic Sea station BY31 for (a) abundance and (b) biomass (in wet weight). Protozoans only include the species *Radiosperma* sp. Zooplankton data include the period 2007–2018 available at (https.//sharkweb.smhi. se/). The temporal resolution of the underlying sampling is monthly during winter (Nov–Feb), weekly during the spring bloom (Mar–Apr) and bi-weekly during the remaining season. Data are daily linearly interpolated between observation days and smoothed using a kernel density estimate. Biomass data for *Radiosperma* sp. are not available.

occur in lower abundances (data not shown). Phytoplankton concentrations are low and food limitation affects zooplankton abundances. Cladoceran and rotifer abundances in the water column typically drop to zero, but *Bosmina, Synchaeta* and *Keratella* may occur in very low abundances below 0.03 ind. L^{-1} in some winters (Fig. 8.7a). These organisms are known to produce fertilized diapausing eggs at declining temperature and food availability, and eggs of both groups are observed in the sediment (Viitasalo and Katajisto 1994).

Some zooplankton species overwinter as active stages in the water column, but in low abundances such as for the copepods *Acartia* spp. (about 0.5 ind. L^{-1}), or *Temora, Eurytemora* and *Pseudocalanus* (with about 0.3 ind. L^{-1}) (Fig. 8.7b). This suggests continuous reproduction by a few individuals at low rate. Most copepods produce overwintering resting eggs and calanoid copepods are the most common eggs identified in Baltic Sea sediments (Viitasalo and Katajisto 1994), belonging to *Acartia* spp. and *Eurytemora. Temora*, on the other hand, is thought to hibernate in the water column during the winter as immature and adult stage. Occurrence of copepods in the sediment is also confirmed by metabarcoding analysis of Baltic Sea sediment showing that copepods, particularly *Eurytemora*, dominate DNA sequences in the northern Baltic Proper sediment, although *Temora* dominates in the southern Baltic Proper sediment (Broman et al. 2019). Copepod DNA in the sediment may derive from buried eggs, sinking remains from copepods such as carcasses, and faecal pellets.

Lipid reserves of overwintering copepod species are generally lower for the small-sized copepods in the Baltic Sea compared to the larger copepods of polar regions (Lee et al. 2006, Peters et al.

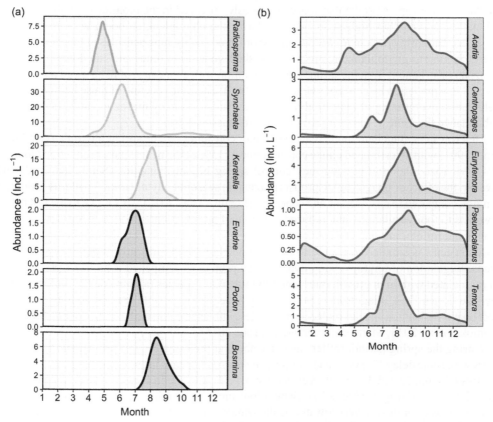

Figure 8.7: Average multi-year seasonal dynamics of zooplankton species abundances at the Baltic Sea station BY31. The left panel, (a) includes the protist *Radiosperma*, the rotifers *Synchaeta* and *Keratella* and cladocerans *Bosmina, Evadne* and *Podon*, the right panel (b) are copepod species. Zooplankton data include juvenile and adult stages over the period 2007–2018 available at (https.//sharkweb.smhi.se/). Data are daily linearly interpolated between observation days and smoothed using a kernel density estimate.

2013). *Temora* accumulates lipid reserves in autumn that may buffer against starvation during winter; however, the reserves (in the form of triacylglycerols) may suffice only for a few days or weeks. This suggests that copepods in the Baltic Sea either depend on food availability through the winter or rely on resting egg strategies, with lipid reserves playing a minor role for overwintering and reproduction (Peters et al. 2013), except for *Limnocalanus* (Dahlgren et al. 2012).

8.7.2 Stratification Onset and the Spring Phytoplankton Bloom Period

Phytoplankton spring blooms appear in March–April at low temperature (below 5°C) (Figs. 8.4 and 8.5) at which development rates and generation times of zooplankton are long. Zooplankton develop in response to food availability and warmer temperature. Protozoans, represented by heterotrophic flagellates and ciliates, are the first ones that build up zooplankton biomass in spring (Arndt 1991) and can make up about 85% of the biomass of the zooplankton spring community (Johansson 2004). Bacterial heterotrophic production increases with the phytoplankton bloom (Bunse et al. 2019), followed by protozoans with brief delays and high abundances reached prior to the establishment of mesozooplankton (Arndt 1991). Protozoans depend on phytoplankton as food resource, either directly or indirectly via grazing on bacteria on phytoplankton. The small-sized protozoans are susceptible to much the same grazing pressures as the phytoplankton and abundances decrease with increase in mesozooplankton (Johansson 2004). Given that protozoan densities are close to zero over winter, the population likely rejuvenates from hatching of pelagic or benthic resting stages.

Radiosperma sp., a protist that was continuously observed in the monitoring, reflect this spring peak pattern and reaches highest abundances during the phytoplankton boom with up to 8 ind. L^{-1} (Fig. 8.7a). The spring peak in *Radiosperma* is short-lived and by June densities of this protist are close to zero. Similar to phytoplankton, protists have short generation times and can double within days. Protozoans also benefit from the spring-bloom condition with abundance of nano-sized phytoplankton prey and low predation by copepods (Johansson 2004). The sharp decline of these herbivores is most likely due to overgrazing of their food resource and suppression by mesozooplantkers that increase in abundance this time of the year (Arndt 1991, Johansson 2004).

Heterotrophic flagellates and ciliates are major consumers of the spring phytoplankton biomass (Arndt 1991) with an estimated ciliate consumption of 15% of the net primary production (Johansson 2004). Microzooplankton grazing may control spring bloom dynamics (Schmoker et al. 2013); however, the effect of phagotrophic protists grazing on seasonal phytoplankton patterns is thought to be minor, compared to cladocerans and copepods (Sommer et al. 2012). This is due to the fast response of protists to food availability and reduced potential of phytoplankton to escape top-down control, a prerequisite for blooms and subsequent crashes. In addition, the diet breadth of many phagotrophic protists species is smaller compared to mesozooplankton, and thus only suppresses specific prey groups. Regulation of phytoplankton dynamics by protists is dampened if predator-prey cycles are fast (associated with high growth rates) and phytoplankton species fall outside the preferred protists prey size. Regardless, phagotrophic protists may have a considerable grazing activity and a strong effect on phytoplankton species replacement, but less influence compared to mesozooplankton grazing (Schmoker et al. 2013, Menden-Deuer and Kiørboe 2016).

During the spring period, rotifers increase about at the same time as protozoans but their peak occurs with some delay. The dominant rotifer *Syncheata* spp. peaks in June, reaching average monthly densities of about 35 ind. L^{-1}, while *Keratella* spp. peak later in the season with average densities of up to 20 ind. L^{-1} in August (Fig. 8.7a). Generations most likely rejuvenate from hatching of benthic resting eggs as densities are very low during the winter season. The two most abundant rotifer species have different feeding behaviour. *Synchaeta* spp. is a predator and feeds mainly on dinoflagellates and ciliates, while *Keratella* spp. is a suspension feeder and preys on degraded detritus material but also ingests protists (Arndt 1993) and ciliates (Weisse and Frahm 2002).

The observed spring phytoplankton declines before the increase in dominating mesozooplankton grazers suggest that the clear water phase is mainly driven by microzooplankton grazing, nutrient limitation and subsequent acceleration of sinking particles. The transition period between the spring and summer phytoplankton bloom is characterized by a diverse community dominated by dinoflagellates, the mixotrophic ciliate *Myrionecta rubra* (formerly *Mesodinium rubrum*) (Hansen et al. 2012, Kim et al. 2016) and Prymnesiophytes (Fig. 8.5).

8.7.3 Stratified Summer Period and the Dominance of Cyanobacteria Bloom

The biomass increases of slower growing cladocerans and copepods are delayed until the stratification period between June and August, with highest abundances of crustacean zooplankton coinciding with the period of the cyanobacteria bloom (Figs. 8.5 and 8.6). Cladoceran abundances are low during the winter months and it is thought that the population rejuvenate from hatching of benthic resting eggs triggered by increasing temperature, light and oxygen or a combination of them (Kankaala 1983). Cladoceran taxa prevail during the summer period and reach densities up to 8 ind. L^{-1} or 40 mg m^{-3} in wet weight biomass (Fig. 8.6). *Evadne* spp. and *Podon* spp. are the first ones to increase in May–June with highest abundances during July with about 2 ind. L^{-1} and decline during August to low abundances (Fig. 8.7a). Abundances of *Bosmina* reveal a seasonal pattern with increase in July, and maxima in August up to 8 ind. L^{-1}, and decline towards September. The introduced carnivorous *Cercopagis* appears only sporadically during June and July at station BY31 (data not shown), likely due to its preference for warm waters.

Copepods exhibit a slow numerical growth response to increasing phytoplankton and temperature in spring, and typically peak during the summer months at BY31 with up to 17 ind. L^{-1} (including all copepodite stages and adults) and dominating the zooplankton biomass with up to 160 mg m^{-3} wet weight (Fig. 8.6). Their population increases coincide with an increase in temperature above about 10°C. The copepods *Eurytemora* and *Temora* are the most dominant taxa of this group followed by *Centropages* and *Acartia* in the upper 30 m, while *Pseudocalanus* is dominant below 30 m, in the 30–60 m depth layer (Fig. 8.7b).

Temora persists throughout the year in the water column and develops slowly in spring. The development of the first generation is initiated by egg production of overwintering females, and nauplii typically dominate during the spring bloom in April (Dutz et al. 2010). Spring egg production is associated with low levels of storage lipid in *Temora*, and the triacylglycerol reserves that increase in autumn probably serve as a relatively short-term buffer against starvation during winter (Peters et al. 2013). Hence, *Temora* is an income breeder relying on phytoplankton for egg production (Dutz et al. 2012). At BY31, *Temora* abundances typically increase rapidly during June, reach peak abundance in July with about 5 ind. L^{-1} and the species remains in the water column throughout the summer (Fig. 8.7b). The number of generations produced by *Temora* depends on environmental conditions. In the southern Baltic Sea (Bornholm Basin), there are five to six generations a year as deduced from stage structure, copepodite length and stage duration (Dutz et al. 2012) (Fig. 8.8). Secondary maxima may occur in autumn dominated by nauplii or copepodite stages, typical for areas where *Temora* produces four to six generations per year (Digby, PSB 1950). Egg production of *Temora* is strongly affected by the low salinity in the Baltic Sea, and maximum production rates of 12 eggs female^{-1} are about 3–5 times lower compared to areas with higher salinity (> 30) (Holste et al. 2009). *Temora* is a suspension feeder primarily targeting non-motile prey (Tiselius and Jonsson 1990).

The copepod *Eurytemora* is thought to produce diapause eggs in autumn in the Baltic Sea that overwinter in the sediment (Tiselius and Jonsson 1990). At BY31, abundances of this species increase in June with peaks in August of about 6 ind. L^{-1} and densities decline to low abundances by November (Fig. 8.7b). *Eurytemora* shows a selective feeding behaviour on rotifers (Feike and Heerkloss 2009) suggesting that the relatively short period of seasonal durations is caused by this specialised interaction.

Acartia, particularly *Acartia bifolosa*, produces resting eggs in the Baltic Sea in the form of quiescence, with no obligatory diapausing phase (Katajisto 2003). Egg hatching is thought to occur throughout the year in the Baltic Sea and to depend on bottom temperature (that peaks in June),

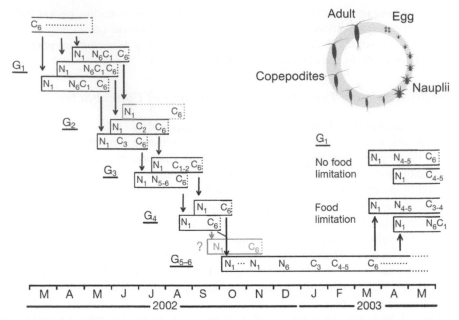

Figure 8.8: Schematic and hypothetical life cycle of *Temora longicornis* in the Bornholm Basin, Baltic Sea during March 2002 to May 2003 based on observed occurrence of nauplii and projected cohort development (after Dutz et al. 2010). The life cycle of copepods and its life stages are illustrated in the insert. Different generations or cohorts are indicated. The projection suggests that the species produced five to six generations in 2002. N = nauplii stage N1 to N6, C = copepodite stage C1–C5 and adult stage C6, G = generation 1–6. Adapted from: Dutz, J., V. Mohrholz and J. van Beusekom. 2010. Life cycle and spring phenology of *Temora longicornis* in the Baltic Sea. Mar. Ecol. Prog. Ser. 406: 223–238, with permission from Inter-Research.

which affects the development rate of the eggs. At BY31, *Acartia* spp. (mainly *A. bifolosa*) occur all year round in the water column and show smaller peaks (around 2 ind. L⁻¹) after the spring bloom and maximum peaks during the warm temperature season in late summer (August) with up to 3.5 ind. L⁻¹. Maximum densities of *A. longiremis* are typically in June, while *A. bifolosa*, the most abundant *Acartia* species, and *A. tonsa* develop peaks in July and August. Densities decline slowly after October and remain low during the rest of the year and at about 0.5 ind. L⁻¹ over the winter (Fig. 8.7b). *Acartia* has a prey switching behaviour and is capable of both raptorial and filter feeding, consuming a broad size spectrum of prey from both the classical and microbial food web (Kiørboe et al. 1996, Engstrom 2000).

   Pseudocalanus individuals are found in the water column year-round and abundances increase in June and peak during August with average densities of 0.5 ind. L⁻¹ in the upper 30 m water column (data not shown) and higher values of 1 ind. L⁻¹ in the depth stratum of 30–60 m at BY31 (Fig. 8.7b). Given the wide distribution of this species in the North Sea and Arctic, and its Arctic origin, it is assumed that *Pseudocalanus* is a glacial relict in the Baltic Sea (Ojaveer 1998). Seasonality of stage composition and growth measurements suggest that this copepod produces only one cohort over the annual cycle (Renz et al. 2007). Females show high seasonal variability in size and lipid content (Peters et al. 2006), suggesting that females mature from overwintering copepodite stages in early spring, a development and growth possibly fueled by storage lipids. This species is thought to have the same life cycle as *Pseudcalanus* spp. in the Arctic (McLaren et al. 1989) with high production in spring and the later developmental stages accumulating storage lipids in the form of wax ester, followed by overwintering and then maturation in early spring. Low abundances in late winter and early spring suggest considerable mortality associated with this pelagic overwintering. Seasonal fatty acid analysis indicates an opportunistic feeding behaviour and high dominance of ciliates in the diet (Peters et al. 2006) as well as feeding on sinking detritus particles.

Population densities of the larger-sized (ca. 1.4 mm) copepod *Centropages* increase after the spring bloom, reaching highest density of about 3 ind. L^{-1} during end of July and early August (Fig. 8.7b). Densities drop quickly to below 1 ind. L^{-1} after mid-August. This calanoid copepod produces resting eggs in other areas, such as the North Sea (Viitasalo 1992), which might also be its overwintering strategy in the Baltic Sea, as suggested by the very low abundances in the pelagic during winter and early spring. *Centropages* is omnivorous eating phyto- and microzooplankton as well as other copepods (Calbet et al. 2007), and it can, similar to *Acartia*, select prey and switch between suspension feeding and ambush predation (Tiselius and Jonsson 1990). Given the diverse diet of *Centropages*, it is suggested that food is not the limiting driver of its population dynamics (Tiselius and Jonsson 1990). Its relatively large body size may however make it vulnerable to visual predation by fish.

Among meroplankton taxa, Bivalvia larvae are the most abundant plankton component. Abundances increase with warming waters through the summer and reach peak abundances in June with about 0.8 ind. L^{-1}, prior to the peaks of the other dominant mesozooplankton taxa (data not shown). Given that *Macoma baltica* is the dominant benthic bivalve, most of the larvae likely belong to this species. Cirripedia larvae appear between June and October (data not shown), albeit in very low abundances, probably due to the brackish water conditions of the Baltic Sea.

8.7.4 Termination of Stratification, Decline in Plankton Productivity and Onset of Autumn Mixing

The stratification period starts to break down with water temperature decrease in September and mixing into deeper water layers (Fig. 8.4). Dissolved inorganic nutrients are redistributed in the water column and concentrations increase again in the upper water layer in October. This coincides with a decline in phytoplankton biomass (Fig. 8.5), although diatom peaks of *Coscinodiscus* spp. may occur after inorganic nutrients are redistributed in the water column. Zooplankton experience declines in all taxa, and all species with resting eggs are absent from the pelagic by November and the copepods (except *Acartia* and *Pseudocalanus*) have reached low abundances by then (Fig. 8.7). This suggests that the production of the large-sized diatom in autumn is not much grazed upon by pelagic species and rather sinks out and becomes food for benthic species.

8.8 Interannual Variability in Seasonal Zooplankton Succession and Phenology

Changes in abiotic and biotic factors and variation in environmental factors can modify phenology and population size fluctuations, affecting the timing and duration of peak abundances and magnitude, as well as the timing of production and emergence of resting stages (Mackas et al. 2012). Interannual variation in zooplankton abundance, timing and duration is likely related to temperature, food availability and predation pressure (Winder and Schindler 2004a, Winder et al. 2009b). Temperature is a key parameter affecting physiological rates in ectotherms (Cushing 1990, Forster and Hirst 2012), and zooplankton population growth fluctuates strongly with seasonal temperature variation. Temperature affects metabolic and vital rates, and increasing temperature within the tolerance range of individual organisms accelerates both growth and developmental rates given sufficient resources. However, population dynamics of zooplankton will depend not only on the direct effects of temperature on vital rates but also on the synchronization of key life stages with food availability (Cushing 1990, Winder and Schindler 2004b). This is particularly important for pelagic herbivores in temperate regions where quantity and quality of phytoplankton, their major food resource, is highly variable on a seasonal basis (Sommer et al. 2012). For instance, the onset of the phytoplankton bloom in the Baltic Sea BY31 station varies between mid and end April, and peak phytoplankton bloom can range from about 50 to over 100 µgC L^{-1} (Hjerne et al. 2019).

In addition to bottom-up processes driving the seasonal production cycle, predation by higher trophic levels can be important for structuring zooplankton populations (Fig. 8.2) and is expected to account for 67–75% of total copepod mortality (Hirst and Kiørboe 2002). In the Baltic Sea, the planktivorous clupeid fish species, sprat and herring, are dominant predators on zooplankton, and the predation pressure varies through the year and between years (Möllmann 2000, Möllmann et al. 2008). For instance, Möllmann (2002) described how the increase of the sprat population through the 1990s may have impacted zooplankton mortality and thereby the population dynamics of zooplankton, *Pseudocalanus* and *Temora* in particular. Fish predation on zooplankton is expected to cascade down to primary producers, suggesting a closely interlinked food web (Casini et al. 2008).

Zooplankton peak abundances, timing of seasonal peak and duration fluctuate substantially from year to year in some taxa, particularly for microzooplankton (Fig. 8.9). For example, average summer densities of the protozoa *Radiosperma* range from close to zero to 6 ind^{-1}. Maximum densities of this protozoa occur between April and September with seasonal durations of about 110 days. For the rotifer *Synchaeta*, summer peak averages are around 7 ind. L^{-1}, while *Keratella* reaches lower densities with averages around 2 ind. L^{-1}. Peaks of both rotifer species appear at a narrow time window with both having short seasonal durations of about 30 days and *Synchaeta* peaking end of June and *Keratella* early August. Given that these zooplankton species fluctuate strongly, sampling frequency may contribute to some of the interannual variability.

For the cladoceran *Podon* and *Evadne*, annual average and maximum densities are in a narrower range with average and maximum values ranging from 0.5 to 1.5 ind. L^{-1}, seasonal peaks typically occurring in July and short seasonal durations of less than 50 days (Fig. 8.9). *Bosmina* shows higher interannual variability with averages from 0.5 to 8 ind. L^{-1}. Bloom timing and duration of this cladoceran is more variable, ranging between mid-July and mid-August with a short median seasonal

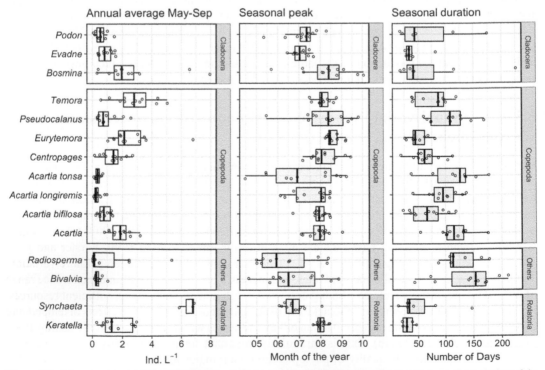

Figure 8.9: Interannual variability of zooplankton multi-year average summer (May–Sept) abundance (left), timing of the seasonal peak (middle) and seasonal duration (right) in the northern Baltic proper station BY31. Plots indicate the median (vertical line), the 25th and 75th percentile (box), the upper and lower whisker (horizontal bars), and individual observations (dots). Seasonal peak is the year day with highest abundance; seasonal duration is the number of days between the 25th and 75th percentile of seasonal abundance.

duration of less than 50 days. The invasive cladoceran *Cercopagis* was only observed sporadically between April and October with average summer abundances below 2 ind. L^{-1} (data not shown).

Within the copepods, *Temora* and *Eurytemora* reach highest summer abundances with interannual variation ranging from averages of 0.5 to 6 ind. L^{-1} (Fig. 8.9). Bloom timing occurs consistently between early and mid-August with *Eurytemora* having short seasonal durations of about 50 days and *Temora* up to 100 days. This confirms earlier observations showing that *Eurytemora* is confined to the warm water season and forms large transitory population peaks in late summer (Möllmann 2000). Average summer densities of *Pseudocalanus* and *Centropages* are around 1.5 ind. L^{-1} with seasonal peaks between July and August. *Centropages* peaks are relatively short (around 75 days), while *Pseudocalanus* peak over a longer time period (more than 100 days).

Among *Acartia*, *A. bifolosa* reaches highest summer densities with about 1.5 ind. L^{-1}. This species shows a narrow bloom timing and typically peaks in early August and with a seasonal duration from 40 to 75 days (Fig. 8.9). Timing of seasonal peak and seasonal duration of *A. tonsa* and *A. longiremis* is more variable, occurring between early June and end of August and with seasonal durations up to 125 days and more. Annual averages of Bivalvia larvae are typically below 0.3 ind. L^{-1} and peaks occur as early as end of April and as late as end of September with longer seasonal duration of about 150 days (Fig. 8.9). This indicates that Bivalvia reproduction is variable between years.

8.9 Spatial Variation in Seasonal Zooplankton Succession and Phenology

Spatial and latitudinal gradients in environmental and biotic factors lead to patterns in zooplankton distributions and community composition, as well as spatial variability in within-species traits and dynamics (Hays et al. 2005, Daase et al. 2013). In the Baltic Sea, copepods typically peak about one to two months earlier in the southern station compared to the central and northern most sampling stations with peaks occurring in June compared to August (Fig. 8.10). In contrast, rotifers and cladocerans vary less across the latitudinal gradient.

Zooplankton species composition in the southern Baltic Proper is similar to the BY31 station; however, species dominance changes, particularly within copepods. In comparison, the northern Baltic Sea comprises more brackish and freshwater species such as *Limnocalanus*. In addition to spatial variability in salinity, food availability and predation are the most pronounced environmental variables structuring the zooplankton assemblage in the Baltic Sea. Zooplankton bloom timing and duration varies across the stations (Fig. 8.11), which for some taxa may lead to a difference in timing of the seasonal peak by up to a month. Some taxa appear considerably earlier in the south whereas differences in timing between the two more northern stations are less clear. Given that BY31 has high sampling frequency (weekly to bi-weekly during the spring and summer period) compared to the monthly sampling at the other stations, differences in variability could be an artefact of the sampling but needs to be verified.

Synchaeta is the most abundant rotifer species in the southern Bornholm basin, whereas *Keratella* is more abundant in the Bothnian Sea. Bloom timing and duration of rotifers are more variable between years in the southern and northern Baltic Sea (Fig. 8.11). *Synchaeta* typically appear before the copepod and cladoceran peaks, while *Keratella* peaks overlap with mesozooplankton. Given that copepods and cladocerans are strong competitors for food, the seasonal dynamics suggests that *Synchaeta* has a temporal niche before crustaceans increase in abundance.

Seasonal peaks of *Bosmina*, the most abundant cladoceran in the Baltic Sea, is quite consistent across the latitudinal gradient with peaks in early August and relatively short seasonal duration. Limited abundances of this cladoceran species early in the season and during mid-summer could be related to intense fish predation by fish larvae, sprat and herring during the spring and early summer period.

Seasonal peaks of most copepods appear in general about one month earlier in the southern Baltic Sea, typically in July compared to August for the central and northern stations (Fig. 8.11). The

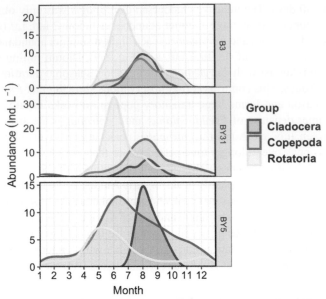

Figure 8.10: Average seasonal abundance of major zooplankton taxonomic groups at the southern (BY5), central (BY31) and northern (B3) sampling stations. Zooplankton data for BY5 include the years 2009–2011, 2013, 2016, 2017, for BY31 the period 2007–2018 and for B3 2012–2017. Data are daily interpolated between observation days and smoothed using a kernel density estimate. Protozoa include only the species *Radiosperma* sp.

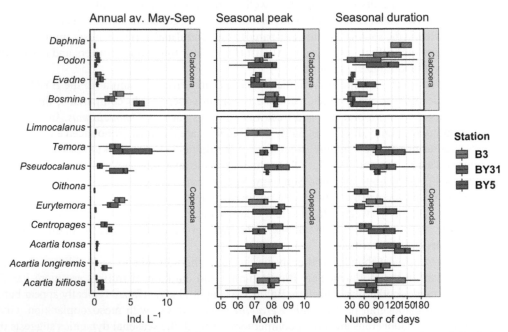

Figure 8.11: Interannual variability of zooplankton multi-year average summer (May–Sep) abundance (left), timing of the seasonal peak (middle) and seasonal duration (right) in a northern (B3), central (BY31) and southern (BY5) sampling location in the Baltic Sea. For description of boxplots, data availability and quantification of plankton phenology see Figs. 8.9 and 8.10.

most abundant copepods, *Temora* and *Pseudocalanus* have a relatively narrow time window of low abundance. The duration of *Pseudocalanus* is about 100 days, while it is more variable for *Temora*. This suggests a narrow prey window for fish species depending on these copepods. *Centropages* is absent in the northernmost station and has similar densities in the central and southern station with seasonal peaks in July in the south and August in the central station. Similarly, *Acartia* spp. copepods typically peak earlier in the southernmost station compared to the central station, except for *Acartia tonsa* that shows high variability in seasonal peak and duration. In the northernmost station, *A. biofolosa* occurs at low abundances. *Eurytemora* is the most abundant copepod in the Bothnian Bay and compared to the central BY31 station has a more variable time window in the northern and southern regions with peaks occurring between May and September. The small-sized *Oithona* species appears only in the southern Baltic Sea with low abundances within a narrow time window in July and seasonal duration below 100 days. Low salinity likely restricts the distribution of this oceanic species.

These latitudinal patterns of zooplankton phenology indicate that the general seasonal succession and species replacement is relatively consistent across the Baltic Sea latitudinal gradient; however, timing of the seasonal peak and duration vary across the gradient. This suggests that environmental conditions of temperature, food availability and predation pressure affect seasonal dynamics across the Baltic Sea and likely affect the interannual and decadal variability in abundance and phenology.

8.10 Plankton Interactions under Changing Environmental Conditions in the Baltic Sea

Zooplankton growth and reproduction rates are sensitive to temperature and food abundance, including the duration of the feeding season, with direct implications for population dynamics (McCauley and Murdoch 1990). For ectotherms in particular, climate warming thus leads to elevated turnover rates, increased number of generations per year, greater population variability, and altered phenology and life history traits, such as timing of dormancy and smaller body sizes (Daufresne et al. 2009, Winder et al. 2009b, Garzke et al. 2015). Moreover, climate change alters the density gradient of the water column and consequently the relative strength of mixing and stratification. Mixing processes are usually accompanied by changes in phytoplankton dynamics that in turn affect the seasonal dynamics of consumers (Winder et al. 2009b). As a result, climate may indirectly affect population dynamics and life histories of zooplankton through its effect on seasonality of resource availability and other components of the ecosystem, such as the extent of the growing season. Such modifications in the environment are expected to affect life cycle responses in zooplankton, particularly in copepods (Drake 2005), given the plasticity of their life histories and their extended longevity compared to cladocerans and rotifers (Allan 1976).

In the Baltic Sea, phytoplankton spring bloom timing, magnitude and composition display high variability from year to year with the peak biomass of the bloom occurring between end of March and end of April at station BY31 (Hjerne et al. 2019). Spring bloom timing occurred about 1–2 weeks earlier over the last 20 years driven by less clouds and less wind. Furthermore, due to warming, the magnitude of spring bloom diatoms decreased, while summer associated cyanobacteria increased (Wasmund and Uhlig 2003, Kahru and Elmgren 2014, Kahru et al. 2016). These changes are expected to affect carbon cycling within the pelagic system and export rates to the benthos.

In addition to temperature, changes in salinity are expected for coastal systems as climate also affects precipitation patterns and freshwater inflow (Käyhkö et al. 2015). For the Baltic Sea, a decrease in salinity was observed and a further decrease is expected for the coming decades due to increasing freshwater inflow and reduced marine water inflow through the narrow entrance (The BACC Teach 2015). Because some zooplankton species are physiologically constrained by salinity, changes in salinity are affecting abundance and distribution. For example, *Pseudocalanus*, an important prey item for larval and planktivorous feeding fish, and as such considered a key species (Möllmann et al. 2003) experienced multi-year declines that coincide with decrease in North Sea water inflow and decreasing salinity in the Baltic Sea. This species is considered a mediator between climate change and herring growth, with consequences for the fisheries (Möllmann et al. 2003, Casini et al. 2008). Similarly, *Temora* serves as the major diet for sprat *Sprattus sprattus* and herring *Clupea harengus*.

The effect of top-down control on *Temora* population dynamics by fish predation is, however, uncertain (Köster et al. 2003, Dutz et al. 2010), and it is thought that hydrographic conditions explain most of the variation in the fluctuations of this copepod species. *Temora* decreased in coastal waters over the last decades with a general trend of decreasing salinity and increase in freshwater inflow in the Baltic Sea (Ojaveer 1998, Vuorinen 1998), and its distribution was shifted westwards into areas of higher salinity.

The light regime of the pelagic domain also determines multiple processes, both photosynthesis and primary production as well as predator-prey interactions involving visually searching predators. Many processes impact the light regime of Baltic Sea waters, including freshwater inflow and the amount of suspended matter, eutrophication and the shading through algae blooms near the surface, and snow-covered sea ice. These processes all have pronounced spatial variability, both with respect to distance from the coast and along latitude, such as for suspended matter (Kyryliuk and Kratzer 2019). Sea ice is more common, and lasts for a longer part of the season, in the northern parts, particularly the Bothnian Bay (Haapala et al. 2015, Hjerne et al. 2019). The amount of sea ice varies considerably between years and has declined over time (Haapala et al. 2015). Runoff processes are also highly dynamic but with no or weaker trends over time (Käyhkö et al. 2015). The net effects on water clarity and light regime are challenging to predict and highly likely to be dependent on region. From other aquatic systems, it is known that increased runoff from land leads to murkier waters and changing ecological interactions (Aksnes et al. 2009, Opdal et al. 2019). Less sea ice leads to more light and may be of particular importance in areas where the sea ice cover historically has lasted well into the well-lit spring period. Such sea ice changes would be expected to have similar implications to those observed and predicted in polar seas (Clark et al. 2013, Langbehn and Varpe 2017). For parts of the Baltic Sea where the net effect is more light for a longer part of the season, we would expect species well adapted to avoid visually searching predators to increase, and selection pressures to shift strategies towards smaller body size, more transparency, and increased diel vertical migration (DVM).

Other human drivers, such as eutrophication, anoxia or fishing that are prominent in the Baltic Sea (Reusch et al. 2018) also have direct effects on the plankton community and likely affect species composition, carbon flow within the food web and seasonal dynamics of zooplankton (Arndt 1991, Casini et al. 2008, 2016). In comparison, many plankton organisms of the Baltic Sea are quite resistant to ocean acidification (Rossoll et al. 2013, Lischka et al. 2017). Given that the Baltic Sea has strong seasonal changes in pH that exceed the predictions for open ocean systems by the end of this century (Reusch et al. 2018), it is assumed that species are adapted to high pH variation (Thomsen et al. 2017). Given that jellyfish are tolerant to low pH levels, ocean acidification and warming may give this group a competitive advantage in the Baltic Sea (Winder et al. 2017a).

Multi-year changes in the annual zooplankton cycle are not well studied in the Baltic Sea but changes can be expected given alterations in phytoplankton seasonal dynamics and abiotic factors (Wasmund and Uhlig 2003, Hjerne et al. 2019). Given that the degree to which individual species respond to changing temperature and salinity varies, it is likely that climate change can significantly alter trophic flows in unpredictable ways. The effect of climate change further depends on the local adaptions of life history traits, as has been shown for *Eurytemora* in the Baltic Sea (Karlsson et al. 2018). This suggests that the extent of physical changes and the potential for species to adapt to changing environmental conditions will greatly influence food web dynamics as future climate warms and becomes more variable.

8.11 Conclusions

Understanding the seasonal succession of zooplankton, as well as their responses to changing environmental conditions, requires detailed knowledge of their population dynamics through the annual cycle as well as their interactions with food, competitors and predators. In this chapter, we have described temporal interactions between primary producers and consumers in plankton food webs of the Baltic Sea. Observed dynamics and patterns have been discussed in relation to driving mechanisms and environmental variables. For our Baltic Sea case, ciliate and rotifer microzooplankton species are the first ones to appear after the spring phytoplankton bloom that occurs around April when water temperatures are low. Mesozooplankton, including cladocerans and copepods, peaks are temporally decoupled from the spring bloom and occur during the cyanobacteria dominated summer blooms in August. Given the strong seasonal overlap of diverse zooplankton, specialized feeding behaviour may allow coexistence of copepod and cladoceran species during the summer season. The absence of some groups of larger and non-visual predatory zooplankton from the Baltic Sea suggests that fish are the main predators on mesozooplankton. Their visual search for food lead to strong selection pressures for anti-predator strategies such as through diel vertical migration, transparency or small body size. The timing of peak abundance of zooplankton species is highly variable from year to year, except for a few species, such as *Eurytemora affinis* that appear during a narrow and similar time window each year. Population peaks in the southern Baltic Sea typically appear one month earlier compared to central and northern stations where the seasonal dynamics of mesozooplankton are more condensed within the summer months. Multi-year changes in zooplankton assemblages are mainly related to abundance declines of some key copepod species driven by salinity declines. However, given alterations in phytoplankton seasonal dynamics and abiotic factors, multi-year changes in zooplankton phenology are expected and the complex interactions with other climate change effects can significantly alter trophic flows.

Acknowledgements

We thank all of those who have contributed to the plankton monitoring programme of the Baltic Sea, particularly the Stockholm University for data from BY31, Umeå University for B3 and Swedish Meteorological and Hydrological Institute for B5 station. The use of the data from the National Swedish Monitoring Program, supported by the Swedish EPA, and supported by SMHI is gratefully acknowledged. We appreciate discussions on the manuscript with Jakob Walve and Andreas Novotny, and thank Erik Swedberg for providing a map of the Baltic Sea. We thank the Wenner-Gren Foundation and The University Centre in Svalbard for supporting ØV's sabbatical stay at the Department of Ecology, Environment and Plant Sciences (DEEP), Stockholm University.

References

Aksnes, D., N. Dupont, A. Staby, Ø. Fiksen, S. Kaartvedt and J. Aure. 2009. Coastal water darkening and implications for mesopelagic regime shifts in Norwegian fjords. Mar. Ecol. Prog. Ser. 387: 39–49.

Allan, J.D. 1976. Life history patterns in zooplankton. Am. Nat. 110: 165–180.

Andersson, A., Tamminen, Timo, Lehtinen, Sirpa, Jurgens, Klaus, Labrenz, Matthias and Viitasalo, Markku. 2017. The pelagic food web. pp. 281–332. *In*: P. Snoeijs-Leijonmalm, H. Schubert and T. Radziejewska [eds.]. Biological Oceanography of the Baltic Sea, Springer.

Arndt, H. 1991. On the importance of planktonic protozoans in the eutrophication process of the Baltic Sea. Int. Rev. Gesamten Hydrobiol. Hydrogr. 76: 387–396.

Arndt, H. 1993. Rotifers as predators on components of the microbial web (bacteria, heterotrophic flagellates, ciliates) —A review. Hydrobiologia 255: 231–246.

Atkinson, A. 1998. Life cycle strategies of epipelagic copepods in the Southern Ocean. J. Mar. Syst. 15: 289–311.

Barnes, C., D. Maxwell, D.C. Reuman and S. Jennings. 2010. Global patterns in predator–prey size relationships reveal size dependency of trophic transfer efficiency. Ecology 91: 222–232.

Basedow, S.L., N.A.L. de Silva, A. Bode and J. van Beusekom. 2016. Trophic positions of mesozooplankton across the North Atlantic estimates derived from biovolume spectrum theories and stable isotope analyses. J. Plankton Res. 38: 1364–1378.

Behrenfeld, M.J. and E.S. Boss. 2014. Resurrecting the ecological underpinnings of ocean plankton blooms. Annu. Rev. Mar. Sci. 6: 167–194.

Bjærke, O., T. Andersen and J. Titelman. 2014. Predator chemical cues increase growth and alter development in nauplii of a marine copepod. Mar. Ecol. Prog. Ser. 510: 15–24.

Boyce, D.G., K.T. Frank and W.C. Leggett. 2015. From mice to elephants overturning the 'one size fits all' paradigm in marine plankton food chains. Ecol. Lett. 18: 504–515.

Broman, E., M. Brüsin, M. Dopson and S. Hylander. 2015. Oxygenation of anoxic sediments triggers hatching of zooplankton eggs. Proc. R. Soc. B Biol. Sci. 282: 20152025.

Broman, E., C. Raymond, C. Sommer, J.S. Gunnarsson, S. Creer and F.J.A. Nascimento. 2019. Salinity drives meiofaunal community structure dynamics across the Baltic ecosystem. Mol. Ecol. 28: 3813–3829.

Bunse, C., S. Israelsson, F. Baltar, M. Bertos-Fortis, E. Fridolfsson, C. Legrand et al. 2019. High frequency multi-year variability in Baltic Sea microbial plankton stocks and activities. Front. Microbiol. 9: 3296.

Burian, A., J.M. Nielsen and M. Winder. 2020. Food quantity–quality interactions and their impact on consumer behavior and trophic transfer. Ecol. Monogr. 90: e01395.

Calbet, A. and M.R. Landry. 2004. Phytoplankton growth, microzooplankton grazing, and carbon cycling in marine systems. Limnol. Oceanogr. 49: 51–57.

Calbet, A., F. Carlotti and R. Gaudy. 2007. The feeding ecology of the copepod *Centropages typicus* (Kröyer). Prog. Oceanogr. 72: 137–150.

Carstensen, J., J.H. Andersen, B.G. Gustafsson and D.J. Conley. 2014. Deoxygenation of the Baltic Sea during the last century. Proc. Natl. Acad. Sci. 111: 5628–5633.

Casini, M., J. Lovgren, J. Hjelm, M. Cardinale, J.C. Molinero and G. Kornilovs. 2008. Multi-level trophic cascades in a heavily exploited open marine ecosystem. Proc. R. Soc. B 275: 1793–1801.

Casini, M., F. Käll, M. Hansson, M. Plikshs, T. Baranova, O. Karlsson et al. 2016. Hypoxic areas, density-dependence and food limitation drive the body condition of a heavily exploited marine fish predator. R. Soc. Open Sci. 3: 160416.

Chinnery, F. and J. Williams. 2003. Photoperiod and temperature regulation of diapause egg production in *Acartia bifilosa* from Southampton Water. Mar. Ecol. Prog. Ser. 263: 149–157.

Clark, G.F., J.S. Stark, E.L. Johnston, J.W. Runcie, P.M. Goldsworthy, B. Raymond et al. 2013. Light-driven tipping points in polar ecosystems. Glob. Change Biol. 19: 3749–3761.

Cloern, J.E., P.C. Abreu, J. Carstensen, L. Chauvaud, R. Elmgren, J. Grall et al. 2016. Human activities and climate variability drive fast-paced change across the world's estuarine-coastal ecosystems. Glob. Change Biol. 22: 513–529.

Conley, K.R., F. Lombard and K.R. Sutherland. 2018. Mammoth grazers on the ocean's minuteness a review of selective feeding using mucous meshes. Proc. R. Soc. B Biol. Sci. 285: 20180056.

Conover, R.J. 1988. Comparative life histories in the genera *Calanus* and *Neocalanus* in high latitudes of the northern hemisphere. Hydrobiologia 167: 127–142.

Continuous Plankton Recorder Survey Team. 2004. Continuous Plankton Records. Plankton Atlas of the North Atlantic Ocean (1958–1999). II. Biogeographical charts. Mar. Ecol. Prog. Ser. Suppl. 11–75.

Cushing, C.E. 1990. Plankton production and year-class strength in fish populations an update of the match/mismatch hypothesis. Adv. Mar. Biol. 26: 249–292.

Daase, M., S. Falk-Petersen, Ø. Varpe, G. Darnis, J.E. Søreide, A. Wold et al. 2013. Timing of reproductive events in the marine *Calanus glacialis* a pan-Arctic perspective. Can. J. Fish. Aquat. Sci. 70: 871–884.

Dahlgren, K., B.R. Olsen, C. Troedsson and U. Bamstedt. 2012. Seasonal variation in wax ester concentration and gut content in a Baltic Sea copepod *Limnocalanus macrurus* (Sars 1863). J. Plankton Res. 34: 286–297.

Danks, H.V. 2002. The range of insect dormancy responses. Eur. J. Entomol. 99: 127–142.

Daufresne, M., K. Lengfellner and U. Sommer. 2009. Global warming benefits the small in aquatic ecosystems. Proc. Natl. Acad. Sci. 106: 12788–12793.

De Meester, L. 1993. Genotype, fish-mediated chemicals, and phototactic behavior in Daphnia magna. Ecology 74: 1467–1474.

Diekmann, A.B.S., C. Clemmesen, M.A. St John, M. Paulsen and M.A. Peck. 2012. Environmental cues and constraints affecting the seasonality of dominant calanoid copepods in brackish, coastal waters a case study of *Acartia, Temora and Eurytemora* species in the south-west Baltic. Mar. Biol. 159: 2399–2414.

Digby, P.S.B. 1950. The biology of the small planktonic copepods of Plymouth. J. Mar. Biol. Assoc. UK 29: 393–438.

Drake, J.M. 2005. Population effects of increased climate variation. Proc. Biol. Sci. 272: 1823–1827.

Durbin, E. and J. Kane. 2007. Seasonal and spatial dynamics of *Centropages typicus* and *C. hamatus* in the western North Atlantic. Prog. Oceanogr. 72: 249–258.

Dutz, J., V. Mohrholz and J. van Beusekom. 2010. Life cycle and spring phenology of *Temora longicornis* in the Baltic Sea. Mar. Ecol. Prog. Ser. 406: 223–238.

Dutz, J., J. van Beusekom and R. Hinrichs. 2012. Seasonal dynamics of fecundity and recruitment of *Temora longicornis* in the Baltic Sea. Mar. Ecol. Prog. Ser. 462: 51–66.

Edwards, M. and A.J. Richardson. 2004. Impact of climate change on marine pelagic phenology and trophic mismatch. Nature 430: 881–884.

Ejsmond, M.J., J.M. McNamara, J. Søreide and Ø. Varpe. 2018. Gradients of season length and mortality risk cause shifts in body size, reserves and reproductive strategies of determinate growers. Funct. Ecol. 32: 2395–2406.

Engstrom, J. 2000. Feeding interactions of the copepods *Eurytemora affinis* and *Acartia bifilosa* with the cyanobacteria *Nodularia* sp. J. Plankton Res. 22: 1403–1409.

Falkowski, P.G. and M.J. Oliver. 2007. Mix and match how climate selects phytoplankton. Nat. Rev. Microbiol. 5: 813–819.

Falkowski, R., T. Barber and V. Smetacek. 1998. Biogeochemical controls and feedbacks on ocean primary production. Science 281: 200–206.

Feike, M. and R. Heerkloss. 2009. Does *Eurytemora affinis* (Copepoda) control the population growth of *Keratella cochlearis* (Rotifera) in the brackish water Darss-Zingst Lagoon (southern Baltic Sea)? J. Plankton Res. 31: 571–576.

Feró, O., P.A. Stephens, Z. Barta, J.M. McNamara and A.I. Houston. 2008. Optimal annual routines new tools for conservation biology. Ecol. Appl. 18: 1563–1577.

Flynn, K.J., A. Mitra, K. Anestis, A.A. Anschütz, A. Calbet, G.D. Ferreira et al. 2019. Mixotrophic protists and a new paradigm for marine ecology where does plankton research go now? J. Plankton Res. 41: 375–391.

Forster, J. and A.G. Hirst. 2012. The temperature-size rule emerges from ontogenetic differences between growth and development rates. Ontogenetic differences between growth and development rates. Funct. Ecol. 26: 483–492.

Galloway, A.W.E. and M. Winder. 2015. Partitioning the relative importance of phylogeny and environmental conditions on phytoplankton fatty acids. PLoS One 10: e0130053–23.

Garzke, J., S.M.H. Ismar and U. Sommer. 2015. Climate change affects low trophic level marine consumers warming decreases copepod size and abundance. Oecologia 177: 849–860.

Gasol, J.M., P.A. del Giorgio and C.M. Duarte. 1997. Biomass distribution in marine planktonic communities. Limnol. Oceanogr. 42: 1353–1363.

Griffiths, J.R., M. Kadin, F.J.A. Nascimento, T. Tamelander, A. Törnroos, S. Bonaglia et al. 2017. The importance of benthic-pelagic coupling for marine ecosystem functioning in a changing world. Glob. Change Biol. 23: 2179–2196.

Haapala, J.J., I. Ronkainen, N. Schmelzer and M. Sztobryn. 2015. Recent Change—Sea Ice. Second Assessment of Climate Change for the Baltic Sea Basin, The BACC II Author Team, Ed., Springer International Publishing, 145–153.

Hagen, W. 1999. Reproductive strategies and energetic adaptations of polar zooplankton. Invertebrate. Reprod. Dev. 36: 25–34.

Hansen, B., P.K. Bjornsen and P.J. Hansen. 1994. The size ratio between planktonic predators and their prey. Limnol. Oceanogr. 39: 395–403.

Hansen, P., M. Moldrup, W. Tarangkoon, L. Garcia-Cuetos and Ø. Moestrup. 2012. Direct evidence for symbiont sequestration in the marine red tide ciliate *Mesodinium rubrum*. Aquat. Microb. Ecol. 66: 63–75.

Hays, G.C., A.J. Richardson and C. Robinson. 2005. Climate change and marine plankton. Trends Ecol. Evol. 20: 337–344.

Heino, M. and V. Kaitala. 1999. Evolution of resource allocation between growth and reproduction in animals with indeterminate growth. J. Evol. Biol. 12: 423–429.

HELCOM. 2007. Climate Change in the Baltic Sea Area—HELCOM Thematic Assessment in 2007.

HELCOM. 2009a. Eutrophication in the Baltic Sea. An integrated thematic assessment of the effects of nutrient enrichment in the Baltic Sea region.

HELCOM. 2009b. Biodiversity in the Baltic Sea. An integrated thematic assessment on biodiversity and nature conservation in the Baltic Sea.

HELCOM. 2014. Manual for Marine Monitoring in the COMBINE Programme.

Hirst, A. and T. Kiørboe. 2002. Mortality of marine planktonic copepods global rates and patterns. Mar. Ecol. Prog. Ser. 230: 195–209.

Hjerne, O., S. Hajdu, U. Larsson, A.S. Downing and M. Winder. 2019. Climate driven changes in timing, composition and magnitude of the Baltic Sea phytoplankton spring bloom. Front. Mar. Sci. 6: 482.

Holm, M.W., T. Kiørboe, P. Brun, P. Licandro, R. Almeda and B.W. Hansen. 2018. Resting eggs in free living marine and estuarine copepods. J. Plankton Res. 40: 2–15.

Holste, L., M.A. St. John and M.A. Peck. 2009. The effects of temperature and salinity on reproductive success of *Temora longicornis* in the Baltic Sea a copepod coping with a tough situation. Mar. Biol. 156: 527–540.

Hutchinson, G.E. 1967. A Treatise on Limnology, Vol. II. Introduction to Lake Biology and the Limnoplankton, Wiley, Ed. John Wiley.

Irigoien, X. 2004. Some ideas about the role of lipids in the life cycle of *Calanus finmarchicus*. J. Plankton Res. 26: 259–263.

Ji, R., M. Edwards, D.L. Mackas, J.A. Runge and A.D. Thomas. 2010. Marine plankton phenology and life history in a changing climate current research and future directions. J. Plankton Res. 32: 1355–1368.

Ji, R., M. Jin and Ø. Varpe. 2013. Sea ice phenology and timing of primary production pulses in the Arctic Ocean. Glob. Change Biol. 19: 734–741.

Johansson, F., R. Stoks, L. Rowe and M. De Block. 2001. Life history plasticity in a damselfly effects of combined time and biotic constraints. Ecology 82: 1857–1869.

Johansson, M., E. Gorokhova and U. Larsson. 2004. Annual variability in ciliate community structure, potential prey and predators in the open northern Baltic Sea proper. J. Plankton Res. 26: 67–80.

Kaartvedt, S. 2000. Life history of *Calanus finmarchicus* in the Norwegian Sea in relation to planktivorous fish. ICES J. Mar. Sci. 57: 1819–1824.

Kahru, M. and R. Elmgren. 2014. Multidecadal time series of satellite-detected accumulations of cyanobacteria in the Baltic Sea. Biogeosciences 11: 3619–3633.

Kahru, M., R. Elmgren and O.P. Savchuk. 2016. Changing seasonality of the Baltic Sea. Biogeosciences 13: 1009–1018.

Kankaala, P. 1983. Resting eggs, seasonal dynamics, and production of *Bosmina longispina maritima* (P.E. Müller) (Cladocera) in the northern Baltic proper. J. Plankton Res. 5: 53–69.

Karlsson, K., S. Puiac and M. Winder. 2018. Life-history responses to changing temperature and salinity of the Baltic Sea copepod *Eurytemora affinis*. Mar. Biol. 165.

Katajisto, T., M. Viitasalo and M. Koski. 1998. Seasonal occurrence and hatching of calanoid eggs in sediments of the northern Baltic Sea. Mar. Ecol. -Prog. Ser. 163: 133–143.

Katajisto, T. 2003. Development of *Acartia bifilosa* (Copepoda. Calanoida) eggs in the northern Baltic Sea with special reference to dormancy. J. Plankton Res. 25: 357–364.

Käyhkö, J., E. Apsite, A. Bolek, N. Filatov, S. Kondratyev, J. Korhonen et al. 2015. Recent change—river run-off and ice cover. Second Assessment of Climate Change for the Baltic Sea Basin, The BACC II Author Team, Ed., Springer International Publishing, 99–116.

Kim, G.H., J.H. Han, B. Kim, J.W. Han, S.W. Nam, W. Shin et al. 2016. Cryptophyte gene regulation in the kleptoplastidic, karyokleptic ciliate *Mesodinium rubrum*. Harmful Algae 52: 23–33.

Kiørboe, T. 1993. Turbulence, phytoplankton cell size, and the structure of pelagic food webs. Adv. Mar. Biol. 29: 1–72.

Kiørboe, T., E. Saiz and M. Viitasalo. 1996. Prey switching behaviour in the planktonic copepod *Acartia tonsa*. Mar. Ecol. Prog. Ser. 143: 65–75.

Klais, R., M. Lehtiniemi, G. Rubene, A. Semenova, P. Margonski, A. Ikauniece et al. 2016. Spatial and temporal variability of zooplankton in a temperate semi-enclosed sea implications for monitoring design and long-term studies. J. Plankton Res. 38: 652–661.

Koski, M., W. Klein Breteler and N. Schogt. 1998. Effect of food quality on rate of growth and development of the pelagic copepod *Pseudocalanus elongatus* (Copepoda, Calanoida). Mar. Ecol. Prog. Ser. 170: 169–187.

Koski, M., M. Viitasalo and H. Kuosa. 1999. Seasonal development of meso-zooplankton biomass and production on the SW coast of Finland. Ophelia 50: 69–91.

Köster, F.W., H.-H. Hinrichsen, D. Schnack, M.A. St. John, B.R. Mackenzie, J. Tomkiewicz et al. 2003. Recruitment of Baltic cod and sprat stocks identification of critical life stages and incorporation of environmental variability into stock-recruitment relationships. Sci. Mar. 67: 129–154.

Kyryliuk, D. and S. Kratzer. 2019. Summer distribution of total suspended matter across the Baltic Sea. Front. Mar. Sci. 5: 504.

Landry, M.R. and A. Calbet. 2004. Microzooplankton production in the oceans. ICES J. Mar. Sci. 61: 501–507.

Langbehn, T.J. and Ø. Varpe. 2017. Sea-ice loss boosts visual search fish foraging and changing pelagic interactions in polar oceans. Glob. Change Biol. 23: 5318–5330.

Lee, C.E. 2016. Evolutionary mechanisms of habitat invasions, using the copepod *Eurytemora affinis* as a model system. Evol. Appl. 9: 248–270.

Lee, R., W. Hagen and G. Kattner. 2006. Lipid storage in marine zooplankton. Mar. Ecol. Prog. Ser. 307: 273–306.

Liblik, T. and U. Lips. 2019. Stratification has strengthened in the Baltic Sea—an analysis of 35 years of observational data. Front. Earth Sci. 7: 174.

Lima-Mendez, G., K. Faust, N. Henry, J. Decelle, S. Colin, F. Carcillo et al. 2015. Ocean plankton. Determinants of community structure in the global plankton interactome. Science 348: 1262073–1262073.

Lischka, S., L.T. Bach, K.-G. Schulz and U. Riebesell. 2017. Ciliate and mesozooplankton community response to increasing CO_2 levels in the Baltic Sea insights from a large-scale mesocosm experiment. Biogeosciences 14: 447–466.

Litchman, E., P. de Tezanos Pinto, K.F. Edwards, C.A. Klausmeier, C.T. Kremer and M.K. Thomas. 2015. Global biogeochemical impacts of phytoplankton: a trait-based perspective. J. Ecol. 103: 1384–1396.

Loick-Wilde, N., I. Fernández-Urruzola, E. Eglite, I. Liskow, M. Nausch, D. Schulz-Bull et al. 2019. Stratification, nitrogen fixation, and cyanobacterial bloom stage regulate the planktonic food web structure. Glob. Change Biol. 25: 794–810.

Mackas, D.L., W. Greve, M. Edwards, S. Chiba, K. Tadokoro, D. Eloire et al. 2012. Changing zooplankton seasonality in a changing ocean: comparing time series of zooplankton phenology. Prog. Oceanogr. 97-100: 31–62.

Marcus, N. 1996. Ecological and evolutionary significance of resting eggs in marine copepods past, present, and future studies. Hydrobiologia 320: 141–152.

McCauley, E. and W.W. Murdoch. 1990. Predator-prey dynamics in environments rich and poor in nutrients. Nature 343: 455–457.

McLaren, I.A., E. Laberge, C.J. Corkett and J.-M. Sévigny. 1989. Life cycles of four species of *Pseudocalanus* in Nova Scotia. Can. J. Zool. 67: 552–558.

McNamara, J.M. and A.I. Houston. 2008. Optimal annual routines behaviour in the context of physiology and ecology. Philos. Trans. R. Soc. B Biol. Sci. 363: 301–319.

Menden-Deuer, S. and T. Kiørboe. 2016. Small bugs with a big impact linking plankton ecology with ecosystem processes. J. Plankton Res. 38: 1036–1043.

Möllmann, C. 2000. Long-term dynamics of main mesozooplankton species in the central Baltic Sea. J. Plankton Res. 22: 2015–2038.

Mollmann, C. 2002. Population dynamics of calanoid copepods and the implications of their predation by clupeid fish in the Central Baltic Sea. J. Plankton Res. 24: 959–978.

Möllmann, C., G. Kornilovs, M. Fetter, F.W. Koster and H.-H. Hinrichsen. 2003. The marine copepod, *Pseudocalanus elongatus*, as a mediator between climate variability and fisheries in the Central Baltic Sea. Fish. Oceanogr. 12: 360–368.

Möllmann, C., B. Muller-Karulis, G. Kornilovs and M.A. St. John. 2008. Effects of climate and overfishing on zooplankton dynamics and ecosystem structure regime shifts, trophic cascade, and feedback coops in a simple ecosystem. ICES J. Mar. Sci. 65: 302–310.

Ojaveer, E. 1998. Highlights of zooplankton dynamics in Estonian waters (Baltic Sea). ICES J. Mar. Sci. 55: 748–755.

Ojaveer, H., A. Jaanus, B.R. MacKenzie, G. Martin, S. Olenin, T. Radziejewska et al. 2010. Status of biodiversity in the Baltic Sea. PLoS One 5: e12467.

Ok, J.H., H.J. Jeong, A.S. Lim and K.H. Lee. 2017. Interactions between the mixotrophic dinoflagellate Takayama helix and common heterotrophic protists. Harmful Algae 68: 178–191.

Opdal, A.F., C. Lindemann and D.L. Aksnes. 2019. Centennial decline in North Sea water clarity causes strong delay in phytoplankton bloom timing. Glob. Change Biol. 25: 3946–3953.

Paerl, H.W. and J. Huisman. 2008. Blooms like it hot. Science 320: 57–58.

Peters, J., J. Renz, J. van Beusekom, M. Boersma and W. Hagen. 2006. Trophodynamics and seasonal cycle of the copepod *Pseudocalanus acuspes* in the Central Baltic Sea (Bornholm Basin) evidence from lipid composition. Mar. Biol. 149: 1417–1429.

Peters, J., J. Dutz and W. Hagen. 2013. Trophodynamics and life-cycle strategies of the copepods *Temora longicornis* and *Acartia longiremis* in the Central Baltic Sea. J. Plankton Res. 35: 595–609.

Pijanowska, J. and G. Stolpe. 1996. Summer diapause in *Daphnia* as a reaction to the presence of fish. J. Plankton Res. 18: 1407–1412.

Priyadarshi, A., S.L. Smith, S. Mandal, M. Tanaka and H. Yamazaki. 2019. Micro-scale patchiness enhances trophic transfer efficiency and potential plankton biodiversity. Sci. Rep. 9: 17243.

Record, N.R., R. Ji, F. Maps, Ø. Varpe, J.A. Runge, C.M. Petrik et al. 2018. Copepod diapause and the biogeography of the marine lipidscape. J. Biogeogr. 45: 2238–2251.

Renz, J., J. Peters and H.-J. Hirche. 2007. Life cycle of *Pseudocalanus acuspes* Giesbrecht (Copepoda, Calanoida) in the Central Baltic Sea. II. Reproduction, growth and secondary production. Mar. Biol. 151: 515–527.

Reusch, T.B.H., J. Dierking, H.C. Andersson, E. Bonsdorff, J. Carstensen, M. Casini et al. 2018. The Baltic Sea as a time machine for the future coastal ocean. Sci. Adv. 4: eaar8195.

Romagnan, J.-B., L. Legendre, L. Guidi, J.-L. Jamet, D. Jamet, L. Mousseau et al. 2015. Comprehensive model of annual plankton succession based on the whole-plankton time series approach. PLoS ONE 10: e0119219.

Rossoll, D., U. Sommer and M. Winder. 2013. Community interactions dampen acidification effects in a coastal plankton system. Mar. Ecol. Prog. Ser. 486: 37–46.

Rudstam, L.G., K. Danielsson, S. Hansson and S. Johansson. 1989. Diel vertical migration and feeding patterns of *Mysis mixta* (Crustacea, Mysidacea) in the Baltic Sea. Mar. Biol. 101: 43–52.

Saage, A., O. Vadstein and U. Sommer. 2009. Feeding behaviour of adult *Centropages hamatus* (Copepoda, Calanoida). Functional response and selective feeding experiments. J. Sea Res. 62: 16–21.

Sainmont, J., K.H. Andersen, Ø. Varpe and A.W. Visser. 2014. Capital versus income breeding in a seasonal environment. Am. Nat. 184: 466–476.

Sandberg, J., A. Andersson, S. Johansson and J. Wikner. 2004. Pelagic food web structure and carbon budget in the northern Baltic Sea potential importance of terrigenous carbon. Mar. Ecol. Prog. Ser. 268: 13–29.

Santer, B. and W. Lampert. 1995. Summer diapause in cyclopoid copepods adaptive response to a food bottleneck? J. Anim. Ecol. 64: 600–613.

Schmoker, C., S. Hernández-León and A. Calbet. 2013. Microzooplankton grazing in the oceans impacts, data variability, knowledge gaps and future directions. J. Plankton Res. 35: 691–706.

Sherr, E.B. and B.F. Sherr. 2002. Significance of predation by protists in aquatic microbial food webs. Antonie Van Leeuwenhoek 81: 293–308.

Sieburth, J.McN., V. Smetacek and J. Lenz. 1978. Pelagic ecosystem structure. Heterotrophic compartments of the plankton and their relationship to plankton size fractions 1. Limnol. Oceanogr. 23: 1256–1263.

Snoeijs-Leijonmalm, P., H. Schubert and T. Radziejewska. 2017. Biological Oceanography of the Baltic Sea. Springer Netherlands.

Sommer, U. 1989. Plankton Ecology. Succession in Plankton Communities. Springer-Verlag Berlin Heidelberg, 369 p.

Sommer, U., H. Stibor, A. Katechakis, F. Sommer and T. Hansen. 2002. Pelagic food web configurations at different levels of nutrient richness and their implications for the ratio fish production primary production. Hydrobiologia 484: 11–20.

Sommer, U., R. Adrian, L. De Senerpont Domis, J.J. Elser, U. Gaedke, B. Ibelings et al. 2012. Beyond the Plankton Ecology Group (PEG) model. Mechanisms driving plankton Succession. Annu. Rev. Ecol. Evol. Syst. 43: 429–448.

Sommer, U., E. Charalampous, S. Genitsaris and M. Moustaka-Gouni. 2017. Benefits, costs and taxonomic distribution of marine phytoplankton body size. J. Plankton Res. 39: 494–508.

Stearns, S.C. 1992. The Evolution of Life Histories. Oxf. Univ. Press Lond., 249 p.

Steinberg, D.K. and M.R. Landry. 2017. Zooplankton and the ocean carbon cycle. Annu. Rev. Mar. Sci. 9: 413–444.

Stibor, H., O. Vadstein, S. Diehl, A. Gelzleichter, T. Hansen, F. Hantzsche et al. 2004. Copepods act as a switch between alternative trophic cascades in marine pelagic food webs. Trophic cascades in marine plankton. Ecol. Lett. 7: 321–328.

Stoecker, D., M. Johnson, C. de Vargas and F. Not. 2009. Acquired phototrophy in aquatic protists. Aquat. Microb. Ecol. 57: 279–310.

Sverdrup, H.U. 1953. On conditions for the vernal blooming of phytoplankton. ICES J. Mar. Sci. 18: 287–295.

Telesh, I., L. Postel, R. Heerkloss, E. Mironova and S. Skarlato. 2008. Zooplankton of the open Baltic Sea. Atlas. Institute for Baltic Sea Research, Marine Science Reports.

The BACC Teach. 2015. Second Assessment of Climate Change for the Baltic Sea Basin. Springer.

Thomsen, J., L.S. Stapp, K. Haynert, H. Schade, M. Danelli, G. Lannig et al. 2017. Naturally acidified habitat selects for ocean acidification–tolerant mussels. Sci. Adv. 3: e1602411.

Tiselius, P. and P. Jonsson. 1990. Foraging behaviour of six calanoid copepods observations and hydrodynamic analysis. Mar. Ecol. Prog. Ser. 66: 23–33.

Varpe, Ø., Ø. Fiksen and A. Slotte. 2005. Meta-ecosystems and biological energy transport from ocean to coast: the ecological importance of herring migration. Oecologia 146: 443–451.

Varpe, Ø., C. Jørgensen, G.A. Tarling and Ø. Fiksen. 2009. The adaptive value of energy storage and capital breeding in seasonal environments. Oikos 118: 363–370.

Varpe, Ø. and Ø. Fiksen. 2010. Seasonal plankton–fish interactions: light regime, prey phenology, and herring foraging. Ecology 91: 311–318.

Varpe, Ø. 2012. Fitness and phenology: annual routines and zooplankton adaptations to seasonal cycles. J. Plankton Res. 34: 267–276.

Varpe, Ø. 2017. Life history adaptations to seasonality. Integr. Comp. Biol. 57: 943–960.

Varpe, Ø. and M.J. Ejsmond. 2018. Semelparity and Iteroparity. pp. 97–124. *In*: G.A. Wellborn and M. Thiel [eds.]. Natural History of Crustacea, Life Histories, Vol. 5., Oxford University Press.

Viitasalo, M., I. Vuorinene and E. Ranta. 1990. Changes in crustacean mesozooplankton and some environmental parameters in the Archipelago Sea (Northern Baltic) in 1976–1984. Ophelia 31: 207–217.

Viitasalo, M. 1992. Calanoid resting eggs in the Baltic Sea implications for the population dynamics of *Acartia bifilosa* (Copepoda). Mar. Biol. 114: 397–405.

Viitasalo, M. and T. Katajisto. 1994. Mesozooplankton resting eggs in the Baltic Sea identification and vertical distribution in laminated and mixed sediments. Mar. Biol. 120: 455–466.

Viitasalo, M., T. Katajisto and I. Vuorinen. 1994. Seasonal dynamics of *Acartia bifilosa* and *Eurytemora affinis* (Copepods, Calanoida) in relation to abiotic factors in the northern Baltic Sea. Hydrobiologia 292/293: 415–422.

Viitasalo, M., I. Vuorinen and S. Saesmaa. 1995. Mesozooplankton dynamics in the northern Baltic Sea implications of variations in hydrography and climate. J. Plankton Res. 17: 1857–1878.

Vuorinen, I. 1998. Proportion of copepod biomass declines with decreasing salinity in the Baltic Sea. ICES J. Mar. Sci. 55: 767–774.

Ward, B.A. and M.J. Follows. 2016. Marine mixotrophy increases trophic transfer efficiency, mean organism size, and vertical carbon flux. Proc. Natl. Acad. Sci. 113: 2958–2963.

Wasmund, N. and S. Uhlig. 2003. Phytoplankton trends in the Baltic Sea. ICES J. Mar. Sci. 60: 177–186.

Weisse, T. and A. Frahm. 2002. Direct and indirect impact of two common rotifer species (*Keratella* spp.) on two abundant ciliate species (*Urotricha furcata, Balanion planctonicum*). Freshw. Biol. 47: 53–64.

Weithoff, G. and B.E. Beisner. 2019. Measures and approaches in trait-based phytoplankton community ecology—from freshwater to marine ecosystems. Front. Mar. Sci. 6: 40.

Winder, M. and D.E. Schindler. 2004a. Climatic effects on the phenology of lake processes. Glob. Change Biol. 10: 1844–1856.

Winder, M. and D.E. Schindler. 2004b. Climate change uncouples trophic interactions in a lake ecosystem. Ecology 85: 2100–2106.

Winder, M., J.E. Reuter and S.G. Schladow. 2009a. Lake warming favours small-sized planktonic diatom species. Proc. R. Soc. B Biol. Sci. 276: 427–435.

Winder, M., D.E. Schindler, T.E. Essington and A.H. Litt. 2009b. Disrupted seasonal clockwork in the population dynamics of a freshwater copepod by climate warming. Limnol. Oceanogr. 54: 2493–2505.

Winder, M. and J.E. Cloern. 2010. The annual cycles of phytoplankton biomass. Philos. Trans. R. Soc. B Biol. Sci. 365: 3215–3226.

Winder, M. and U. Sommer. 2012. Phytoplankton response to a changing climate. Hydrobiologia 698: 5–16.

Winder, M., J.-M. Bouquet, J. Rafael Bermúdez, S.A. Berger, T. Hansen, J. Brandes et al. 2017a. Increased appendicularian zooplankton alter carbon cycling under warmer more acidified ocean conditions. Limnol. Oceanogr. 62: 1541–1551.

Winder, M., J. Carstensen, A.W.E. Galloway, H.H. Jakobsen and J.E. Cloern. 2017b. The land-sea interface. A source of high-quality phytoplankton to support secondary production. Limnol. Oceanogr. 62: S258–S271.

Winkler, G.S., P.C. Souissi and V. Castric. 2011. Genetic heterogeneity among *Eurytemora affinis* populations in Western Europe. Mar. Biol. 158: 1841–1856.

CHAPTER 9

Visualizing and Exploring Zooplankton Spatio-Temporal Variability

Todd D. O'Brien* and *Stephanie A. Oakes*

9.1 Introduction

The distribution and abundance of zooplankton varies over time and space in response to physical and biological drivers operating at different temporal (e.g., seasonal, interannual and decadal) and spatial (e.g., local, regional, global) scales. As a result, zooplankton are often described as having patchy distributions and boom-or-bust abundance cycles. Zooplankton data are collected with different frequencies (e.g., sub-weekly to once-per-season), sampling methods (e.g., nets, pumps, and optical systems), and measurement techniques ranging from total biomass to full microscope-based taxonomic identification and enumeration. The resulting time series data are often complex and require data usage and application caveats, as do the advanced statistics used for their analysis. Whether one is a statistics and programming expert, or a student analyzing data with a spreadsheet, the first step in any analysis should be to explore the data with basic visual and statistical techniques.

The overall aim of this chapter is to introduce some simple techniques for visualizing and exploring zooplankton variability, patterns, and trends over time and space. The chapter discusses how zooplankton time series provide an ideal starting point to explore the impacts of changing climate and environmental conditions on ocean food webs. A historical overview provides background on international efforts to compile and synthesize global ocean time series data and develop standard visualization tools. Simple and more complex visualization techniques for seasonal, annual and interannual cycles, both within and between sampling sites and regions, are then provided. An example of broad scale spatio-temporal synthesis of North Atlantic sampling program time series data illustrates how to combine and compare data with different sampling methods and frequencies. Finally, the chapter ends with recommendations for how to proceed with advanced statistical analysis, and future challenges are discussed.

National Oceanic & Atmospheric Administration (NOAA) Fisheries, Office of Science & Technology/Marine Ecosystems Division, 1315 East-West Highway, Silver Spring, Maryland, USA 20910.
Email: stephanie.oakes@noaa.gov
* Corresponding author: todd.obrien@noaa.gov

9.1.1 The Need for Data Exploration

Knowledge about the world's oceans is becoming increasingly data-driven. From satellite observation to more sophisticated *in situ* sensors, the volume of data produced and accessible to researchers today is ever growing, allowing them to explore and describe more complex systems. Advanced approaches to address "big data" with artificial intelligence and machine learning are at the forefront of cutting edge data science today, but scientists have always struggled with ways to interpret and analyze complex data. In their paper "Applied Statistics in Ecology: Common Pitfalls and Simple Solutions", Steel et al. (2013) start their list of pitfalls with "failure to explore the data" and suggest a very simple solution: "plot the data early and often". Constructing basic plots can help highlight data issues such as skewed data or an abundance of zero-values or large outliers, allowing further investigation to distinguish natural phenomena from equipment or data transcription errors. Even a simple histogram can quickly display the skew and spread (numeric range) of the data.

Some other common pitfalls mentioned in Steel et al. (2013) involve not defining and dealing appropriately with zeros (e.g., was it a true absence or just below detection), and applying unnecessary transformations. Finally, they warn of the dangerous pitfalls of using an advanced statistical tool without understanding the statistics, data prerequisites, or correct interpretation of its results. In short, one should not dismiss or ignore the step of first applying simple statistical and visualization methods to explore the properties, patterns, and trends present in their data sets. Many of these identified pitfalls are applicable to zooplankton time series data. Zooplankton, though relatively abundant in the ocean, exist over large spatial scales and tend to occur in patches (Mackas et al. 1985, Mackas and Beaugrand 2010). Unlike temperature and chlorophyll, with global satellite-based measurements and networks of drifting floats and buoys, zooplankton are still predominantly sampled shipboard or from small boats, and often identified via microscope examination. This means zooplankton data often have limited spatial and temporal coverage, which influences choice of data exploration and analysis approaches.

9.1.2 Why Zooplankton

An Ideal Study Subject

Zooplankton are a key component of marine ecosystems, and serve a broad spectrum of trophic roles: grazers, detritivores, predators, and prey. They help transfer primary production to the upper trophic levels (e.g., fish, birds, marine mammals) and the microbial community (e.g., via fecal pellets and nitrogen excretion), and transfer carbon and organic materials to the benthic communities and deep ocean (Ruhl and Smith 2004, Richardson 2008, Steinberg and Landry 2017).

Zooplankton can be categorized by their attributes including trophic level, feeding habits, preferred environment, life history and size. The majority of historical zooplankton surveys and long-running time series sample only the top 200 meters of water using 200–333 µm mesh nets, most commonly with a 200 µm mesh UNESCO Working Party 2 net (WP2), a 270 µm silk Continuous Plankton Recorder, or a 333 µm mesh bongo net (Reid et al. 2003, O'Brien 2005, O'Brien et al. 2013). Zooplankton size classes caught by this range of mesh sizes are generally defined as micro- (20–200 µm), meso- (200 µm to 20 mm), and macro-zooplankton (20–200 mm). Using 200–333 µm mesh nets, the mesozooplankton size class is generally the best sampled, both quantitatively and qualitatively, as microzooplankton[1] tend to pass through larger mesh sizes and

---

[1] Microzooplankton include both unicellular phagotrophic protists (e.g., flagellates, ciliates, foraminifera, radiolarians) as well as smaller species and early life stages of multicellular metazooplankton. Quantitative sampling of these smaller taxa requires bottle, pump, or fine mesh sampling methods and as such generally fall within the scope of the microbial and phytoplankton time series community (O'Brien et al. 2012).

macrozooplankton may actively avoid the nets or be inaccessible in the shallower sampling depths (e.g., some euphausiid species). Additionally, many gelatinous zooplankton are damaged in nets and difficult to identify and enumerate. For these reasons, the majority of readily available zooplankton time series data sets are dominated by less delicate taxa in the mesozooplankton size class, which is often dominated by copepods. From here onward, the term "zooplankton" is used to refer to 200–333 um mesh net-caught, primarily non-gelatinous, mesozooplankton.

Several characteristics of marine zooplankton make them especially suitable for studying variability and trends at seasonal to interannual to decadal time scales, especially in relation to climate change. Zooplankton growth and reproduction are highly sensitive to temperature (Mauchline 1998), and with few exceptions (e.g., some euphausiids) most are short-lived (~ 1 year) with relatively short generation times (weeks to months)[2] often resulting in a tight coupling between climate and environmental stresses on zooplankton biomass and community structure (Hays et al. 2005, Chiba et al. 2018). With short generation times, zooplankton populations are minimally affected by multi-year carryover of individuals (Hayes et al. 2005, Mackas and Beaugrand 2010), which means their population size closely tracks seasonal-to-interannual changes in environmental conditions with relatively little time lag (Mackas and Beaugrand 2010). Zooplankton can also exhibit dramatic changes in geographic distribution, responding to changes in temperature and oceanic current systems by expanding and contracting their biogeographic ranges (Hays et al. 2005, Richardson 2008, Mackas and Beaugrand 2010).

Zooplankton are relatively abundant in the ocean, can be quantified with relatively inexpensive and simple sampling methods (e.g., net tows and in some cases automated optical methods), and in many regions their interseasonal and interannual variability can be sufficiently tracked with only a monthly-to-seasonal sampling frequency (Mackas and Beaugrand 2010). Since zooplankton are not commercially harvested (with the exception of some euphausiid species and recently *Calanus* copepods off Norway), changes in abundance and community composition can be attributed to environmental changes or predator prey interactions, independent from fishing pressure or exploitation. The trends and state of the zooplankton may be a useful leading indicator for commercial fish stocks, as the youngest life stages of and some adult fish are often heavily dependent on zooplankton as a food source (Mackas et al. 2012a).

Multiple studies and programs from regions around the globe have shown that multi-year zooplankton time series can provide useful insights into interactions between climate and marine ecosystems (Perry et al. 2004, Hays et al. 2005, Chiba et al. 2006, Richardson 2008, Mackas and Beaugrand 2010, Mackas et al. 2012a, Hays et al. 2018). Several global and regional comparisons of marine ecosystem variability have also been conducted using zooplankton time series and monitoring data (Rombouts et al. 2010, Mackas et al. 2012a, O'Brien et al. 2012, 2013, Beaugrand et al. 2014, O'Brien et al. 2017).

Beacons for a Changing Ocean

The world's oceans are warming, with many regions seeing unprecedented high or even record temperatures (IPCC 2019). Based on results from an ensemble of ocean models (IPCC 2013), this warming is already having and will continue to have detrimental impacts on marine ecosystems (Richardson 2008, Hoegh-Guldberg and Bruno 2010, Doney et al. 2011, Wernberg et al. 2016, Frölicher and Laufkötter 2018). While these IPCC predictions are dire enough, Cheng et al. (2019) recently reported that newly improved models, run by four independent groups, predict even larger and faster increases than predicted by the 2013 IPCC AR5 model ensemble.[3] Ensemble modeling

---

[2] These generation times are for mesozooplankton. Within the microzooplankton, phagotrophic protists can have generation times of days to hours (Tillmann 2004).

[3] The IPCC AR6 has a planned release date of June 2022.

results may be abstract to some but the real world impact of warming oceans combined with large scale and cyclic environmental drivers are making readily understandable headlines (e.g., the melting Arctic, Marine Heat Waves, coral bleaching; Hoegh-Guldberg 1999, IPCC 2013, Frölicher et al. 2018, Oliver et al. 2019).

Zooplankton can manifest changes in their physical and biological environment (e.g., temperature or food availability) through changes in their population abundance and composition, geographic distribution (biogeography), and seasonal timing (phenology). For example, changes of just a few degrees in water temperature have been associated with measurable changes in zooplankton abundance and composition (McGowan et al. 1998, Mackas et al. 2001, Beaugrand et al. 2002, 2003, Beaugrand 2003, 2004, Hays et al. 2005, Chiba et al. 2006, Mackas et al. 2007, Valdés et al. 2007, Mackas and Beaugrand 2010, Mackas et al. 2012a, b, Beaugrand and Kirby 2018).

Biogeographic and community composition changes in zooplankton have been documented, and in many cases involve poleward migration of cold water species, to cooler waters, and replacement by warmer water species (Johns et al. 2001, Beaugrand et al. 2002, Beaugrand 2003, Richardson 2008, Beaugrand et al. 2009, Batten and Walne 2011, Keister et al. 2011, Mackas et al. 2012a, Chust et al. 2013, Nelson et al. 2014, Chiba et al. 2015, Villarino et al. 2015). Biogeographic changes have also been reported in the phytoplankton (Rosseaux and Gregg 2015, Barton et al. 2016, Neukermans et al. 2018).

Phenological changes in some zooplankton have also occurred, especially when seasonal temperature changes coincided with the weeks-to-months preceding the dominant zooplankton's growing seasons. Zooplankton with a spring or summer maximum abundance exhibit an "occur earlier when warmer" response (Mackas et al. 1998, Edwards and Richardson 2004, Greve et al. 2005, Chiba et al. 2006, Mackas et al. 2007), while zooplankton with a late summer maximum tend to exhibit an "occur later when warmer" response (Edwards and Richardson 2004, Conversi et al. 2009, Mackas and Beaugrand 2010). Phenological changes have also been reported in the phytoplankton (Thackeray et al. 2008, Racault et al. 2012, Winder et al. 2012, Henson et al. 2018, Friedland et al. 2018, Wasmund et al. 2019, Winder and Varpe 2021, this book, Chapter 8).

Zooplankton's ability to respond rapidly to environmental changes makes them a good 'early warning' beacon for changes in ecosystem productivity that can often be detected before these impacts are seen further up the food web (e.g., in the fish, birds, marine mammals, and human fisheries catch). Yet, prior to 2004, zooplankton-climate interactions are mostly examined within a single ocean or region (e.g., the North Atlantic or the California Current). There is a need for and value of doing a global comparison of zooplankton time series trends and variability in relation to each other and climate (Perry et al. 2004).

9.1.3 Working Group 125 (and its Successors)

In 2004, the Scientific Committee on Oceanic Research (SCOR) formed an international working group titled and focused-on Global Comparisons of Zooplankton Time Series (WG125). The primary focus of SCOR WG125 was to compare synchrony and trends between marine zooplankton time series, and to explore how zooplankton interannual variability is related to environmental conditions in the upper ocean and climate variability (Mackas et al. 2012a). SCOR WG125 had three main goals: (1) compile a collection of globally representative zooplankton time series of greater than 10 years in length; (2) develop shared protocols to summarize time series within regional spatio-temporal scales; and (3) examine modes of interannual variability by comparing global zooplankton time series across different regions and oceans.

The working group recognized that the analytical methods available to principal investigators had grown larger and more complicated over time, and were not always available to researchers that did not have programming skills or access to (often expensive) statistical software. One of the Terms of Reference for SCOR WG125 was to develop a set of common analytical protocols and data visualization tools for examining seasonal and interannual variability within and between zooplankton time series. These tools and methods had to be robust enough to handle times series that differed in method and sampling frequency (from weekly to once-per-season), and had to be accessible to individuals without access to expensive software, and with limited (if any) programming skills. During the four years of SCOR WG125, a collection of visualization tools and methods began growing through its collaboration with the National Oceanic and Atmospheric Administration (NOAA) Coastal and Oceanic Plankton Ecology, Production, and Observation Database (COPEPOD; O'Brien 2005, 2017). Shortly after the completion of SCOR WG125, these tools were made available to the public as an online time series toolkit, called COPEPOD's Interactive Time-series Explorer, or COPEPODITE (Box 9.1).

Even after SCOR WG125 completed its four-year course, the capabilities and components of this toolkit continued to expand over the next 15 years, as COPEPOD continued similar collaborative work with multiple other time series working groups (Table 9.1). This suite of analytical and visualization tools was expanded and used within these groups, and components of it are featured in many of their

---

Box 9.1: Copepodite: A Plankton and Ecosystem Time Series Toolkit.

COPEPODITE: A Plankton and Ecosystem Time Series Toolkit

COPEPOD's *Interactive Time-series Explorer* (COPEPODITE, http://copepodite. org) is an online tool collection that allows users to visualization time series data using a variety of standard graphics and analysis tools that have been collaboratively developed with the marine ecological time series research community over the last 15 years. With nothing more than a web browser, a user can upload and explore their own data, or extract data from satellite and model sources (e.g., sea surface temperature, chlorophyll, surface winds, mixed-layer-depth, net primary production, and more). Using only a simple comma-separated-values (CSV) file format and a web browser, this tool requires no programming skills or expensive software. The simple, statistical methods and visualizations of this toolkit are described in the online toolkit documentation and in multiple applied-use technical publications (O'Brien et al. 2012, 2013, 2017).

With the exception of the "Trends-n-Triangles Maps" shown in Figs. 9.10 and 9.11, all of the analyses and figures shown in this chapter can be generated with the COPEPODITE toolkit, accessible online at http://copepodite.org.

Table 9.1: Summary and foci of international plankton time series working groups that have collaborated with the Coastal and Oceanic Plankton Ecology, Production and Observation Database (COPEPOD), and inspired and co-developed elements of its time-series analyses and graphical visualizations.

Time series working group	Focus variable(s)	Focus region(s)
SCOR WG125	Zooplankton	Global (primarily open ocean)
SCOR WG137 which later became IOC-UNESCO TrendsPO	Phytoplankton and Chlorophyll	Global (primarily coastal and estuarine systems)
ICES WGZE	Zooplankton	ICES North Atlantic Area
ICES WGPME	Phytoplankton and Microbes	ICES North Atlantic Area
IOC-UNESCO IGMETS	All plankton groups, pigments, nutrients, temperature, salinity, oxygen, pH, alkalinity, carbon	Global

Group Acronyms and Websites: Global Comparisons of Zooplankton Time Series (WG125, http://oceantimeseries. net/wg125/); Global Patterns of Phytoplankton Dynamics in Coastal Ecosystems (WG137, http://wg137.net); Climate Change and Global Trends of Phytoplankton in the Ocean (TrendsPO, http://trendspo.net); Working Group on Zooplankton Ecology (WGZE, http://wgze.net); Working Group on Phytoplankton and Microbial Ecology (WGPME, http://wgpme. net); International Group for Marine Ecological Time Series (IGMETS, http://igmets.net).

reports and publications (Mackas et al. 2012a, O'Brien et al. 2012, 2013, Paerl et al. 2015, O'Brien et al. 2017). With free online access and an easy-to-use graphical interface, these tools continue to be used by both researchers and students, to visualize and explore data, and to produce graphics used in publications and scientific presentations.

9.2 Visualization of Patterns and Variability

This section demonstrates how to visualize and explore patterns and variability within and between variables at a single time series site, looking at within-year (e.g., the seasonal cycle) and between-year (e.g., interannual variability) elements. Each subsection provides a basic introduction to the concept and then an example of visualization and interpretation using real time series data along with graphics generated by the COPEPODITE toolkit.[4] The analytical examples in these sections are not meant to be comprehensive, only to introduce how to quickly visualize and explore time series data.

9.2.1 Seasonal Cycles

Introduction

Primary production in the ocean is influenced by seasonal fluctuations in environmental parameters (e.g., temperature, nutrient and light availability) that drive growth and reproduction. Water temperature varies with changes in solar radiation leading to higher temperature during the summer months with increased radiative input and lower temperatures during the winter when sunlight is limited. Nutrient levels are influenced by seasonal factors (e.g., stratification, wind driven mixing, upwelling) and are often depleted during summer phytoplankton growth cycle and replenished by winter mixing that deepens mixed layer depth. The strength of these seasonal cycles may be modulated by different drivers with latitude, for example by greater extremes in seasonal light availability at high latitudes or variability in wind speed and direction in low latitudes.

The mid and high latitude oceans tend toward strong seasonal cycles that drive large spring phytoplankton blooms and in some regions smaller autumn blooms (Findlay et al. 2006). Spring blooms result from a combination of bottom-up processes, including increased nutrient levels due to winter mixing, and the onset of stratification at a time when light levels are increasing to drive photosynthesis and zooplankton populations are limited. Blooms often progress until nutrients are depleted in the mixed layer and zooplankton populations increase to levels high enough to exert top-down control through grazing (Fasham 1993). Conversely, autumn phytoplankton blooms are often driven by the relaxation of vertical stratification and increased mixing, that replenishes nutrient concentration in the upper mixed layer when light levels are still sufficient to support photosynthesis and effectively dilutes the zooplankton population releasing grazing pressure (Findlay et al. 2006). For a more comprehensive review of phytoplankton seasonal cycles and bloom hypotheses, see Behrenfeld and Boss (2018).

Zooplankton that feed primarily on phytoplankton tend to have seasonal cycles that follow the timing of phytoplankton blooms and are subject to both prey limitation (bottom-up forcing) and predation (top-down forcing) by carnivorous zooplankton, fish, birds and marine mammals. As with their phytoplankton food sources, the strength of zooplankton seasonal cycles can vary with latitude and tend to exhibit larger seasonal ranges (maximums and minimums) in abundance or biomass in the mid to high latitudes (Ji et al. 2010, Mackas et al. 2012a). Latitudinal gradients in the timing

---

[4] All of the graphics and analyses presented here can be created using the free, online COPEPODITE time series toolkit (http://copepodite.org/toolkit).

of these peaks are also influenced by shorter growth seasons at higher latitudes, often with earlier seasonal peaks in spring and summer zooplankton species (Fanjul et al. 2017). Yet, even in the lower latitude oceans, measurable seasonal variation in zooplankton is also observed (Sheridan and Landry 2004, Ivory et al. 2019).

Simple Graphical Visualization

Zooplankton seasonal cycles become readily apparent when visualized over time and there are many ways to display the data graphically. Copepod abundance data from the Stonehaven time series located off the coast of Scotland in the northern North Sea are sampled as part of a long-term monitoring project operated by Marine Scotland Science since 1997 (Bresnan et al. 2015). The seasonal cycle of these data can be visualized in several ways.

A simple plot of raw copepod abundance values (small gray dots) overlaid with a solid line indicating the monthly-binned, median abundances for each month (e.g., the open squares on the solid line) is shown in Fig. 9.1a. This treatment captures the monthly-median and general spread of values across the seasonal cycle, but it does not provide any information on the relative density of these data values (e.g., where do the majority of the individual data values fall within each month).

A slightly more sophisticated way to visualize copepod abundance uses some basic statistical properties of the data. The same copepod abundance data can be represented as a Box and Whisker plot (Fig. 9.1b), where the center line within the box represents the median, the ends of the box represent the 25th and 75th percentiles, and the whiskers denote the 5th and 95th percentiles[5] (e.g., 50 percent of the copepod abundance values fall within the area of this box). This treatment captures the monthly-median, spread, and density of values across the seasonal cycle.

Finally, a Contoured Box and Whisker plot (Fig. 9.1c) presents an interpolated version of the Box and Whisker plot, where the darker shaded area represents the 25th to 75th percentile and the lighter shaded areas represent the 5th and 95th percentile, centered on the median (dashed line). The contoured box and whisker plot works best when data are present for multiple contiguous months (e.g., all months from March through September with no gaps).

Comparing Seasonal Cycles

Environmental variables (e.g., temperature, nutrients) in the ocean and predator-prey dynamics that influence the seasonal cycles of zooplankton are also seen in time series data. Visualizations of these time series from Stonehaven can be compared to copepod abundance to explore the relationships between them, such as how they impact one another and whether there is evidence for leading or lagging indicators.

Water temperature is an important physical variable that affects nutrient concentration, and phytoplankton and zooplankton growth. At Stonehaven (Bresnan et al. 2015), surface water temperature (Fig. 9.2a) varies from seasonal summer maximums of ~ 14°C to winter minimums of ~ 5°C (Bresnan et al. 2015). During the summer, the surface waters heat and when combined with calm weather conditions (e.g., low wind), the usually well mixed waters tend to stratify setting up a temporary thermocline. These changes in water column stability affect the seasonal total nitrate concentration (Fig. 9.2b)—an important nutrient that promotes phytoplankton growth. During the winter months, heightened storm activity increases water column mixing and re-suspension of sediments leading to higher nitrate concentration (~ 8 μM) in the surface waters (Fig. 9.2b) at a time when light limits phytoplankton growth and prey availability limits copepod growth and abundance.

---

[5] Some other commonly used criteria for setting the whiskers include using the minimum and maximum values, using the 2nd and 98th percentiles, or using +/– 1.5 times the interquartile range (IQR) as defined by the Tukey rule.

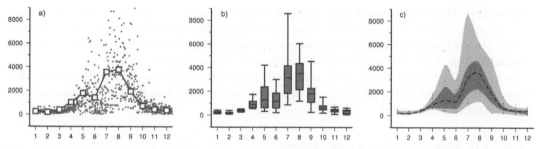

Figure 9.1: Visualization of total copepod abundance (individuals m⁻³) using three different "seasonal cycle" plotting methods. [Data from Stonehaven time series, western North Sea, Scottish coast.] From left to right: (a) Plot of monthly-binned monthly medians (solid line with squares) on a background of unbinned values; (b) box and whisker plot of monthly-binned data; (c) contoured box and whisker plot.

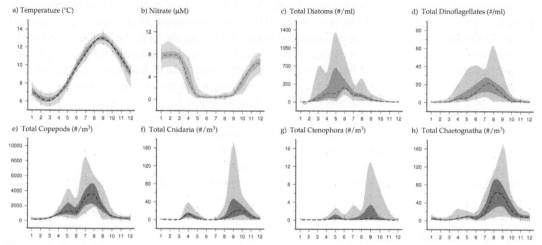

Figure 9.2: Seasonal cycle (contoured box and whisker plots) of sampled variables at the Stonehaven time series site (western North Sea, Scottish coast). From left to right: (a) temperature (°C) ; (b) nitrate (µM)—a dissolved inorganic nutrient; (c–d) diatoms and dinoflagellates (cells L⁻¹)—phytoplankton primary producers; (e) copepods (individuals m⁻³)—zooplankton grazers; and (f–h) cnidaria, ctenophora and chaetognatha (individuals m⁻³)—zooplankton carnivores.

As spring progresses, increased light levels and replenished nutrient concentrations promote the spring diatom bloom (Fig. 9.2c), which at the Stonehaven site is often dominated by *Chaetoceros, Thalassiosira,* and *Skeletonema* spp. (Bresnan 2012). In some years, diatom abundance increases even when water temperatures are still at winter minimums (~ 6 C). Summer sees an increase in smaller photosynthetic dinoflagellate abundance (Fig. 2d), which tends to lag behind that of the diatoms as nutrient levels decrease (Sarthou et al. 2005) to near zero in the surface waters and water temperatures continue to increase.

As spring evolves into summer, phytoplankton-dependent copepod abundances (Fig. 9.2e) increase in response to heightened food availability provided by the spring phytoplankton bloom and generally coincide with warming water temperatures (~ 8 C). At Stonehaven, the copepod community is generally dominated by *Pseudocalanus* spp. during the spring bloom and *Acartia clausi* in the summer and autumn, with *Calanus helgolandicus* and *C. finmarchicus* present during both seasons (Bresnan et al. 2015). The relative dominance of *Calanus* spp. are influenced by differences in their overwintering strategies and thermal tolerance (Bresnan et al. 2015). Phytoplankton growth is sustained further into summer by nitrate and ammonia generated through the microbial loop and zooplankton excretion (Cook 2013).

Copepod abundances begin to decrease in late summer and early autumn following increases in the abundance of carnivorous zooplankton (Fig. 9.2f, g, h), which coincide with peak surface water temperatures. This increase in copepod predators, such as cnidarians, ctenophores, and chaetognaths (Brenan et al. 2015), exerts top-down control on copepod populations at a time when declining diatom blooms limit food availability (bottom-up pressure), which also leads to declines in copepod abundance (Fig. 9.2e). The release of copepod grazing pressure combined with increased mixing leads to a smaller autumn bloom that in some years include large diatoms such as *Rhizosolenia* and *Pseudo-nitzschia* spp. (Bresnan 2012). Finally, winter storms and water column mixing set up a repeat of the seasonal cycle.

Changes in Seasonal Patterns Over Time

In the example plots of this subsection, each seasonal cycle plot was created using data from the entire duration of the time series. In a changing ocean, it may also be useful to visualize and compare seasonal cycles between two different time periods (e.g., before and during a period of record water temperatures). In a recent paper by Wasmund et al. (2019), the seasonal cycles of phytoplankton were compared between the start and end of a 29-year Baltic Sea time series. They found significant and visually obvious differences in the two seasonal cycles, including shifts in the start and end of the growing season (e.g., an earlier start and a later end), which expanded the length of the growing season by over 125 days since the start of the time series. They attributed these differences to increasing light levels in the early spring and an extension of warmer waters across the growing season.

The seasonal cycle plot is generally used for examining within-year patterns and interactions, or perhaps for comparing seasonal cycle changes between a two time periods (Wasmund et al. 2019). The next subsection presents ways to look at year-by-year and multi-year patterns. Data visualization and exploration should involve looking at the same data in multiple different ways. For example, the discovery of Wasmund et al. (2019) is not visually evident if only a single seasonal cycle plot or an Annual Anomaly Plot were used (Fig. 9.4 next section). However, the differing seasonal patterns are evident in an Interseasonal Matrix Plot (Fig. 9.5 next section).

9.2.2 Interannual Variability

Introduction

In addition to recurrent, month-to-month seasonal variation (previous section), zooplankton can also exhibit cyclical and non-cyclical, year-to-year (interannual) variations, with up and down (positive and negative) patterns that can last from a few years to decades. Some of this interannual variation is caused by the same environmental variables that drive seasonal growth and production (e.g., temperature and nutrient availability), through the carryover (or loss) of zooplankton and phytoplankton populations from the previous year. For example, an especially productive season may increase the number of individuals that survive into the next year, which can lead to increased carryover into the following years. Likewise, a less productive year could decrease population carryover.

Interannual changes in the zooplankton are often influenced by seasonal and interannual productivity of their food sources. Nutrient availability in turn heavily influences annual phytoplankton production through stratification, wind-mixing and upwelling, driven by water temperature and surface winds. Short-term weather events, such as an unusually windy or a warm-and-calm summer, can increase or decrease phytoplankton production (through mixing or stratification and nutrient availability) which may influence zooplankton productivity and year-to-year population sizes.

Both phytoplankton and the zooplankton that consume them are subject to bottom-up forcing. Yet, zooplankton also exert top-down pressure on phytoplankton populations. In any given year, extreme zooplankton grazing pressure can lead to phytoplankton bloom collapse (Thackeray et al. 2012) but this must be considered along with nutrient drawdown and other weather and climate related physical drivers that can diminish phytoplankton populations (Winder and Cloern 2010).

Short-term weather impacts may punctuate longer-term and large-scale influences (e.g., climate oscillations) that create persistent conditions that favor one set of dominant plankton over another. For example, warmer water conditions (e.g., often associated with calmer winds, increased stratification and decreased nutrients) may favor gelatinous zooplankton over other species (Richardson et al. 2009, Codon et al. 2013, Winder et al. 2017).

Large-scale Drivers of Interannual Variability

The cyclical nature of multi-year climate oscillations can also affect zooplankton interannual variation on different timescales (Beaugrand and Reid 2003, Perry et al. 2004, Pershing et al. 2004, Hays et al. 2005, Chiba et al. 2006, Beaugrand et al. 2014, Harris et al. 2014). Climate oscillations operate on timescales that range from several years to decades (Table 9.2), and their effect can be seen across entire oceans or hemispheres. The El Niño Southern Oscillation (ENSO) variation between the El Niño "warm phase" and La Niña "cool phase" can alternate over a 2–7 year period, while the temperature and wind effects of the Pacific Decadal Oscillation (PDO) and Atlantic Multidecadal Oscillation (AMO) can persist for decades (Kerr 2000, Mantua and Hare 2002, McPhaden et al. 2006, Newman et al. 2016, Zhang et al. 2019).

Climate oscillations are often quantified by the use of an index. For example, the AMO index and PDO index are calculated from the spatial average of sea surface temperature (SST) anomalies in the North Atlantic and North Pacific, respectively (Mantua and Hare 2002, Trenberth and Shea 2006). In contrast, the North Atlantic Oscillation (NAO) index is calculated as the difference in sea level pressure between a point located near Iceland and a point located off Portugal (Hurrell 1995). Multiple indices are used to describe ENSO including those based on sea level pressure (Southern Oscillation Index, SOI), SST (Nino 3.4), and even trade wind strength (Trenberth and Caron 2000, McPhaden et al. 2006, Wang and Fiedler 2006).

The strength and effects of these climate oscillations vary even within their respective regions of influence. For example, when the PDO Index is positive (warm phase), the waters along the west coast of North America tend to be warmer with lower productivity and the waters in the north central

Table 9.2: Examples of selected Northern Hemisphere climate oscillations indices, illustrating their average phase and period durations, index calculation base variable, and primary region of influence.

Index name	Phase duration (years)	Period duration (years)	Index base	Primary region of influence	Introductory references
El Niño Southern Oscillation (ENSO): El Niño (warm phase), La Niña (cold phase)	0.75–1 (sometimes up to two years)	2–7	Sea Surface Temperature, Atmospheric Pressure, or Trade Winds strength	Central and eastern tropical Pacific Ocean (and beyond)	McPhaden et al. 2006, Wang and Fiedler 2006
North Atlantic Oscillation (NAO)	1–10	(variable)	Sea Level Pressure	Northern North Atlantic	Hurrell 1995, Kerr 1997
Pacific Decadal Oscillation (PDO)	20–30 (sometimes less)	40–60	Sea Surface Temperature	North Pacific	Mantua et al. 1997, Mantua and Hare 2002, Newman et al. 2016
Atlantic Multi-decadal Oscillation (AMO)	30–40	60–80	Sea Surface Temperature	North Atlantic	Kerr 2000, Zhang et al. 2019

Phase Duration is the average length of a positive or negative anomaly phase (e.g., the length of the warm or cool phase). Period Duration is the average length of a full positive-and-negative cycle. The values provided here are the commonly reported ranges but can be variable and considerably shorter. The Index Base is the measured variable used to calculate the index (i.e., they are not directly measurable).

Pacific Ocean tend to be cooler (Newman et al. 2016). In the North Atlantic, the effects of the positive and negative NAO differ between northwest and southeast North Atlantic regions (Hurrell 1995, Kerr 1997). Through teleconnections, large climate oscillations like the PDO and AMO can also affect weather patterns and ocean conditions in regions far from their location (e.g., monsoon strength in the Indian Ocean, or water temperatures in the Gulf of Mexico; Krishnamurthy and Krishnamurthy 2014, Poore et al. 2009). Recent work suggests the AMO may force changes in North Pacific atmospheric circulation and surface winds (Wu et al. 2019) that drive the PDO. Modeling studies also forecast that some of these indices may weaken in strength and increase in frequency (e.g., have shorter phases or periods) due to climate change (Fang et al. 2014).

Climate oscillations can have significant impacts on marine ecosystems by altering both the abiotic and biotic environments (McGowan et al. 1998, Winder and Sommer 2012, Rosseaux and Gregg 2014, Racault et al. 2017). Zooplankton time series should show some level of correlation with regional climate oscillations (e.g., PDO/SOI in the North Pacific, AMO/NAO in the North Atlantic). SCOR WG125 found that many zooplankton time series had strong correlations with the PDO or AMO indices, while few had statistically significant correlations with the NAO or SOI indices (O'Brien, unpublished data). Correlation strength appears to depend on the index's calculation base-variable (e.g., was it derived from SST or sea level pressure; Table 9.2, Index Base). The WG125 time series often have clear correlations with SST-based indices (e.g., AMO or PDO), but rarely with pressure-derived indices (e.g., NAO or SOI). There is a strong relationship between zooplankton and temperature, so an SST-based index should also have a relationship with the zooplankton variable (e.g. abundance or biomass). In pressure-based indices, even the relationships between an index itself and SST are not always clear or consistent (Wang et al. 2004, Trenberth and Caron 2000), and may result in a weaker or inconsistent correlation between the index and the zooplankton variable.

The causal mechanisms behind zooplankton and climate oscillation relationships (or lack of relationship) are often complex, and complicate understanding the processes by which they actually affect zooplankton populations. For example, warm and cool phases of the PDO are actually influenced by several factors, including the strength of the Aleutian low (e.g., the North Pacific winter wind impact on upwelling), SST anomalies (e.g., ENSO), and strength and location of currents (e.g., the Kuroshio) or wind, all of which can influence the productivity in the California Current (Mantua and Hare 2002, Newman et al. 2016). This is one reason why ENSO has multiple index base types (Table 9.2, SST, atmospheric pressure, or wind), as the ENSO phenomenon itself cannot be directly measured.

The causal mechanisms behind zooplankton and climate oscillation relationships (or lack of relationship) are often complex, and complicate understanding the processes by which they actually affect the zooplankton populations. For example, the warm and cool phases of the PDO are actually influenced by several factors, including the strength of the Aleutian low (e.g., the North Pacific winter wind impact on upwelling), SST anomalies (e.g., ENSO), and the strength and location of currents (e.g., the Kuroshio) or wind, all of which can influence the productivity in the California Current (Mantua and Hare 2002, Newman et al. 2016). In Table 9.2, this is one reason why ENSO has multiple index base types (e.g., SST, atmospheric pressure, or wind), as the ENSO phenomenon itself cannot be directly measured.

When trying to resolve the complex interactions and correlations between zooplankton and cyclical climate patterns, capturing several cycles of the climate pattern is ideal (Mackas et al. 2012a, Valdes and Lomas 2017). Unfortunately, this suggests that to comprehensively study the impacts of a climate driver with a 30–60 year cycle would require more than 120 years of data. While data of this length may exist for water temperature, the longest zooplankton time series are currently less than 70 years in length (e.g., Continuous Plankton Recorder (CPR) and California Cooperative Oceanic Fisheries Investigations (CalCOFI)), with the majority being less than 30 years in length (Mackas et al. 2012a, O'Brien 2017). Generally, at least 30 years of zooplankton data are needed to begin distinguishing large scale climate effects from other drivers of variability, but in some cases 20 years were found sufficient (Batchelder et al. 2012, Mackas et al. 2012a). This is in line with satellite-derived chlorophyll, where Henson et al. (2010, 2016) found that 20–40 years of data were needed, with fewer years generally needed in equatorial regions.

9.2.3 Simple Graphical Visualization

The seasonal cycle plots (Figs. 9.1 and 9.2) present by month the average seasonal state of a variable measured over time, but do not convey information about the interannual state of that variable. For decades, several variables including nutrients (e.g., nitrate and phosphate), phytoplankton production (e.g., diatoms) and mesozooplankton production (e.g., copepods) have been measured at the Helgoland Roads monitoring site in the North Sea off the coast of Germany (Wiltshire et al. 2008, Boersma et al. 2015). The sampling site is visited multiple times a week providing a long-running, high-frequency, parameter-rich time series, and is an excellent time series for visualizing interannual variability within and between data sets. A plot of the raw copepod abundance data from 1975–2017 (Fig. 9.3a) exhibits much scatter and is difficult to interpret, especially with so many data points. Presenting the same data as a monthly average (Fig. 9.3b) or annual average (Fig. 9.3c) generally makes changes over time easier to interpret but does not necessarily present a clear picture as to how variable the value for a given month or year is relative to the past. One commonly used method for displaying interannual variability, patterns and trends is to calculate and plot the data as annual anomalies by year (Fig. 9.3d).

Annual Anomaly Plots

Annual Anomaly Plots are relatively easy to create and quickly convey information on the interannual states, trends and patterns in one or more variables (Fig. 9.3d, Fig. 9.4a–e). An annual anomaly plot is typically a bar graph where each bar represents the annual value of a variable relative to the long-

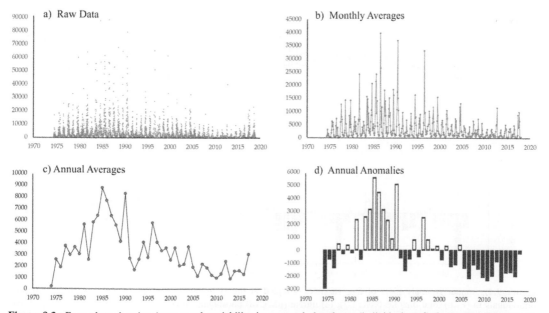

Figure 9.3: Four plots showing interannual variability in copepod abundance (individuals m^{-3}) from 1974–2017 at the Helgoland Roads time series: (a) raw, sub-weekly-sampled data; (b) monthly averaged data; (c) annually averaged data; and (d) annual averages plotted as difference anomalies from the climatological average (ClimAvg).

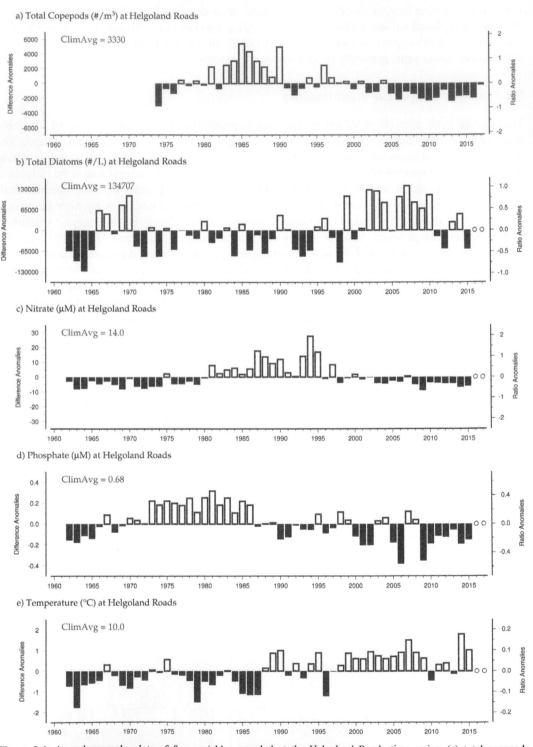

Figure 9.4: Annual anomaly plots of five variables sampled at the Helgoland Roads time series: (a) total copepods (individuals m^{-3}); (b) total diatoms (cells L^{-1}); (c) nitrate (μM); (d) phosphate (μM); and (e) temperature (°C). Open bars indicate positive anomalies, years with an annual average that was above the climatological average (ClimAvg). Solid bars indicate negative anomalies, or years with an annual average below the climatological average.

term average of all annual values (e.g., the climatological average or climatology). Bars that extend above the zero value on the Y-axis represent years that were above the long-term average (a positive anomaly) and bars below zero represent years that were below the long-term average (a negative anomaly). The upward or downward height of each bar indicates the magnitude of the difference between the annual value and the long-term average.

There are multiple methods for calculating anomalies, often with advantages and disadvantages (O'Brien et al. 2012, 2013, 2017, https://copepodite.org/methods). The simplest method is to calculate the annual anomaly as the difference between the annual value and the climatological average of all years. In this method, the original annual value can be recalculated by summing the anomaly with the climatological average. For example, copepods at Helgoland Roads (Fig. 9.4a) have a climatological average of 3,300 copepods m^{-3}, and a positive anomaly of approximately 5,000 copepods m^{-3} in the year 1990, which when summed returns the original annual average value (e.g., ~ 8,300 copepods m^{-3}). Annual anomaly calculations can also be based on standard deviation or percentiles, but those treatments are not considered here (https://copepodite.org/methods).

By presenting the annual values in relation to the climatology, annual anomaly plots (Fig. 9.4) quickly convey the interannual state (e.g., above or below average) and trend over time (e.g., increasing or decreasing). Visualizing multiple variables on a synchronized X-axis also allows for rapid visual intercomparison of trends and patterns between variables. For example, comparing the Helgoland Roads copepods (Fig. 9.4a) to the annual temperature anomaly (Fig. 9.4e) over the time period 1975–1989, there is a negative relationship such that when the temperature is cooler than average (negative temperature anomalies), copepod abundance is higher (positive copepod abundance anomalies). The temperature and copepod anomalies switch signs (e.g., positive and negative) from 1998–2015. Analyses of copepod anomalies at Helgoland Roads show that the positive anomalies are driven by several species such as *Acartia* spp., *Temora* spp., *Pseudo/Paracalanus* spp., and *Centropages* spp. (Boersma et al. 2015). Copepod anomalies (Fig. 9.4a) also show a negative relationship with total diatom anomalies (Fig. 9.4b) over the same two time periods which is an unexpected result (e.g., more diatoms should support higher copepod abundance). Unlike temperature and diatom biomass anomalies, total nitrogen and phosphorus anomalies show a positive relationship with the copepod anomaly (Fig. 9.4a, d).

Deeper analysis of the annual anomalies in the Helgoland Roads data is outside the scope of this chapter, but is presented in multiple publications (Greve et al. 2005, Boersma et al. 2007, Wiltshire et al. 2008, 2010, Kraberg et al. 2011, Boersma et al. 2015). Many mechanisms for change in the Helgoland Roads time series data have been discussed in the literature. Earlier studies suggest a novel ctenophore copepod predator (Boersma et al. 2007) or cell size increases making diatoms difficult to graze (Wiltshire et al. 2008, 2010, Kraberg et al. 2011) impacted copepod biomass. More recently, it has been suggested that the unexpected relationship between the copepod and diatom anomalies may result from decreasing quality of phytoplankton as a food source under nutrient limitation (Boersma et al. 2015). Non-calanoid copepod species (not shown) showed decreases in biomass with a time lag, perhaps because they were more omnivorous consumers (Boersma et al. 2015). Trends for dinoflagellates (not presented here) that may compete with (or eat) the diatoms or impact zooplankton grazers at Helgoland Roads are equivocal (Wiltshire et al. 2008, Kraberg et al. 2011, O'Brien et al. 2012, Boersma et al. 2015). Comparisons of these anomalies taken together suggest regime shifts occurred in this ecosystem, potentially related with changes in oceanic water influx (O'Brien et al. 2012, Boersma et al. 2015), reduced land based nutrient loading (van Beuesekom 2004, van Beusekom et al. 2006) and changes in diatoms species composition (Wiltshire et al. 2008).

Interseasonal Matrix Plots

The Interseasonal Matrix Plot is a more advanced visualization method that presents all months and years of a given time series variable. In this plot, monthly mean values are plotted in a table structure (a matrix) with each cell corresponding to the sampling year and month, and the cells' colors represent the values of these means. Interseasonal Matrix Plots (Fig. 9.5) can help visualize interannual patterns at the monthly-level, and often highlight related seasonal mechanisms (e.g., high winter nutrients, low summer biomass).

Some of the Helgoland Roads variables shown in Fig. 9.4 are displayed as Interseasonal Matrix Plots in Fig. 9.5. In the matrices, light to dark blues represent lower values, green to yellow intermediate values, and orange to red higher values. Additionally, contoured box and whisker plots of the seasonal cycles (Figs. 9.1d and 9.2) are aligned with the monthly axis of matrix plots to illustrate the relationship between the seasonal cycle plots and the interseasonal matrix plots.

Many of the interseasonal matrix plots seen in Fig. 9.5 demonstrate a visual clustering of same-colored matrix cells. This clustering illustrates changes in the intensity and duration of the copepod and diatom growing seasons (Figs. 9.5a and 9.5b), multi-year periods with strong seasonality in the winter or early spring concentrations of total nitrogen and phosphate (Figs. 9.5c and 9.5d), and the seasonal cycle in temperature over decades (Fig. 9.5e). Comparing the seasonal cycle plot and interseasonal matrix plot for copepods makes clear the highest seasonal peaks dominated in the 1975 to 1990 time frame and have generally diminished since then (Fig. 9.5a). The characteristics of this 1975 to 1990 time period are also visually present in the copepod annual anomaly plot (Fig. 9.4a). By itself, or in conjunction with the annual anomaly plot, the interseasonal matrix plot visualizes information that is not always discernable in the seasonal cycle or annual anomaly plots alone.

Discerning Climate Change from Climate Variability

Discerning cyclical climate variability from climate change in a time series requires careful consideration of data limitations. These include the length of the time series and the spatial scale over which data are sampled. With time series on the order of 30 years (Mackas et al. 2012a), the ability to visualize and compare long term data can identify areas of interest to address climate related questions and data sets that would benefit from more sophisticated statistical treatment. For example, comparing decadal-scale positive and negative annual anomalies present in Helgoland Roads data (Fig. 9.4a–d), some are in phase and some are out of phase, suggesting the influence of larger scale forcing. Comparing time series for water temperature and copepod abundance from the southeastern North Sea against the dominant climate index (AMO) can also be revealing. The Helgoland Roads time series demonstrates that copepod abundance has decreased for approximately 40 years (Fig. 9.4a) and *in situ* water temperature (Fig. 9.4e) shows a multi-decadal increase over that same time period.

While long-term trends of warming water and decreasing copepod abundance demonstrate concerning changes in the ecosystem, they are not necessarily a result of climate change. For example, comparing the Helgoland Roads temperature (Fig. 9.4e) and abundance (Fig. 9.4a) to the AMO Index (Fig. 9.6a), it may appear these 1962–2017 trends are only part of the natural 60–80 year AMO climate cycle. However, SST-based climate indices, like the AMO and PDO, are intentionally detrended to remove the climate warming signal (Mantua et al. 1997, Kerr 2000). Using the Hadley Centre Sea Ice and Sea Surface Temperature data set (HadISST, https://www.metoffice.gov.uk/hadobs/hadisst/),[6] sea surface temperatures around Helgoland Roads can be extended back to the year 1900. In this extended-years temperature anomaly plot (Fig. 9.6b), any relationship with the AMO is less clear, and it turns out water temperatures over the last 30 years (e.g., 1988–2017) have repeatedly been the warmest seen in the last 118 years (1900–2017).

The AMO index (and PDO) is also the spatial average of the entire North Atlantic (North Pacific for PDO) SST values, and may not fully reflect local SST trends. Depending on whether the

---

[6] HadISST is one of many standard satellite/model/global data sets automatically extracted by COPEPODITE based on user-provided geographic coordinates (e.g., in this case the latitude and longitude of Helgoland Roads time series site). This technique and method are featured in the ICES Plankton Status Reports (O'Brien et al. 2012, 2013) and in IGMETS (O'Brien et al. 2017).

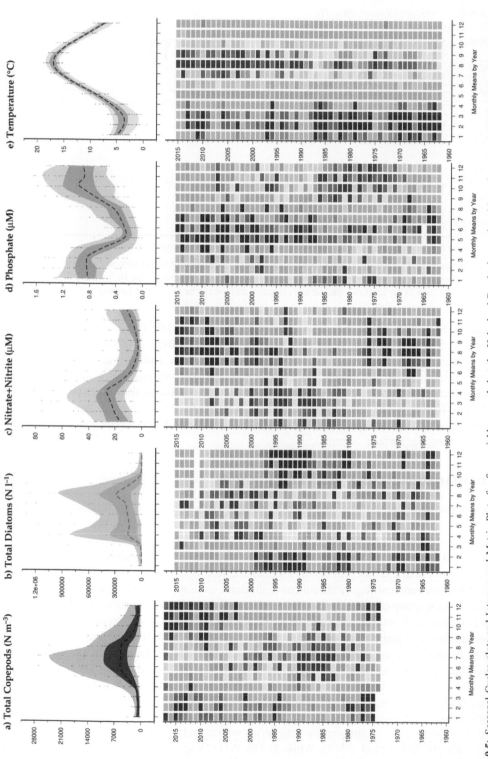

Figure 9.5: Seasonal Cycles plots and Interseasonal Matrix Plots for five variables sampled at the Helgoland Roads time series: (a) total copepods (individuals m⁻³); (b) total diatoms (cells L⁻¹); (c) nitrate (μM); (d) phosphate (μM); and (e) temperature (°C). Colors in the matrix rows and columns represent average monthly values by year: dark to light blues represent lower values, green and yellow intermediate values, and orange to red higher values.

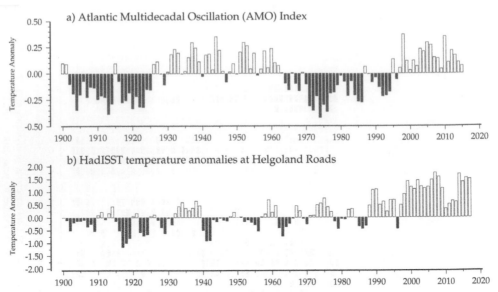

Figure 9.6: Annual anomaly plots of (a) the Atlantic Multi-decadal Oscillation (AMO) index, and (b) Hadley Centre Sea Ice and Sea Surface Temperature data set (HadISST) corresponding to the location of the Helgoland Roads time series. (Explanation of the HadISST data set and extraction method can be found in O'Brien et al. 2013.)

AMO and the SST index are in the same phase or opposite, the AMO can either enhance or mask the increase in SST in the North Atlantic region (Edwards et al. 2013). This tendency is especially applicable where time series are sampled at limited spatial scales (Edwards et al. 2010), such as the single station Helgoland Roads time series. Local or shorter-cycle processes can also mask the effect of large-scale climate forcing. For example, SST at Helgoland Roads shows periodic inversions (e.g., 1940–1970; Fig. 9.6b) that are not seen in the AMO (Fig. 9.6a). These differences could be counter effects of the shorter-cycle NAO or other local conditions acting on the SST (e.g., changes in wind, upwelling, or surface currents).

9.3 Comparisons Across Sites and Regions

As demonstrated with the seasonal and interannual data visualizations, it is clear that zooplankton respond to changes in the physical, chemical, and biological components of their environment at both local and regional scales. While global efforts such as SCOR WG125 focused on zooplankton time series, and SCOR WG137 focused on phytoplankton time series, neither working group delved deeply into the physical and biochemical time series parameters that connect these plankton functional groups to their environment (e.g., nutrient concentrations, salinity, oxygen, wind and stratification). In 2014, a working group of the Intergovernmental Oceanographic Commission (IOC) of the United Nations Educational, Scientific and Cultural Organization (UNESCO) was formed to look at ship-based biogeochemical and marine ecological time series. The IOC-UNESCO International Group for Marine Ecological Time Series (IGMETS) was an expansion and compilation of previous time series workings groups, such as SCOR WG125 and SCOR WG137, that now included time series data of ocean and atmospheric physical data (e.g., temperature, salinity, oxygen, wind), biogeochemical data (e.g., nutrients, carbon, alkalinity), and lower trophic-level biology (e.g., zoo-, phyto- and other microbial-plankton). Similar to SCOR WG125, the main goals of IGMETS included compiling data and developing tools to summarize time series within and across regions at different spatio-temporal scales (O'Brien et al. 2017). While SCOR WG125 spawned many of the individual site seasonal and interannual visual graphics shown in the previous sections, IGMETS inspired new visual intercomparison tools for exploring its global collection of over 350 time series.

The analyses developed for IGMETS (O'Brien 2017) addressed the following challenges: (1) How to compare time series with different collection methods or measurement units; (2) How to compare time series of different years in length; and (3) How to spatially visualize trends and patterns across multiple time series, variables, and/or regions.

9.3.1 Comparing Time Series with Different Methods and Varying Lengths

The time series collection assembled by IGMETS varied greatly in their methodologies and years length. This required developing approaches for comparing considerably different time series either in a quantitative or qualitative way. For example, one challenge was to compare the 20-year Stonehaven and 45-year Helgoland Roads time series with other existing long running (60-year) time series such as the North Atlantic Continuous Plankton Recorder survey (Reid et al. 2003). While all three contain "total copepods" data, the three programs differ in the number of years sampled and sampling methodology used. The IGMETS approach was simple: compare relative values (e.g., ratio-based anomalies) from the time windows of overlapping years (e.g., in this case the 20 years shared in common).

Calculation of Ratio-based Anomalies and Trends

Since the initial work of SCOR WG125, ratio-based anomalies have been used to compare time series variables with different methods and/or measurement units (Mackas et al. 2012a, O'Brien et al. 2012, 2013, 2017). The original SCOR WG125 methodology used log-transformed zooplankton and nutrient concentration data, both to reduce the impact of outliers and to minimize skew in the data to satisfy parametric statistics assumptions and requirements (Mackas and Beaugrand 2010, Mackas et al. 2012a). With the start of IGMETS, non-parametric statistics were used, which did not require log-transformation to address skew (see O'Brien 2017 for a full methodology). Three additional features of the IGMETS non-parametric (non-log-transformed) approach are: (1) zero value counts data do not need special handling (e.g., log "x+1") because the data are not being log transformed; (2) the resulting anomaly and trend values do not need to be transformed back to non-log values for interpretation; and (3) the annual anomalies can be shown using a dual-Y-axis plot that present the anomalies in both original units and relative units (Fig. 9.7).

When working with non-log-transformed data (e.g., the IGMETS method), the simple difference anomalies (Fig. 9.2 and Fig. 9.7) retain their original units (e.g., copepods m^{-3}, μM nitrate). When these difference anomalies are divided by their climatology, the result is a unitless ratio of the anomaly value relative to the long-term average (e.g., climatology). In this ratio-based anomaly, a value of

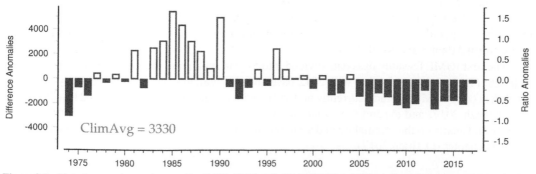

Figure 9.7: Plot of copepod annual anomalies (1974–2017) at the Helgoland Roads time series, plotted on a dual Y-axis. The left Y-axis provides difference anomaly values, with original units preserved (e.g., individuals m^{-3}). The right Y-axis provides ratio-based anomaly values relative to the climatological average (ClimAvg). In this treatment, a ratio anomaly value of 1.0 corresponds to a difference anomaly of 3,330 copepods m^{-3} (i.e., 1.0 * ClimAvg).

1.5 indicates the difference anomaly was 1.5 times greater than the climatological average, while a value of –0.5 indicates the difference anomaly was value 0.5 times less than the climatological average. For example, in Fig. 9.7, the anomaly bar for year 1990 has a difference anomaly value of approximately 5,000 (left Y-axis), a ratio-based anomaly of around 1.5 (right Y-axis), and the climatological average (ClimAvg) is 3,330 copepods m^{-3}. In 1990, copepod abundance was 1.5 times larger than the long-term average (3,330 copepods $m^{-3)}$. The 1990 difference anomaly (~ 5,000 copepods m^{-3}) can be read from the left Y-axis, or calculated by multiplying the ratio-based anomaly by the climatological average (1.5 * 3,330 = 4,995 or ~ 5,000).

With non-log-transformed data, the linear trend of ratio-based anomalies is also ratio-based, and its slope represents the rate of change relative to the climatological average. For example, a ratio-based annual anomaly trend with a slope of –0.10 would indicate a decreasing trend of 10 percent per year, and a positive slope of 0.25 would indicate an increasing trend of 25 percent per year. By working with ratio-based anomalies and ratio-based trends, it is possible to compare time series with different methods, different value ranges and units, or even completely different variables, because the comparison is based on relative amounts and relative change (e.g., "copepods decreased by 10 percent while diatoms increased by 25 percent").

When working with log-transformed data (e.g., the SCOR WG125 method), the calculations and interpretation of the visualizations are quite different. First, ratio-based anomalies are already a default product of the difference anomaly calculation due to the mathematical properties of logarithms (e.g., $\log(x) – \log(y) = \log(x/y)$). Second, while the trend slopes of log-transformed annual anomalies still represent relative change per year, they are based on log-transformed values, which can complicate interpretation of the annual anomalies and slopes. Finally, because of the non-linearity of log-transformed anomalies, it is not possible to create a dual-Y-axis plot (Fig. 9.7) that shows both original unit and relative (ratio-based) values. The log-transformed anomalies do have a higher tolerance for data with large outliers, and still offer an alternate way to visualize the data. The COPEPODITE toolkit offers multiple analysis and visualization options using both the SCOR WG125 and IGMETS approaches. Online documentation for the toolkit (http://copepodite. org/methods) also provides additional information on methods for calculating anomalies and trends, and discusses methods for handling large-value outliers, zeros and trace-values.

Time Windows

When comparing time series, it is not meaningful to directly compare a 30-year time series with a 10-year time series, but it is possible to compare a common 10-year overlapping Time Window (TW). Using this approach, the first IGMETS study (O'Brien et al. 2017) analyzes and compares time series ranging from 5 to 50+ years in length. The shorter time window analyses includes more time series and has greater spatial coverage (e.g., participating sites were present in most regions). In contrast, the longer time window analyses provides better temporal resolution of interannual variability, but more limited spatial coverage in fewer regions (e.g., the majority of the longest time series are found in the North Atlantic and North Pacific).

The first IGMETS study uses data divided into 5-year expanding increments from 5 to 30 years in length (Table 9.3) terminating with 2012 based on the available data[7] when the study started in 2014. A second IGMETS study currently underway, with a planned completion in 2021, will include data through 2017, and expand time windows as far back as 50 years in length (e.g., "TW50" in Table 9.3). Details on the calculation and assignment of time windows are presented in the IGMETS methods chapter (O'Brien 2017).

---

[7] When the IGMETS study started in 2014, many of the biological time series were still processing their 2012 and 2013 year samples. While some variables like temperature data can be measured real-time, microscope-based analysis of plankton often take much longer to process and enumerate.

Table 9.3: Time Window (TW) year minimums and spans used by the first International Group for Marine Ecological Time Series (IGMETS) study (O'Brien et al. 2017) and the in-progress, second IGMETS study. The first IGMETS study only examined time windows through TW30.

Time Window (TW)	Minimum years to qualify	IGMETS (1st study) year span	IGMETS (2nd study) year span
TW05	4	2008–2012	2013–2017
TW10	8	2003–2012	2008–2017
TW15	12	1998–2012	2003–2017
TW20	16	1993–2012	1998–2017
TW25	20	1988–2012	1993–2017
TW30	24	1983–2012	1988–2017
TW35	28	n/a	1983–2017
TW40	32	n/a	1978–2017
TW45	36	n/a	1973–2017
TW50	40	n/a	1968–2017

In the IGMETS methodology (O'Brien 2017), a time series qualifies for a given time window if it has data in at least 80 percent of the years in that time window (Minimum Years to Qualify column in Table 9.3). For example, a time series must have data present in at least 12 of the 15 years to qualify for a TW15 time window. In the longer time windows (e.g., TW25-TW50), an additional qualification criterion is added that requires that any gaps of missing years must be less than 5 consecutive years in length. For example, TW30 requires at least 24 years to be present, which means up to six years can be missing. The presence of a two- and four-year gap would be acceptable, but a consecutive five- or six-year gap would disqualify this time series from the TW30 time window.

Time Window-based Estimates of Relative Change

In the time window-based analysis, the climatology,[8] anomalies and trends are independently calculated for each and every qualifying time window in a time series variable. For example, Table 9.4 shows results from eight time-window-based, linear trends calculated on the difference and ratio-based anomalies of copepod abundances at the 45-year Helgoland Roads time series. This table provides calculated trend slopes for the difference anomalies (DIFFanom) and ratio-based anomalies (RATIOanom), as well as their level of statistical significance (e.g., Signif), and then provides a simplified Trend Symbol that represents both trend direction and statistical significance.

Even when statistical trend is non-significant, the direction can still provide useful information on the temporal or spatial patterns in a time series. O'Brien (2017) discusses statistical significance of linear trends in relation to the IGMETS analysis, which are influenced by several factors including time series length, strength of the trend (i.e., magnitude of the trend slope), and level of variance (noise) in the time series variable. In general, shorter time windows often need a stronger trend to be statistically significant, while longer time windows can be statistically significant even with a very small magnitude trend. Further, relatively low variance variables like temperature or salinity tend to need fewer years (to have significant trends) than "noisier" plankton and other biologically influenced (e.g., nutrients and pigments) data. For example, time series of nitrate and chlorophyll needed almost twice as many years of data as needed for SST to distinguish climate change trends from natural variability (Henson et al. 2016).

---

[8] Each time window has its own long-term average (climatology), calculated only from the years within that time window. Ratio-based anomalies are calculated using this time window-specific climatology.

Table 9.4: Time Window (TW) based linear trend statistics for copepod abundances in the Helgoland Roads time series, showing trends and relative change for eight time windows.

Time Window	Year span	Trend slope			Trend symbol	Relative change	
		DIFFanom	**RATIOanom**	**Signif**		**per year**	**[over TW]**
TW10	2008–2017	−5.917	−0.005	non-sig	▽	−0.5%	[−5%]
TW15	2003–2017	−39.183	−0.028	non-sig	▽	−2.8%	[−42%]
TW20	1998–2017	−71.307	−0.044	p < 0.05	▼	−4.4%	[−88%]
TW25	1993–2017	−87.918	−0.046	p < 0.01	▼	−4.6%	[−115%]
TW30	1988–2017	−92.132	−0.041	p < 0.01	▼	−4.1%	[−123%]
TW35	1983–2017	−137.207	−0.048	p < 0.01	▼	−4.8%	[−168%]
TW40	1978–2017	−99.291	−0.033	p < 0.01	▼	−3.3%	[−132%]
TW45	1973–2017	−81.323	−0.027	p < 0.01	▼	−2.7%	[−122%]

This table features statistics from eight time-window-based, linear trend slopes calculated on the difference and ratio-based anomalies of copepod abundances at the 45-year Helgoland Roads time series. The difference anomalies (DIFFanom) trend slopes have units of number of copepods $m^{-3}\,yr^{-1}$, while the Ratio-based anomaly (RATIOanom) trends slopes are unitless values of relative change per year (relative to the long-term average copepod abundance from each time window). The statistical significance of each trend is shown in the "Signif" column and represented in the Trend Symbol column. Values in the Relative Change column summarize the relative rate of change per year and average total changes across the entire time window.

As the statistical trends are calculated on annual anomalies, the trend slopes represent a "per year" rate of change, averaged across the time window. Difference anomaly trend slopes retain their original units (e.g., copepods $m^{-3}\,yr^{-1}$), while the ratio-based anomaly trend slopes provide a relative change per year (e.g., −0.005 per year or −0.5% per year). Multiplying the trend slope by the number of years in the time window gives a rough estimate of the total change that occurred across the entire time window. For example, in a 20-year time window (TW20), if copepod abundance decreased at a rate of −4.4% per year, they decreased by approximately 88 percent over that entire 20 year period.

These simple estimates of relative change allow for comparisons within and between different time series (e.g., copepods decreased by 10 percent while diatoms increased by 25 percent). In the next section, the gray shaded trend information columns in Table 9.4 have been combined with dual-Y-axis anomaly plots (Fig. 9.7), creating a simple visual and statistical tool for exploring and comparing trends and patterns between multiple time series.

9.3.2 *Visualizing Trends and Patterns Across Time and Space*

A Four Site Example (North Sea Copepods)

Earlier sections in this chapter introduced the Stonehaven (*Seasonal Cycle* section) and Helgoland Roads (*Interannual Variability* section) time series, located in the western and southeastern North Sea (Fig. 9.8b). Another zooplankton-sampling time series program from this region is the CPR Survey (Reid et al. 2003, http://cprsurvey.org) which samples over much of the North Atlantic and the European shelf (Fig. 9.8a). In contrast to station- or transect- based time series, the spatially massive CPR data are often averaged across standard areas, producing spatially-averaged month-by-year time series for each area. For example, the CPR Survey uses this method to divide the North Sea into six standard areas (Fig. 9.8b), with the Helgoland Roads site falling within CPR area D1 and the Stonehaven site falling within CPR area C2. These two CPR standard areas will be used in a simple comparison between four North Sea zooplankton time series.

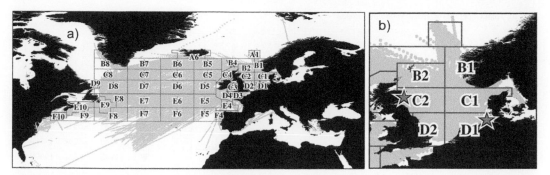

Figure 9.8: Map of Continuous Plankton Recorder (CPR) Survey standard areas in the (a) North Atlantic with a zoomed inset of the (b) North Sea. Light gray dots indicate transects sampled by the CPR survey. Labels and box-outlines indicate the geographic coverage and names of each CPR standard area. The star symbols in the North Sea inset indicate the locations of the Stonehaven (western North Sea, area C2) and Helgoland Roads (southeast North Sea, area D1) time series sites.

Directly comparing these four time series poses a challenge, as they differ in the number of years sampled (e.g., ranging from 20 years to over 50 years) and in the types of sampling gear employed (e.g., from standard bongo net to the small-aperture CPR sampler).[9] Incorporation of time windows based analysis (to address the different year lengths) and ratio-based anomaly trends (to address the different sampling methods) creates the ability to compare and contrast relative change in copepod abundances within and between these four time series. This is done by combining elements from the dual-Y-axis anomaly plot (Fig. 9.7) and the simple trend statistics from Table 9.4.

Figure 9.9 is a standard COPEPODITE-created comparison plot that features annual anomaly plots and time-window-based trend statistics for copepod abundances, generated from four North Sea example time series. The four time series in Fig. 9.9 are ordered: (a) Helgoland Roads, (b) its encompassing CPR standard area D1, (c) Stonehaven, and (d) its encompassing CPR standard area C2, such that Fig. 9.9a and 9.9b represent copepod abundances from the southeast North Sea area and Fig. 9.9c and 9.9d represent copepod abundances from the western North Sea (Fig. 9.8b). The left side of each figure panel is a dual-Y-axis anomaly plot (Fig. 9.7) with matched right-Y-axis ratio-based anomaly value ranges (e.g., +/−1.75). The right side of each figure panel summarizes statistics of relative change (per year and over the entire time window), based on ratio-based anomaly trends in each qualifying time window (Table 9.4).

The shape of the annual anomalies in Fig. 9.9 (left subpanels) demonstrates long-term decreases in copepod abundance at three of the four sites. These trends are supported by corresponding statistical summaries (right subpanel) having significant negative trends indicated in many of the longer time windows (e.g., TW20-TW45). The remaining site, Stonehaven (Fig. 9.9c), with only ~ 20 years of sampling, also appears to have slight decrease in abundance but the statistical summary shows non-significant trends in its three qualifying time windows (TW10-TW20). Trends for the shortest time windows (e.g., TW10-TW15) are non-significant in all four sites, which is likely due, in part, to these being higher variance biological data with a limited number of sampling years (see statistical significance discussion in previous section and O'Brien 2017). Trends at CPR area C2 (Fig. 9.9d), for time windows T10-TW30, are all non-significant, and the trend directions switch from positive (TW10-TW25) to negative (TW30-TW45). While the longest time windows in the CPR and Helgoland Roads sites show significant decreases in copepod abundance, the more recent years (especially TW10-TW15) exhibit considerable interannual variability at both the Stonehaven and CPR area C2 sites, as seen in the annual anomalies within that time period. Some of the variability at Stonehaven

---

[9] Web links for additional information on these four time series (and their methodologies) can be found in the Acknowledgements section.

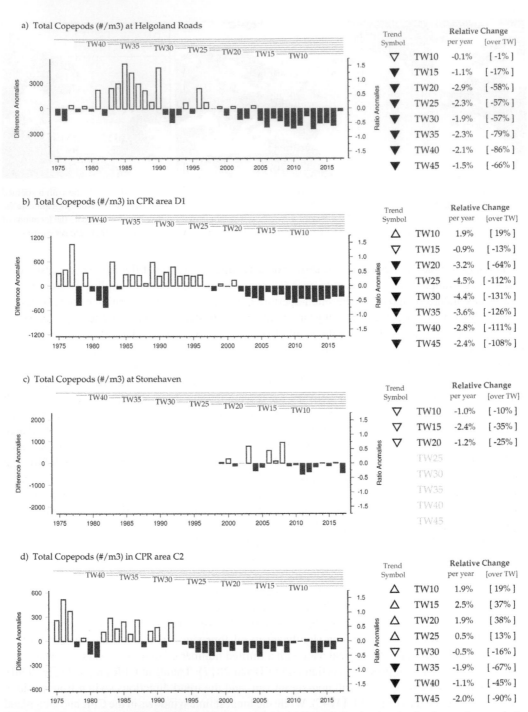

Figure 9.9: Example of a combined Annual Anomalies and Statistical Trends plot showing the abundance of copepods from the (a) Helgoland Roads time series, (b) Helgoland-Roads-adjacent CPR standard area "D1", (c) Stonehaven time series, and (d) Stonehaven-adjacent CPR standard area "C2". Horizontal lines above each anomaly plot mark the year-span of anomalies included in each time window (e.g., TW10 = 2008–2017, TW20 = 1998–2017). Trend Symbols in the adjacent column indicate the direction and statistical significance of linear trends calculated across each time window. Filled triangles indicate a significant trend (p < 0.05), and empty rows indicate time windows in which the time series was too short to qualify. Values in the Relative Change tables present slopes of ratio-based anomalies linear trends, provided as percentage change per year and over the entire time window period. For example, a rate of 1.9% change per year in a 10-yr time window would have an estimated 19% total change during those 10 years.

is likely influenced by currents moving north-south along the coastline and mixing of different water masses and their associated species (Bresnan et al. 2015).

The average yearly rates of change (i.e., trend slopes) are listed in the Relative Change per Year column of the right-side statistical summary subpanels of Fig. 9.9. Comparing all sites and time windows, rates of change range from –4.5% to +2.5% per year, with none of the positive rates having statistically significant trends. Trends for the longest time windows (e.g., TW35-TW45), when present, are all significant and negative, with rates ranging from –1.1% to –4.8% per year.

The Helgoland Roads TW35 time window rate of –2.3% (per year) estimates that copepod populations were on average decreasing by 2.3% for each of the 35 years. Multiplying this rate by the number of years in the time window, the copepod population decreases by 79% over the 1983–2017 time period. As this is based on a linear trend, it is only an estimate of total change, but it still allows for comparison of relative change between the different sites and variables. For example, the Helgoland Roads encompassing CPR area D1 copepods have a TW35 decrease of 126%, while the Stonehaven encompassing CPR area C2 copepods have a TW35 decrease of 67% (Fig. 9.9b, d statistical summary table). Based on these estimated amounts, CPR area D1 decreases almost two times faster than CPR area C2 over the same 35-year time period.

A simple visualization like Fig. 9.9 can provide basic statistics and summary information as well as help focus the direction of further investigation. For example, the reasons for the large decreases in the longer time windows, or the greater relative rates of decrease in the southeastern North Sea (CPR area D1 and Helgoland Roads) in comparison to the western North Sea (CPR area C2) might be considered. This example analysis could be expanded by adding additional CPR standard areas in the North Sea (Fig. 9.8b, areas B1, C1, B2, D2), or by additional time series variables (e.g., diatom and dinoflagellate abundance, temperature, and/or chlorophyll concentrations).

In this simple example, data visualization is provided for only four sites, and additional graphics are required to view more time series sites or additional variables. Working with many sites and multiple variables (e.g., to compare trends across an entire region or ocean) using these graphical techniques is challenging. A more complex visualization is needed for larger, multi-site comparisons, but the basic analytical concepts remain the same, with ratio-based anomalies and time-window-based trends being calculated for each variable at every site. This collection of statistical results is stored in compilation files, indexed by time series name, geographic location, and analyzed time window. These data compilations are fed into more advanced statistical analyses and correlation tools or plotted in spatio-temporal visualizations. For example, using geographic locations and simple Trend Symbols (the triangles shown in Fig. 9.9 and Table 9.4), a map of trends can be created for an entire region or ocean.

The Trends-n-Triangles Map (A Sixty-one Site Example)

The Trends-n-Triangles Map displays statistical trends for a single variable and time window (e.g., trends in total copepod abundance over the last 15 years, TW15), by plotting the Trend Symbol (i.e., triangle symbol used in Table 9.4 and Fig. 9.9) at the time series' geographical location on a map (Fig. 9.10). In this example, the shading-darkness of the triangles represent the statistical significance of the copepod abundance trend at each site (e.g., darker = p < 0.05 or p < 0.01, light gray = non-significant). Similar coloration approaches can be used to represent the relative rate of change (e.g., ratio-based trend slopes in Table 9.4) or the correlation direction and strength between two variables (e.g., are copepods at this location positively or negatively correlated with SST or chlorophyll). Optionally, a gray circle can be drawn at the locations of time series that were too short for the current time window, or that did not have the variable currently plotted. The gray circles close to the North American and European coastlines most often indicate locations of estuarine monitoring programs that do not sample copepod data (Fig. 9.10).

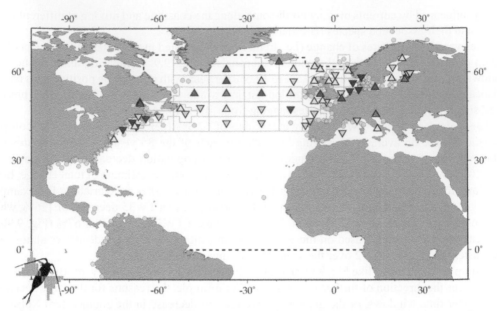

Figure 9.10: Trends-n-Triangles Map of IGMETS 15-year (1998–2012) statistical trends from 61 North Atlantic copepod abundance time series (triangles symbols). Trend direction is represented by upward facing (positive trend) and downward facing (negative trend) triangles. Symbol coloration represent statistical significance of the trends (darkest = p < 0.01, medium = p < 0.05, light gray = non-significant).

Background Spatio-Temporal Trend Fields

The top panel of Fig. 9.11 is a color version of Fig. 9.10, with the addition of blue and green background shading in the ocean regions. This blue and green shading indicates trends in satellite-based chlorophyll concentration. The colors in this background field represent both trend direction and strength and have the same time window as the copepod trend triangles (e.g., TW15). Green colors indicate an increase in chlorophyll over the last 15 years, and blue colors indicate a decrease, with darker colors indicating a stronger increase (or decrease) in chlorophyll concentration at the location.

The chlorophyll spatio-temporal trend fields in Fig. 9.11 were developed for IGMETS (see O'Brien 2017 for a detailed methodology), along with spatio-temporal fields of SST trends and Chlorophyll-vs-SST correlations. Created from global-coverage satellite and model products, these fields were used to provide spatio-temporal background information in regions where *in situ* time series were sparse or not available (e.g., the southern central North Atlantic in Figs. 9.10 and 9.11). Since the first IGMETS study (O'Brien et al. 2017), additional spatio-temporal fields were created, including trends fields of surface winds, mixed layer depth, and net primary production. As discussed earlier in this chapter, mixing or stratification can influence the availability of nutrients for primary production, and thus affect the upper trophic levels (e.g., total copepod abundance).

A Visual Summary of 15-year Copepod Trends in the North Atlantic

Exploring Fig. 9.11, in the North Atlantic open ocean regions there is a positive correlation between trend directions of copepod abundances and satellite chlorophyll concentrations. In the northwest North Atlantic, increasing copepod abundances (e.g., green upward pointing triangles south of Greenland and Iceland) are often found in regions of increasing satellite chlorophyll concentrations (green background fields). In the open ocean region west of Spain, decreasing copepod abundances are often found in regions of decreasing chlorophyll concentrations. While this relationship holds

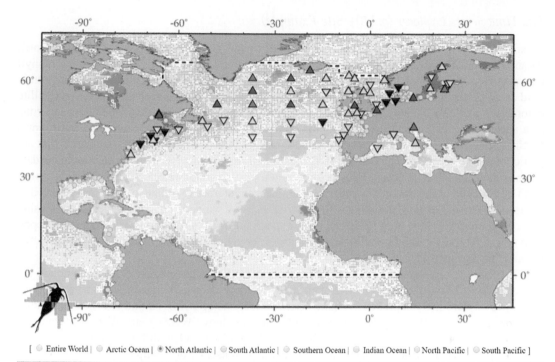

[○ Entire World | ○ Arctic Ocean | ◉ North Atlantic | ○ South Atlantic | ○ Southern Ocean | ○ Indian Ocean | ○ North Pacific | ○ South Pacific]

Time-series Symbol Layer: *Choose a variable below* ○ No Symbols (*turn off triangles symbol layer*)									Time Windows 15 year (1998-2012) & Oceanic Realms		Background Fields		
Hydrography				**Chemistry / Carbon**				**Biology**				**TWin**	**Realm**	**Select Background**	**BackFade**
	Trend	*CorrT*	*CorrC*		*Trend*	*CorrT*	*CorrC*		*Trend*	*CorrT*	*CorrC*	○ 05-yr	Ocean ◉	*Field* □	NONE ○
Temp	○	●	○	N-NO3	○	●	○	Copepods	•	●	○	○ 10-yr	Estuarine ○	SST *trends* ○	FADED ○
PSal	○	●	○	P-PO4	○	●	○	ZooMass	○	●	○	◉ 15-yr	All-TS ○	CHL *trends* ◉	LITE ○
D-Oxy	○	●	○	Si-SiO4	○	●	○	ZooCombo	○	●	○	○ 20-yr			MED ○
SST(R)	●			CHL(S)	○	●	○	Diatoms	○	●	○	○ 25-yr		CHL-vs-SST *correlations* ○	DARK ◉
PSAL(H)	○	●		Chl-a	○	●	○	Dinoflags	○	●	○	○ 30-yr			DARK! ○
WIND(I)	○			Chl+PCI	○	●	○	DiaDinRatio	○	●					

Figure 9.11: Trends-n-Triangles Map of 15-year (1998–2012) statistical trends from 61 North Atlantic copepod abundance time series (triangle symbols) on a background map of 15-year trends in satellite-derived chlorophyll concentration. Green coloration indicates increasing copepod abundance (triangles) or chlorophyll concentration (background color) over time. Blue coloration indicates decreasing copepod abundance (inverted triangles) or chlorophyll concentration. The lower panel of this figure shows the selection menu from the IGMETS Time Series Explorer (http://igmets.net/explorer), allowing the user to select from 18 different time series variables, six Time Windows, three background fields, and eight ocean regions. Via this selection menu, the triangles plotted on the map can show statistical trends over time ("Trend") or correlation direction and strength between the selected variable (e.g., copepod abundance) and either sea surface temperature ("CorrT") or satellite chlorophyll ("CorrC"). Explanations of the various plotting options and variable acronyms (e.g., "Temp" = *in situ* temperature, "SST (R)" = remotely-sensed satellite sea surface temperature) are provided interactively in the Explorer interface by holding the mouse cursor over any item or option in question. See O'Brien (2017) for full methods and additional information.

up well in many of the open ocean regions, copepod-vs-chlorophyll trends along the coastlines do not always match, often having opposite-direction trends. The satellite-derived chlorophyll estimates used are not always precise in shallow, optically complex coastal areas, or areas with exceptionally high concentrations of colored dissolved organic matter, such as the Baltic Sea (Kratzer and Moore 2018, Wozniak et al. 2014). Further discussions on trends in copepods and other variables in the North Atlantic are presented in the North Atlantic chapter of the first IGMETS status report (Bode et al. 2017) and in the ICES plankton status reports (O'Brien 2012, 2013).

The Time Series Explorer (a 340+ Site Example)

Presented here is a single figure (Fig. 9.11, upper panel), a map of 15-year trends in North Atlantic copepod abundance, visualized on a background of 15-year trends in satellite chlorophyll concentration. To examine how copepod trends correlate with water temperatures, this background field could be changed to show 15-year trends in SST, or even 15-year correlations between satellite chlorophyll and SST. Finally, similar maps could be created for diatoms, dinoflagellates, nutrients, or a variety of other variables (Fig. 9.11, lower variable selection menu). Figure 9.11 only shows 61 time series, from one ocean, out of the multi-ocean collection of over 340 time series that are participating in IGMETS. The full IGMETS catalog of trend maps is expansive and includes data visualizations for all seven oceans (plus a global view), six time windows, and 18 time series variables. With a collection of thousands[10] of image possibilities, an interactive online interface was created to let users select and explore trend maps for any region, time window, and variable of interest using a clickable map selection menu (Fig. 9.11 lower panel). For example, the North Atlantic copepod map (Fig. 9.11) was created by selecting the buttons for "North Atlantic" and "Copepods" and "15 yr" and "background CHL trends".

The Time Series Explorer interface and visualization tools were created as a companion to the printed IGMETS Status Report (O'Brien et al. 2017), to allow study participants (and later the general public) to explore the comprehensive time series collections of IGMETS and two ICES plankton working groups, the Working Group on Zooplankton Ecology (WGZE) and Working Group on Phytoplankton and Microbial Ecology (WGPME). The ability to select different variables, regions and time periods makes the tool useful for a broad range of users. For example, someone looking at North Pacific climate variability might be interested in the longer time windows in the North Pacific map, while someone looking at recent marine heat wave events may be interested in the last 5-10 years, viewed using a global and/or individual ocean maps.

9.3.3 *Advanced Statistical Methods (Where to Start)*

Steel et al. (2013) encouraged starting every analysis with basic visualizations of the data ("plot the data early and often"). The simple visualizations presented in this chapter were designed to allow for quick exploration of time series within and between different variables and different regions. This initial information can then be used to guide, focus, or expand further analysis, possibly including the application of more advanced statistics and procedures.

When moving forward into more advanced methods, three things may become quickly apparent. First, the data format and prerequisites of these advanced methods (or software) may require careful preparation of the data, discussion of additional topics (e.g., missing data interpolation, smoothing and filtering, autocorrelation, statistical biases), and might limit which sites can actually be used (e.g., spectral analysis and phenology studies quite often require gap-free, higher sampling frequency, and/or long-running data sets). Second, specialized software is often required, which may have a steep learning curve and require additional programming skills. Finally, many of these methods return pages of results, tables and figures that sometimes only a statistical expert can safely interpret. Steel et al. (2013) warned of the dangerous pitfalls of using an advanced statistical tool without understanding the statistics, data prerequisites, or correct interpretation of its results. So, what is the next step?

If looking for advanced techniques to address a specific question or statistical result, a good starting point is to look for similar analyses in the scientific literature. The methods sections of these publications should summarize the exact statistical methods and software used in the study, and a well-written results section should provide a real-data example of how to interpret the generated

---

[10] Multiplying 7+1 regions x 6 Time Windows x 18 variables x 3 background fields = 2,592 map configurations.

statistical results. The exact methods (and software) used will vary greatly, depending on whether the researcher is studying timing changes in phenology, detecting regime shifts, analyzing changes in community structure, or looking for spatial scales of coherent temporal synchrony. Attempting to provide examples of the literature here would be incomplete and quickly become outdated. The reader is therefore encouraged to conduct their own literature search to find the most recent and relevant publications for the statistical methods to apply (e.g., search for "phenology analysis methods plankton time series" science academic databases and search engines, such as Web of Science or Scopus).

If looking for basic statistical techniques and background, hundreds of books have been written on this topic at a variety of technical levels. Selecting a book is a matter of preference and existing statistical background. For those that like to learn with hands-on examples, "Time Series Analysis and Its Applications: With R Examples" (Shumway and Stoffer 2017) provides a brief statistical background within each analysis example, and provides hands-on data and programming examples that can be run in the free R statistical software package (URL https://www.R-project.org). This book offers a way to learn time series analysis methods, and R, and is part of collection of related books published by the same authors. As such, it can be a good steppingstone for pursuing more advanced statistics and/or R statistical tools.

9.4 Final Thoughts

The impacts of decades of climate change on marine ecosystems and their many inhabitants are observable and measurable, most readily in the zooplankton. Warming marine waters impact seasonal as well as interannual cycles in the oceans. While these impacts are often discussed in terms of global averages, different regions of the planet (e.g., especially the higher latitudes) are experiencing warming at rates much higher than the global planet, and may suffer additional localized extreme events (e.g., marine heat waves). The stress on these systems (e.g., breakdown of cycles, timing, and production) reverberate throughout the food web and impact prey and predator species alike (e.g., fish, birds, marine mammals). While distinguishing natural variability from climate change is still a challenge in some data sets and variables, these differences should be easier to distinguish as the ocean time series get longer.

While not exhaustive, the graphical examples and techniques presented in this chapter show the usefulness of even basic time series visualization and exploration. This crucial first step helps visualizing past and present state of the oceans, as well as aid in directing continued, more advanced analyses. Global ocean efforts like COPEPOD and IGMETS will continue to support and collaborate with the time series communities, creating new tools for data visualization and exploration along the way. Simple visualization tools such as COPEPODITE offer encouragement to scientists (and the public) to "plot the data early and often" and explore ocean data.

Acknowledgements

The example figures and analyses in this chapter used time series data provided by the Alfred Wegener Institute (AWI) Helgoland Roads time series (https://igmets.net/sites/?id=de-30201), the Marine Scotland Science—Scottish Coastal Observatory Stonehaven time series (https://igmets.net/sites/?id=uk-30101), and the Marine Biological Association (MBA) Continuous Plankton Recorder (CPR) Survey (https://igmets.net/sites/?id=uk-40131). In addition to their contribution to this chapter, these time series programs are participants in the IGMETS and ICES WGZE/WGPME plankton working groups, whose ongoing work has helped make the tools and visualizations in this chapter possible. As the work of COPEPOD, IGMETS, and the ICES plankton working groups continues, new data and capabilities will be added (including next generations of the Time Series Explorer and new tools in the Time Series Toolkit). Information on the work of these groups can be found on their respective web pages (e.g., http://igmets.net, http://wgze.net, http://wgpme.net). Information and access to the time-series related tools, publications, and Explorers can be found via the COPEPOD web pages (http://copepod.org or https://www.st.nmfs.noaa.gov/copepod). We would also like to thank the reviewers and editors, whose guidance (and patience) helped this chapter become a reality.

References

Barton, A.D., A.J. Irwin, Z.V. Finkel and C.A. Stock. 2016. Anthropogenic climate change drives shift and shuffle in North Atlantic phytoplankton communities. P. Natl. Acad. Sci. USA 113(11): 2964–2969.

Batchelder, H.P., D.L. Mackas and T.D. O'Brien. 2012. Spatio-temporal scales of synchrony in marine zooplankton biomass and abundance patterns: A world-wide comparison. Prog. Oceanogr. 97: 15–30.

Batten, S.D. and A.W. Walne. 2011. Variability in northwards extension of warm water copepods in the NE Pacific. J. Plankton Res. 33: 1643–1653.

Beaugrand, G., P.C. Reid, F. Ibanez, J.A. Lindley and M. Edwards. 2002. Reorganization of North Atlantic marine copepod biodiversity and climate. Science 296: 1692–1694.

Beaugrand, G. 2003. Long-term changes in copepod abundance and diversity in the north-east Atlantic in relation to fluctuations in the hydroclimatic environment. Fish Oceanogr. 12: 270–283.

Beaugrand, G., K.M. Brander, J.A. Lindley, S. Souissi and P.C. Reid. 2003. Plankton effect on cod recruitment in the North Sea. Nature 426(6967): 661.

Beaugrand, G. 2004. The North Sea regime shift: evidence, causes, mechanisms and consequences. Prog. Oceanogr. 60(2-4): 245–262.

Beaugrand, G., C. Luczak and M. Edwards. 2009. Rapid biogeographical plankton shifts in the North Atlantic Ocean. Glob. Change Biol. 15(7): 1790–1803.

Beaugrand, G., A. Conversi, S. Chiba, M. Edwards, S. Fonda-Umani, C. Greene et al. 2014. Synchronous marine pelagic regime shifts in the Northern Hemisphere. Philos. Trans. R. Soc. B Biol. Sci. 370(1659): 20130272.

Beaugrand, G. and R.R. Kirby. 2018. How do marine pelagic species respond to climate change? Theories and observations. Ann. Rev. Mar. Sci. 10: 169–197.

Behrenfeld, M.J. and E.S. Boss. 2018. Student's tutorial on bloom hypotheses in the context of phytoplankton annual cycles. Glob. Change Biol. 24(1): 55–77.

Bode, A., H.W. Bange, M. Boersma, E. Bresnan, K. Cook, A. Goffart et al. 2017. North Atlantic Ocean. pp. 55–82. *In*: T.D. O'Brien, L. Lorenzoni, K. Isensee and L. Valdés [eds.]. What are Marine Ecological Time Series Telling us About the Ocean? A Status Report, IOC-UNESCO, IOC Technical Series, No. 129.

Boersma, M., A.M. Malzahn, W. Greve and J. Javidpour. 2007. The first occurrence of the ctenophore Mnemiopsis leidyi in the North Sea. Helgoland Mar. Res. 61(2): 153.

Boersma, M., K.H. Wiltshire, S. Kong, W. Greve and J. Renz. 2015. Long-term change in the copepod community in the southern German Bight. J. Sea Res. 101: 41–50.

Bresnan, E. 2012. Phytoplankton and microbial plankton of the North Sea and english channel: Stonehaven. pp. 80–82. *In*: T.D. O'Brien, W.K.W. Li and X.A.G. Morán [eds.]. ICES Phytoplankton and Microbial Plankton Status Report 2009/2010. ICES Cooperative Research Report No. 313.

Bresnan, E., K. Cook, S.L. Hughes, S.J. Hay, K. Smith, P. Walsham et al. 2015. Seasonality of the plankton community at an east and west coast monitoring site in Scottish waters. J. Sea Res. 105: 16–29.

Chiba, S., K. Tadokoro, H. Sugisaki and T. Saino. 2006. Effects of decadal climate change on zooplankton over the last 50 years in the western subarctic North Pacific. Glob. Chang. Biol. 12: 907–920.

Chiba, S., S.D. Batten, T. Yoshiki, Y. Sasaki, K. Sasaoka, H. Sugisaki et al. 2015. Temperature and zooplankton size structure: climate control and basin-scale comparison in the North Pacific. Ecol. Evol. 5(4): 968–978.

Chiba, S., S. Batten, C.S. Martin, S. Ivory, P. Miloslavich and L.V. Weatherdon. 2018. Zooplankton monitoring to contribute towards addressing global biodiversity conservation challenges. J. Plankton Res. 40(5): 509–518.

Chust, G., C. Castellani, P. Licandro, L. Ibaibarriaga, Y. Sagarminaga and X. Irigoien. 2013. Are *Calanus* spp. shifting poleward in the North Atlantic? A habitat modelling approach. ICES J. Mar. Sci. 71(2): 241–253.

Condon, R.H., C.M. Duarte, K.A. Pitt, K.L. Robinson, C.H. Lucas, K.R. Sutherland et al. 2013. Recurrent jellyfish blooms are a consequence of global oscillations. Proc. Natl. Acad. Sci. USA 110: 1000–1005.

Conversi, A., T. Peluso and S. Fonda-Umani. 2009. Gulf of Trieste: a changing ecosystem. J. Geophys. Res. 114(C3).

Cook, K.B. 2013. Zooplankton of the North Sea and english channel: Stonehaven. pp. 119–122. *In*: T.D. O'Brien, P.H. Wiebe and T. Falkenhaug [eds.]. ICES Zooplankton Status Report 2010/2011. ICES Cooperative Research Report No. 318.

Doney, S.C., M. Ruckelshaus, J.E. Duffy, J.P. Barry, F. Chan, C.A. English et al. 2011. Climate change impacts on marine ecosystems. Ann. Rev. Mar. Sci. 4: 11–37.

Edwards, M. and A.J. Richardson. 2004. Impact of climate change on marine pelagic phenology and trophic mismatch. Nature 430: 881–884.

Edwards, M., G. Beaugrand, G.C. Hays, J.A. Koslow and A.J. Richardson. 2010. Multi-decadal oceanic ecological datasets and their application in marine policy and management. Trends Ecol. Evol. 25(10): 602–610.

Edwards, M., G. Beaugrand, P. Helaouët, J. Alheit, and S. Coombs. 2013. Marine ecosystem response to the Atlantic Multidecadal Oscillation. PloS One 8(2): e57212.

Fang, C., L. Wu and X. Zhang. 2014. The impact of global warming on the Pacific Decadal Oscillation and the possible mechanism. Adv. Atmos. Sci. 31(1): 118–130.

Fanjul, A., F. Villate, I. Uriarte, A. Iriarte, A. Atkinson and K. Cook. 2017. Zooplankton variability at four monitoring sites of the Northeast Atlantic Shelves differing in latitude and trophic status. J. Plankton Res. 39(6): 891–909.

Fasham, M.J.R. 1993. Modelling the marine biota. pp. 457–504. *In*: M. Heimann [ed.]. The Global Carbon Cycle, Springer-Verlag, New York, USA.

Findlay, H.S., A. Yool, M. Nodale and J.W. Pitchford. 2006. Modelling of autumn plankton bloom dynamics. J. Plankton Res. 28(2): 209–220.

Friedland, K.D., C.B. Mouw, R.G. Asch, A.S.A. Ferreira, S. Henson, K.J. Hyde et al. 2018. Phenology and time series trends of the dominant seasonal phytoplankton bloom across global scales. Global Ecol. Biogeogr. 27(5): 551–569.

Frölicher, T.L. and C. Laufkötter. 2018. Emerging risks from marine heat waves. Nature Comm. 9(1): 650.

Greve, W., S. Prinage, H. Zidowitz et al. 2005. On the phenology of North Sea ichthyoplankton. ICES J. Mar. Sci. 62: 1216–1223.

Harris, V., M. Edwards and S.C. Olhede. 2014. Multidecadal Atlantic climate variability and its impact on marine pelagic communities. J. Marine Syst. 133: 55–69.

Hays, G.C., A.J. Richardson and C. Robinson. 2005. Climate change and marine plankton. Trends Ecol. Evol. 20(6): 337–344.

Henson, S.A., J.L. Sarmiento, J.P. Dunne, L. Bopp, I. Lima, S.C. Doney et al. 2010. Detection of anthropogenic climate change in satellite records of ocean chlorophyll and productivity. Biogeosciences 7(2): 621–640.

Henson, S.A., C. Beaulieu and R. Lampitt. 2016. Observing climate change trends in ocean biogeochemistry: when and where. Glob. Change Biol. 22(4): 1561–1571.

Henson, S.A., H.S. Cole, J. Hopkins, A.P. Martin and A. Yool. 2018. Detection of climate change-driven trends in phytoplankton phenology. Glob. Change Biol. 24(1): e101–e111.

Hoegh-Guldberg, O. 1999. Climate change, coral bleaching and the future of the world's coral reefs. Mar. Freshwater Res. 50(8): 839–866.

Hoegh-Guldberg, O. and J.F. Bruno. 2010. The impact of climate change on the world's marine ecosystems. Science 328(5985): 1523–1528.

Hurrell, J.W. 1995. Decadal trends in the North Atlantic Oscillation: regional temperatures and precipitation. Science 269(5224): 676–679.

IPCC. 2013. Climate Change 2013: The Physical Science Basis. *In*: T.F. Stocker, D. Qin, G.K. Plattner, M. Tignor, S.K. Allen, J. Boschung et al. [eds.]. Fifth Assessment Report of the Intergovernmental Panel on Climate Change Cambridge University Press, Cambridge, United Kingdom and New York, NY, USA.

IPCC. 2019. Summary for Policymakers. *In*: H.O. Pörtner, D.C. Roberts, V. Masson-Delmotte, P. Zhai, M. Tignor, E. Poloczanska et al. [eds.]. IPCC Special Report on the Ocean and Cryosphere in a Changing Climate. In press.

Ivory, J.A., D.K. Steinberg and R.J. Latour. 2019. Diel, seasonal and interannual patterns in mesozooplankton abundance in the Sargasso Sea. ICES J. Mar. Sci. 76(1): 217–231.

Ji, R., M. Edwards, D.L. Mackas, J.A. Runge and A.C. Thomas. 2010. Marine plankton phenology and life history in a changing climate: current research and future directions. J. Plankton Res. 32(10): 1355–1368.

Johns, D.G., M. Edwards and S.D. Batten. 2001. Arctic boreal plankton species in the Northwest Atlantic. Can. J. Fish Aquat. Sci. 58: 2121–2124.

Keister, J.E., E. Di Lorenzo, C.A. Morgan, V. Combes and W.T. Peterson. 2011. Zooplankton species composition is linked to ocean transport in the Northern California Current. Glob. Chang. Biol. 17: 2498–2511.

Kerr, R.A. 1997. A new driver for the Atlantic's moods and Europe's weather? Science 275(5301): 754–755.

Kerr, R.A. 2000. A North Atlantic pacemaker for the centuries. Science 288: 1984–1986.

Kraberg, A.C., N. Wasmund, J. Vanaverbeke, D. Schiedek, K.H. Wiltshire and N. Mieszkowska. 2011. Regime shifts in the marine environment: the scientific basis and political context. Mar. Pollut. Bull. 62: 7–20.

Kratzer, S. and G. Moore. 2018. Inherent optical properties of the baltic sea in comparison to other seas and oceans. Remote Sens-Basel. 10(3): 418.

Krishnamurthy, L. and V. Krishnamurthy. 2014. Influence of PDO on South Asian summer monsoon and monsoon–ENSO relation. Clim. Dynam. 42(9-10): 2397–2410.

Mackas, D.L., K.L. Denman and M.R. Abbott. 1985. Plankton patchiness: biology in the physical vernacular. Bulletin of Marine Science 37(2): 652–674.

Mackas, D.L., R. Goldblatt and A.G. Lewis. 1998. Interdecadal variation in developmental timing of Neocalanus plumchrus populations at Ocean Station P in the subarctic North Pacific. Can. J. Fish Aqua. Sci. 55: 1878–1893.

Mackas, D.L., R.E. Thomson and M. Galbraith. 2001. Changes in the zooplankton community of the British Columbia continental margin, 1985–1999 and their covariation with oceanographic conditions. Can. J. Fish Aqua. Sci. 58: 685–702.

Mackas, D.L., S. Batten and M. Trudel. 2007. Effects on zooplankton of a warmer ocean: recent evidence from the Northeast Pacific. Prog. Oceanogr. 75: 223–252.

Mackas, D.L. and G. Beaugrand. 2010. Comparisons of zooplankton time series. J. Mar. Sys. 79(3-4): 286–304.

Mackas, D.L., P. Pepin and H. Verheye. 2012a. Interannual variability of marine zooplankton and their environments: within-and between-region comparisons. Prog. Oceanogr. 97: 1–14.

Mackas, D.L., W. Greve, M. Edwards, S. Chiba, K. Tadokoro, D. Eloire et al. 2012b. Changing zooplankton seasonality in a changing ocean: Comparing time series of zooplankton phenology. Prog. Oceanogr. 97: 31–62.

Mantua, N.J., S.R. Hare, Y. Zhang, J.M. Wallace and R.C. Francis. 1997. A Pacific interdecadal climate oscillation with impacts on salmon production. B Am. Meteorol. Soc. 78(6): 1069–1080.

Mantua, N.J. and S. Hare. 2002. Pacific-Decadal Oscillation (PDO). Encyclopedia of Global Environmental Change 1: 592–594.

Mauchline, J. 1998. The biology of calanoid copepods. Adv. Mar. Biol. 33: 1–710.

McPhaden, M.J., S.E. Zebiak and M.H. Glantz. 2006. ENSO as an integrating concept in earth science. Science 314(5806): 1740–1745.

McGowan, J.A., D.R. Cayan and L.M. Dorman. 1998. Climate–ocean variability and ecosystem response in the Northeast Pacific. Science 281: 210–217.

Nelson, R.J., C.J. Ashjian, B.A. Bluhm, K.E. Conlan, R.R. Gradinger, J.M. Grebmeier et al. 2014. Biodiversity and biogeography of the lower trophic taxa of the Pacific Arctic region: sensitivities to climate change. pp. 269–336. *In*: The Pacific Arctic Region. Springer, Dordrecht.

Neukermans, G., L. Oziel and M. Babin. 2018. Increased intrusion of warming Atlantic water leads to rapid expansion of temperate phytoplankton in the Arctic. Glob. Change Biol. 24(6): 2545–2553.

Newman, M., M.A. Alexander, T.R. Ault, K.M. Cobb, C. Deser, E. Di Lorenzo et al. 2016. The Pacific decadal oscillation, revisited. J. Climate 29(12): 4399–4427.

O'Brien, T.D. 2005. COPEPOD: A global plankton database. NOAA Technical Memorandum NMFS-F/SPO 73: 1–19.

O'Brien, T.D., W.K.W. Li and X.A.G. Morán [eds.]. 2012. ICES Phytoplankton and Microbial Plankton Status Report 2009/2010. ICES Cooperative Research Report No. 313.

O'Brien, T.D., P.H. Wiebe and T. Falkenhaug [eds.]. 2013. ICES Zooplankton Status Report 2010/2011. ICES Cooperative Research Report No. 318.

O'Brien, T.D. 2017. Methods and visualizations. pp. 19–35. *In*: T.D. O'Brien, L. Lorenzoni, K. Isensee and L. Valdés [eds.]. What are Marine Ecological Time Series telling us About the Ocean? A Status Report, IOC-UNESCO, IOC Technical Series, No. 129.

O'Brien, T.D., L. Lorenzoni, K. Isensee and L. Valdés [eds.]. 2017. What are Marine Ecological Time Series telling us About the Ocean? A Status Report. IOC-UNESCO, IOC Technical Series, No. 129.

Oliver, E.C.J., M.T. Burrows, M.G. Donat, A.S. Gupta, L.V. Alexander, S.E. Perkins-Kirkpatrick et al. 2019. Projected marine heatwaves in the 21st century and the potential for ecological impact. Front. Mar. Sci. 6: 734.

Paerl, H.W., K. Yin and T.D. O'Brien. 2015. SCOR Working Group 137: Global Patterns of Phytoplankton Dynamics in Coastal Ecosystems: An introduction to the special issue of Estuarine, Coastal and Shelf Science. Estuar. Coast. Shelf S. 162: 1–3.

Perry, R.I., H.P. Batchelder, D.L. Mackas, E. Durbin, W. Greve, S. Chiba et al. 2004. Identifying global synchronies in marine zooplankton populations: issues and opportunities. ICES J. Mar. Sci. 61: 445–456.

Pershing, A.J., C.H. Greene, B. Planque and J.M. Fromentin. 2004. The influence of climate variability on North Atlantic zooplankton populations. pp. 59–94. *In*: N.C. Stenseth, G. Ottersen, J.W. Hurrell and A. Belgrano [eds.]. Ecological Effects of Climate Variations in the North Atlantic. Oxford, UK: Oxford University Press.

Poore, R.Z., K.L. DeLong, J.N. Richey and T.M. Quinn. 2009. Evidence of multidecadal climate variability and the Atlantic Multidecadal Oscillation from a Gulf of Mexico sea-surface temperature-proxy record. Geo-Mar. Lett. 29(6): 477–484.

Racault, M.F., c. Le Quéré, E. Buitenhuis, S. Sathyendranath and T. Platt. 2012. Phytoplankton phenology in the Global Ocean. Ecol. Indic. 14(1): 152–163.

Racault, M.F., S. Sathyendranath, R.J. Brewin, D.E. Raitsos, T. Jackson and T. Platt. 2017. Impact of El Niño variability on oceanic phytoplankton. Front. Mar. Sci. 4: 133.

Reid, P.C., J.M. Colebrook, J.B.L. Matthews, J. Aiken and C.P.R. Team. 2003. The continuous plankton recorder: concepts and history, from Plankton Indicator to undulating recorders. Prog. Oceanogr. 58(2-4): 117–173.

Richardson, A.J. 2008. In hot water: zooplankton and climate change. ICES J. Mar. Sci. 65(3): 279–295.

Richardson, A.J., A. Bakun, G.C. Hays and M.J. Gibbons. 2009. The jellyfish joyride: causes, consequences and management responses to a more gelatinous future. Trends Ecol. Evol. 24(6): 312–322.

Rombouts, I., G. Beaugrand, F. Ibañez, S. Gasparini, S. Chiba and L. Legendre. 2010. A multivariate approach to large-scale variation in marine planktonic copepod diversity and its environmental correlates. Limnol. Oceanogr. 55: 2219–2229.

Rousseaux, C.S. and W.W. Gregg. 2014. Interannual variation in phytoplankton primary production at a global scale. Remote Sens-Basel. 6(1): 1–19.

Rousseaux, C.S. and W.W. Gregg. 2015. Recent decadal trends in global phytoplankton composition. Global Biogeochem. Cy. 29(10): 1674–1688.

Ruhl, H.A. and K.L. Smith. 2004. Shifts in deep-sea community structure linked to climate and food supply. Science 305: 513–515.

Sarthou, G., K.R. Timmermans, S. Blain and P. Tréguer. 2005. Growth physiology and fate of diatoms in the ocean: A review. J. Sea Res. 53(1-2): 25–42.

Sheridan, C.C. and M.R. Landry. 2004. A 9-year increasing trend in mesozooplankton biomass at the Hawaii Ocean Time-series Station ALOHA. ICES J. Mar. Sci. 61: 457–463.

Shumway, R.H. and D.S. Stoffer. 2017. Time Series Analysis and its Applications: with R Examples. Springer.

Steel, E.A., M.C. Kennedy, P.G. Cunningham and J.S. Stanovick. 2013. Applied statistics in ecology: common pitfalls and simple solutions. Ecosphere 4(9): 115.

Steinberg, D.K. and M.R. Landry. 2017. Zooplankton and the ocean carbon cycle. Ann. Rev. Mar. Sci. 9: 413–444.

Thackeray, S.J. 2012. Mismatch revisited: What is trophic mismatching from the perspective of the plankton? J. Plankton Res. 34: 1001–1010.

Tillmann, U. 2004. Interactions between planktonic microalgae and protozoan grazers. J. Eukaryot. Microbiol. 51(2): 156–168.

Trenberth, K.E. and J.M. Caron. 2000. The Southern Oscillation revisited: Sea level pressures, surface temperatures, and precipitation. J. Climate 13(24): 4358–4365.

Trenberth, K.E. and D.J. Shea. 2006. Atlantic hurricanes and natural variability in 2005. Geophys. Res. Lett. 33(12).

Valdés, L., A. López-Urrutia, J. Cabal, M. Alvarez-Ossorio, A. Bode, A. Miranda et al. 2007. A decade of sampling in the Bay of Biscay: What are the zooplankton time series telling us? Prog. Oceanogr. 74: 98–114.

Valdés, L. and M.W. Lomas. 2017. New light for ship-based time series. pp. 11–17. *In*: T.D. O'Brien, L. Lorenzoni, K. Isensee and L. Valdés [eds.]. What are Marine Ecological Time Series telling us About the Ocean? A Status Report, IOC-UNESCO, IOC Technical Series, No. 129.

van Beusekom, J.E.E. 2004. A historic perspective on Wadden Sea eutrophication. Helgoland Mar. Res. 59: 45–54.

van Beusekom, J.E.E., S. Weigelt-Krenz and P. Martens. 2008. Long term variability of winter nitrate concentrations in the Northern Wadden Sea driven by freshwater discharge, decreasing riverine loads and denitrification. Helgoland Mar. Res. 62: 49–57.

Villarino, E., G. Chust, P. Licandro, M. Butenschön, L. Ibaibarriaga, A. Larrañaga et al. 2015. Modelling the future biogeography of North Atlantic zooplankton communities in response to climate change. Mar. Ecol. Prog. Ser. 531: 121–142.

Wang, W., B.T. Anderson, R.K. Kaufmann and R.B. Myneni. 2004. The relation between the North Atlantic Oscillation and SSTs in the North Atlantic basin. J. Climate 17(24): 4752–4759.

Wang, C. and P.C. Fiedler. 2006. ENSO variability and the eastern tropical Pacific: A review. Progr. Oceangr. 69(2-4): 239–266.

Wasmund, N., G. Nausch, M. Gerth, S. Busch, C. Burmeister, R. Hansen et al. 2019. Extension of the growing season of phytoplankton in the western Baltic Sea in response to climate change. Mar. Ecol. Prog. Ser. 622: 1–16.

Wernberg, T., S. Bennett, R.C. Babcock, T. De Bettignies, K. Cure, M. Depczynski et al. 2016. Climate-driven regime shift of a temperate marine ecosystem. Science 353(6295): 169–172.

Wiltshire, K.H., A.M. Malzahn, K. Wirtz, W. Greve, S. Janisch, P. Mangelsdorf et al. 2008. Resilience of North Sea phytoplankton spring bloom dynamics: An analysis of long-term data at Helgoland Roads. Limnol. Oceanogr. 53(4): 1294–1302.

Wiltshire, K.H., A. Kraberg, I. Bartsch, M. Boersma, H.D. Franke, J. Freund et al. 2010. Helgoland roads, North Sea: 45 years of change. Estuar. Coasts 33(2): 295–310.

Winder, M. and J.E. Cloern. 2010. The annual cycles of phytoplankton biomass. Philos. T Roy Soc. B 365(1555): 3215–3226.

Winder, M. and U. Sommer. 2012. Phytoplankton response to a changing climate. Hydrobiologia 698(1): 5–16.

Winder, M., S.A. Berger, A. Lewandowska, N. Aberle, K. Lengfellner, U. Sommer et al. 2012. Spring phenological responses of marine and freshwater plankton to changing temperature and light conditions. Mar. Biol. 159: 2491–2501.

Winder, M., J.M. Bouquet, J. Rafael Bermúdez, S.A. Berger, T. Hansen, J. Brandes et al. 2017. Increased appendicularian zooplankton alter carbon cycling under warmer more acidified ocean conditions. Limnol. Oceanogr. 62(4): 1541–1551.

Winder, M. and Ø. Varpe. 2021. Interactions in Plankton Food Webs: Seasonal Succession and Phenology of Baltic Sea Zooplankton. pp. 162–191. *In*: M.A. Teodósio and A.B. Barbosa [eds.]. Zooplankton Ecology. CRC Press.

Woźniak, M., K.M. Bradtke and A. Krężel. 2014. Comparison of satellite chlorophyll a algorithms for the Baltic Sea. J. Appl. Remote Sens. 8(1): 083605.

Wu, C.R., Y.F. Lin and B. Qiu. 2019. Impact of the Atlantic Multidecadal Oscillation on the Pacific North Equatorial Current bifurcation. Sci. Rep-UK. 9(1): 2162.

Zhang, W., X. Mei, X. Geng, A.G. Turner and F.F. Jin. 2019. A nonstationary ENSO–NAO relationship due to AMO modulation. J. Climate 32(1): 33–43.

Part 3

Advanced Techniques to Study Zooplankton

CHAPTER 10

New Approaches to Study Jellyfish

From Autonomous Apparatus to Citizen Science

Catarina Magalhães,[1,2,3,*] *Alfredo Martins*[4,5] and
Antonina dos Santos[1,6]

10.1 Introduction

Knowledge on dynamics of gelatinous zooplankton populations is very limited worldwide and mainly restricted to specific highly-impacted regions, like the Mediterranean basin and Black Sea (Boero 2013), Seto Inland Sea of Japan (Uye et al. 2003), and East Asian Marginal Seas (Uye 2008), but still with very sparse information available, and focus on specific bloom phenomena. The scarcity of studies addressing this zooplankton group is responsible for the still unresolved questions related with jellyfish biology, its global distribution and community dynamics, the environmental controls of its populations, as well as its pulses of temporal and spatial variability. In this sense, there is an urgent need on developing sustained long-term jellyfish monitoring programs to expand current time series observations, and support research with consistent long-term data sets, at both regional and global scales (Condon et al. 2013). These monitoring strategies will be highly-relevant to elucidate if there is a worldwide increasing trend of jellyfish populations, and if this tendency is driven by natural interannual variability of environmental variables or caused by global warming and the progressive deterioration of coastal areas due to human industry intensification. Sustained data sets are also essential to validate predictive models of jellyfish dynamics and blooms and understand the causes and effects of increased bloom frequency and extension (Purcell et al. 2009, Condon et al. 2013, Graham et al. 2014).

In this context, this chapter aims to revise new technological developments available to acquire information on jellyfish worldwide. This chapter starts with a brief description of the life history and ecology of jellyfish, followed by an outline of their growing ecological relevance and impact

---

[1] CIIMAR – Interdisciplinary Center of Marine and Environmental Research, University of Porto, Portugal.
[2] FCUP – Faculty of Sciences of University of Porto, Porto, Portugal.
[3] School of Science, University of Waikato, Hamilton, New Zealand.
[4] INESC TEC – INESC Technology and Science, Porto Portugal.
[5] ISEP – School of Engineering, Polytechnic Institute of Porto, Porto, Portugal.
[6] IPMA – Portuguese Sea and Atmosphere Institute, Lisbon, Portugal.
  Emails: alfredo.martins@inesctec.pt; antonina@ipma.pt
* Corresponding author: cmagalhaes@ciimar.up.pt

on marine ecosystems. Then, the chapter also goes on to explore emerging technologies, such as the coupling of distributed autonomous systems with new real-time data processing methods, as well as the usefulness of autonomous *in situ* environmental DNA (eDNA) samplers. Citizen science programs are presented as an important tool to improve and complement the existing monitoring programs by increasing spatial and temporal resolution.

10.2　The Life History, Biology and Ecology of Jellyfish

The term "jellyfish" is generally used to refer the medusa phase of some cnidarians that have attracted much attention in recent years, mainly due to the proliferation of some species in coastal areas (Attrill et al. 2007), especially in the Mediterranean Sea (Boero 2013). However, "gelatinous zooplankton" is the best term to refer to the group of transparent and fragile marine zooplanktonic organisms (Haddock 2004). It is important to clearly identify which groups are considered as gelatinous plankton. Usually, Cnidaria (Cubozoa, Hydrozoa and Scyphozoa), Ctenophora (Beroida and Tentaculata) and Tunicata (Salpida) are representatives of the gelatinous plankton (Boero 2013). Gelatinous zooplankton are also often defined as planktonic organisms with tissues containing a high percentage of water (Madinand and Harbison 2001). This definition would also include representatives of Radiolaria, Heteropoda, Pteropoda, some Polychaeta, and some Holothuria, among others. Kiørbe (2013) used an extensive database of a wide range of zooplankton representatives to examine how body composition varies as a function of individual size, how would groups be separated, and to verify if the water content varies between common gelatinous groups (Cnidaria, Ctenophora and Tunicata), with very different ecology and taxonomy. According to this study, gelatinous and non-gelatinous organisms were clearly separated. The former shows an average dry matter content of ±4–5% of the live weight and a carbon content of ±0.5% of the live weight, whereas the latter presents average dry matter content of ±15–25% and carbon content of ±5–10% of the live weight. Therefore, according to Kiørbe (2013), gelatinous zooplankton are the Cubozoa, Hydrozoa and Scyphozoa (Cnidaria), the Beroida and Tentaculata (Ctenophora) and the Appendicularia, Doliolida, Salpida and Pyrosomatida (Tunicata).

Considering all phases of the life cycle and colonial stages, gelatinous zooplankton have a wide range of sizes, from a few millimeters to several centimeters, and representatives in three phyla with different ecological traits and life histories. Although gelatinous bodies are fragile structures, they survive under low food concentrations, therefore prevailing in the open ocean. All gelatinous zooplankton have many unique adaptations that enable them to reach high population densities when facing favorable environmental conditions, and quickly adapt to unfavorable conditions in the marine pelagic ecosystem.

Zooplankton is traditionally sampled using plankton net hauls. These methods are not adequate to sample gelatinous zooplankton due to their fragility, and these organisms are destroyed or seriously damaged when captured in a plankton haul. As a consequence, jellyfish are poorly studied with respect to other zooplankton groups, but they are gaining much attention due to their outbreaks and related interferences with human activities (Condon et al. 2012).

10.2.1　Cnidaria

Cnidarians are commonly known as the true jellyfish, and are represented in zooplankton by Cubozoa, Hydrozoa and Scyphozoa. The Hydrozoa are mostly marine components of mesozooplankton and distributed in all oceans and seas around the world. Under favorable environmental conditions, they are able to rapidly bloom and spread over large areas and, when resources are limiting, cnidarians are able to enter into a resting phase. Cnidarians are carnivorous, eating all kinds of zooplankton, such as copepods, eggs and fish larvae and other gelatinous organisms, depending on their size. All cnidarians possess stinging cells, armed with cnidocysts (nematocysts in the case of the colonial

Hydrozoa), which are small structures with the capacity to inject venom that is used to capture prey or as a protection against predators. Both, Hydrozoa and Scyphozoa, have species capable of producing bioluminescence (Haddock et al. 2010). The medusae move by jet propulsion propelled by the rhythmic contractions of the umbrellas, and colonial species that float at the ocean surface are mostly driven by wind and surface currents (Ferrer and Pastor 2017, Pires et al. 2018).

Cnidarian life cycles are mostly dimorphic, with a polyp and a medusae phase. In meroplanktonic species, the polyp is a benthic phase that produces the ephyra by budding. In the case of Scyphozoa, the ephyra is produced by strobilation and, in the case of Cubozoa, by polyp metamorphosis into medusae (Boero 2013). The ephyra is a planktonic immature medusae which develops into a medusa. The medusae reproduces sexually, with females releasing eggs into the ocean where they are fertilized by sperm liberated by male medusae. Eggs develop to become a planula larva, which will metamorphose and settle as a polyp in the benthic substrate. The asexual reproduction is by budding of the young medusa stage or by longitudinal fission. Hydrozoa can be solitary or colonial. In the case of the holoplanktonic solitary species, the planula stage develops directly into an ephyra larval stage. Colonial species from the order Siphonophorae, which are mostly holoplanktonic, are formed by specialized individuals, with specific functions in the colony, the zooids, which are not able to survive by themselves. Siphonophorae are integrated into the macrozooplankton, since mature colonies are long and float or swim in the ocean (e.g., *Physalia physalis*, the Portuguese man o'war). Sexually mature individuals from both sexes, designated gonophores, are released by the colony, except in some species, as *Physalia physalis*, which releases gonophores of only one sex (Munro et al. 2019). When completely mature, they will release the gametes for external fertilization, leading to the production of the planula larval stage, that develop into a calyconula/siphonula larva, and eventually into a new colony.

10.2.2 Ctenophora

Ctenophora is an entirely marine phylum of invertebrate organisms that are commonly known as comb jellies and are mostly pelagic. Ctenophores have a gelatinous body covered by comb rows of fused ciliary plates, which are used for locomotion. Most ctenophores are bioluminescent, especially the deep-sea species (Haddock et al. 2010). Ctenophores include two relevant groups, Beroida whose species lack tentacles, and Tentaculata. The latter presents a pair of tentacles, that usually carry colloblasts, sticky adhesive cells used to capture the prey (Haddock 2007). Adults can have a wide variety of sizes, from a few millimeters to more than one meter. Ctenophores are carnivorous, feeding on other zooplankton but also on marine snow, and Beroida also feed on gelatinous zooplankton (Licandro and Lindsay 2017). Studies of *Mnemiopsis leidyi*, a ctenophore from western Atlantic, well known as an invasive species in northern European waters and Mediterranean Sea, demonstrate that survival is possible under a broad range of sea temperatures and without food ingestion (Gambill et al. 2015, Jasper et al. 2015). Most ctenophores are simultaneous hermaphrodites, producing at the same time eggs and sperm that are released through separate gonoducts. In some groups, as Beroida, the resultant cydippid larvae is very similar to the adult; in other groups, larvae undergo changes during the development till the adult phase.

10.2.3 Tunicata

The pelagic Tunicata include Appendicularia and Thaliacea, the latter comprising the orders Doliolida, Salpida and Pyrosomatida. All pelagic tunicates have a transparent gelatinous body, with complex life cycles having solitary (all appendicularians) and, in the case of thaliaceans, a colonial life stage. Pelagic tunicates are mostly holoplanktonic and inhabit open ocean but also coastal waters, especially in the case of appendicularians. Tunicates are mucous-mesh grazers, feeding on phytoplankton, bacteria, viruses and detritus, and exhibit higher feeding rates and much shorter generation times than other marine planktonic grazers. The muscles of their body contract and relax to force the seawater passing through a large buccal siphon, creating currents and moving it forward and backward, to be expelled by the posterior siphon. The muscle bands are also used for propelling the organisms through the ocean (Conley et al. 2018).

Appendicularians are hermaphrodites, except *Oikopleura dioica*, with external fertilization and direct development (Deibel and Lowen 2012). Salpids display an obligatory alternation of sexually and asexually reproducing generations, and the oozooid reproduces asexually by budding releasing discrete blocks, presumably genetic clones, originating a colonial phase. The buds develop into a free-living form, the blastozooid, which is fertilized by an older male blastozooid, with internal fertilization. Doliolids have a similar life-cycle of salpids, only more complex, since the buds develop into three different types of blastozooids, depending on their position on the dorsal spur: trophozooids are gastric-feeding units, responsible for nourishing the colony; and phorozooids develop into a gonozooid stage and reproduce sexually (Deibel and Lowen 2012). The development of pyrosomatids, described only for *Pyrosoma atlanticum*, is considered similar to other thaliaceans, but without a larval stage (Licandro and Brunetta 2017).

10.3 Jellyfish Proliferation and Interactions with Human Activities

10.3.1 *Natural and Anthropogenic Drivers of Population Dynamics*

Jellyfish are adapted to different environments and can live in a wide range of temperatures and salinities, which make them successfully widely distributed in every ocean. Jellyfish have an important ecological value due to its high diversity (over 1200 species; Daly et al. 2007, Milles 2011) and role in marine carbon pump (Lebrato et al. 2011), through carbon transport across water column (Doyle et al. 2014), to benthic coastal systems and the deep ocean (Sweetman and Chapman 2011, Sweetman et al. 2014). In fact, recent studies of jellyfish carbon transfer to the deep ocean layers demonstrated the important role of massive jellyfish sink events to local and regional biogeochemical cycles (Lebrato et al. 2019). In addition, jellyfish are the main source of food for specialized predators, can be opportunistically consumed by a high range of marine predators (Cardona et al. 2012, Hays et al. 2018), and are commercially exploited for human food production (Torri et al. 2020).

Despite the recognized ecological roles of jellyfish, global spatial and temporal dynamics of these organisms across marine environments are still not well understood (Mills et al. 2001, Purcell 2005, Condon et al. 2013). The limited availability of long-term records of gelatinous zooplankton, since the 18th century, and the recent yearly records showed a large temporal scale periodicity of jellyfish blooms (Goy et al. 1989). Current evidences point to a regional increase in the frequency of gelatinous plankton blooms in different oceanic basins, with a negative impact on human industries (Boero 2013). Examples of the most affected regions by jellyfish blooms include the Mediterranean and Black Seas (Bernard et al. 1988, Goy et al. 1989, Boero 2013), the Black, Azov and Caspian seas (Shiganova et al. 2003), Gulf of Mexico (Graham et al. 2003), the Seto Inland Sea of Japan (Uye et al. 2003), and the East Asian Marginal Seas (Uye 2008).

In some of these regions, a regional intensification tendency was reported for blooms of Ctenophore and medusae (Mills et al. 2001). Jellyfish populations are sensitive to seasonal changes in the environment, at annual and interannual time scales (Stone 2016), usually following planktonic food availability (Mills et al. 2001). Region-specific raises in jellyfish densities can result from increases in specific native or introduced species. The progressive dominance of *Chrysaora melanaster*, in the Bering Sea since 1990s, and *Chrysaora hysoscella*, as well as *Aequorea aequorea*, in the Benguela Current since 1970s, are examples of regional changes on the dominant species of native jellyfish populations (Mills et al. 2001). Examples of proliferations of non-native jellyfish species include *Rhopilema nomadica* in the Mediterranean Sea, and the estuarine hydromedusae *Maeotias marginata*, in San Francisco Bay and Chesapeake Bay, amongst others (Mills et al. 2001).

Although in some specific regions there is evidence of an increase in jellyfish bloom intensity and frequency (Mills et al. 2001), the analysis of long-term datasets of jellyfish abundance, from worldwide coastal areas, showed no evidence of a tendency of global jellyfish proliferation across marine environments, for the period 1874 to 2011 (Condon et al. 2013). Instead, long term population size fluctuations, with approximately a 20-y periodicity, were detected (Condon et al. 2013). These findings highlight the urgent need for sustained long term monitoring of jellyfish occurrences and abundance to understand global trends in population dynamics.

Higher incidences of localized jellyfish blooms (Mills et al. 2001, Graham et al. 2014) and blooms affecting the Large Marine Ecosystem (LMEs) of the world's coastal regions (Brotz and Cheung 2012) has been associated with the impacts of anthropogenic activities. These impacts include: (i) the eutrophication of coastal systems, with water column oxygen depletion, favoring for example the proliferation of *Aurelia aurita* in some harbors and bays (Elefsis Bay, Greece: Papathanassiou et al. 1987; Tokyo Bay, Japan: Ishii et al. 2008); (ii) the introduction of non-native species, for example in San Francisco Bay (Rees and Gershwin 2000) or the Mediterranean and Black Sea (Holland et al. 2004, Dawson et al. 2005, Graham and Bayha 2007); and (iii) overfishing, as reported in the Black Sea (Daskalov et al. 2007), with a direct impact on the jellyfish eating mackerel (Zaitsev and Mamaev 1997). There are also evidences that aquaculture, using floating raft ropes, favor jellyfish by increasing the eutrophic status of the coastal waters and providing increased artificial substrate for polyp colony formation (Lo et al. 2008).

The few long-term records of jellyfish available also suggest that global warming may affect spatial and temporal distribution of gelatinous zooplankton organisms (Purcell 2005). For example, periodic monitoring of *Mnemiopsis leidyi* in the East coast of the USA (Sullivan et al. 2001) revealed that an increase in temperature of 2°C promoted high abundances and prolonged occurrence periods. Warm spring temperatures in the Black Sea and Chesapeake Bay were also associated with the occurrence of large ctenophores (Purcell et al. 2001). In addition, high abundance of the cosmopolitan *Aurelia* spp. in the last 20 years, in Tokyo Bay, was associated with the increase of minimum winter temperature, from 1.5°C to 11°C (Uye and Ueta 2004).

All these evidences together suggest that several of the observed regional long term increases in jellyfish populations could be linked to human activities (Richardson 2009), which also induce global warming, but the mechanisms involved in those shifts are still far from being fully understood (Brotz and Cheung 2012). However, based on a systematic review of 365 published studies on jellyfish blooms, Pitt et al. (2018) concluded that there aren't reliable evidences that anthropogenic stressors could enhance jellyfish blooms, recommending researchers the need to qualify the statements about the jellyfish in order to accurately describe the state of the knowledge.

10.3.2 *Impacts of Jellyfish Blooms on Human Activities and Marine Ecosystems*

Expanded temporal and spatial distribution of jellyfish blooms is a concern due to their potential negative impacts on several human activities. Jellyfish blooms have been reported to have enormous ecological, economic, and societal impacts, and implicated in the decline of commercial fisheries, economic losses in aquaculture, tourism and power plants maintenance.

Because jellyfish blooms tend to overlap with important worldwide fisheries areas (Graham et al. 2014), they directly impact fisheries by clogging and damaging fishing nets and causing mortality of commercial fish species. There is a considerable documented impact on the fishing industry of Northern Adriatic Sea (Palmieri et al. 2014), Japan, South Korea and East China Sea (Kawahara et al. 2006, Uye 2011, Kim et al. 2012), Gulf of Mexico (Quiñones et al. 2013) and Black Sea (Boero 2013). Also, an indirect impact has been recognized to occur due to the fact that fish and jellyfish may compete for the same prey, including ichthyoplankton (fish eggs and larvae) and other zooplankton (Kawahara et al. 2006, Condon et al. 2013), being responsible for consuming large quantities of these planktonic organisms jeopardizing fisheries' sustainability (Graham et al. 2014). Jellyfish blooms have been shown to drive fundamental ecological impacts by trapping large amounts of carbon (C) and changing C transfer across food webs, favoring the flow of C to microbial food web instead to higher trophic-level predators (Condon et al. 2011), thereby limiting fisheries.

Jellyfish can also cause damage in aquaculture industry (Purcell et al. 2013) due to intensification of aquaculture operations in coastal areas. There are reports of Cnidarian jellyfish impact on North Europe salmonids fish farms, being implicated in gill damage and fish kill events (Rodgen et al. 2011), and considered a potential vector of the bacterial fish pathogen *Tenacibaculum maritimum*. The sessile phase of hydrozoans is also a main cause of biofouling in aquaculture structures, with implications on hydrodynamics and water quality (Carl et al. 2010).

Massive jellyfish blooms block seawater intakes in power plants and desalination plants, increasing prevention costs. There were several shut down situations reported in the United States, Israel, Scotland, Japan, and the Philippines due to the clogging of the seawater filters used for the water intake to cool down the reactors (Graham et al. 2014). Tourism is one of the most affected industries by jellyfish outbreaks, with important economic losses documented for the Mediterranean Sea (Ghermandi et al. 2015) and warm waters of north Australia and Florida with venomous jellyfish incidences (Graham et al. 2014).

In face of the detrimental impacts of jellyfish blooms on marine ecosystems and human activities, it is undoubtedly important to improve our ability to predict and manage jellyfish incidences. Although there is still a lack of studies focusing on these events to properly identify bloom drivers, and there are a few modeling studies for forecasting jellyfish abundance in coastal areas (Brown et al. 2013, Alekseenko et al. 2019).

10.4 Emerging Technologies for Jellyfish Monitoring

10.4.1 *Remote Sensing Technologies*

Zooplankton and jellyfish monitoring in particular can be performed using a multitude of methods and tools, ranging from remote sensing imaging, from satellite and aircraft (Houghton et al. 2006), to trawl plankton nets (Wiebe and Benfield 2003), using acoustics for biomass characterization (Moline et al. 2015), optical imaging at the micro- or macro level (Cowen and Guigand 2008, Reisenbichler et al. 2016), and individual organism tracking (Fossette et al. 2016).

Not only new technology developments, such as new plankton imaging sensors, eDNA *in situ* samplers, and advanced robotic underwater vehicles, reaching Technology Readiness Levels (TRL), but also technical advances in artificial intelligence techniques and computational hardware have been providing new ways of data gathering and processing. These sensors and systems make the transition between classic methods, performed in the laboratory from sampling and measurements obtained during research surveys, and *in situ* sampling and automated processing with advanced tools such as autonomous underwater vehicles (AUV). In the case of jellyfish studies, due to the delicacy and transparency of the individuals, there is a growing interest in using indirect and automated sampling techniques (Graham et al. 2003), such as acoustic target strength or animal detailed images.

Remote sensing and satellite observations, in particular, provide global scale information on water properties such as surface water temperature, suspended particulate matter, and chlorophyll concentration (Table 10.1). Hyperspectral imaging can also discriminate the composition of suspended particulate matter and phytoplankton (Mouw et al. 2017, Fossum et al. 2019), which could be relevant for developing large scale models of jellyfish blooms. Short-distance observations with aircrafts and unmanned aerial vehicles can also be used to provide finer scale measurements for large organisms, such as jellyfish, even allowing for individual identification and monitoring (Houghton et al. 2006).

Individual jellyfish tagging and tracking has been limited due to their gelatinous nature and can only be applied to larger specimens. Developments in miniaturization of monitoring tags allow for the use of these in increasingly smaller marine animals (Viana et al. 2018). With small tags, less than two centimeters long (Moriarty et al. 2012), and new fixation developments, it has been possible to track jellyfish (Fossette et al. 2016).

High frequency acoustic sensors provide a set of tools for zooplankton studies (Wiebe and Benfield 2003). Acoustic backscatter information can be used for biomass estimation and characterization of

Table 10.1: Sensing technologies used for the analysis of planktonic organisms, with information on the sensor deployment platform, scope of application (depth level, organism size), biological information retrieved, advantages, limitations and examples of application. In some cases, both the vertical sensor distance range ("range"), and the sensor depth-ranging (i.e., maximum sensor working depth, "depth") are provided. DCP: acoustic doppler vertical profiler; AUV: autonomous underwater vehicle; ROV: remotely operated vehicle.

Sensor	Deployment	Depth level	Organism size	Biological information	Advantages	Limitations	Example
Spectro-radiometer	Satellite Aircraft	Surface	1–200 µm phytoplankton	Chlorophyll Phytoplankton composition	Very large area coverage	Low temporal/spatial resolution Ocean "skin"	Fossum et al. 2019, Mouw et al. 2019
Imaging systems	Aircraft	Few meters	10 cm–few meters	Abundance macro-megazooplankt Aerial coverage	Medium area coverage Fast coverage	Limited range Sea surface or near surface only High cost	Houghton et al. 2006
Scientific echosounders	Moored, towed; Ship, AUV/Glider, Robotic lander	Range: 10,000 m Depth: 3,000 m	> 1 mm	Biomass, size (target strength)	Water column scans Measurements from surface	Calibration issues Noise sensitivity Difficult analysis, limited information	Moline et al. 2015
Imaging sonar	Moored, towed; ROV, AUV, Robotic lander	Range: 100 m Depth: 4,000 m	> 10 cm	Imaging (low resolution) Identification Abundance	Versatility Long range imaging Additional backscatter data Robustness to water visibility	Low resolution Large organisms Difficult to detect jellyfish (low reflection) High power consumption	Makabe et al. 2012
ADCP (Backscatter)	Moored (surface to bottom), ROV, AUV, Robotic lander	Range: 100 m Depth: 6,000 m	All sizes	Biomass	Common installed sensor	Low accuracy in biomass estimation	Geoffrey et al. 2016
Conventional cameras	All	Range: 5–10 m Depth: 10,000	> 1 cm	Identification, size	Low cost Very common Versatile deployment Possibly high-quality data	Water visibility (limited range) Illumination issues/power hungry	Robison et al. 2010
Plankton counters	Towed, AUV	Depth: 1,000 m	<40 mm	Biomass, size	Fast *In situ* biomass estimation	Noise/other particles Only very small particles Limited application	Pederson et al. 2010

Table 10.1 contd. ...

...Table 10.1 contd.

Sensor	Deployment	Depth level	Organism size	Biological information	Advantages	Limitations	Example
Plankton imaging—line scan	Towed, AUV	Depth: 500 m	100 mm–2 m (one dimension)	Biomass, identification	Combines detailed imaging for very small and large organisms	Requires constant velocity/flow Image formation issues Limited deployment Visibility	Samson et al. 2001
Plankton imaging-matrix sensor	Mooring, ROV, AUV	Depth: 6,000 m	2–20 cm	Biomass, identification	Ease of application of image processing	Optical setup limits organism size/detail Visibility, range High power consumption	Picheral et al. 2010
eDNA samplers	Moored, Towed, ROV, AUV (large), Robotic lander	Depth: 1000 m	< 1 mm	Genetic information	*In situ* genetic info – allow partial transfer of lab work to site	Cost Current limited analysis capabilities Sample recovery Microbial organisms Limited direct application to jellyfish	Ribeiro et al. 2019

animal distributions in the water column (Foote and Stanton 2000). For these purposes, frequencies between 38 kHz and 1 MHz are commonly used. Acoustic sensors can provide two types of information: volume backscatter or echo integration of the energy returned by all the individuals in a given volume, or target strength corresponding to a particular eco from an individual (Table 10.1).

With a single beam, only volume backscatter can be obtained thus providing a total estimation of sound reflection in a given volume. By using multiple frequencies, such as with Acoustic Zooplankton Fish Profiler (AZFP) (Ludvigsen et al. 2018), it is possible to estimate animal size distribution. This method uses a prior model of sound backscatters from different animals, depending on frequency. Multi-beam acoustic system allows the acquisition of individual target strength. The most commonly used acoustic biomass profilers are split beam or dual beam echosounders. In dual-beam sonar, wide and narrow beam transducers are used, and sound emitted in the narrow beam is received by both. Differences on the receiving echoes can be used to determine individual targets. Split-beam applies a similar approach, where intensity and phase information are used from four 90° sectors and integrated to obtain target strength (TS). These sonars used for biomass estimation and characterization are usually designated as scientific echosounders or calibrated echosounders, since they require precise calibration in order to model the returns and thus the estimation performed (Moline et al. 2015).

There is also a vast diversity of other acoustic sensors, not developed specifically for biomass or zooplankton studies, but nevertheless relevant. Acoustic Doppler profilers use multiple beams and the Doppler frequency shift in echo return to provide velocity information of either water volumes (currents) or velocity against the sea bottom Doppler Velocity Log (DVL). There is an increasing interest of using the acoustic backscatter information of these common sensors for zooplankton studies (Powell and Ohman 2012). Multibeam sonars, designed for three-dimensional mapping and for imaging purposes (in this case providing an intensity of the echo along time and for different directions), can also be used for direct large animal imaging utilizing sound (Båmstedt et al. 2003).

10.4.2 Optical Systems

Optical based sensors provide one of the more relevant tools for the study of zooplankton and jellyfish, in particular (Lombard et al. 2019). These sensors are used for characterization of zooplankton, providing direct counting of particles or individuals and image data that can be used for the identification of different species and their developmental stages. Optical sensors can be divided into conventional imaging cameras, such as video or digital stills cameras, commonly included in multiple marine observing services, and specific systems designed with particular measuring or imaging purposes, such as plankton counting (Herman 2004), imaging (Martins et al. 2016) or holographic recording (Sun et al. 2008; see Table 10.1). For a more detailed treatment of optical sensors for plankton studies, the reader is referred to Lombard et al. (2019).

Gelatinous zooplankton image and video observations, and their use for long term studies, suffer from two main problems. One is the nature of these organisms, with low image contrast due to their transparency, and a wide range of sizes, from tens of micrometers (e.g., microzooplankton) up to meters (e.g., large jellyfish). Another problem is that for extended temporal and spatial video/imaging surveys, the amount of generated data is very large requiring automated processing for increased efficiency. The first problem is also encountered in the design of plankton counting/imaging systems, as for short range high magnification optics required by microorganisms, the volume of imaged water is necessarily small. On the other hand, large organisms require different instrumentation, such as: the *In Situ* Ichthyoplankton Imaging System (ISIIS) (Cowen and Guigand 2008), with telecentric optics and 68 μm-pixel resolution; the Jellycam (Graham et al. 2003) that addresses jellyfish over two centimeters in size; and other scientific cameras, with conventional perspective lenses but much wider fields of view and imaging volume dimensions mainly limited by available illumination, dependent on specific optic configuration.

The optical sensors designed for plankton studies can be classified as plankton counters, flow cytometry devices or plankton imaging systems. Plankton counters (Herman 2004) are designed to count the number of particles or plankton organisms of a specific size range, within a known volume of seawater. Thus, they are mostly used for estimation of abundance and biomass and, in general, do not provide information on species identification. Flow cytometry devices automatically measure different optical properties of individual cells in suspension, including light scatter, at multiple angles, and fluorescence intensity, at multiple wavelengths, including both natural autofluorescence (phytoplankton) or artificially induced fluorescence by pre-labeling the sample with secondary fluorescent stains. Flow cytometry can also provide images and identification of individual cells, when combined with imaging systems (microscopy) and artificial intelligence technology. For example, the Imaging FlowCytobot is an *in-situ* automated advanced submersible system that provides information on the abundance and composition of phytoplankton and phagotrophic protists (Olson and Sosik 2007, Doucette et al. 2018, Glibert et al. 2018). However, since flow cytometry devices are tailored to acquire information on unicellular planktonic organisms, representatives of lower trophic levels have limited use for jellyfish studies.

Plankton imaging sensors are designed to acquire images of plankton in a predetermined water volume. These sensors are not only used for counting and biomass estimation, but also for species identification. Some examples of these systems are the Video Plankton Recorder (VPR) (Ohman et al. 2019), Underwater Vision Profiler (UVP) (Picheral et al. 2010), ISIIS (Cowen and Guigand 2008), Shadowed Image Particle Profiling and Evaluation Recorder (SIPPER) (Samson et al. 2001) and JellyCam (Graham et al. 2003). Most of these sensors address micro- and mesozooplanktonic organisms (only JellyCam is designed for larger organisms), since larger plankton can, in many cases, be imaged using traditional cameras. From an optical design point of view, the use of telecentric lenses is frequent, as it preserves the object size in the image independent of the distance to the camera, thus facilitating identification. This is the case of ISIIS or MarinEye camera (Martins et al. 2016). Other systems, such as the VPR or UVP, use conventional perspective lens.

One problem associated with the design of plankton imaging sensors is the need to achieve good resolution for small organisms (e.g., microzooplankton, in the scale of tenths of micrometers) and, simultaneously, image large volumes of water. At high optical magnification, the depth-of-field is reduced as well as the field of view. Many of these systems (e.g., VPR, ISIIS, JellyCam) are thus designed to be operated in towfish configuration, with sensor motion providing a scan over large volumes of water. This advantageous configuration provides motion free images, and energy for strong illumination required for reducing shutter speeds. Many of these sensors are associated with other sensors, that measure seawater properties (e.g., CTD, dissolved oxygen probe, fluorometer) in the same instrument (e.g., towfish).

One relevant example of optical sensors is MarinEye system (Martins et al. 2016), an autonomous *in situ* system that incorporates an imaging camera designed for micro- and mesozooplankton, a biosampler system able to collect and concentrate different size classes of planktonic microbes, using *in situ* filtration (Ribeiro et al. 2019), different probes for measuring seawater properties, acoustic sensors for imaging and biomass estimation, and acoustic Doppler for water current measurement (Fig. 10.1).

10.4.3 *Environmental DNA Based Technologies*

Environmental DNA (eDNA), defined as the genetic material obtained directly from environmental samples, is becoming an important tool for biodiversity monitoring (Deiner et al. 2017), with some, but still few, applications in the study of spatial-temporal distribution of jellyfish (Minamoto et al. 2017). DNA extracted from environmental samples enables a comprehensive DNA-based taxonomic analysis, using metabarcoding or metagenomic approaches (Deiner et al. 2017, Rey et al. 2021, this book, Chapter 11) for detection and quantification of jellyfish (Minamoto et al. 2017) and other planktonic organisms (Ribeiro et al. 2019, Matos et al. 2021, this book, Chapter 12). Unfortunately, up to date genomic sensors are not available, although different initiatives have been undertaken to develop eDNA autonomous samplers (Trembani et al. 2012, Yamahara et al. 2019, Ribeiro et al. 2019)

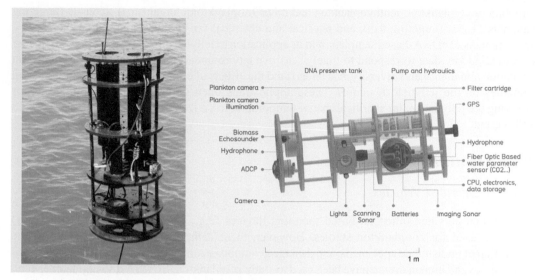

Figure 10.1: MarinEye prototype, an *in situ* autonomous biosampler with multiple technological components for integrated ocean observation. Left panel: sensor prototype deployed from ship (reproduced from Ribeiro et al. 2019); Right panel: schematic representation of system components (MarinEye website, http://marineye.ciimar.up.pt/).

or systems for identifying particular marine pelagic organisms (Scholin et al. 2009, 2017). Traditional jellyfish monitoring approaches involve highly costly and time-consuming sampling procedures, limiting sampling frequency and data availability. Moreover, conventional sampling programs are mostly limited to target regions, and a large spatial and temporal gap in data availability do exist worldwide (Condon et al. 2013). Thus, the autonomous sampling of *in situ* planktonic assemblages, for posterior genomic and metagenomic analysis (see Matos et al. 2021, this book Chapter 12), might represent an important improvement for monitoring jellyfish, enabling the increase of sampling spatial and temporal resolution, especially critical for remote or hardly accessible aquatic ecosystems.

For example, the previously referred autonomous biosampler, MarinEye (Ribeiro et al. 2019; Table 10.1), allows *in situ* filtration and concentration of different size classes of planktonic microbes (using standard filters Sterivex TM-GV 0.22/0.45 μm, PVDF) and preservation eDNA, with high-performance in eDNA quality, yield and replicability. eDNA is later used for in-depth taxonomic analysis, based on megasequencing approaches (Ribeiro et al. 2019). The capability of *in situ* water filtration minimizes the risk of contamination and eDNA degradation, when compared to manual sample collection. This autonomous biosampler can also integrate a user-friendly application for programming sampling definitions (e.g., number of replicates, sampling depth, volume of water to be filtered). This system can be used in open ocean studies, but also in coastal, estuarine and aquaculture environments, to successfully increase monitoring resolution of aquatic plankton in general and, jellyfish assemblages in particular. The Environmental Sample Processor (ESP), developed by the Monterey Bay Aquarium Research Institute, MBARI (Scholin et al. 2006, 2009), is also an example of an autonomous instrument that collects and filters water samples, for further laboratory analysis. This system can also carry out real-time *in situ* molecular detection of specific microbial groups and/or biological toxins, with environmental relevance, and has been applied to the quantification of mixotrophic species of phytoplankton that should also be functionally considered members of zooplankton assemblages (Scholin et al. 2006, 2009, Doucette et al. 2018, Glibert et al. 2018).

eDNA systems based on *in situ* filtration allow the concentration of large volumes of water into a final sample, with a reduced volume, being flexible in terms of operation mode and integration on different fixed and mobile (e.g., AUV) aquatic observation systems. eDNA samplers can also be integrated in multitrophic observatory systems, like MarinEye (Fig. 10.1), by combining imaging (targeting phytoplankton, ichthyoplankton and other zooplanktonic organisms), acoustic (targeting mammals), sonar (targeting fish), and physical and chemical sensors (targeting biogeochemistry). Developments of eDNA-based samplers, and its application to jellyfish organisms, are highly relevant for providing key data to respond to timely sensitive environmental issues, as changes in jellyfish population abundance and diversity, and understand the effects of climate change and environmental anomalies on the jellyfish dynamics. These applications open up remarkable opportunities for observing the jellyfish realm at critical spatial-temporal scales, a feature which has been out of reach until present.

10.4.4 Robotic Platforms

Remotely Operated Vehicles

Conventional nets and other net-based sampling devices (e.g., Continuous Plankton Recorder) have been used for zooplankton studies. However, these sampling methods do not allow the observation of undisturbed specimens, and damage gelatinous zooplankton. Since the 1950s, manned submersibles and human diving have been used to study zooplankton. SCUBA diving (Hamner et al. 1975) and Human Occupied Vehicles, HOVs (Hidaka-Umetsu and Lindsay 2018), allow the direct observation of undisturbed marine organisms. However, these strategies present strong spatial and temporal limitations, due to constraints in the range of human diving and the cost of human operated vehicles.

The high cost of human operated vehicles and deep-sea submersibles, in particular, has promoted the use of remotely operated vehicles (ROVs) and other robotic systems for the ocean exploration (Allen et al. 1997, Manley 2008), in multiple science fields. ROVs (Youngbluth et al. 2008, Smith et al. 2019) and autonomous submersible vehicles (Eriksen et al. 2001, Hobson et al. 2007) allow for a very diverse set of sensing abilities and more efficient data collection, relevant aspects in face of the limitations accessing specific ocean areas (Yoerger et al. 1986) and the need to survey large areas of oceans.

ROVs are unoccupied submersible remotely controlled vehicles that vary in size, depth-rating and application (see Table 10.2). Small ROVs, with dimensions on the order of tens of centimeters and weighing a few kilograms, have limited sensor payloads and are mostly used in shallow marine systems (Wang and Clark 2006). By contrast, large ROVs, with dimensions of meters and weighing thousands of kilograms, are used in deep sea systems (Richmond and Rock 2006). ROVs accommodate a variety of common sensors, including cameras and acoustic systems, but also allow the manipulation of objects and collection of samples (although for many soft bodied organisms, such as jellyfish, this can be problematic). ROVs, and deep ROVs in particular, allow the use of a very complete set of payload sensors and their typical scientific use is to provide a remote acquisition of data. The real time data access from the surface (e.g., as video feeds) allows the scientists to directly control and readjust the sampling process. The main limitation of ROVs is the umbilical cable, which restricts the operational area and requires heavy surface vessel support.

ROVs are typically used to obtain images of the environment but, since the imaging systems usually available (video and stills cameras) are designed for large fields of view, they tend to be used more frequently for benthic systems. Although multiple optical systems have been developed targeting micro- and mesozooplankton, these systems are not typically used in ROV systems due to limitations in both the operating area and mission duration, as many of the plankton imaging systems are designed for continuous motion operation.

ROVs can also accommodate acoustic sensors, ranging from calibrated echosounders for biomass estimation, to sonar imaging systems, such as scanning or multibeam sonars due to their relatively large payload capability (depending on ROV system size) and power availability. Also, the large payload common in deep sea ROVs easily allows the integration of sensors such as laser optical counters, specific plankton- sensors, as eDNA samplers (Scholin et al. 2006, Ribeiro et al.

Table 10.2: Robotic platforms potentially available for jellyfish sensing, with information on their size, weight, speed and autonomy, along with maximum working depth, operational limitations, overall sensing payload capacity, biology sensing payload and examples of specific platforms.

Robotic platform	Size/weight	Speed (knots)	Autonomy	Depth (m)	Payload capacity	Biology sensing payload	Example systems
Surface Robots	2–4 m/100–200 kg	1–10	Variable (hrs to months)	0	Medium 40–100 kg	Acoustic Video Sampling	Waveglider ROAZ ASV
AUVs (small)	Diam 20–30 cm | 1.5–3 m/ 50–250 kg	3–4	10–24 hr	300 –1,000	Small 5–7 kg	Acoustic Video Conventional cameras	REMUS 100/600 Iver3
AUVs (large)	Diam 0.5–1 to 5 m/400–800 kg	3–4	Days 25–60 hr	4,500–6,000	Medium 30–50 kg	Acoustic Video Plankton imaging	Dorado Autosub Hugin
Gliders	Diam 20 cm | 1.5–2 m/ 70 kg	0.7	Months (4–18) 3,000–13,000 km	1,000	Small, custom made (few kg)	Acoustic	Slocum Seaglider Spray
AUV/ Glider Hybrid	Diam 0.5–0.7 m / 500 kg	1	Days/weeks 2,000 km	6,000	Medium 30 kg	Acoustic Video Plankton imaging	Tethys Autosub LR
ROV (Large)	2 × 2 × 2 m/ 2–5 ton	2	Hours (limited by operation conditions)	4,000–6,000	Large 250 kg	Acoustic Video Plankton imaging Sampling	Doc Ricketts Victor6000 Hyperdolphin Luso
Robotic landers	1.5 × 1.5 ×1.5 m/600–1,500 kg	2	Weeks-months	4,000	Large 100–200 kg	Acoustic Video Plankton imaging Sampling	Turtle
Rovers	1.5 × 1 ×1/ 400 kg	1	Months	6,000	Large 100 kg	Acoustic Video Plankton imaging Sampling	Bathybot Wally Benthic rover

2019), and large water sampling devices. Large ROV systems, such as the MBARI Doc Ricketts (USA), NMRF Aglantha (NO), IFREMER Victor6000 (FR), JAMSTEC Kayko, Dolphin3K (JP) or IPMA Luso (PT), have the capacity to include a large set of sensors, such as video, acoustic, water parameter probes, and dedicated micro- and macro-particle sensors (counters or imaging) largely benefiting field observations.

Vertical profiling is a common operation mission for ROVs in zooplankton studies. Deep sea ROVs have been used to acquire plankton data using imaging systems such as UVP (Picheral et al. 2010, Hosia et al. 2017), in vertical dives in multiple locations in the North Atlantic, in conjunction with net sampling. During these studies, ROVs collect video images and samples, and measure environmental variables such as depth, temperature, salinity, oxygen and chlorophyll *a* concentration. Counting and classification of zooplankton specimens is typically performed offline with human experts (Lindsay et al. 2015) from video/image recorded data. According to Hosia et al. (2017), net trawl samples and other net systems only caught ca. 21% of gelatinous macrozooplankton taxa sampled during the study, mainly due to the damage inflicted by the nets sampling systems on ctenophores. Further, Raskoff et al. (2010) also reported differences in the composition of Arctic gelatinous zooplankton between net and ROV collections, at the level of organism size, fragility and/or transparency but, due to the bias of the two methods, the authors considered that "When used together, they paint a more complete picture of the community as a whole". However, the vertical distribution of zooplankton (including gelatinous) in a Norwegian fjord was consistent using dedicated mesozooplankton (UVP) imaging or conventional imaging methods (for larger specimens) from samples obtained either with ROV mounted samplers (open or suction samplers) or with plankton nets (Youngbluth and Båmstedt 2001).

Hidaka-Umetsu and Lindsay (2018) observed jellyfish blooms in a deep-sea caldera, at 790 m deep offshore Tokyo, with the Dolphin 3 K ROV and the Shinkai 2000 HOV. Video information recorded with the ROV standard camera was compared with online classification performed onboard the HOV. More individuals and species were identified by the HOV, due to a higher resolution of the human eye (with direct observation), and the relatively low resolution of the ROV imaging system (standard video). Comparing the two systems, HOVs are considered more suitable tools for biodiversity surveys, and ROVs more suitable for impact assessment, due to possibility of repeating the missions. This difference currently is much more attenuated due to high video definition (full HD 1080p at 60fps or higher such as 4 K), and state of art imaging capabilities of ROV systems.

Traditional video footage taken from ROVs during multiple years, either during vertical dives or horizontal transects (Robison et al. 2010, Smith et al. 2019), provides information on spatial distribution and temporal variability, at both intra- and interannual scales, of gelatinous plankton (Smith et al. 2019). Integration of ROV dives imaging information with environmental data collected from other sources, such as instrumented Landers and bottom crawlers (Henthorn et al. 2010), provides a more coherent view of jellyfish dynamics and their interactions in marine ecosystems. During a three-year period study of zooplankton in the abyssal Northeast Atlantic, using ROV dives (video imaging), time lapse image stills (Canon 5Diii camera with 22.3Mpx resolution) from fixed Lander and current measurements from a bottom crawler. Smith et al. (2019) identified a large presence of jellyfish in the Benthic Boundary Layer (BBL). Thus, water column moving robotic systems are relevant, but the possibility of using benthic rovers (Henthorn et al. 2010, Thomsen et al. 2015) provides mobility at the bottom, with spatial coverage and long-term permanence.

One of the main characteristics of deep-sea research ROVs, and their imaging systems applied for jellyfish monitoring, is mainly the use of video recording from standard or high definition cameras. ROVs do not have the energy limitations of other systems, such as AUVs, and usually possess multiple sources of illumination of considerable power. For example, Dolphin3K (Hidaka-Umetsu and Lindsay 2018) includes six 400 W halogen lights plus 3 gas arc 250 W illuminators, allowing for a large illumination area. Currently, there are already multiple large sized (e.g., full frame) cameras, with resolutions of tens of Mpixels, at relatively low cost. These advances provide higher resolution and detail, thus facilitating species identification using imaging analysis. Manual identification of relevant organisms in a large dataset has high costs and an error prone process. For ROVs, a key concern is the need for processing large amounts of video data, which requires appropriate logging and data management tools. Video annotation tools, like VARS (Schlining and Stout 2006), facilitate human data processing and also pave the way for automation in data processing (Reisenbichler et al. 2016).

Large scale ROV systems, like the examples referred above, have a high operational cost thus limiting their widespread use for zooplankton studies. Although being able to reach the deep ocean, and to carry a much-diversified payload, their spatial and temporal coverage is limited by the tethered operation and global costs. Smaller scale ROV systems have much lower operational costs but are very limited in propulsion, range and available payload. These ROVs are not suitable for these types of studies, since they have a much tighter set of limitations (depth, range and payload) and do not bring advantages over other systems, such as moorings, deployments from ROVs or autonomous vehicles.

Autonomous Underwater Vehicles

The use of autonomous underwater vehicles (AUVs), in detriment of ROVs, has been increasing in all areas of ocean research and in jellyfish studies as well (Rife and Rock 2003, Ura et al. 2008).

The possibility of unsupervised missions, with a potentially larger spatial coverage, which glider vehicles are a good example of, increased energy availability and recent technological advancements, leading to sensor miniaturization, allow for a growing use of new sensors with AUVs and for their application on studies of jellyfish ecology.

Single or multiple AUV missions allow data acquisition of jellyfish distribution and behavior in large areas, together with data on ocean phenomena and processes, such as upwelling and oceanic fronts. AUVs can carry traditional oceanographic sensors (e.g., CTDs, fluorometers, chemical probes), but also dedicated jellyfish sensors, such as calibrated echosounders or plankton imaging systems. Depending on their size and type, AUV systems can also accommodate water sampling systems (Harvey et al. 2012), as in the MBARi AUV Dorado Gulper, or even dedicated eDNA sensors, such as the Environmental Sampling System (see Section 10.4.4; Scholin et al. 2006) installed onboard the long-range Tethys AUV (Hobson et al. 2012).

There are a large variety of AUV systems, from small shallow water systems (Moline et al. 2005, Cruz and Matos 2008) to large deep sea robots with standard propeller propulsion (Bowen et al. 2008, McPhail 2009), gliders (Webb et al. 2001), and a recent hybrid class (Hobson et al. 2012) that combines the characteristics of conventional propelled AUVs with the long range of gliders and with a dock station for recharge. For marine research, oceanography, biology and jellyfish studies, the AUVs usually are designed to cover large ocean areas, thus favoring forward motion with typical torpedo shaped designs.

The first class of small AUVs, of which REMUS (Allen et al. 1997) is a typical example, corresponds to easy operation and deployment robots (one- to two-man deployment from a rib boat; Table 10.2). These AUVs have the advantage of a much lower system and operation cost, when compared with deep sea AUVs, and can navigate with relative precision. The main limitations are their size (limiting scientific payload) and autonomy (energy availability) that restrict its use to half to one day operation. The longer-range robots have the advantage of reaching considerable depths (up to 6000 m deep) and being able to carry a more substantial sensor payload (McPhail 2009, Pedersen et al. 2010). Thus, long-range AUVs are very relevant alternatives to deep sea ROVs, in cases where the direct real-time, human control is not required (Bowen et al. 2008, Reisenbichler et al. 2016). These AUVs are much more cost efficient in terms of deployment and operation and can cross much larger areas of the ocean.

Gliders, either traditional underwater gliders (Powell and Ohman 2015) or surface bound ones (Manley and Willcox 2010), are the most common types of autonomous vehicles used for monitoring large ocean areas (Table 10.2). Underwater gliders move without propeller, "gliding" in the water column in a yo-yo vertical motion, using a variable buoyancy mechanism and displacement of their center of gravity (achieved through displacement of internal batteries position). This allows for a much reduced energy consumption, thus extending autonomy. However, glider displacement speed is relatively low (usually below 0.25 m s^{-1}), due to the small amount of buoyancy displacement, resulting in a relatively uncontrolled path subject to ambient conditions, with emphasis on water currents. Due to its motion control, albeit somewhat limited, gliders that change their buoyancy, and thus execute vertical profiling motions, are the next step for measurements made with drifter buoys, such as ARGO floats.

The glider limited energy availability also limits the possible sensor payload and curbs the use of high-power optical sensors or acoustic measurements. In terms of size, cost and deployment, underwater gliders have similar requirements as the class of small AUVs. Recently, there has been a growing interest in combining the navigation and trajectory control of standard propelled AUVs with the long endurance of glider vehicles. One example is the Tethys AUV (Hobson et al. 2012), developed by MBARI (U.S.A.), or the Autosub LR from NOC (U.K.) (Table 10.2). These vehicles extend endurance from days to weeks, thus combining long term permanence with deep sea access and relevant payload capability. By taking advantage of the innovations in embedded computation, sensor miniaturization and vehicle design, these long-range AUVs allow the use of sensors previously possible only for short ROV or AUV missions, with a much wider spatial and temporal coverage. One good example of increased capabilities of these AUVs is the incorporation of the third generation ESP, an automated sampling device and analytical molecular biology laboratory (see Section 10.4.4; Scholin et al. 2006), onboard the AUV Tethys in long term missions.

Environmental data (water parameters) and zooplankton acoustic measurements acquired with AUVs, namely gliders, have mostly been used indirectly for jellyfish studies. On board processing and mission control provide AUVs with new methods of ocean observation, including adaptive sampling (Zhang et al. 2016, Fossum et al. 2019), maximizing the collected information and allowing for tracking of relevant ocean phenomena such as upwelling or ocean fronts and jellyfish behavior.

Since AUVs and gliders are commonly equipped with Acoustic Doppler Current Profilers (ADCP), there is interest in using the acoustic backscatter for plankton biomass estimation. When comparing with calibrated split-beam or dual beam echosounders, the quality of the results is inferior since ADCPs usually operate at a single frequency (limiting size discrimination) and their calibration is much coarser. However, the availability of these sensors in many AUV missions makes them interesting for zooplankton studies. For example, Powell and Ohman (2012) evaluated zooplankton biomass in the upper 500 m in Santa Barbara basin using measurements of acoustic backscatter of the 750 kHz ADCP, installed in a Spray glider, in conjunction with conventional net sampling. According to this study, the results obtained using acoustic backscatter were proportional to those derived from net samples, collected in the vicinity of the ADCP. Later, Powel and Ohman (2015) evaluated zooplankton biomass in upwelling fronts, during a six-year glider study, combining ADCP with CTD and chlorophyll-*a* concentration, based on over 23000 vertical profiles. An 614 kHz ADCP installed on the REMUS600 AUV was also used to evaluate the distribution of zooplankton biomass in Arctic pelagic scattering layers, being able to discriminate single targets as small as 3 mm (Geoffroy et al. 2017).

Plankton counting or imaging sensors can be also used with AUVs. Although many of the configurations of these systems are tailored for towfish operation, whenever the size and configuration allow, these sensors can be integrated in AUVs or gliders. Larger AUVs, such as Hugin (Table 10.2), are able to integrate (Pedersen et al. 2010) systems like the Laseroptical Plankton Counter (LOPC) (Herman 2004). The system is installed in the exterior of the vehicle, so that the submersible motion does not affect the water flow in the sensor. A similar mounting option is taken by Ohman et al. (2019) with a Spray glider carrying a Zoocam plankton imaging system externally.

Large AUVs (Reisenbichler et al. 2016) also have the possibility of carrying enough energy to allow illumination of the environment and gathering of video or conventional imaging. This is relevant for the study and monitoring of larger jellyfish. Synchronous illumination and laser based structured light can also be used, in short range (up to 5 m, depending on water turbidity), for efficient imaging and 3 D morphology information (e.g., EVA AUV; Martins et al. 2018).

Autonomous surface vehicles, such as the Waveglider, move taking advantage of the waves, have the possibility of recharging the batteries with solar panels, and can thus have longer endurance (Table 10.2). Recently, calibrated split-beam echosounders operating at 70 kHz and 333 kHz were used during a six-month Waveglider mission off the Norwegian shelf (Pedersen et al. 2019). Another example, reported by Ludvigsen et al. (2018), is the use of a jetkayak ASV, equipped with a hyperspectral irradiance sensor and an AZFP acoustic multi-frequency biomass profiler, to study zooplankton vertical distribution and migration to evaluate the vertical zooplankton migration in the Norwegian shelf.

Robotic Autonomous Landers and Long-Term Robotic Permanence at Sea

Robotic deep-sea autonomous landers are a type of autonomous vehicle that combines characteristics of a traditional fixed lander with the mobility of an AUV (Table 10.2). These vehicles can stay for long periods in the sea bottom and autonomously move to another position (Silva et al. 2016). Equipped with a relatively large variable buoyancy system (VBS)- 10s of kg when comparing with

100 s of grams in traditional gliders- their payload is considerable (100 s of kg). The TURTLE class of robotic autonomous deep-sea Landers has onboard processing and navigation sensors as a deep sea AUV and combines that with a Lander form and large modular energy autonomy (Silva et al. 2016). The VBS allows for efficient ascent and diving, and the autonomous vehicle navigation provides reduced operation cost (when compared with large deep sea ROVs) in deployment and recovery, along with the capability of precise bottom positioning. These vehicles can have multiple roles in marine sciences and jellyfish studies, in particular. Deep sea landers can carry a very diverse set of sensors, such as those used in moorings, fixed observatories, or in deep sea ROVs and AUVs. Sensors comprise traditional plankton samplers, imaging systems, video, multiple acoustic sensors, and the conventional water parameter probes such as CTDs, dissolved oxygen, fluorescence, amongst others (Martins et al. 2016, Ribeiro et al. 2019). Robotic Landers can be used as an excellent carrier for marine biology integrating systems such as MarinEye (Fig. 10.1; Martins et al. 2016). The bottom positioning and capability of moving in the water column makes these systems particularly useful for studies of the BBL. Similarly, to benthic rovers, these systems can take multiple measurements and samples (water, organisms and bottom sediments), from multiple locations. The combination of permanence at sea bottom with sensor payload and mobility makes these systems valuable tools for sampling benthic jellyfish stages, as they are usually under-studied due to the limitations on access and observation at deep sea.

Robotic Landers can integrate vast and heterogeneous autonomous sampling networks (Curtin et al. 1993), combining cabled fixed observatories, AUVs, gliders and surface vehicles, not only in the science data gathering process but also providing navigation support for other robotic vehicles and extending their permanence at sea by providing recharging or data transfer points. AUV docking (Singh et al. 2001, Palomeras et al. 2018) for energy recharging and communications is a critical development step in order to have long term oceanic presence of robotic vehicles and to extend their efficacy and efficiency. Robotic Landers, due to their size, available energy and placement at the sea bottom, can be equipped with AUV docking systems (Hobson et al. 2007).

10.4.5 *Image Processing and Classification*

Identification of zooplankton through image data (either obtained from dedicated plankton imaging systems or conventional video surveys) is a relevant mission. In the case of plankton specific imaging sensors, the recording of all frames (either containing the organisms or not) during the mission is, in many cases, impossible due to local disk size limits. Thus, image processing algorithms are not only needed in an offline process but also in real time, in order to decide what to record or not (Geraldes et al. 2019).

Image processing can be performed using traditional techniques (Szeliski 2011) such as color correction, equalization and contrast enhancement and segmentation. Yet, machine learning techniques, including Support Vector Machines (Olson and Sosik 2007) and, more recently, convolutional neural networks (Dai et al. 2016, Geraldes et al. 2019), have also been applied to identify and classify planktonic organisms. One recent development in the area is the availability of dedicated hardware for neural network processing. Both dedicated computing systems, such as the Intel Movidius Neural Compute Stick (NCS) and the Google Edge Tensor Processing Unit (TPU), or power efficient graphic processors, like the NVIDIA Tegra line, allow real time classification, with modest energy requirements. For example, a neural network-based classification, with hardware implementation, is used for classification and identification of plankton in the MarinEye plankton imaging camera (Geraldes et al. 2019).

The coupling of real time processing with classification allows increased data processing automation, new modalities of observation with active targeting the organisms of interest, reduced cost of observation (both financial and human), and simplification of data integration, thus coping with the increasing volumes of data produced by a more comprehensive and extensive ocean observation.

10.5 Citizen Science as a Valuable Instrument for Jellyfish Research

Citizen science is the contribution of non-professionals to the development of scientific research (Silvertown 2009, Miller-Rushing et al. 2012). This contribution of amateurs can be of various degrees, (i) contributory citizen science, when collecting data for research designed by scientists, (ii) collaborative science, when citizens collect data for the research designed by scientists but also contribute to analyze the data, refine the project design and disseminate the findings of the research, and (iii) co-created science, when the research is designed by scientists together with citizens with at least some of the non-professionals participating in all phases of the development of the research project (Miller-Rushing et al. 2012). In fact, science started with non-professionals and the scientist as a paid profession begun only in the later part of the 19th century (Silvertown 2009). However, even then the citizen scientist continued to exist, especially in areas such as archaeology, astronomy and natural history, where observation and acquisition of data are valuable for long periods of time and space. One of the first citizen science projects that is still running is the Christmas Bird Count (https://www.audubon.org/conservation/science/christmas-bird-count#) that started in December 1900 in the U.S.A. Nowadays, thousands of research projects have a component of citizen science, with amateurs collecting, categorizing and analyzing data and helping to communicate the findings of research to a broader community. Most of these projects were enhanced by the internet and the possibility to submit data to online databases, allowing the participants to analyze and categorize photos and videos on various projects, like the Zooniverse (Bonney et al. 2014). However, data quality is a critical issue for any citizen science project (Bonney et al. 2009). Ensuring that the public can gather and submit accurate data depends on providing clear data collection protocols, simple and logical data forms, and being able to give support to participants on the proceedings for submitting information (Bonney et al. 2009). Results from citizen science projects are often published in peer-reviewed journals (Rapacciuolo et al. 2017).

Citizen science is extremely important in collecting large quantities of data across a multitude of locations over long periods of time, being especially well suited to contribute to long-term biodiversity surveys, species conservation and climate change impact assessments. Citizen science can be used to respond to crises and inform management actions (McKinley et al. 2017) and, therefore, is being used to monitor jellyfish populations around the world. One of the first and well known citizen science projects on jellyfish is the Jellywatch (https://jellywatch.org/), whose objective is to monitor the status of jellyfish populations around the world, using information on sightings submitted online by citizens worldwide. Another important citizen science project dedicated to jellyfish monitoring, aiming to future forecast jellyfish outbreaks in Mediterranean waters, is Med-Jellyrisk (http://jellyrisk.com/). This project was launched in 2013, and assembles several national citizen science programs in Mediterranean countries, such as Italy, Spain, Malta and Tunisia. For Portuguese waters (Portugal mainland, Madeira and Azores), the citizen science project GelAvista (http://gelavista.ipma.pt/) was created in February 2016. One of the most recent citizen science projects, The Great British Jellywatch Weekend (https://www.gbjellywatchweekend.com/), was created during 24th–26th August 2019 to allow citizens sending information on their sightings around the United Kingdom during this specific weekend.

In the case of the GelAvista, as in the other citizen science projects mentioned, the objectives are: (i) gather long-term data on gelatinous organisms; (ii) understand biodiversity, distribution, dynamics and the jellyfish role in the ecosystems; (iii) model and predict bloom occurrences; (iv) promote marine ecosystem's literacy; (v) inform the population regarding hazards and precautions; and (vi) engage the population in the science developed nationally. The citizens were requested to report information on their sightings when near the sea, at the beach and doing any nautical activity, by taking a photo or video of the gelatinous organism spotted with an object of known size (to be used as a scale), and reporting the date, hour and location where the photo was taken, together with information on the number of jellyfish individuals spotted. Citizens were also requested to report on the absence of gelatinous organisms. All the information is then gathered in a database where the identification of the specimens is confirmed, and a confidence level is attributed to each report. Additionally, some observers were requested, whenever possible and necessary, to provide samples of the organisms sighted for later molecular analysis, with the objective to create a barcode library of the jellyfish in Portugal. The protocols for sampling as well as for reporting the sightings are well disseminated, and training is given regularly upon request. Annual meetings, open to all citizens, are organized to provide information on the latest results, and information on the occurrences of jellyfish is regularly posted on the GelAvista social media, as well as disseminated through press releases.

The information being gathered by the citizen science projects dedicated to jellyfish allowed the detection of new species occurrences (Boero 2011, Deidun et al. 2017, Guerrero et al. 2019, Langeneck et al. 2019), monitoring invasive alien species (Mannino and Balistreri 2018), description of new species to science (Manko et al. 2017), description of jellyfish seasonal and geographical patterns (Kienberger and Prieto 2018), and monitoring jellyfish for assessing their impacts on human activities (Lucas et al. 2013) and marine food webs (Nordstrom et al. 2019). The use of citizen science information, in conjunction with data from plankton surveys, also enhances our knowledge on the main drivers of jellyfish transport at sea (Ferrer and Pastor 2017, Pires et al. 2018).

10.6 Concluding Remarks

New sensor technologies can bring relevant knowledge to marine biology research and thus to jellyfish studies. For example, advances in imaging sensors with relevant improvements in underwater imaging, coupled with new imaging technologies such as light field techniques can address some of the current limitations of underwater vision with potential impact on jellyfish studies. The use of robotic platforms and autonomous systems has reduced data gathering costs, simultaneously providing larger spatial and temporal coverage. Many of these systems, such as AUVs, still have a few uses in jellyfish studies mainly due to limitations on sensor payload. With the miniaturization of the sensors and the consequent need for lower energy requirements, it is expected that in the near future there will be a substantial increase in the use of AUVs and gliders in the monitoring of jellyfish. Robotic landers, a very recent class of vehicles, combining sensor payload, mobility and access to deep sea bottom, are very promising tool for jellyfish studies, especially the benthic stages. The widespread use of robotic platforms, increased sources of data, and high resolution imagery bring new challenges to big volume data processing. Recent tools and artificial intelligence algorithms are proving to be valuable, not only for traditional analysis but also for detecting new patterns and insights on zooplankton distributions. These algorithms also play a key role in citizen science initiatives by exploiting the ubiquity of personal computing devices with communication and imaging capabilities.

Cooperative international programs on jellyfish monitoring, integrating different observation and sampling techniques, are imperative. The involvement of common people in jellyfish observation, through citizen science programs where they choose to participate in the scientific process of understanding jellyfish community dynamics, is a promising approach that is being applied in jellyfish monitoring programs worldwide. Knowledge derived from consistent long term monitoring of gelatinous zooplankton would be of great relevance to predict jellyfish blooms and develop early warning systems to manage and regulate the most affected industries (e.g., tourism, aquaculture, fisheries, power plants) and, consequently, to minimize and mitigate the impacts of jellyfish bloom events.

Acknowledgements

AdS acknowledge financial support from projects PLANTROF (Programa Operacional Mar 2020 Portaria n° 118/2016), HiperSea (Portugal 2020, Aviso 03/SI/2017, Projecto 33889), EMODnet (EASME/EMFF/2016/1.3.1.2-Lot5/SI2.750022-Biology) and Jellyfisheries (PTDC/MAR-BIO/0440/2014) and the ICES Working Group on Zooplankton Ecology (WGZE). CM acknowledges financial support from EMSO-PT (PINFRA/22157/2016). All authors are grateful for the help provided by the editors throughout the process.

References

Alekseenko, E., M. Baklouti and F. Carlotti. 2019. Main factors favoring *Mnemiopsis leidyi* individual's growth and population outbreaks: A modelling approach. J. Mar. Sys. 196: 14–35.

Allen, B., R. Stokey, T. Austin, N. Forrester, R. Goldsborough, M. Purcell et al. 1997. REMUS: a small, low cost AUV; system description, field trials and performance results. pp. 994–1000. *In*: Oceans '97. MTS/IEEE Conference Proceedings, IEEE.

Attrill, M.J., J. Wright and M. Edwards. 2007. Climate-related increases in jellyfish frequency suggest a more gelatinous future for the North Sea. Limnol. Oceanogr. 52(1): 480–485.

Båmstedt, U., S. Kaartvedt and M. Youngbluth. 2003. An evaluation of acoustic and video methods to estimate the abundance and vertical distribution of jellyfish. J. Plankton Res. 25(11): 1307–1318.

Barnes, C.R., M.M.R. Best, F.R. Johnson, L. Pautet and B. Pirenne. 2013. Challenges, benefits, and opportunities in installing and operating cabled ocean observatories: Perspectives from NEPTUNE Canada. IEEE J. Oceanic Eng. 38(1): 144–157.

Bernard, P., F. Couasnon, J.-P. Soubiran and J.-F. Goujon. 1988. Surveillance estivale de la méduse *Pelagia noctiluca* (Cnidaria, Scyphozoa) sur les côtes Méditerranéennes Françaises. Annales de l'Institut océanographique, Paris 64: 115–125.

Boero, F. 2011. New species are welcome, but ... what about the old ones? Ital. J. Zool. 78(1): 1–2.

Boero, F. 2013. Review of jellyfish blooms in the Mediterranean and Black Sea. Studies and Reviews. General Fisheries Commission for the Mediterranean. No. 92. Rome, FAO, 53 p.

Bonney, R., C.B. Cooper, J. Dickinson, S. Kelling, T. Phillips, K.V. Rosenberg et al. 2009. Citizen science: A developing tool for expanding science knowledge and scientific literacy. BioScience 59(11): 977–984.

Bonney, R., J.L. Shirk, T.B. Phillips, A. Wiggins, H.L. Ballard, A.J. Miller-Rushing et al. 2014. Next steps for citizen science. Science 343(6178): 1436–1437.

Brotz, L., W.W. Cheung, K. Kleisner, E. Pakhomov and D. Pauly. 2012. Increasing jellyfish populations: trends in large marine ecosystems. Hydrobiol. 690: 3–20.

Brown, C.W., R.R. Hood, W. Long, J. Jacobs, D.L. Ramers, C. Wazniak et al. 2013. Ecological forecasting in Chesapeake Bay: Using a mechanistic–empirical modeling approach. J. Mar. Sys. 125: 113–125.

Cardona, L., I. Álvarez de Quevedo, A. Borrell and A. Aguilar. 2012. Massive consumption of gelatinous plankton by Mediterranean apex predators. PLoS ONE 7: e31329.

Carl, C., J. Guenther and L.M. Sunde. 2010. Larval release and attachment modes of the hydroid *Ectopleura larynx* on aquaculture nets in Norway. Aquac. Res. 42: 1056–1060.

Condon, R.H., D.K. Steinberg, P.A. del Giorgio, T.C. Bouvier, D.A. Bronk, W.M. Graham et al. 2011. Jellyfish blooms result in a major microbial respiratory sink of carbon in marine systems. PNAS 108(25): 10225–10230.

Condon, R.H., W.M. Graham, C.M. Duarte, K.A. Pitt, C.H. Lucas, S.H.D. Haddock et al. 2012. Questioning the rise of gelatinous zooplankton in the World's Oceans. BioScience 62(2): 160–169.

Condon, R.H., C.M. Duarte, K.A. Pitt, K.L. Robinson, C.H. Lucas, K.R. Sutherland et al. 2013. Recurrent jellyfish blooms are a consequence of global oscillations. Proc. Natl. Acad. Sci. USA 110(3): 1000–1005.

Conley, K.R., F. Lombard and K.R. Sutherland. 2018. Mammoth grazers on the ocean's minuteness: a review of selective feeding using mucous meshes. Proc. R. Soc. B 285: 20180056.

Cowen, R.K. and C.M. Guigand. 2008. *In situ* ichthyoplankton imaging system (ISIIS): system design and preliminary results. Limnol. Oceanogr-Meth. 6(2): 126–132.

Cruz, N.A. and A.C. Matos. 2008. The MARES AUV, a modular autonomous robot for environment sampling. In IEEE/MTS OCEANS 2008: 1–6.

Curtin, T.B., J.G. Bellingham, J. Catipovic and D. Webb. 1993. Autonomous oceanographic sampling networks. Oceanography 6: 86–94.

Dai, J., R. Wang, H. Zheng, G. Ji and X. Qiao. 2016. ZooplanktoNet: Deep convolutional network for zooplankton classification. In OCEANS 2016—Shanghai. Institute of Electrical and Electronics Engineers Inc.

Daly, M., M.R. Brugler, P. Cartwright, A.G. Collins, M.N. Dawson, D.G. Fautin et al. 2007. The phylum Cnidaria: A review of phylogenetic patterns and diversity 300 years after Linneus. Zootaxa 1668: 127–182.

Daskalov, G.M., A.N. Grishin, S. Rodionov and V. Mihneva. 2007. Trophic cascades triggered by overfishing reveal possible mechanisms of ecosystem regime shifts. Proc. Natl. Acad. Sci. USA 104: 10518–23.

Deibel, D. and B. Lowen. 2012. A review of the life cycles and life-history adaptations of pelagic tunicates to environmental conditions. ICES J. Mar. Sci. 69: 358–369.

Deidun, A., J. Sciberras, A. Sciberras, A. Gauci, P. Balistreri, A. Salvatore et al. 2017. The first record of the white-spotted Australian jellyfish *Phyllorhiza punctata* von Lendenfeld, 1884 from Maltese waters (western Mediterranean) and from the Ionian coast of Italy. BioInvasions Rec. 6: 119–124.

Deiner, K., H.M. Bik, E. Mächler, M. Seymour, A. Lacoursière-Roussel, F. Altermatt et al. 2017. Environmental DNA metabarcoding: Transforming how we survey animal and plant communities. Mol. Ecol. 26: 5872–5895.

Doucette, G.J., L.K. Medlin, P. McCarron and P. Hess. 2018. Detection and surveillance of harmful algal bloom species and toxins. pp. 39–113. *In*: S.E. Shumway, J.M. Burkholder and S.L. Morton [eds.]. Harmful Algal Blooms: A Compendium Desk Reference, John Wiley & Sons.

Doyle, T.K., G.C. Hays, C. Harrod and J.D.R. Houghton. 2014. Ecological and societal benefits of jellyfish. pp. 105–127. *In*: K.A. Pitt and C.H. Lucas [eds.]. Jellyfish Blooms. Dordrecht, the Netherlands: Springer.

Eriksen, C.C., T.J. Osse, R.D. Light, T. Wen, T.W. Lehman, P.L. Sabin et al. 2001. Seaglider: A long-range autonomous underwater vehicle for oceanographic research. IEEE J. Oceanic. Eng. 26(4): 424–436.

Ferrer, L. and A. Pastor. 2017. The Portuguese man-of-war: Gone with the wind. Reg. Stud. Mar. Sci. 14: 53–62.

Foote, K.G. and T.K. Stanton. 2000. Acoustical methods. pp. 223–258. *In*: ICES Zooplankton Methodology Manual, Elsevier.

Fossette, S., K. Katija, J.A. Goldbogen, S. Bograd, W. Patry, M.J. Howard et al. 2016. How to tag a jellyfish? A methodological review and guidelines to successful jellyfish tagging. J. Plankton Res. 38(5): 1347–1363.

Fossum, T.O., G.M. Fragoso, E.J. Davies, J.E. Ullgren, R. Mendes, G. Johnsen et al. 2019. Toward adaptive robotic sampling of phytoplankton in the coastal ocean. Sci. Robot. 4(27): eaav3041.

Gambill, M., L.F. Møller and M.A. Peck. 2015. Effects of temperature on the feeding and growth of the larvae of the invasive ctenophore *Mnemiopsis leidyi*. J. Plankt. Res. 37: 1001–1005.

Geoffroy, M., F.R. Cottier, J. Berge and M.E. Inall. 2017. AUV-based acoustic observations of the distribution and patchiness of pelagic scattering layers during midnight sun. ICES J. Mar. Sci. 74(9): 2342–2353.

Geraldes, P., J. Barbosa, A. Martins, A. Dias, C. Magalhães, S. Ramos et al. 2019. *In situ* real-time Zooplankton Detection and Classification (pp. 1–6). Institute of Electrical and Electronics Engineers (IEEE).

Ghermandi, A., B. Galil, J. Gowdy and P.A.L.D. Nunes. 2015. Jellyfish outbreak impacts on recreation in the Mediterranean Sea: welfare estimates from a socio-economic pilot survey in Israel. Ecosyst. Serv. 11: 140–147.

Glibert, P.M., G.C. Pitcher, S. Bernard and M. Li. 2018. Advancements and continuing challenges of emerging technologies and tools for detecting harmful algal blooms, their antecedent conditions and toxins, and applications in predictive models. pp. 339–357. *In*: P.M. Glibert, E. Berdalet, M.A. Burford, G.C. Pitcher and M. Zhou [eds.]. Global Ecology and Oceanography of Harmful Algal Blooms, Springer.

Goy, J., P. Morand and M. Etienne. 1989. Long-term fluctuations of *Pelagia noctiluca* (Cnidaria, Scyphomedusa) in the western Mediterranean Sea. Prediction by climatic variables. Deep-Sea Res. 36: 269–279.

Graham, W.M., D.L. Martin and J.C. Martin. 2003. *In situ* quantification and analysis of large jellyfish using a novel video profiler. Mar. Ecol. Prog. Ser. 254: 129–140.

Graham, W.M., S. Gelcich, K.L. Robinson, C.M. Duarte, L. Brotz, J.E. Purcell et al. 2014. Linking human well-being and jellyfish: ecosystem services, impacts and societal responses. Front. Ecol. Environ. 12: 515–523.

Guerrero, E., K. Kienberger, A. Villaescusa, J.-M. Gili, G. Navarro and L. Prieto. 2019. First record of beaching events for a calycophoran siphonophore: *Abylopsis tetragona* (Otto, 1823) at the Strait of Gibraltar. Mar. Biodivers. 49: 1587–1593.

Haddock, S.H.D. 2004. A golden age of gelata: past and future research on planktonic ctenophores and cnidarians. Hydrobiol. 530/531: 549–556.

Haddock, S.H.D. 2007. Comparative feeding behavior of planktonic ctenophores. Integr. Comp. Biol. 47: 847–853.

Haddock, S.H.D., M.A. Moline and J.F. Case. 2010. Bioluminescence in the Sea. Ann. Rev. Mar. Sci. 2: 443–493.

Hamner, W.M., L.P. Madin, A.L. Alldredge, R.W. Gilmer and P.P. Hamner. 1975. Underwater observations of gelatinous zooplankton: Sampling problems, feeding biology, and behavior. Limnol. Oceanogr. 20(6): 907–917.

Harvey, J.B.J., J.P. Ryan, R. Marin, C.M. Preston, N. Alvarado, C.A. Scholin et al. 2012. Robotic sampling, *in situ* monitoring and molecular detection of marine zooplankton. J. Exp. Mar. Biol. Ecol. 413: 60–70.

Hays, G.C., T.K. Doyle and J.D.R. Houghton. 2018. A paradigm shift in the trophic importance of jellyfish? Trends Ecol. Evol. 33: 874–884.

Henthorn, R.G., B.W. Hobson, P.R. McGill, A.D. Sherman and K.L. Smith. 2010. MARS Benthic Rover: *In-situ* rapid proto-testing on the monterey accelerated research system. In MTS/IEEE Seattle, OCEANS 2010.

Herman, A.W. 2004. The next generation of Optical Plankton Counter: the Laser-OPC. J. Plankton Res. 26(10): 1135–1145.

Hidaka-Umetsu, M. and D.J. Lindsay. 2018. Comparative ROV surveys reveal jellyfish blooming in a deep-sea caldera: The first report of *Earleria bruuni* from the Pacific Ocean. J. Mar. Biol. Assoc. UK 98: 2075–2085.

Hobson, B.W., R.S. McEwen, J. Erickson, T. Hoover, L. McBride, F. Shane et al. 2007. The development and ocean testing of an AUV docking station for a 21" AUV. In Oceans Conference Record (IEEE).

Hobson, B.W., J.G. Bellingham, B. Kieft, R. McEwen, M. Godin and Y. Zhang. 2012. Tethys-class long range AUVs—extending the endurance of propeller-driven cruising AUVs from days to weeks. IEEE/OES Autonomous Underwater Vehicles (AUV) (pp. 1–8).

Hosia, A., T. Falkenhaug, E.J. Baxter and F. Pagès. 2017. Abundance, distribution and diversity of gelatinous predators along the northern Mid-Atlantic Ridge: A comparison of different sampling methodologies. PLoS One 12(11): e0187491.

Houghton, J., T. Doyle, J. Davenport and G. Hays. 2006. Developing a simple, rapid method for identifying and monitoring jellyfish aggregations from the air. Mar. Ecol. Prog. Ser. 314: 159–170.

Ishii, H., T. Ohba and T. Kobayashi. 2008. Effects of low dissolved oxygen on planula settlement, polyp growth and asexual reproduction of *Aurelia aurita*. Plankton Benthos Res. 3(Suppl.): 107–13.

Jaspers, C., L.F. Møller and T. Kiørbe. 2015. Reproduction rates under variable food conditions and starvation in *Mnemiopsis leidyi*: significance for the invasion success of a ctenophore. J. Plankt. Res. 37: 1011–1018.

Kawahara, M., S.-I. Uye, K. Ohtsu and H. Iizumi. 2006. Unusual population explosion of the giant jellyfish *Nemopilema nomurai* (Scyphozoa: Rhizostomeae) in East Asian waters. Mar. Ecol. Prog. Ser. 307: 161–173.

Kienberger, K. and L. Prieto. 2018. The jellyfish *Rhizostoma luteum* (Quoy & Gaimard, 1827): not such a rare species after all. Mar. Biodiv. 48: 1455–1462.

Kim, D.-H., J.-N. Seo, W.-D. Yoon and Y.-S. Suh. 2012. Estimating the economic damage caused by jellyfish to fisheries in Korea. Fisheries Sci. 78: 1147–1152.

Kiørbe, T. 2013. Zooplankton body composition. Limnol. Oceanogr. 58(5): 1843–1850.

Langeneck, J., F. Crocetta, N. Doumpas, I. Giovos, S. Piraino and F. Boero. 2019. First record of the non-native jellyfish *Chrysaora* cf. *achlyos* (Cnidaria: Pelagiidae) in the Mediterranean Sea. BioInvasions Records 8.

Lebrato, M. and D.O.B. Jones. 2011. Expanding the oceanic carbon cycle: jellyfish biomass in the biological pump. Biochem. Evolution 33: 35–39.

Lebrato, M., M. Pahlow, J.R. Frost, M. Kuter, P.J. Mendes, J.-C. Molinero et al. 2019. Sinking of gelatinous zooplankton biomass increases deep carbon transfer efficiency globally. Global. Biogeochem. Cy. 33(12): 1764–1783.

Licandro, P. and M. Brunetta. 2017. Chordata: Thaliacea. pp. 584–598. *In*: C. Castellani and M. Edwards [eds.]. Marine Plankton, Oxford University Press.

Licandro, P. and D.J. Lindsay. 2017. Ctenophora. pp. 251–263. *In*: C. Castellani and M. Edwards [eds.]. Marine Plankton, Oxford University Press.

Lindsay, D., M. Umetsu, M. Grossmann, H. Miyake and H. Yamamoto. 2015. The gelatinous macroplankton community at the hatoma knoll hydrothermal vent. In Subseafloor Biosphere Linked to Hydrothermal Systems: TAIGA Concept (pp. 639–666). Springer Japan.

Lo, W.T., J.E. Purcell, J.J. Hung, H.M. Su and P.K. Hsu. 2008. Enhancement of jellyfish (*Aurelia aurita*) populations by extensive aquaculture rafts in a coastal lagoon in Taiwan. ICES J. Mar. Sci. 65: 453–61.

Lombard, F., E. Boss, A.M. Waite, M. Vogt, J. Uitz, L. Stemmann et al. 2019. Globally consistent quantitative observations of planktonic ecosystems. Front. Mar. Sci. 6: 196.

Lucas, C.H., S. Gelcich and S.-I. Uye. 2013. Living with Jellyfish: Management and adaptation strategies. pp. 129–150. *In*: K.A. Pitt and C.H. Lucas [eds.]. Jellyfish Blooms, Springer Science.

Ludvigsen, M., J. Berge, M. Geoffroy, J.H. Cohen, P.R. De La Torre, S.M. Nornes et al. 2018. Use of an autonomous surface vehicle reveals small-scale diel vertical migrations of zooplankton and susceptibility to light pollution under low solar irradiance. Sci. Adv. 4(1): eaap9887.

Lüskow, F. 2020. Importance of environmental monitoring: Long-term record of jellyfish (*Aurelia aurita*) biomass in a shallow semi-enclosed cove (Kertinge Nor, Denmark). Reg. Stud. Mar. Sci. 34: 100998.

Madinand, L.P. and J.R. Harbison. 2001. Gelatinous Zooplankton. pp. 1120–1130. *In*: Encyclopedia of Ocean Sciences, Vol. 2, Elsevier Ltd.

Makabe, R., T. Kurihara and S.-I. Uye. 2012. Spatio-temporal distribution and seasonal population dynamics of the jellyfish *Aurelia aurita* s.l. studied with Dual-frequency IDentification SONar (DIDSON). J. Plankton Res. 34(11): 936–950.

Manko, M.K., A. Weydmann and G.M. Mapstone. 2017. A shallow-living Benthic Rhodaliid siphonophore: Citizen science discovery from Papua New Guinea. Zootaxa 4324: 189–194.

Manley, J.E. 2008. Unmanned surface vehicles, 15 years of development. OCEANS 2008 IEEE, pp. 1–4.

Manley, J. and S. Willcox. 2010. The wave glider: A persistent platform for ocean science. OCEANS'10 IEEE Sydney, OCEANSSYD.

Mannino, A.M. and P. Balistreri. 2018. Citizen science: a successful tool for monitoring invasive alien species (IAS) in Marine Protected Areas. The case study of the Egadi Islands MPA (Tyrrhenian Sea, Italy). Biodiversity 19: 1–7.

Martins, A., A. Dias, E. Silva, H. Ferreira, I. Dias, J.M. Almeida et al. 2016. MarinEye—A tool for marine monitoring. In OCEANS 2016—SHANGHAI. IEEE.

Martins, A., J. Almeida, C. Almeida, B. Matias, S. Kapusniak and E. Silva. 2018. EVA a hybrid ROV/AUV for underwater mining operations support. In 2018 OCEANS—MTS/IEEE Kobe Techno-Oceans, OCEANS-Kobe 2018. Institute of Electrical and Electronics Engineers Inc.

Matos, A., J.-B. Ledoux, D. Domínguez-Pérez, D. Almeida and A. Antunes. 2021. Omics advances in the study of zooplankton: Big data for small drifting organisms. pp. 264–277. *In*: M.A. Teodósio and A.B. Barbosa [eds.]. Zooplankton Ecology. CRC Press.

McKinley, D.C., A.J. Miller-Rushing, H.L. Ballard, R. Bonney, H. Brown, S.C. Cook-Patton et al. 2017. Citizen science can improve conservation science, natural resource management, and environmental protection. Biol. Conserv. 208: 15–28.

McPhail, S. 2009. Autosub6000: A deep diving long range AUV. J. Bionic. Eng. 6(1): 55–62.

Miller-Rushing, A., R. Primack and R. Bonney. 2012. The history of public participation in ecological research. Front. Ecol. Environ. 10(6): 285–290.

Mills, C.E. 2001. Jellyfish blooms: Are populations increasing globally in response to changing ocean conditions? Hydrobiol. 451: 55–68.

Minamoto, T., M. Fukuda, K.R. Katsuhara, A. Fujiwara, S. Hidaka, S. Yamamoto et al. 2017. Environmental DNA reflects spatial and temporal jellyfish distribution. PLoS ONE 12: e0173073.

Moline, M.A., K. Benoit-Bird, D. O'Gorman and I.C. Robbins. 2015. Integration of scientific echo sounders with an adaptable autonomous vehicle to extend our understanding of animals from the surface to the bathypelagic. J. Atmos. Ocean. Tech. 32(11): 2173–2186.

Moriarty, P., K. Andrews, C. Harvey and M. Kawase. 2012. Vertical and horizontal movement patterns of scyphozoan jellyfish in a fjord-like estuary. Mar. Ecol. Prog. Ser. 455: 1–12.

Mouw, C.B., N.J. Hardman-Mountford, S. Alvain, A. Bracher, R.J.W. Brewin, A. Bricaud et al. 2017. A consumer's guide to satellite remote sensing of multiple phytoplankton groups in the Global Ocean. Frontiers in Marine 4: 41.

Munro, C., Z. Vue, R.R. Behringer and C.W. Dunn. 2019. Morphology and development of the Portuguese man of war, *Physalia physalis*. bioRxiv: 645465. D.

Nordstrom, B., M.C. James, K. Martin and B. Worm. 2019. Tracking jellyfish and leatherback sea turtle seasonality through citizen science observers. Mar. Ecol. Prog. Ser. 620: 15–32.

Ohman, M.D., R.E. Davis, J.T. Sherman, K.R. Grindley, B.M. Whitmore, C.F. Nickels et al. 2019. Zooglider: An autonomous vehicle for optical and acoustic sensing of zooplankton. Limnol. Oceanogr: Meth. 17(1): 69–86.

Olson, R.J. and H.M. Sosik. 2007. A submersible imaging-in-flow instrument to analyze nano-and microplankton: Imaging FlowCytobot. Limnol. Oceanogr. Meth. 5(6): 195–203.

Palmieri, M.G., A. Barausse, T. Luisetti and K. Turner. 2014. Jellyfish blooms in the northern Adriatic Sea: fishermen's perceptions and economic impacts on fisheries. Fish. Res. 155: 51–58.

Palomeras, N., G. Vallicrosa, A. Mallios, J. Bosch, E. Vidal, N. Hurtos et al. 2018. AUV homing and docking for remote operations. Ocean Eng. 154: 106–120.

Papathanassiou, E., P. Panayotidis and K. Anagnostaki. 1987. Notes on the biology and ecology of the jellyfish *Aurelia aurita* Lam. in Elefsis Bay (Saronikos Gulf, Greece). P. S. Z. N. I.: Mar. Ecol. 8: 49–58.

Pedersen, G., D. Peddie, S. Falk-Petersen, K. Dunlop, L. Camus, M. Daase et al. 2019. Autonomous surface vehicles for persistent acoustic monitoring of zooplankton in a highly productive shelf area. In IEEE/MTS Oceans 2019 Marseille (pp. 1–7). Institute of Electrical and Electronics Engineers (IEEE).

Pedersen, O.P., F. Gaardsted, P. Lågstad and K.S. Tande. 2010. On the use of the HUGIN 1000 HUS Autonomous Underwater Vehicle for high resolution zooplankton measurements. J. Oper. Oceanogr. 3(1): 17–25.

Picheral, M., L. Guidi, L. Stemmann, D.M. Karl, G. Iddaoud and G. Gorsky. 2010. The Underwater Vision Profiler 5: An advanced instrument for high spatial resolution studies of particle size spectra and zooplankton. Limnol. Oceanogr. Meth. 8(9): 462–473.

Pires, R.F.T., N. Cordeiro, J. Dubert, A. Marraccini, P. Relvas and A. Dos Santos. 2018 Untangling *Velella velella* (Cnidaria: Anthoathecatae) transport: a citizen science and oceanographic approach. Mar. Ecol. Prog. Ser. 591: 241–251.

Pitt, K.A., C.H. Lucas, R.H. Condon, C.M. Duarte and B. Stewart-Koster. 2018. Claims that anthropogenic stressors facilitate jellyfish blooms have been amplified beyond the available evidence: A systematic review. Front. Mar. Sci. 5: 451.

Powell, J.R. and M.D. Ohman. 2012. Use of glider-class acoustic Doppler profilers for estimating zooplankton biomass. J. Plankton Res. 34(6): 563–568.

Powell, J.R. and M.D. Ohman. 2015. Covariability of zooplankton gradients with glider-detected density fronts in the Southern California Current System. Deep-Sea Res. Pt II: Topical Studies in Oceanography 112: 79–90.

Purcell, J.E. and M.N. Arai. 2001. Interactions of pelagic cnidarians and ctenophores with fishes: a review. Hydrobiol. 451(Dev. Hydrobiol. 155): 27–44.

Purcell, J.E. 2009. Extension of methods for jellyfish and ctenophore trophic ecology to large-scale research. Hydrobiol. 616: 23–50.

Purcell, J.E., E.J. Baxter and V.L. Fuentes. 2013. Jellyfish as products and problems of aquaculture. pp. 404–430. *In*: G. Allan and G. Burnell [eds.]. Advances in Aquaculture Hatchery Technology. Woodhead Publishing, Cambridge. Woodhead Publishing Series in Food Science, Technology and Nutrition No. 242.

Rapacciuolo, G., J.E. Ball-Damerow, A.R. Zeilinger and V.H. Resh. 2017. Detecting long-term occupancy changes in Californian odonates from natural history and citizen science records. Biodivers. Conserv. 26(12): 2933–2949.

Raskoff, K.A., R.R. Hopcroft, K.N. Kosobokova, J.E. Purcell and M. Youngbluth. 2010. Jellies under ice: ROV observations from the Arctic 2005 hidden ocean expedition. Deep-Sea Res. Pt II: Topical Studies in Oceanography 57(1-2): 111–126.

Rees, J.T. and L.A. Gershwin. 2000. Non-indigenous hydromedusae in California's upper San Francisco Estuary: life cycles, distribution, and potential environmental impacts. Sci. Mar. 64(Suppl. 1): 73–86.

Reisenbichler, K.R., M.R. Chaffey, F. Cazenave, R.S. McEwen, R.G. Henthorn, R.E Sherlock et al. 2016. Automating MBARI's midwater time-series video surveys: The transition from ROV to AUV. In OCEANS 2016 MTS/IEEE Monterey, OCE 2016. Institute of Electrical and Electronics Engineers Inc.

Rey, A., J. Corell and N. Rodríguez-Ezpeleta. 2021. Metabarcoding to study zooplankton diversity. pp. 252–263. *In*: M.A. Teodósio and A.B. Barbosa [eds.]. Zooplankton Ecology. CRC Press.

Ribeiro, H., A. Martins, M. Gonçalves, M. Guedes, M.P. Tomasino, N. Dias et al. 2019. Development of an autonomous biosampler to capture in situ aquatic microbiomes. PLoS ONE 14(5): e0216882.

Richardson, A.J., A. Bakun, G.C. Hays and M.J. Gibbons. 2009. The jellyfish joyride: causes, consequences and management responses to a more gelatinous future. Trends in Ecology and Evolution 24(6): 312–322.

Rife, J. and S.M. Rock. 2003. Segmentation methods for visual tracking of deep-ocean jellyfish using a conventional camera. IEEE Journal of Oceanic Engineering 28(4): 595–608.

Robison, B.H., R.E. Sherlock and K.R. Reisenbichler. 2010. The bathypelagic community of Monterey Canyon. Deep-Sea Res. Pt II: Topical Studies in Oceanography 57(16): 1551–1556.

Rodger, H.D., L. Henry and S.O. Mitchell. 2011. Non-infectious gill disorders of marine salmonid fish. Rev. Fish Biol. Fisheries 21: 423–440.

Samson, S., T. Hopkins, A. Remsen, L. Langebrake, T. Sutton and J. Patten. 2001. A system for high-resolution zooplankton imaging. IEEE J. Oceanic Eng. 26(4): 671–676.

Schlining, B.M. and N.J. Stout. 2006. MBARI's Video Annotation and Reference System. In OCEANS 2006. IEEE Computer Society.

Scholin, C., S. Jensen, B. Roman, E. Massion, R. Marin, C. Preston et al. 2006. The Environmental Sample Processor (ESP)—An autonomous robotic device for detecting microorganisms remotely using molecular probe technology. In OCEANS 2006.

Scholin, C., G. Doucette, S. Jensen, B. Roman, D. Pargett, R. Marin III et al. 2009. Remote detection of marine microbes, small invertebrates, harmful algae, and biotoxins using the Environmental Sample Processor (ESP). Oceanography 22: 158–167.

Scholin, C.A., J. Birch, S. Jensen, R. Marin III, E. Massion, D. Pargett et al. 2017. The quest to develop ecogenomic sensors: A 25-Year history of the Environmental Sample Processor (ESP) as a case study. Oceanography 30(4): 100–113.

Silva, E., A. Martins, J.M. Almeida, H. Ferreira, A. Valente, M. Camilo et al. 2016. TURTLE—A robotic autonomous deep sea lander. In OCEANS 2016 MTS/IEEE Monterey.

Silvertown, J. 2009. A new dawn for citizen science. Trends in Ecology & Evolution 24(9): 467–471.

Singh, H., J.G. Bellingham, F. Hover, S. Lemer, B.A. Moran, K. von der Heydt et al. 2001. Docking for an autonomous ocean sampling network. IEEE J. Oceanic Eng. 26(4): 498–514.

Smith, K.L., C.L. Huffard, P.R. McGill, A.D. Sherman, T.P. Connolly, S. Von Thun et al. 2019. Gelatinous zooplankton abundance and benthic boundary layer currents in the abyssal Northeast Pacific: A 3-yr time series study. Deep-Sea Res. Pt II: Topical Studies in Oceanography.

Stone, J.P. 2016. Population Dynamics of Gelatinous Zooplankton in the Chesapeake Bay and Sargasso Sea, and Effects on Carbon Export. Master Thesis, Virginia Institute of Marine Science.

Sun, H., P.W. Benzie, N. Burns, D.C. Hendry, M.A. Player and J. Watson. 2008. Underwater digital holography for studies of marine plankton. Philos. A. Math. Phys. Eng. Sci. 366(1871): 1789–1806.

Sweetman, A.K. and A. Chapman. 2011. First observations of jelly-falls at the seafloor in a deep-sea fjord. Deep-Sea Res. I 58: 1206–11.

Sweetman, A.K., C.R. Smith, T. Dale and D.O.B. Jones. 2014. Rapid scavenging of jellyfish carcasses reveals the importance of gelatinous material to deep-sea food webs. Proc. R. Soc. B 281: 20142210.

Szeliski, R. 2011. Computer Vision: Algorithms and Applications. Springer.

Thomsen, L., A. Purser, J. Schwendner, A. Duda, S. Flogen, T. Kwasnitschka et al. 2015. Temporal and spatial benthic data collection via mobile robots: Present and future applications. In OCEANS 2015—Genova (pp. 1–5). IEEE.

Titelman, J., L. Riemann, T. Sørnes, T. Nilsen, P. Griekspoor and U. Båmstedt. 2006. Turnover of dead jellyfish: stimulation and retardation of microbial activity. Mar. Ecol. Prog. Ser. 325: 43–58.

Torria, L., F. Tuccilloa, S. Bonelliab, S. Pirainoc and A. Leoned. 2020. The attitudes of Italian consumers towards jellyfish as novel food. Food Qual. Prefer. 79: 103782.

Trembani, A.C., C. Cary, V. Schmidt, D. Clark, T. Crees and E. Jackson. 2012. Modular Autonomous Biosampler (MAB): A prototype system for distinct biological size-class sampling and preservation. Oceans, Hampton Roads, VA, 2012, pp. 1–6.

Ura, Tamaki, Y. Yamada, B. Thorton, Y. Nose and T. Sakamaki. 2008. 2A1-A12 Development of the AUV'T-pod'for Catching of Deep Sea Jellyfish. In The Proceedings of JSME annual Conference on Robotics and Mechatronics (Robomec) 2008, pp. _2A1-A12_1. The Japan Society of Mechanical Engineers.

Uye, S.-I., N. Fujii and H. Takeoka. 2003. Unusual aggregations of the scyphomedusa *Aurelia aurita* in coastal waters along western Shikoku, Japan. Plankt. Biol. Ecol. 50: 17–21.

Uye, S. and Y. Ueta. 2004. Recent increase of jellyfish populations and their nuisance to fisheries in the Inland Sea of Japan. Bull. Jpn. Soc. Fish Oceanogr. 68: 9–19 (in Japanese with English abstract).

Uye, S. 2008. Blooms of the giant jellyfish *Nemopilema nomurai*: a threat to the fisheries sustainability of the East Asian Marginal Seas. Plankton and Benthos Research 3(Suppl.): 125–131.

Uye, S.I. 2011. Human forcing of the copepod–fish–jellyfish triangular trophic relationship. Hydrobiol. 666: 71–83.

Viana, N., P. Guedes, D. Machado, D. Pedrosa, A. Dias, J.M. Almeida et al. 2018. Underwater Acoustic Signal Detection and Identification Study for Acoustic Tracking Applications. In OCEANS 2018 MTS/IEEE Charleston. Institute of Electrical and Electronics Engineers Inc.

Wang, W. and C.M. Clark. 2006. Modeling and Simulation of the VideoRay Pro III Underwater Vehicle. pp. 1–7. *In*: IEEE/MTS OCEANS 2006—Asia Pacific.

Webb, D.C., P.J. Simonetti and C.P. Jones. 2001. SLOCUM: an underwater glider propelled by environmental energy. IEEE Journal of Oceanic Engineering 26(4): 447–452.

Wiebe, P.H. and M.C. Benfield. 2003. From the Hensen net toward four-dimensional biological oceanography. Progress in Oceanography. Elsevier Ltd.

Yamahara, K.M., C.M. Preston, J. Birch, K. Waltz, R. Marin III, S. Jensen et al. 2019. *In situ* autonomous acquisition and preservation of marine environmental DNA using an autonomous underwater vehicle. Front. Mar. Sci. 6: 373.

Youngbluth, M.J. and U. Båmstedt. 2001. Distribution, abundance, behavior and metabolism of *Periphylla periphylla*, a mesopelagic coronate medusa in a Norwegian fjord. pp. 321–333. *In*: Jellyfish Blooms: Ecological and Societal Importance, Springer Netherlands.

Youngbluth, M., T. Sørnes, A. Hosia and L. Stemmann. 2008. Vertical distribution and relative abundance of gelatinous zooplankton, *in situ* observations near the Mid-Atlantic Ridge. Deep-Sea Res. Pt II: Topical Studies in Oceanography 55(1-2): 119–125.

Zaitsev, Y. and V. Mamaev. 1997. Marine biological diversity in the Black Sea: a study of change and decline. United Nations Publications, New York: 208 pp.

Zhang, Y., J.G. Bellingham, J.P. Ryan, B. Kieft and M.J. Stanway. 2016. Autonomous four-dimensional mapping and tracking of a coastal upwelling front by an autonomous underwater vehicle. J. Field Robot 33(1): 67–81.

CHAPTER 11

Metabarcoding to Study Zooplankton Diversity

Anaïs Rey, Jon Corell and *Naiara Rodriguez-Ezpeleta**

11.1 Introduction

Over the past decades, significant advances in molecular technologies have shifted the way ecological studies are conducted (Bourlat et al. 2013). In particular, the advent of high throughout sequencing (HTS) technologies has enabled the exploration of biodiversity at an unprecedented scale as these technologies allow generating massive amounts of genetic data from hundreds of environmental samples simultaneously (Margulies et al. 2005). HTS has been mostly applied to the study of prokaryotic diversity (Sogin et al. 2006), but has been later extensively adopted for studying fungi (Buée et al. 2009) and protists (Amaral-Zettler et al. 2009, De Vargas et al. 2015). Applications to study metazoan diversity using HTS are scarcer, but recent developments of alternative sampling methods, laboratory procedures and data analysis pipelines are contributing to increase the number of studies relying on HTS for ecological studies of metazoan taxa, including freshwater and marine zooplankton (Bucklin et al. 2016).

Traditional studies on zooplankton diversity and ecology have relied on visual taxonomic identification of samples, which is tedious, time consuming and complex due to the wide phylogenetic range of the group, presence of cryptic and sibling species and lack of diagnostic characters for early life stages (Bucklin et al. 2016). For nano- and microzooplankton (unicellular eukaryotes), precise identification is problematic requiring special methods such as silver staining or electron microscopy, which are time consuming and expensive for routine application. Additionally, species are often inadequately described in the literature (McManus and Katz 2009). Over the last years, different omics techniques have been used for the study of zooplankton (see example for ciliates in Santoferrara and McManus 2021, this book, Chapter 5). Among them, DNA barcoding and metabarcoding (Fig. 11.1) have become the most promising ones to improve taxonomic identification of zooplankton samples.

DNA barcoding consists of sequencing a standardized short DNA fragment (the barcode) that is unique to each species and therefore useful for taxonomic assignment (Hebert et al. 2003). The process is applied to a sample composed of tissue material of single species and relies on two short

AZTI, Marine Research Division, Sukarrieta 48395, Bizkaia, Spain.
Emails: anais.rey47@hotmail.fr; jonko_hms@hotmail.com
* Corresponding author: nrodriguez@azti.es

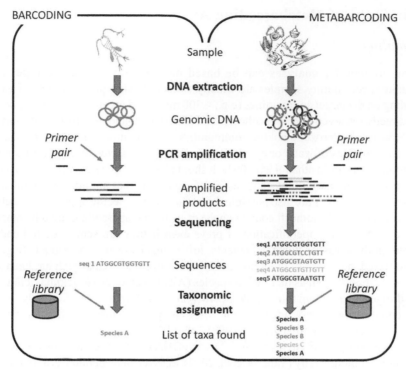

Figure 11.1: Main common steps involved in the barcoding (left) and metabarcoding (right) processes (bold) and the resulting product of each one. Barcoding and metabarcoding starting material is composed by a single or multiple species, respectively. In both cases, the genomic DNA resulting from the extraction is amplified using a primer pair that targets the group(s) of interest. Barcoding products are sequenced using the Sanger technology and metabarcoding products are sequenced using high throughout sequencing technologies. Resulting sequences are, in both cases, compared against a reference database.

DNA sequences flanking the barcode (primers) that allow amplification of the barcode during the Polymerase Chain Reaction (PCR). The resulting sequences are compared to a reference database that contains the correspondence between the barcodes and the species (Hebert et al. 2003, Bourlat et al. 2013). DNA barcoding has been extensively used to study plankton (Webb et al. 2006, Bucklin et al. 2007), revealing its usefulness for accurate taxonomic identification, including cryptic species and larvae.

DNA metabarcoding consists of applying the DNA barcoding procedure to a sample composed of individuals belonging to several species (Taberlet et al. 2012), to simultaneously detect the taxa present in the sample. While the products from DNA barcoding are generally sequenced using the Sanger method (Sanger et al. 1977), the mixed products obtained from metabarcoding can be sequenced thanks to HTS technologies, which allow the simultaneous and individual sequencing of millions of DNA molecules. Provided its cost-effectiveness, allowing the taxonomic identification of hundreds of samples composed of thousands of species simultaneously, DNA metabarcoding is the method of choice for performing taxonomic inventories of zooplankton samples. Yet, the accuracy of this process relies on several factors such as choice of sample type, selection of appropriate barcode and amplification primers, completeness and correctness of reference database, application of appropriate raw data pre-processing, read clustering and taxonomic assignment procedures.

In this chapter, we review each of the steps of a metabarcoding study, examine the aspects to consider during the choices to be made at each step of the process, and discuss the factors that need to be taken into account when drawing ecological conclusions from metabarcoding data. Finally, we provide recommendation for future studies considering the challenges presented.

11.2 Main Steps of Metabarcoding Analyses

11.2.1 Sampling

Zooplankton community analyses can be based on a variety of sample types (Fig. 11.2). Mesozooplankton community samples are usually collected using plankton nets of variable mesh sizes depending on the target communities (e.g., > 200 mm sized organisms). In those cases, samples are usually directly retrieved from the collector and stored in ethanol 96–100% until further use. For nano- (2–20 mm) and microzooplankton communities (20–200 mm), samples are usually collected by filtering large volumes of water or by passing the material collected by the net through filters that are in turn stored frozen or in ethanol 96–100%. It should be noted that both micro- and mesoplankton samples include meroplankton (eggs and larvae of organisms that are not planktonic at their adult phase). Other sample sources are also possible depending on the specific application, such as stomach contents for diet analyses. Stomach content analyses using metabarcoding are advantageous as this approach allows taxonomic identification of preys even if they are semi-digested and/or visually imperceptible. In these cases, stomach contents, full stomachs or full individuals (e.g., larvae) are collected and stored in ethanol or other preservative, or frozen. This approach has been successfully applied to assess zooplankton as diet of fish species (Albaina et al. 2016, Jakubavičiūtė et al. 2017). For bulk and digested zooplankton samples, preservation in formalin, traditionally used for visual taxonomic identification, is not adequate for DNA analysis, as formalin fixation produces DNA degradation and alteration (Williams et al. 1999).

Recently, the analysis of environmental DNA (eDNA) has surged as a novel, game-changing approach to monitor biodiversity (Taberlet et al. 2012). eDNA can be defined as a source of macrobial DNA (from mucus, skin cells, gametes) recovered from an environmental sample such as water, sediment or air without the need of sampling the organisms themselves (Deiner et al. 2017). Besides being non-intrusive and non-destructive, eDNA is also advantageous because it can reduce the cost and time associated with sampling in comparison to traditional collection methods as it only requires sampling water (Borrell et al. 2017). The application of water eDNA metabarcoding for producing inventories of zooplankton diversity at large scale has already started (Lim et al. 2016, Gunther et al. 2018), although most eDNA studies targeting zooplankton communities to date have aimed at monitoring biodiversity and detect new non-indigenous species in coastal waters and maritime traffic environments such as ports and ballast water from commercial vessels (Koziol et al. 2019,

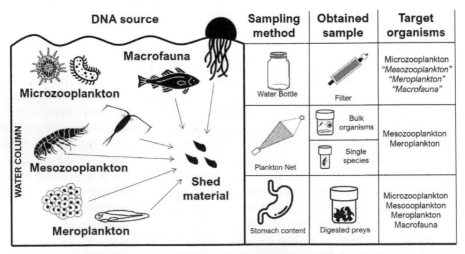

Figure 11.2: Different methods for sampling zooplankton for metabarcoding analyses associated with their DNA source, type of sample obtained and target organisms. Italics indicate that the corresponding target organisms are sampled in form of their environmental DNA.

Rey et al. 2019, Walsh et al. 2019). Sampling eDNA for zooplankton diversity analysis requires collecting a variable number of liters of water (usually 2 to 5) and filtering through filters of < 1 mm pore size. One particularity of using eDNA for zooplankton already explored is the impact of pre-filtration (e.g., 3 μm pre-filter) to remove intact large organisms that may have been caught in the water samples, although this approach has also been associated to a reduced number of target taxa recovered (Djurhuus et al. 2018).

11.2.2 DNA Extraction

Depending on the starting material, DNA extraction procedures will differ. For stomach content and bulk zooplankton samples stored in ethanol, a prior centrifugation to remove ethanol might be required. DNA from both organismal samples and filters (for eDNA and microzooplankton samples) can be extracted using commercial kits or in-house procedures, such as the SDS-chloroform method. In any case, the method of choice should account for the heterogeneity of organismal types (gelatinous organisms, exoskeletal structures) and preservation methods used, yield enough good integrity and purity DNA, and finally, remove compounds that could possibly inhibit subsequent reactions such as PCR (Wilson 1997).

One of the greatest challenges associated with metabarcoding analyses, particularly those based on eDNA samples, is to minimize contaminations (e.g., cross-contamination between samples or contaminations from external sources). For that aim, extreme caution in decontaminating every work surface and material used during the sampling and laboratory analysis is required and the use of negative controls is strongly advised (Deiner et al. 2017). Also, methods of capturing eDNA and of extracting genetic material from the water samples have been extensively explored, with the use of enclosed filters for eDNA capture to avoid contaminations and commercial kits for extraction to ensure reproducibility being increasingly applied (Spens et al. 2017, Djurhuus et al. 2018).

11.2.3 Amplification and Sequencing

The fragments corresponding to the barcode region of choice are usually amplified through Polymerase Chain Reaction (PCR) using a primer pair selected for that aim. The resulting amplicon sequences need to be modified so that they include (i) a unique sequence tag specific for each sample and (ii) the adapters required for sequencing. The amplification of the barcode and addition of index and adapters can be done in a single step or in two steps. Usually, a two-step amplification is performed, the first involving more PCR cycles (which are usually from 10–30, but it is specific to the primer pair selected) and the second, less PCR cycles (8–10). The total number of PCR cycles should be as low as possible to minimize errors and chimeric sequences. Since each sample will have a unique sequence tag that identifies it, samples can be pooled in groups of 96–384 and sequenced together (Bourlat et al. 2016).

Because the PCR step can introduce bias that can skew species abundance representation and over and under detect specific taxa (Piñol et al. 2014), alternative PCR-free approaches are being introduced. Examples of such approaches are the mitogenome skimming (Crampton-Platt et al. 2016) and the extraction of miTags (Logares et al. 2013), which consist of extracting informative mitochondrial or nuclear markers, respectively, from whole genome sequencing. The drawback of these methods is that a large proportion of the sequences are discarded and that there can also exist biases in the extraction of informative sequences from whole metagenomes.

11.2.4 Data Analysis

Sequencing reads produced by the sequencer need to be preprocessed, clustered and taxonomically assigned. Due to the large size of the sequence files and high computer processing power and large memory requirements, these steps are usually done in powerful computers in a UNIX environment. Several programs and tool suites exist to analyze metabarcoding data such as mothur (Schloss et al. 2009), QIIME (Caporaso et al. 2010) and obitools (Boyer et al. 2016), among others. In general, the steps to be performed to go from raw sequencing reads to taxonomic data are: (i) raw sequence quality checking, (ii) discarding low quality sequences or fragments, (iii) merging paired-end reads, (iv) removing primer sequences, (v) removing sequencing errors and chimeras, (vi) clustering reads into Operational Taxonomic Units (OTU) and (vii) taxonomic assignment of OTUs. The order in which some of these steps are done and the specific approaches and parameter values used are dependent on each case study (see next section for considerations on some of them).

11.3　Considerations for Metabarcoding Based Zooplankton Diversity Studies

11.3.1　*Barcode of Choice and Reference Databases*

The choice of the barcode and primers used to amplify it are crucial for any metabarcoding study. The barcode of choice should be variable enough so that species can be distinguished and conserved enough so that primers that allow the simultaneous amplification of a wide range of taxonomic groups can be developed. Additionally, the barcode should be short enough so that it can be amplified from semi-degraded samples and sequenced in HTS platforms, which generate short reads, and long enough so that enough differences among species can be accumulated. Finally, the barcode selected should have a large representation of sequences in reference databases so that it allows confident and accurate taxonomic assignment.

One of the most used molecular markers for metabarcoding metazoan communities are the hypervariable gene regions (particularly the V9 region) of the small subunit ribosomal RNA gene (18S rRNA) (Pearman et al. 2014, Albaina et al. 2016). This marker has the advantage of having a more extensive reference database, but the disadvantage of being too conserved to discriminate further than the genus level (Machida and Knowlton 2012, Tang et al. 2012). Alternatively, the mitochondrial Cytochrome c Oxidase Unit I (COI) (Hebert et al. 2003, Bucklin et al. 2010) has a faster evolutionary rate and allows discriminating to species level (Machida and Knowlton 2012, Hirai et al. 2013). This marker has been extensively used for metabarcoding of zooplankton samples, particularly since the development of a primer pair that amplifies a shorter region than the traditional barcode (Leray et al. 2013), which allows full length sequencing in HTS platforms.

eDNA samples require an even more careful selection of barcode and primer selection as these samples include DNA from a huge variety of organisms, from bacteria to large animals. Numerous eDNA studies have highlighted that only a low proportion of reads obtained from eDNA samples are from macroorganisms when using "universal" primers designed for metazoans (Leray et al. 2013) and that organisms such as bacteria, fungi and algae are also present in the sample (Borrell et al. 2017, Rey et al. 2019, Stat et al. 2017). Consequently, employing a multiple marker approach (different barcodes or different primers for the same barcode) and/or developing group specific primers to decrease amplification bias towards non-target taxa will be one way forward to optimize the use of eDNA samples for zooplankton diversity surveys. In that sense, there is an on-going development of zooplankton taxa group specific primers (e.g., for decapods see Komai et al. 2019; for copepods see Clarke et al. 2017 or Berry et al. 2015; for crustacean see Jeunen et al. 2019 or Berry et al. 2017; for freshwater invertebrates see Klymus et al. 2017), which prompt their further use in eDNA studies.

Reliable taxonomic assignments of metabarcoding reads depend on the availability of a curated and well-populated reference database. Public genetic databases such as GenBank (http://www.ncbi.nlm.nih.gov/nuccore) contain an increasing amount of sequence data available for taxonomic identification of sequence data. However, although inferences based on Genbank data are quite right (Leray et al. 2019), 4 specialized databases for different markers are available that provide aligned sequences and more curated information such as SILVA (Quast et al. 2013) or PR2 (Guillou et al. 2012) for 18S and BOLD (Ratnasingham and Hebert 2007) for COI.

11.3.2 Raw Data Preprocessing and OTU Clustering

HTS platforms produce millions of reads that need to be pre-processed before taxonomic and biodiversity analyses. This pre-processing consists on removing sequences that do not represent reliable information such as low quality reads (those sequences whose nucleotide string inferred from the sequencer is not reliable), singletons (those sequences that appear only once and which are most likely products of PCR or sequencing errors), chimeric sequences (produced during the amplification step and increasing in number as more PCR cycles are used), and non-target sequences (those sequences that do not cover the barcode of choice) and on merging paired-end reads if required. There are standardized protocols to perform these steps (Aylagas and Rodriguez-Ezpeleta 2016), although, in most cases, the steps to perform, the order at which they should be performed and the specific parameters to be used are case-dependent. For example, short or long barcodes will require different data pre-processing; for example, in the first case, forward and reverse reads overlap, whereas in the second, they do not (Fig. 11.3).

After pre-processing, filter passing merged contigs are usually clustered into OTUs, which represent an arbitrary taxonomic level that should ideally be equivalent to ecological species used for estimating biodiversity (Cristescu 2014). Strategies for OTU clustering are diverse (e.g., vsearch (Rognes et al. 2016); swarm (Mahé et al. 2014)) and threshold used for clustering are variable (e.g., 97–99% identity (Jackson et al. 2016) for vsearch or d = 1–13 for swarm (Rey et al. 2019, Siegenthaler

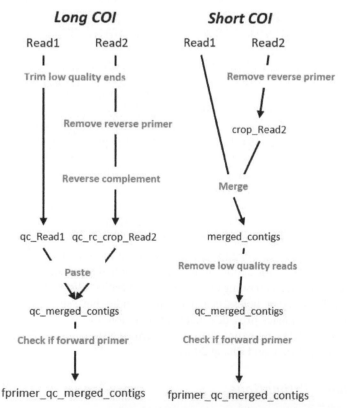

Figure 11.3: Raw read pre-processing for long and short Cytochrome c Oxidase I (COI) barcodes. For both cases, the starting sequences (Read1 and Read2) and the resulting sequences (quality filtered merged contigs that contain the forward primer) are the same, but the intermediate steps (indicated in grey) are specific to each case.

et al. 2019)). Additionally, because these clustering methods are known to produce overestimates of diversity, post clustering algorithms such as LULU (Frøslev et al. 2017) have also been developed. Nonetheless, no optimum universal approach for OTU clustering exists as the clustering algorithm and parameters that provide closest match between OTU and species are dependent on the barcode of choice, the sequencing platform used, the pre-processing approach applied and, most importantly, are specific to each species (Cristescu 2014).

11.3.3 Taxonomic Assignment

Biodiversity analyses can be performed on OTUs without even the need of taxonomically assigning them (Rey et al. 2019), but most studies require information on the taxonomic group each OTU belongs to and some require information on the specific species that are found in a sample. This information is obtained by comparing sequences against reference databases such as BOLD (for COI) or SILVA (for 18S). Several approaches for taxonomic assignment exist, which are often applied to OTU representative sequences (Pearman and Irigoien 2015). BLAST (Altschul et al. 1990) is a popular approach for taxonomic assignment of zooplankton metabarcoding data. This method consists on assigning the query sequence the taxonomy of the most similar sequence on the reference database. Yet, this approach can lead to incorrect classifications due to absence of reference data, to some barcodes not being able to distinguish among closely related species and/or to best-hit not often being the closest phylogenetic neighbour (Koski and Golding 2001). Other methods such as the Naïve Bayesian classifier (Wang et al. 2007) provide a measure of confidence for inclusion of a read into a taxonomic rank and are increasingly applied to metazoans (Porter and Hajibabaei 2018) and, in particular, zooplankton (Pearman and Irigoien 2015, Rey et al. 2019).

11.4 Making Ecological Sense of Metabarcoding Data

11.4.1 Metabarcoding Based Biodiversity Estimates

Numerous studies have focused on translating zooplankton metabarcoding data into ecologically meaningful information (Yang and Zhang 2020), for which the inherent properties of the metabarcoding process and, when relevant, of the use of environmental DNA as opposed to bulk samples should be considered (Piper et al. 2019). Relying on OTUs for performing biodiversity assessments translates into a much higher diversity than that estimated using morphological data, particularly for microzooplankton (De Vargas et al. 2015) and can reveal presence of species otherwise missed (Lindeque et al. 2013). Yet, metabarcoding has disadvantages such as not distinguishing among developmental stages or not distinguishing if a species detected is present in the environment or if it has been transported as prey from another species for example. Additionally, the potential biases accumulated throughout the metabarcoding process (e.g., over and under amplification of species and false negatives for some taxa due to primer bias) might translate into biodiversity estimates that differ from reality. Thus, for some applications (e.g., detection of non-indigenous species), exploratory results obtained thought metabarcoding might need to be verified using alternative techniques such as quantitative PCR (qPCR). Nonetheless, in general, most studies have shown that metabarcoding based zooplankton diversity estimates provide reliable information for assessing community composition (Pearman and Irigoien 2015), revealing seasonal patterns (Casas et al. 2017) or assessing freshwater and marine ecosystems (Yang and Zhang 2020).

11.4.2 Abundance Estimations

Quantitative estimations of taxon abundance are required for performing reliable biodiversity estimates and calculating most biological indices (Aylagas et al. 2018). Yet, obtaining abundance information from metabarcoding data is not trivial as there are biological factors (e.g., varying size of organisms leading to different number of cells per individual, number of DNA copies per cell) and technical factors (e.g., extraction and PCR amplification bias) that make the conversion of number of reads to number of individuals or biomass complex. Yet, despite these difficulties, several studies have shown good correlation between metabarcoding and morphology derived relative abundances (Lindeque et al. 2013, Hirai et al. 2017, Santoferrara and McManus 2021, this book, Chapter 5).

11.4.3 *Understanding Origin, State, Fate and Transport of Environmental DNA*

The amount of eDNA released by organisms and consequently the amount of DNA available in the environment to be sampled is not homogenous among species and life stages and is tightly linked to the abiotic (pH, salinity, temperature, currents, ultraviolet radiation) and biotic (e.g., microbial activity) conditions of the environment (Barnes and Turner 2016). Regarding zooplankton, the potential limitation of DNA shedding rates for invertebrates with exoskeleton may limit their detection from eDNA and warrants further research (Tréguier et al. 2014). Understanding how eDNA persists and ultimately represents the spatial-temporal patterns of species is essential and depends on the type of environment such as marine (coastal, epipelagic or deep ocean) and freshwater (lentic or lotic). For instance, eDNA studies in marine environments have reported a strong localized signal of eDNA, highlighting the great potential of eDNA to represent a contemporaneous community (Jeunen et al. 2019) but also detected the presence of freshwater species in offshore areas which can be riverine input transported over large distance (Yamamoto et al. 2017). Altogether, this suggests that the on-going active research on eDNA will ultimately refine our use of eDNA to help unveil and monitor zooplankton communities.

11.4.4 *Distinguishing Active versus Inactive Organisms by Means of Environmental RNA*

One of the major concerns when working with eDNA is the potential detection of DNA from organisms that are not alive, which can blur the spatial-temporal patterns inferred and can limit its use for, for example, biosecurity purposes (Cristescu 2019). Using environmental RNA (eRNA) instead has been evaluated as a potential, although challenging, solution to overcome this issue (Laroche et al. 2017, Pochon et al. 2017). Compared to DNA, RNA degrades rapidly after cell death; this, together with the increase in RNA copy number depending on gene expression and metabolic activity, has made some authors suggest that eRNA should be a better proxy than DNA for organismal activity. Yet, the first metabarcoding studies based on comparing eDNA and eRNA samples did not observe a clear advantage of using eRNA over eDNA (Pochon et al. 2017, von Ammon et al. 2019); therefore, assessing the potential benefits of eRNA over eDNA for biodiversity studies will require further studies.

11.5 Final Remarks

Despite the numerous factors to consider and the challenges involved in metabarcoding projects aimed at studying zooplankton ecology, current research has provided encouraging data (Bucklin et al. 2019, Yang and Zhang 2020). Certainly, metabarcoding can cost-effectively provide accurate taxonomic identification and relative quantification of complex zooplankton samples, with immense potential for large scale zooplankton monitoring programs. Yet, the quality of the obtained inferences will depend on the rigorous application of each of the steps involved in the project and on the adaptation of the sampling, laboratory and bioinformatic procedures to each case study. Thus, it is strongly advised that there is a fluid and continuous communication between the different members of the research team so that the samples collected, barcode of choice and laboratory procedures, available reference databases and bioinformatic procedures are fit for purpose. Especially when envisaging research in an area or a group of organisms for which no previous studies have been performed, it is strongly advised, when possible, to evaluate alternative sampling types and locations (Rey et al. 2020), completeness of reference database and available amplification primers (Aylagas et al. 2014), and performance of different amplification conditions and primers (Aylagas et al. 2016). Similarly, data preprocessing should be performed using alternative approaches (e.g., using different OTU clustering strategies and/or parameters, removing singletons or not) to check for consistency or for variations that could provide additional ecological information. Finally, data interpretation should be performed considering all the potential biases accumulated during all the steps involved in the project and, when relevant, validated using alternative methods (e.g., validate potential presence of an invasive species detected through metabarcoding through qPCR).

References

Albaina, A., M. Aguirre, D. Abad, M. Santos and A. Estonba. 2016. 18S rRNA V9 metabarcoding for diet characterization: a critical evaluation with two sympatric zooplanktivorous fish species. Ecol. Evol. 6: 1809–1824.

Altschul, S.F., W. Gish, W. Miller, E.W. Myers and D.J. Lipman. 1990. Basic local alignment search tool. J. Mol. Biol. 215: 403–410.

Amaral-Zettler, L.A., E.A. Mccliment, H.W. Ducklow and S.M. Huse. 2009. A method for studying protistan diversity using massively parallel sequencing of V9 hypervariable regions of small-subunit ribosomal RNA genes. PLoS ONE 4: e6372.

Aylagas, E., Á. Borja and N. Rodríguez-Ezpeleta. 2014. Environmental status assessment using DNA metabarcoding: Towards a Genetics Based Marine Biotic Index (gAMBI). PLoS ONE 9: e90529.

Aylagas, E. and N. Rodriguez-Ezpeleta. 2016. Analysis of Illumina MiSeq metabarcoding data: Application to benthic indices for environmental monitoring. *In*: S. Bourlat [ed.]. Marine Genomics Methods in Molecular Biology. New York, NY: Humana Press.

Aylagas, E., Á. Borja, X. Irigoien and N. Rodríguez-Ezpeleta. 2016. Benchmarking DNA metabarcoding for biodiversity-based monitoring and assessment. Front. Mar. Sci. 3: 1–12.

Aylagas, E., Á. Borja, I. Muxika and N. Rodríguez-Ezpeleta. 2018. Adapting metabarcoding-based benthic biomonitoring into routine marine ecological status assessment networks. Ecol. Indicators 95: 194–202.

Barnes, M.A. and C.R. Turner. 2016. The ecology of environmental DNA and implications for conservation genetics. Conserv. Genet. 17: 1–17.

Borrell, Y.J., L. Miralles, H. Do Huu, K. Mohammed-Geba and E. Garcia-Vazquez. 2017. DNA in a bottle—Rapid metabarcoding survey for early alerts of invasive species in ports. PLoS ONE 12: e0183347.

Bourlat, S., Q. Haenel, J. Finnman and M. Leray. 2016. Preparation of amplicon libraries for metabarcoding of marine eukaryotes using Illumina MiSeq: The Dual-PCR method. pp. 197–207. *In*: S. Bourlat [ed.]. Marine Genomics Methods in Molecular Biology. New York, NY, Humana Press.

Bourlat, S.J., A. Borja, J. Gilbert, M.I. Taylor, N. Davies, S.B. Weisberg et al. 2013. Genomics in marine monitoring: new opportunities for assessing marine health status. Mar. Pollut. Bull. 74: 19–31.

Boyer, F., C. Mercier, A. Bonin, Y. Le Bras, P. Taberlet and E. Coissac. 2016. obitools: a unix-inspired software package for DNA metabarcoding. Mol. Ecol. Resour. 16: 176–182.

Bucklin, A., P.H. Wiebe, S.B. Smolenack, N.J. Copley, J.G. Beaudet, K.G. Bonner et al. 2007. DNA barcodes for species identification of euphausiids (Euphausiacea, Crustacea). J. Plankton Res. 29: 483–493.

Bucklin, A., R.R. Hopcroft, K.N. Kosobokova, L.M. Nigro, B.D. Ortman, R.M. Jennings et al. 2010. DNA barcoding of Arctic Ocean holozooplankton for species identification and recognition. Deep Sea Res. II 57: 40–48.

Bucklin, A., P.K. Lindeque, N. Rodriguez-Ezpeleta, A. Albaina and M. Lehtiniemi. 2016. Metabarcoding of marine zooplankton: prospects, progress and pitfalls. J. Plankton Res. 38: 393–400.

Bucklin, A., H.D. Yeh, J.M. Questel, D.E. Richardson, B. Reese, N.J. Copley et al. 2019. Time-series metabarcoding analysis of zooplankton diversity of the NW Atlantic continental shelf. ICES J. Mar. Sci. 76: 1162–1176.

Buée, M., M. Reich, C. Murat, E. Morin, R.H. Nilsson, S. Uroz et al. 2009. 454 Pyrosequencing analyses of forest soils reveal an unexpectedly high fungal diversity. New Phytol. 184: 449–456.

Caporaso, J.G., J. Kuczynski, J. Stombaugh, K. Bittinger, F.D. Bushman, E.K. Costello et al. 2010. QIIME allows analysis of high-throughput community sequencing data. Nat. Meth. 7: 335–336.

Casas, L., J.K. Pearman and X. Irigoien. 2017. Metabarcoding reveals seasonal and temperature-dependent succession of zooplankton communities in the red sea. Front. Mar. Sci. 4: 241.

Crampton-Platt, A., D.W. Yu, X. Zhou and A.P. Vogler. 2016. Mitochondrial metagenomics: letting the genes out of the bottle. GigaScience 5: 15.

Cristescu, M.E. 2014. From barcoding single individuals to metabarcoding biological communities: towards an integrative approach to the study of global biodiversity. Trends Ecol. Evol. 29: 566–571.

Cristescu, M.E. 2019. Can environmental RNA revolutionize biodiversity science? Trends Ecol. Evol. 34: 694–697.

De Vargas, C., S. Audic, N. Henry, J. Decelle, F. Mahé, R. Logares et al. 2015. Eukaryotic plankton diversity in the sunlit ocean. Science 348: 1–12.

Deiner, K., H.M. Bik, E. Mächler, M. Seymour, A. Lacoursière-Roussel, F. Altermatt et al. 2017. Environmental DNA metabarcoding: Transforming how we survey animal and plant communities. Mol. Ecol. 26: 5872–5895.

Djurhuus, A., K. Pitz, N.A. Sawaya, J. Rojas-Márquez, B. Michaud, E. Montes et al. 2018. Evaluation of marine zooplankton community structure through environmental DNA metabarcoding. Limnol. Oceanogr. Methods 16: 209–221.

Frøslev, T.G., R. Kjøller, H.H. Bruun, R. Ejrnæs, A.K. Brunbjerg, C. Pietroni et al. 2017. Algorithm for post-clustering curation of DNA amplicon data yields reliable biodiversity estimates. Nat. Commun. 8: 1188.

Guillou, L., D. Bachar, S. Audic, D. Bass, C. Berney, L. Bittner et al. 2012. The Protist Ribosomal Reference database (PR2): a catalog of unicellular eukaryote Small Sub-Unit rRNA sequences with curated taxonomy. Nucleic Acids Res. 1–8.

Gunther, B., T. Knebelsberger, H. Neumann, S. Laakmann and P. Martinez Arbizu. 2018. Metabarcoding of marine environmental DNA based on mitochondrial and nuclear genes. Sci. Rep. 8: 14822.

Hebert, P.D.N., S. Ratnasingham and J.R. Dewaard. 2003. Barcoding animal life: cytochrome c oxidase subunit 1 divergences among closely related species. Proc. Biol. Sci. 270(Suppl. 1): S96–S99.

Hirai, J., S. Shimode and A. Tsuda. 2013. Evaluation of ITS2-28S as a molecular marker for identification of calanoid copepods in the subtropical western North Pacific. J. Plankton Res. 35: 644–656.

Hirai, J., S. Nagai and K. Hidaka. 2017. Evaluation of metagenetic community analysis of planktonic copepods using Illumina MiSeq: Comparisons with morphological classification and metagenetic analysis using Roche 454. PLoS ONE 12: e0181452–e0181452.

Jackson, M.A., J.T. Bell, T.D. Spector and C.J. Steves. 2016. A heritability-based comparison of methods used to cluster 16S rRNA gene sequences into operational taxonomic units. PeerJ 4: e2341–e2341.

Jakubavičiūtė, E., U. Bergström, J.S. Eklöf, Q. Haenel and S.J. Bourlat. 2017. DNA metabarcoding reveals diverse diet of the three-spined stickleback in a coastal ecosystem. PLoS ONE 12: e0186929.

Jeunen, G.-J., M. Knapp, H.G. Spencer, M.D. Lamare, H.R. Taylor, M. Stat et al. 2019. Environmental DNA (eDNA) metabarcoding reveals strong discrimination among diverse marine habitats connected by water movement. Mol. Ecol. Resour. 19: 426–438.

Koski, L.B. and G.B. Golding. 2001. The closest BLAST hit is often not the nearest neighbor. J. Mol. Evol. 52: 540–542.

Koziol, A., M. Stat, T. Simpson, S. Jarman, J.D. Dibattista, E.S. Harvey et al. 2019. Environmental DNA metabarcoding studies are critically affected by substrate selection. Mol. Ecol. Resour. 19: 366–376.

Laroche, O., S.A. Wood, L.A. Tremblay, G. Lear, J.I. Ellis and X. Pochon. 2017. Metabarcoding monitoring analysis: the pros and cons of using co-extracted environmental DNA and RNA data to assess offshore oil production impacts on benthic communities. PeerJ 5: e3347.

Leray, M., J.Y. Yang, C.P. Meyer, S.C. Mills, N. Agudelo, V. Ranwez et al. 2013. A new versatile primer set targeting a short fragment of the mitochondrial COI region for metabarcoding metazoan diversity: application for characterizing coral reef fish gut contents. Front. Zool. 10: 34.

Leray, M., N. Knowlton, S.-L. Ho, B.N. Nguyen and R.J. Machida. 2019. GenBank is a reliable resource for 21st century biodiversity research. Proc. Natl. Acad. Sci. 116: 22651–22656.

Lim, N.K., Y.C. Tay, A. Srivathsan, J.W. Tan, J.T. Kwik, B. Baloglu et al. 2016. Next-generation freshwater bioassessment: eDNA metabarcoding with a conserved metazoan primer reveals species-rich and reservoir-specific communities. R. Soc. Open Sci. 3: 160635.

Lindeque, P.K., H.E. Parry, R.A. Harmer, P.J. Somerfield and A. Atkinson. 2013. Next generation sequencing reveals the hidden diversity of zooplankton assemblages. PLoS ONE 8: e81327.

Logares, R., S. Sunagawa, G. Salazar, F.M. Cornejo-Castillo, I. Ferrera, H. Sarmento et al. 2013. Metagenomic 16S rDNA Illumina tags are a powerful alternative to amplicon sequencing to explore diversity and structure of microbial communities. Environ. Microbiol. 16: 2659–2671.

Machida, R.J. and N. Knowlton. 2012. PCR primers for Metazoan Nuclear 18S and 28S ribosomal DNA sequences. PLoS ONE 7: e46180.

Mahé, F., T. Rognes, C. Quince, C. De Vargas and M. Dunthorn. 2014. Swarm: robust and fast clustering method for amplicon-based studies. PeerJ 2: e593.

Margulies, M., M. Egholm, W.E. Altman, S. Attiya, J.S. Bader, L.A. Bemben et al. 2005. Genome sequencing in microfabricated high-density picolitre reactors. Nature 437: 376–380.

Mcmanus, G.B. and L.A. Katz. 2009. Molecular and morphological methods for identifying plankton: what makes a successful marriage? J. Plankton Res. 31: 1119–1129.

Pearman, J.K., M.M. El-Sherbiny, A. Lanzén, A.M. Al-Aidaroos and X. Irigoien. 2014. Zooplankton diversity across three Red Sea reefs using pyrosequencing. Front. Mar. Sci. 1: 27.

Pearman, J.K. and X. Irigoien. 2015. Assessment of zooplankton community composition along a depth profile in the central red sea. PLoS ONE 10: e0133487.

Piper, A.M., J. Batovska, N.O.I. Cogan, J. Weiss, J.P. Cunningham, B.C. Rodoni et al. 2019. Prospects and challenges of implementing DNA metabarcoding for high-throughput insect surveillance. GigaScience 8.

Pochon, X., A. Zaiko, L.M. Fletcher, O. Laroche and S.A. Wood. 2017. Wanted dead or alive? Using metabarcoding of environmental DNA and RNA to distinguish living assemblages for biosecurity applications. PLoS ONE 12: e0187636.

Porter, T.M. and M. Hajibabaei. 2018. Automated high throughput animal CO1 metabarcode classification. Sci. Rep. 8: 4226–4226.

Quast, C., E. Pruesse, P. Yilmaz, J. Gerken, T. Schweer, P. Yarza et al. 2013. The SILVA ribosomal RNA gene database project: improved data processing and web-based tools. Nucleic Acids Res. 41: D590–D596.

Ratnasingham, S. and P.D.N. Hebert. 2007. Bold: The Barcode of Life Data System (http://www.barcodinglife.org). Mol. Ecol. Notes 7: 355–364.

Rey, A., K.J. Carney, L.E. Quinones, K.M. Pagenkopp Lohan, G.M. Ruiz, O.C. Basurko et al. 2019. Environmental DNA metabarcoding: A promising tool for ballast water monitoring. Environ. Sci. Technol. 53: 11849–11859.

Rey, A., O. Basurko and N. Rodriguez-Ezpeleta. 2020. Considerations for metabarcoding-based port biological baseline surveys aimed at marine non-indigenous species monitoring and risk-assessments. Ecol. Evol. (in press).

Rognes, T., T. Flouri, B. Nichols, C. Quince and F. Mahé. 2016. VSEARCH: a versatile open source tool for metagenomics. PeerJ 4: e2584–e2584.

Sanger, F., S. Nicklen and A.R. Coulson. 1977. DNA sequencing with chain-terminating inhibitors. Proc. Natl. Acad. Sci. 74: 5463–5467.

Santoferrara, L.F. and G.B. McManus. 2021. Diversity and biogeography as revealed by morphologies and DNA sequences: Tintinnid ciliates as an example. pp. 85–118. *In*: M.A. Teodósio and A.B. Barbosa [eds.]. Zooplankton Ecology. CRC Press.

Schloss, P.D., S.L. Westcott, T. Ryabin, J.R. Hall, M. Hartmann, E.B. Hollister et al. 2009. Introducing mothur: open-source, platform-independent, community-supported software for describing and comparing microbial communities. Appl. Environ. Microbiol. 75: 7537.

Siegenthaler, A., O.S. Wangensteen, C. Benvenuto, J. Campos and S. Mariani. 2019. DNA metabarcoding unveils multiscale trophic variation in a widespread coastal opportunist. Mol. Ecol. 28: 232–249.

Sogin, M.L., H.G. Morrison, J.A. Huber, D.M. Welch, S.M. Huse, P.R. Neal et al. 2006. Microbial diversity in the deep sea and the underexplored "rare biosphere". Proc. Natl. Acad. Sci. 103: 12115.

Spens, J., A.R. Evans, D. Halfmaerten, S.W. Knudsen, M.E. Sengupta, S.S.T. Mak et al. 2017. Comparison of capture and storage methods for aqueous macrobial eDNA using an optimized extraction protocol: advantage of enclosed filter. Meth. Ecol. Evol. 8: 635–645.

Taberlet, P., E. Coissac, F. Pompanon, C. Brochmann and E. Willerslev. 2012. Towards next-generation biodiversity assessment using DNA metabarcoding. Mol. Ecol. 21: 2045–2050.

Tang, C.Q., F. Leasi, U. Obertegger, A. Kieneke, T.G. Barraclough and D. Fontaneto. 2012. The widely used small subunit 18S rDNA molecule greatly underestimates true diversity in biodiversity surveys of the meiofauna. Proc. Natl. Acad. Sci. 109: 16208–16212.

Tréguier, A., J.-M. Paillisson, T. Dejean, A. Valentini, M.A. Schlaepfer and J.-M. Roussel. 2014. Environmental DNA surveillance for invertebrate species: advantages and technical limitations to detect invasive crayfish Procambarus clarkii in freshwater ponds. J. Appl. Ecol. 51: 871–879.

Von Ammon, U., S.A. Wood, O. Laroche, A. Zaiko, S.D. Lavery, G.J. Inglis et al. 2019. Linking environmental DNA and RNA for improved detection of the marine invasive fanworm sabella spallanzanii. Front. Mar. Sci. 6: 621.

Walsh, J.R., M.J. Spear, T.P. Shannon, P.J. Krysan and M.J. Vander Zanden. 2019. Using eDNA, sediment subfossils, and zooplankton nets to detect invasive spiny water flea (Bythotrephes longimanus). Biol. Invasions 21: 377–389.

Wang, Q., G.M. Garrity, J.M. Tiedje and J.R. Cole. 2007. Naive Bayesian classifier for rapid assignment of rRNA sequences into the new bacterial taxonomy. Appl. Environ. Microbiol. 73: 5261–5267.

Webb, K.E., D.K.A. Barnes, M.S. Clark and D.A. Bowden. 2006. DNA barcoding: A molecular tool to identify Antarctic marine larvae. Deep Sea Res. II 53: 1053–1060.

Wilson, I.G. 1997. Inhibition and facilitation of nucleic acid amplification. Appl. Environ. Microbiol. 63: 3741–3751.

Williams, C., F. Pontén, C. Moberg, P. Söderkvist, M. Uhlén, J. Pontén et al. 1999. A high frequency of sequence alterations is due to formalin fixation of archival specimens. Am. J. Pathol. 155: 1467–1471.

Yamamoto, S., R. Masuda, Y. Sato, T. Sado, H. Araki, M. Kondoh et al. 2017. Environmental DNA metabarcoding reveals local fish communities in a species-rich coastal sea. Sci. Rep. 7: 40368.

Yang, J. and X. Zhang. 2020. eDNA metabarcoding in zooplankton improves the ecological status assessment of aquatic ecosystems. Environ. Int. 134: 105230.

CHAPTER 12

Omics Advances in the Study of Zooplankton

Big Data for Small Drifting Organisms

Ana Matos,[1,2] *Jean-Baptiste Ledoux,*[1,3] *Dany Domínguez-Pérez,*[1]
Daniela Almeida[1] and *Agostinho Antunes*[1,2,*]

12.1 Introduction

The world ocean covers 71% of the Earth's surface and provides key ecosystem services ranging from food security to climate regulation. As the largest ecosystem on Earth, the world ocean also harbors a large fraction of biodiversity, estimated as ca. one million species (Hays et al. 2005, Appeltans et al. 2012). While the complex interactions among those species contribute to a proper ecosystem functioning, also impacting related services (Doney et al. 2012), a large fraction of marine-living species, and their respective ecological roles, are not yet characterized (Appeltans et al. 2012). In addition to supporting a remarkable biodiversity and sustaining a large part of worldwide economy, the role of the world ocean on the mitigation of the on-going climate crisis has also recently been emphasized. Accordingly, there is an unquestionable need to improve our knowledge on marine diversity by focusing first on key species (Gattuso et al. 2018). In this context, planktonic organisms, that include viruses, heterotrophic prokaryotes, phytoplankton and zooplankton, have received particular attention in the last decades. Indeed, plankton are recognized as a central component of marine diversity mainly due to their key ecological roles, supporting all the marine life and controlling global biogeochemical cycles, with direct consequences for the Earth climate (Strom 2008, Chavez et al. 2011). The distribution of this vast diversity relies on biotic and abiotic factors, such as water temperature, depth, currents, nutrients and light (Harris et al. 2000). Consequently, environmental

---

[1] CIIMAR/CIMAR, Interdisciplinary Centre of Marine and Environmental Research, University of Porto, Terminal de Cruzeiros do Porto de Leixões, Av. General Norton de Matos, s/n, 4450-208 Porto, Portugal.
[2] Biology Department, Faculty of Sciences, University of Porto, Rua do Campo Alegre, s/n, 4169-007 Porto, Portugal.
[3] Institut de Ciències del Mar CSIC, Passeig Maritim de la Barceloneta 37-49, E-08003, Barcelona, Spain.
  Emails: anabastosmatos@gmail.com; jbaptiste.ledoux@gmail.com; dany.perez@ciimar.up.pt; danielaalmeida23@gmail.com
* Corresponding author: aantunes@ciimar.up.pt

changes, such as ocean warming and acidification, might influence the spatial and temporal distribution of planktonic organisms and their interactions in the food web, with potential feedbacks on the Earth biota (Edwards and Richardson 2004, Chavez et al. 2011, Doney et al. 2012).

Zooplankton are a diverse group, including both phagotrophic protists and metazoans, that remain in the plankton permanently, for their entire life cycles (holoplankton), or only a part of their life cycles (e.g., eggs and larval stages; meroplankton). Therefore, zooplankton comprise a variety of organism sizes, life history traits, life cycles, reproductive strategies (e.g., sexual phase or sexual and asexual phases; Peijnenburg and Goetze 2013), niche requirements and phylogenetic diversity (de Vargas et al. 2015, Lima-Mendez et al. 2015). Zooplankton include representatives of different taxa (e.g., Foraminifera, Cnidaria, Mollusca, Crustace, Tunicata; Harris et al. 2000, Machida et al. 2009), encompassing non-gelatinous and gelatinous species, the latter associated with bloom events (D'Alelio et al. 2019). Thus, zooplankton include important primary consumers and predators (Litchman et al. 2013, Peijnenburg and Goetze 2013) and play a key role in the functioning of marine food webs and ecosystems, linking primary production to higher trophic positions and pelagic to benthic habitats, and shaping global biogeochemical fluxes and cycles (i.e., removal of CO_2 from atmosphere; Jeong et al. 2010, Daewel et al. 2014).

Due to its central ecological role, it is of extreme importance to characterize the diversity patterns and the eco-evolution of zooplanktonic organisms. Traditionally, the methods to identify zooplankton relied solely on morphological characters, assessed using microscopy techniques. However, those strategies have several limitations since different species can share similar morphological traits (e.g., cryptic species, and lack of diagnostic morphological traits). Further, the distinction and species identification in early developmental stages is particularly challenging, as reported for copepods (Bucklin et al. 1999, Lindeque et al. 2006).

Since the introduction of the first molecular biology techniques until the current state-of-the-art "omics" technologies, our ability to characterize the diversity of zooplanktonic organisms, their eco-evolution and functions within world ocean ecosystems has greatly improved. Molecular approaches have been applied, for instance, to refine the estimation of species diversity (e.g., identification of cryptic species, different development stages of the same species, sexual dimorphism), as well as to recognise invasive from non-invasive species (Kiesling et al. 2002, Ghahramanzadeh et al. 2013, Jeon et al. 2018). With the advance of molecular biology, zooplankton identification studies based on Polymerase Chain Reaction (PCR) techniques and Sanger sequencing (Bucklin et al. 1999, Lindeque et al. 2004) following a barcoding approach (see below; Bucklin et al. 2010), have been complementing conventional microscopy-based approaches. DNA barcode databases of specific ecosystems are being developed and used, for example, to identify copepods inhabiting Korean Peninsula and Arctic Ocean (Bucklin et al. 2010, Baek et al. 2016). The recent development and application of metabarcoding, transcriptomics, and metabolomics improve our understanding on the biotic interactions among marine organisms (Evans et al. 2017, Behrendt et al. 2018). The development of high throughput sequencing technologies, such as Illumina and Ion Torrent, opened new avenues for the identification and characterization of zooplankton communities (de Vargas et al. 2015, Hirai et al. 2017a, b, D'Alelio et al. 2019).

This chapter describes how -omics techniques, particularly genomics, transcriptomics and proteogenomics, contribute to: (i) characterize the spatial and temporal patterns of zooplankton; (ii) shed new light on ecology and life cycles of the organisms composing the zooplankton; (iii) understand global change effects on specific zooplanktonic groups; (iv) unravel ecological patterns of tunicates and cnidarians; and (v) create extensive genetic databases.

12.2 Insight into the Spatial and Temporal Patterns of Zooplankton

As stated previously, zooplankton are a diverse group ubiquitously distributed in all marine ecosystems. Yet, zooplankton distribution patterns are strongly influenced by factors, processes and variables such as ocean dynamics (e.g., currents), depth and temperature. Recently, studies using omics technologies have granted a better understanding of those distribution patterns. Genomic approaches have evolved to large-scale analyses, providing knowledge on ocean microbes and zooplankton assemblages, and offering cues on species distribution, nutrient and biogeochemical cycles, and ecological factors that structure the ocean (Kopf et al. 2015, Delmont et al. 2018, Carradec et al. 2018). Genomic-derived techniques, such as metagenomics and metatranscriptomics, have revolutionized the study of the ocean microbiome diversity (Carradec et al. 2018, Salazar et al. 2019). These techniques not only provide large amounts of data from complex communities, in a single sequencing event, but also allow scientists to characterize thousands of genes and their functions and, accordingly, shed new light on the biotic and abiotic processes that shape this central component of marine ecosystems.

The application of omics approaches with traditional methodologies has allowed a better identification of zooplankton species, increasing resolution at both spatial and temporal levels. After the development of molecular-based methodologies, such as the application of PCR followed by Restriction Fragment Length Polymorphism (RFLP), new distributional areas and co-occurrence of species (e.g., copepods) were discovered (Lindeque et al. 2004). Using molecular approaches, the identification of cryptic species, in particular belonging to merozooplankton and gelatinous zooplankton, was reached (Cheng et al. 2014). Moreover, a large-scale study based on metabarcoding analysis of V9 region—18S rRNA of a total of 47 stations, across tropical and temperate surface ocean, unravelled around 75% of eukaryotic ribosomal diversity in the upper-euphotic zone, revealing that most eukaryotic plankton biodiversity is associated with heterotrophic protists (de Vargas et al. 2015).

In connection with variability patterns, for example, temporal-spatial variability of zooplankton assemblages in the Red Sea was investigated using high-throughput sequencing (Pearman and Irigoien 2015). Sequences belonging to Arthropoda phylum were dominant in all sampling points. Yet, a shift in species composition, with increases in Mollusca and Cnidaria sequences, was detected one month after the start of sampling (Pearman and Irigoien 2015). Within Arthropoda phylum, Maxillopoda represented the dominant class, but a depth-dependent distribution was reported, with *Pleuromamma* sequences being mostly detected in deeper samples, whereas *Corycaeus* was mostly detected near the surface (Pearman and Irigoien 2015).

Long-term monitoring studies allow the detection of community compositional shifts over time, at intra- and interannual scales. Continuous monitoring is needed to produce baseline data for zooplankton communities' structure and seasonal turnover, identify changes in species abundance and diversity, linked for instance to global change, and detect the first stages of biological invasions (Edwards and Richardson 2004, Jonkers et al. 2019). For example, zooplankton samples collected in the NW Atlantic continental shelf, during the period 2002–2012, were used for metabarcoding (V9 region of 18S rRNA) and sequenced on Illumina MiSeq NextGen platform. These samples were assigned to 28 taxonomic groups, and the dominant groups revealed similar variability patterns among regions and years (Bucklin et al. 2019). This study illustrates how metabarcoding approaches improve the detection of species difficult to identify (e.g., small fragile gelatinous zooplankton), that could be destroyed due to the application of net samplers (Bucklin et al. 2019).

Meroplankton, composed of pelagic eggs and larval stages, represent an important component of coastal zooplanktonic assemblages, highly dependent on the annual cycle but, for some groups, morphological traits are still understudied. From fish to sessile invertebrates, larval phase is an important stage in the life cycle, being sometimes the only mobile phase (Ershova et al. 2019). Despite its difficult identification from direct microscopic observations, the study of planktonic larvae composition and abundance allows the prediction of species invasions, supporting the study of reproductive life cycles. DNA barcoding protocols targeting meroplankton have been developed and tested, in combination with morphological traits, to characterise their temporal-spatial diversity patterns. For instance, Ershova et al. (2019) evaluated summer meroplankton assemblages in the Pacific arctic, during a 12-year period, using barcoding of bivalve and echinoderm larvae. Meroplankton assemblage revealed a low diversity that could eventually increase if more species were barcoded (i.e., more exhaustive reference database; Ershova et al. 2019). This study highlights the need of adding more sequences to reference libraries in order to improve species identification, including larval stages (Ershova et al. 2019).

12.3 Shedding New Light on the Ecology and Life Cycles of Zooplanktonic Species

Traditionally, morphological tools were the basis of species identification. However, the advance of molecular approaches revolutionized this field supporting life cycle characterization and understanding the ecological factors associated with each species. These techniques created the opportunity to develop approaches for identifying species at any stage of their life cycles, as observed for copepods, one of the most representative zooplankton groups (Lindeque et al. 1999, Kiesling et al. 2002). One of such approaches relies on the use of species-specific molecular probes allowing the identification of target species. For instance, during probe hybridization assays, the amplified DNA is captured by a specific probe and detected through a colorimetric enzymatic system, allowing the discrimination of copepod species and identification of both adult and larval stages (Kiesling et al. 2002).

Real-time quantitative PCR (qPCR) protocols to detect and quantify the abundance of marine invertebrate larvae have also been developed (Vadopalas et al. 2006, Pan et al. 2008, Jensen et al. 2012). As example, a qPCR protocol was applied for identification and quantification of abalone larvae (*Haliotis kamtschatkana*), also assessing its dispersal patterns (Vadopalas et al. 2006). In addition, proteomics has also been tested to identify specific proteins that support the identification of bivalve larvae collected from seawater samples (López et al. 2005).

Moreover, several studies have been focussing on species and community barcoding. DNA barcoding relies on the identification of species through specific sequences applying one or more genetic markers (Hebert and Gregory 2005), whereas metabarcoding involves the DNA identification of multiple species in a complex sample (Cristescu 2014). For further information on the application of metabarcoding to study zooplankton, see Rey et al. (2021) (this book, Chapter 11). Due to their wide distribution, as well as easy sampling, both unicellular phagotrophic protists (Santoferrara and McManus 2021, this book, Chapter 5) and metazooplankton (Yang et al. 2017) have been considered suitable for the application of metabarcoding approaches. Indeed, the identification of cryptic species has been possible through metabarcoding. Copepods, as example, are known to be very difficult to distinguish morphologically, even in adult phases. Thus, studies aiming to develop barcode approaches address cryptic copepod, namely *Acartia tonsa* (Goetze 2010, da Costa et al. 2011).

The availability of sequences in public databases is strongly correlated with the growth of omics technologies. There are studies in which several Operational Taxonomic Units (OTUs) cannot be assigned to any taxonomic entry. To overcome this drawback, some studies constructed a barcode database focusing on a specific geographic area. As an example, Yang et al. (2017) developed a barcode database from zooplankton native specimens of Tai Lake, China (Yang et al. 2017). Similarly, the holozooplankton community of the Arctic Ocean has started to be characterized. Contributing to the first steps of barcode database development, Bucklin et al. (2010) provided DNA sequences of 41 species (82 specimens—cnidarians, crustaceans, chaetognaths, and nemertean species). This study highlights the potential of bulk sampling barcode, in contrast to barcoding studies focusing only on specific zooplanktonic organisms, such as crustaceans, copepods and gastropods (Bucklin et al. 2003, Remigio 2003, Costa et al. 2007).

With the development of reference databases, the so-called metabarcoding approach, has been conducted. Combining this approach with morphological traits, zooplankton community has been studied and characterized, revealing a high diversity level, particularly in copepod group (Djurhuus et al. 2018). Indeed, the use of several genetic markers in metabarcoding potentiates higher amounts of biodiversity recovery (Djurhuus et al. 2018). Thus, the combination of high throughput sequencing and morphology is currently recommended for assessing copepod diversity (Djurhuus et al. 2018).

12.4 How are Omics Methodologies Helping in the Study of Global Changes Effects?

Plankton are key indicators of ocean status and change, thus representing a vulnerable group to climate variability and changes (Rombouts et al. 2009, Eloire et al. 2010, Lewandowska et al. 2014). For instance, detrimental effects of ocean warming and acidification on calcifying planktonic organisms have been reported (Beaugrand et al. 2013). Besides, the impact of global change on zooplankton assemblages has been studied, and the abundances of both meroplankton and holoplankton can decrease under high CO_2 levels (Smith et al. 2016). Yet, some zooplankton species, such as the tunicate *Oikopleura dioica*, can be favored by ocean warming and acidification (Bouquet et al. 2018).

To date, metatranscriptomics are one of the most powerful techniques that allow the screening of differences in the levels of gene expression in microbe communities, from distinct habitats or facing different environmental conditions (Salazar et al. 2019). Metatranscriptomics, which focuses on the analysis of transcripts, thus, can indirectly provide information on the distribution of marine organisms and on-going physiological and metabolic activities (Wang et al. 2014). Based on the analyses of microbial metagenomes and metatranscriptomes, collected from 126 globally distributed sampling stations, Salazar et al. (2019) reported differences in transcript levels between polar or non-polar regions, and across depth levels. This approach allowed the authors to conclude that microbial communities inhabiting warm waters probably have higher acclimation capacity to temperature variations. Besides, Salazar et al. (2019) also suggested that in polar regions, organismal composition will change more than gene regulatory mechanisms as a response to ocean warming.

Compared to metatranscriptomics, shotgun proteomics is performed with a lower cost and computing power. Although the identification of proteins is database-dependent, in most of the cases, the continuous increase of metagenomic and metatranscriptomic data may solve this limitation in a near future. Proteogenomics, an approach based on bioinformatics to perform proteomic raw data searches against genomic/transcriptomic derived protein databases, has recently emerged and may greatly improve our ability to functionally characterize complex marine communities (Wang et al. 2016).

Metaproteomics is a new field within proteogenomics, which aims to unravel protein expression from a complex biological system, as marine environmental samples (Wang et al. 2014). A metaproteomic approach will enhance our understanding on zooplankton or microbial community composition, as well as their ecological functions (Wilmes and Bond 2004, Wang et al. 2014, 2016, Li et al. 2018). Even if metaproteomic is still growing, there is enough evidence to consider it as one of the most promising approaches for studying zooplankton communities (Wang et al. 2014, 2016). Metaproteomics has been applied to assess microbial plankton metabolic activity, during a winter to spring succession, but also to understand the response of heterotrophic bacterioplankton to phytoplankton blooms (Georges et al. 2014, Wöhlbrand et al. 2017).

12.5 Case Studies

The sub-phylum Tunicata is composed of filter feeding marine invertebrates, subdivided into three main classes: Ascidiacea, Thaliacea and Appendicularia. Thaliacea and Appendicularia are pelagic holoplanktonic tunicates, whereas Ascidians include sessile adult tunicates with a pelagic larval phase (Lemaire 2011). Thaliaceans proliferate abruptly in the presence of optimal environmental conditions thus being frequently associated with blooms (Deibel and Lowen 2012, Piette and Lemaire 2015, Holland 2016). Appendicularians are relatively-small solitary organisms, characterized by a structure referred to as a 'house', associated with buoyancy and filter-feeding strategy (Brena et al. 2003, Holland

2016, Conley et al. 2018). This feeding strategy enables the ingestion of particles with different size range, including picoplanktonic prey that are not usually ingested by other filter feeders (Bone 1998). Thus, this group plays an important function in marine food webs, transferring energy from small picoplankton to higher trophic levels (Conley et al. 2018). Once clogged, the appendicularian houses are discarded and sink to the deep ocean, contributing to marine snow formation and ocean carbon pump (Dash et al. 2012), and a new house is secreted (Holland 2016).

Pelagic tunicates are important ubiquitous members of mesozooplankton, either being important in the food web or as bloom producers (Sommer and Stibor 2002). The reproductive cycle of these tunicates, and the increase of nutrients and food sources, usually contribute to the occurrence of blooms. Being a relevant member of zooplankton, their study may support the prediction of bloom events (see Magalhães et al. 2021, this book, Chapter 10). Pelagic tunicates have been traditionally studied using microscopic techniques. However, for Appendicularia, molecular and proteomic techniques have been applied. Real-time quantitative PCR has been applied to understand *O. dioica* feeding strategy (Troedsson et al. 2007). Moreover, proteomics approaches have been applied to characterize the protein composition of appendicularian houses (Hosp et al. 2012). Metagenomic and metatranscriptomic approaches were also used to unravel the composition of microbial communities and related gene expression under specific environmental conditions and changes (Ottesen et al. 2011, Brown et al. 2015).

High abundance of ascidians, as well as their sessile adult phases, contribute to the existence of a higher number of studies focused on ascidians in comparison with the pelagic tunicates. The presence of ascidians cryptic species and the study of non-indigenous ascidians species have been assessed through molecular biology techniques (Turon et al. 2003). DNA barcoding approaches targeting ascidians have been applied (Ali and Tamilselvi 2016). Ascidians larval stages have also been associated with vertical transmission of microorganisms. In the past years, several studies have applied omics technologies to better understand the microbial associations and biotechnological potential associated with microorganisms (Kwan et al. 2012, Hirose 2015). Moreover, the ascidian larval stage arouses curiosity amongst the scientific community due to the presence of chordate characteristics that are lost after metamorphosis into sessile organisms. Besides the larvae study in terms of developmental and evolutionary studies, ascidians are fouling organisms; nonetheless, since they are sessile as adults, the larvae stage contributes to their dispersal (Aldred and Clare 2014). Proteomic techniques have been applied to study embryonic and larval stages of *Ciona intestinalis*, along development (Nomura et al. 2009).

Like most of marine species, cnidarians are planktonic at some point of their life cycle, and their identification may be challenging due to the existence of several cryptic species (Holland et al. 2004, Bucklin et al. 2016). During sampling and preservation, fragile cnidarian colonies may suffer disaggregation or damage, leading to difficulties in their taxonomic identification (Moura et al. 2008, Grossmann et al. 2013). The misidentification is also caused by the attribution of species category to each life stage, pelagic and benthic, of the same species (Moura et al. 2008, Grossmann et al. 2013). In this sense, molecular methods, such as DNA barcoding or metabarcoding, have been applied as a complement to species detection (Carlos et al. 2014, Magalhães et al. 2021, this book, Chapter 10).

However, some limitations associated with the use of molecular markers in Anthozoa have been reported (i.e., lack of polymorphism in mitochondrial DNA) (Shearer et al. 2002, Calderón et al. 2006, McFadden et al. 2011). A recent study tested four molecular markers, cytochrome oxidase I (COI), 16S, ITS1 and ITS2, to assess the intraspecific and interspecific genetic divergence within the anthozoan group Ceriantharia (Stampar et al. 2012). This study demonstrates that COI is a suitable DNA marker in this taxon, with levels of intraspecific and interspecific genetic divergence similar to those observed in Medusozoa (Stampar et al. 2012). COI has been also pointed by some authors to be a good genetic marker to determine species boundaries (Huang et al. 2008, Ortman et al. 2010).

On the other hand, some authors suggest that 16S rRNA is more appropriate for barcoding of Cnidaria than COI, since it is easier to be amplified and has more available sequences in public databases than COI, which may also not be informative at interspecific level (Moura et al. 2008,

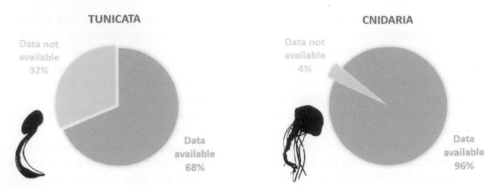

Figure 12.1: Number of species of cnidarian and tunicates deposited in the "Zooplankton Data" section, from COPEPOD database (https://www.st.nmfs.noaa.gov/copepod/atlas/index-atlas.html#zooplankton). Genetic information available for these species was searched from several databases: (1) NCBI "Sequence Set Browser", (2) NCBI nucleotide database, (3) NCBI protein database, (4) DNA Data Bank of Japan (DDBJ); (5) European Molecular Biology Laboratory-European Bioinformatics Institute (EMBL-EBI); and (6) public data portal Barcode of Life Data System (BOLD). All information accessed on 3 December 2019. Sources: (1) https://www.ncbi.nlm.nih.gov/Traces/wgs/?page=1&view=all; (2) https://www.ncbi.nlm.nih.gov/nuccore/?term=; (3) https://www.ncbi.nlm.nih.gov/protein/; (4) http://ddbj.nig.ac.jp/arsa/?lang=en; (5) https://www.ebi.ac.uk/ebisearch/overview.ebi/about; and (6) http://www.boldsystems.org/index.php/Public_BINSearch?searchtype=records.

Cantero et al. 2010, Zheng et al. 2014, Lindsay et al. 2015). Another advantage of this marker is the resolution of cryptic species (Moura et al. 2008). 16S rRNA barcoding gene approach has also been applied to study the introduction and geographical distribution of hydrozoan species (Moura et al. 2008, Grossmann et al. 2013, Miglietta et al. 2015). This approach was also useful for finally linking two different life stages of the same species, the free-living sexual unit (eudoxid stage, *Eudoxia macra*) and its polygastric colony (*Lensia cossack*), formerly classified as two different species (Grossmann et al. 2013).

To overcome the problems of mitochondrial genes, other genes have been tested as possible cnidarian barcodes as the case of 18S rRNA. Ayala et al. (2018), using next generation sequencing (NGS), analysed large particulate organic material (POM) and eel larvae gut contents, finding a dominance of Hydrozoan gene sequences and the presence of Cnidaria in all samples (Ayala et al. 2018). Cnidarians were also highly detected in the POM fraction, being the second most representative group of eukaryotes (Ayala et al. 2018). Another study applied DNA metabarcoding, 18S rRNA, to assess the presence of gelatinous zooplankton in albatross diet concluding that one of the common prey items of albatross are scyphozoans (McInnes et al. 2017).

The application of both taxonomic and molecular approaches is being adopted by several studies. For example, taxonomic observations, COI barcoding, and phylogenetic analyses of 16S rRNA and 18S rRNA genes revealed a new species of the genus *Clytia* (He et al. 2015). Further, despite body disintegration, the application of DNA barcoding with COI genetic marker allowed the identification of *Aurelia aurita* (Keskin and Atar 2013). Both genetic markers, COI and 16S rRNA, are considered useful for Zoantharia identification (Sinniger et al. 2008).

Overall, the genetic data available for the two abovementioned groups of gelatinous zooplankton, Cnidaria and Tunicata, is quite distinct (Fig. 12.1). There is a much higher amount of genetic data deposited in databases for Cnidaria (49 species), in comparison with Tunicates (19 species), reflecting the greater number of studies applying omics methodologies to assess the ecology of Cnidarians.

12.6 Integrative Assessments of Data, Databases and Inferences

In order to construct a custom protein database aiming protein identification in a proteogenomic study of zooplankton (Fig. 12.2), researchers must gather sequence data from specific own-generated

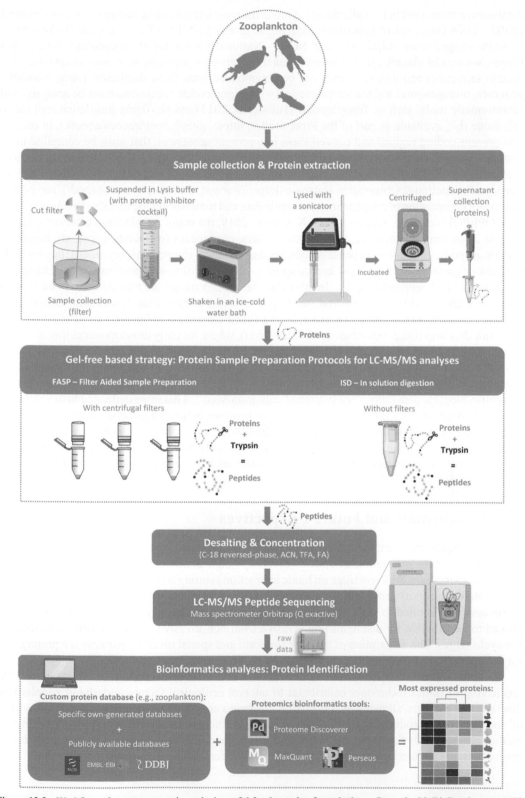

Figure 12.2: Workflow of a proteogenomic analysis useful for the study of zooplankton. Created with BioRender.com (with permission).

databases or from a publicly available one. Among the most known public databases, we recommend: DDBJ—DNA Data Bank of Japan (http://ddbj.nig.ac.jp/arsa/), EMBL-EBI—European Bioinformatics Institute (https://www.ebi.ac.uk/), and NCBI—National Center for Biotechnology Information (https://www.ncbi.nlm.nih.gov/). Sequence data from genomes, metagenomes, nucleotide sequences, protein sequences and transcriptomes can be downloaded from these databases. Later, assembled genomes, metagenomes and transcriptomes, as well as nucleotide sequences, must be analysed with bioinformatic tools, such as TransDecoder (Haas et al. 2013) and six-frame translation tool (script "sixframe.rb"; available as part of the Protk toolkit, https://github.com/iracooke/protk), in order to find protein-coding regions and convert them into protein sequences, that must be compiled into a complex protein custom database for proteogenomic characterization of zooplanktonic samples. Currently, a useful and publicly available database, COPEPOD—The Coastal & Oceanic Plankton Ecology, Production & Observation Database (https://www.st.nmfs.noaa.gov/copepod/), can be used to access a compilation of zooplankton taxonomic data and their worldwide distribution. According to the COPEPOD database, accessed on 21 November 2019, the major zooplankton taxonomic groups include cnidarians, ctenophores, nemerteans, rotifers, crustaceans (ostracods, copepods, barnacles, mysids, isopods, amphipods, euphausiids, decapods), annelids, sipunculans, bryozoans, brachiopods, mollusks (gastropods, bivalves, cephalopods), chaetognaths, echinoderms, hemichordates, urochordates and cephalochordates. In this database, species names are associated with each major taxonomic group, which can be a first step to guide further construction of a custom database, for a particular zooplankton group. An example of a specific database is PdumBase—The *Platynereis dumerilii* database (http://pdumbase.gdcb.iastate.edu), where the early developmental transcriptome of the zooplanktonic *Platynereis dumerilii* can be found. In addition, metatranscriptomic studies have provided useful data comprising hundreds of transcriptomes and genomes of zooplanktonic organisms (Carradec et al. 2018). Another example is Aniseed–ascidian network for *in situ* expression and embryological data (https://www.aniseed.cnrs.fr/aniseed/), a database dedicated to development and evolutionary studies of tunicates with a genome browser, annotated genomes, taxonomic section, gene expression and function dedicated section (Brozovic et al. 2018). For metabarcoding, one of the most useful databases is the Barcode of Life Data Systems (BOLD). BOLD is a freely available web platform (http://www.barcodinglife.org) that provides an integrated environment for the assembly and use of DNA barcode data.

12.7 Conclusions and Future Perspectives

Omics techniques are essential for the characterization of zooplankton assemblages. These approaches enable an accurate estimation of species diversity (e.g., distinguish cryptic species), but also contribute to increase the knowledge on biotic interactions among zooplanktonic organisms, through the characterization of genes and their functions. The assessment of gene/protein levels started to be successfully applied to study dynamics of zooplankton assemblages. Omics technologies have proved to be valuable for distinguishing invasive from non-invasive species and characterizing all life cycle stages. The description of baseline temporal and spatial diversity patterns is a prerequisite for the prediction of community shifts caused by on-going environmental changes.

Despite some limitations, associated with the dependence on well curated databases, the application of omics techniques contributes to unravel ecological factors underlying zooplankton diversity and patterns, allowing the study of all life stages and, thus, promoting a detailed knowledge of zooplankton reproductive life cycles and ecology. Integrative approaches applying traditional methodologies with functional genomic and other genomics, transcriptomics and proteomics approaches, are revolutionizing our knowledge on the overwhelming zooplankton diversity.

Acknowledgments

Ana Matos was funded by a Ph.D grant (SFRH/BD/126682/2016) from the Portuguese Foundation for Science and Technology (Fundação para a Ciência e Tecnologia, FCT). Agostinho Antunes was partially supported by the Strategic Funding UIDB/04423/2020 and UIDP/04423/2020 through national funds provided by FCT and European Regional Development Fund (ERDF) in the framework of the program PT2020, and the FCT projects PTDC/MAR-BIO/0440/2014 and PTDC/CTA-AMB/31774/2017 (POCI-01-0145-FEDER/031774/2017).

References

Aldred, N. and A.S. Clare. 2014. Mini-review: impact and dynamics of surface fouling by solitary and compound ascidians. Biofouling 30: 259–70.

Ali, H.A.J. and M. Tamilselvi. 2016. Classification of Ascidians. pp. 13–18. *In*: Ascidians in Coastal Water. Springer International Publishing, Cham.

Appeltans, W., S.T. Ahyong, G. Anderson, M.V. Angel, T. Artois, N. Bailly et al. 2012. The magnitude of global marine species diversity. Curr. Biol. 22: 2189–2202.

Ayala, D.J., P. Munk, R.B.C. Lundgreen, S.J. Traving, C. Jaspers, T.S. Jørgensen et al. 2018. Gelatinous plankton is important in the diet of European eel (*Anguilla anguilla*) larvae in the Sargasso Sea. Sci. Reports 8: 1–10.

Baek, S.Y., K.H. Jang, E.H. Choi, S.H. Ryu, S.K. Kim, J.H. Lee et al. 2016. DNA barcoding of metazoan zooplankton copepods from South Korea. PLoS ONE 11: e0157307.

Beaugrand, G., A. McQuatters-Gollop, M. Edwards and E. Goberville. 2013. Long-term responses of North Atlantic calcifying plankton to climate change. Nat. Clim. Chang. 3: 263–267.

Behrendt, L., J.-B. Raina, A. Lutz, W. Kot, M. Albertsen, P. Halkjaer-Nielsen et al. 2018. *In situ* metabolomic- and transcriptomic-profiling of the host-associated cyanobacteria *Prochloron* and *Acaryochloris marina*. ISME J. 12: 556–567.

Bouquet, J.-M., C. Troedsson, A. Novac, M. Reeve, A.K. Lechtenborger, W. Massart et al. 2018. Increased fitness of a key appendicularian zooplankton species under warmer, acidified seawater conditions. PLoS ONE 13: e0190625.

Brena, C., F. Cima and P. Burighel. 2003. The highly specialised gut of Fritillariidae (Appendicularia: Tunicata). Mar. Biol. 143: 57–71.

Brown, B.L., R.V. LePrell, R.B. Franklin, M.C. Rivera, F.M. Cabral, H.L. Eaves et al. 2015. Metagenomic analysis of planktonic microbial consortia from a non-tidal urban-impacted segment of James River. Stand Genomic Sci. 10: 65.

Brozovic, M., C. Dantec, J. Dardaillon, D. Dauga, E. Faure, M. Gineste et al. 2018. ANISEED 2017: extending the integrated ascidian database to the exploration and evolutionary comparison of genome-scale datasets. Nucleic Acids Res. 46: D718–D725.

Bucklin, A., B. Frost, J. Bradford-Grieve, L. Allen and N. Copley. 2003. Molecular systematic and phylogenetic assessment of 34 calanoid copepod species of the Calanidae and Clausocalanidae. Mar. Biol. 142: 333–343.

Bucklin, A., M. Guarnieri, R.S. Hill, A.M. Bentley and S. Kaartvedt. 1999. Taxonomic and systematic assessment of planktonic copepods using mitochondrial COI sequence variation and competitive, species-specific PCR. Hydrobiologia 401: 239–254.

Bucklin, A., R.R. Hopcroft, K.N. Kosobokova, L.M. Nigro, B.D. Ortman, R.M. Jennings et al. 2010. DNA barcoding of Arctic Ocean holozooplankton for species identification and recognition. Deep Sea Res. II 57: 40–48.

Bucklin, A., P.K. Lindeque, N. Rodriguez-Ezpeleta, A. Albaina and M. Lehtiniemi. 2016. Metabarcoding of marine zooplankton: prospects, progress and pitfalls. J. Plankton Res. 38: 393–400.

Bucklin, A., H.D. Yeh, J.M. Questel, D.E. Richardson, B. Reese, N.J. Copley et al. 2019. Time-series metabarcoding analysis of zooplankton diversity of the NW Atlantic continental shelf. ICES J. Mar. Sci. 76: 1162–1176.

Calderón, I., J. Garrabou and D. Aurelle. 2006. Evaluation of the utility of COI and ITS markers as tools for population genetic studies of temperate gorgonians. J. Exp. Mar. Biol. Ecol. 336: 184–197.

Cantero, Á.L.P., V. Sentandreu and A. Latorre. 2010. Phylogenetic relationships of the endemic Antarctic benthic hydroids (Cnidaria, Hydrozoa): What does the mitochondrial 16S rRNA tell us about it? Polar. Biol. 33: 41–57.

Carlos, C., D.B.A. Castro and L.M.M. Ottoboni. 2014. Comparative metagenomic analysis of coral microbial communities using a reference-independent approach. PLoS ONE 9: e111626.

Carradec, Q., E. Pelletier, C. Da Silva, A. Alberti, Y. Seeleuthner, R. Blanc-Mathieu et al. 2018. A global ocean atlas of eukaryotic genes. Nat. Commun. 9: 373.

Chavez, F.P., M. Messié and J.T. Pennington. 2011. Marine primary production in relation to climate variability and change. Annu. Rev. Mar. Sci. 3: 227–260.

Cheng, F., M. Wang, C. Li and S. Sun. 2014. Zooplankton community analysis in the Changjiang River estuary by single-gene-targeted metagenomics. Chin. J. Oceanol. Limnol. 32: 858–870.

Conley, K.R., F. Lombard and K.R. Sutherland. 2018. Mammoth grazers on the ocean's minuteness: a review of selective feeding using mucous meshes. Proc. R. Soc. B 285: 20180056.

Costa, F.O., J.R. DeWaard, J. Boutillier, S. Ratnasingham, R.T. Dooh, M. Hajibabaei et al. 2007. Biological identifications through DNA barcodes: the case of the Crustacea. Can. J. Fish Aquat. Sci. 64: 272–295.

Cristescu, M.E. 2014. From barcoding single individuals to metabarcoding biological communities: Towards an integrative approach to the study of global biodiversity. Trends Ecol. Evol. 29: 566–571.

D'Alelio, D., D. Eveillard, V.J. Coles, L. Caputi, M. Ribera d'Alcala and D. Ludicone. 2019. Modelling the complexity of plankton communities exploiting omics potential: From present challenges to an integrative pipeline. Curr. Opin. Syst. Biol. 13: 68–74.

Da Costa, K.G., M. Vallinoto and R.M. Da Costa. 2011. Molecular identification of a new cryptic species of *Acartia tonsa* (Copepoda, Acartiidae) from the northern coast of Brazil, based on mitochondrial COI gene sequences. J. Coast. Res. 64: 359–363.

Daewel, U., S.S. Hjøllo, M. Huret, R. Ji, M. Maar, S. Niiranen et al. 2014. Predation control of zooplankton dynamics: a review of observations and models. ICES J. Mar. Sci. 71: 254–271.

de Vargas, C., S. Audic, N. Henry, J. Decelle, F. Mahé, R. Logares et al. 2015. Eukaryotic plankton diversity in the sunlit ocean. Science 348: 1261605–1261605.

Deibel, D. and B. Lowen. 2012. A review of the life cycles and life-history adaptations of pelagic tunicates to environmental conditions. ICES J. Mar. Sci. 69: 358–369.

Delmont, T.O., C. Quince, A. Shaiber, O.C. Esen, S.T. Lee, M.S. Rappé et al. 2018. Nitrogen-fixing populations of Planctomycetes and Proteobacteria are abundant in surface ocean metagenomes. Nat. Microbiol. 3: 804–813.

Djurhuus, A., K. Pitz, N.A. Sawaya, J. Rojas-Márquez, B. Michaud, E. Montes et al. 2018. Evaluation of marine zooplankton community structure through environmental DNA metabarcoding. Limnol. Oceanogr. Methods 16: 209–221.

Doney, S.C., M. Ruckelshaus, J.E. Duffy, J.P. Barry, F. Chan, C.A. English et al. 2012. Climate change impacts on marine ecosystems. Annu. Rev. Mar. Sci. 4: 11–37.

Edwards, M. and A.J. Richardson. 2004. Impact of climate change on marine pelagic phenology and trophic mismatch. Nature 430: 881–884.

Eloire, D., P.J. Somerfield, D.V.P. Conway, C. Halsband-Lenk, R. Harris and D. Bonnet. 2010. Temporal variability and community composition of zooplankton at station L4 in the Western Channel: 20 years of sampling. J. Plankton Res. 32: 657–679.

Ershova, E.A., R. Descoteaux, O.S. Wangensteen, K. Iken, R.R. Hopcroft, C. Smoot et al. 2019. Diversity and distribution of meroplanktonic larvae in the pacific arctic and connectivity with adult benthic invertebrate communities. Front. Mar. Sci. 6: 490.

Evans, J.S., P.M. Erwin, N. Shenkar and S. López-Legentil. 2017. Introduced ascidians harbor highly diverse and host-specific symbiotic microbial assemblages. Sci. Reports 7: 11033.

Gattuso, J.-P., A.K. Magnan, L. Bopp, W.W.L. Cheung, C.M. Duarte, J. Hinkel et al. 2018. Ocean solutions to address climate change and its effects on marine ecosystems. Front. Mar. Sci. 5: 337.

Georges, A.A., H. El-Swais, S.E. Craig, W.K. Li and D.A. Walsh. 2014. Metaproteomic analysis of a winter to spring succession in coastal northwest Atlantic Ocean microbial plankton. ISME J. 8: 1301–1313.

Ghahramanzadeh, R., G. Esselink, L.P. Kodde, H. Duistermaat, J.L.C.H.V. Valkenburg, S.H. Marashi et al. 2013. Efficient distinction of invasive aquatic plant species from non-invasive related species using DNA barcoding. Mol. Ecol. Resour. 13: 21–31.

Goetze, E. 2010. Species discovery in marine planktonic invertebrates through global molecular screening. Mol. Ecol. 19: 952–967.

Grossmann, M.M., D.J. Lindsay and A.G. Collins. 2013. The end of an enigmatic taxon: Eudoxia macra is the eudoxid stage of *Lensia cossack* (Siphonophora, Cnidaria). Syst. Biodivers 11: 381–387.

Haas, B.J., A. Papanicolaou, M. Yassour, M. Grabherr, P.D. Blood, J. Bowden et al. 2013. De novo transcript sequence reconstruction from RNA-seq using the Trinity platform for reference generation and analysis. Nat. Protoc. 8: 1494–1512.

Harris, R.P., P.H. Wiebe, J. Lenz, H.R. Skjoldal and M. Huntley (eds.). 2000. ICES Zooplankton Methodology Manual. Elsevier.

Hays, G., A. Richardson and C. Robison. 2005. Climate change and marine plankton. Trends Ecol. Evol. 20: 337–344.

He, J., L. Zheng, W. Zhang, Y. Lin and W. Cao. 2015. Morphology and molecular analyses of a new *Clytia* species (Cnidaria: Hydrozoa: Campanulariidae) from the East China Sea. J. Mar. Biol. Assoc. UK 95: 289–300.

Hebert, P.D.N. and T.R. Gregory. 2005. The promise of DNA barcoding for taxonomy. Syst. Biol. 54: 852–859.

Hirai, J., S. Katakura, H. Kasai and S. Nagai. 2017a. Cryptic zooplankton diversity revealed by a metagenetic approach to monitoring metazoan communities in the coastal waters of the Okhotsk Sea, Northeastern Hokkaido. Front. Mar. Sci. 4: 379.

Hirai, J., S. Nagai and K. Hidaka. 2017b. Evaluation of metagenetic community analysis of planktonic copepods using Illumina MiSeq: Comparisons with morphological classification and metagenetic analysis using Roche 454. PLoS ONE 12: e0181452.

Hirose, E. 2015. Ascidian photosymbiosis: Diversity of cyanobacterial transmission during embryogenesis. Genesis 53: 121–131.

Holland, B.S., M.N. Dawson, G.L. Crow and D.K. Hofmann. 2004. Global phylogeography of *Cassiopea* (Scyphozoa: Rhizostomeae): molecular evidence for cryptic species and multiple invasions of the Hawaiian Islands. Mar. Biol. 145: 1119–1128.

Holland, L.Z. 2016. Tunicates. Curr. Biol. 26: R146–R152.

Hosp, J., Y. Sagane, G. Danks and E.M. Thompson. 2012. The evolving proteome of a complex extracellular matrix, the Oikopleura house. PLoS ONE 7: e40172.

Huang, D., R. Meier, P.A. Todd and L.M. Chou. 2008. Slow mitochondrial COI sequence evolution at the base of the metazoan tree and its implications for DNA barcoding. J. Mol. Evol. 66: 167–174.

Jensen, P.C., M.K. Purcell, J.F. Morado and G.L. Eckert. 2012. Development of a real-time Pcr assay for detection of planktonic red king crab. *Paralithodes camtschaticus* (Tilesius 1815). Larvae. J. Shellfish Res. 31: 917–924.

Jeon, D., D. Lim, W. Lee and H.Y. Soh. 2018. First use of molecular evidence to match sexes in the Monstrilloida (Crustacea: Copepoda), and taxonomic implications of the newly recognized and described, partly Maemonstrilla -like females of *Monstrillopsis longilobata* Lee, Kim & Chang, 2016. PeerJ 6: e4938.

Jeong, H.J., Y. du Yoo, J.S. Kim, K.A. Seong, N.S. Kang and T.H. Kim. 2010. Growth, feeding and ecological roles of the mixotrophic and heterotrophic dinoflagellates in marine planktonic food webs. Ocean Sci. J. 45: 65–91.

Jonkers, L., H. Hillebrand and M. Kucera. 2019. Global change drives modern plankton communities away from the pre-industrial state. Nature 570: 372–375.

Keskin, E. and H.H. Atar. 2013. DNA barcoding commercially important aquatic invertebrates of Turkey. Mitochondrial DNA 24: 440–450.

Kiesling, T.L., E. Wilkinson, J. Rabalais, P.B. Ortner, M.M. McCabe and J.W. Fell. 2002. Rapid identification of adult and naupliar stages of copepods using DNA hybridization methodology. Mar. Biotechnol. 4: 30–39.

Kopf, A., M. Bicak, R. Kottmann, J. Schnetzer, I. Kostadinov, K. Lehmann et al. 2015. The ocean sampling day consortium. GigaScience 4: 27.

Kwan, J.C., M.S. Donia, A.W. Han, E. Hirose, M.G. Haygood and E.W. Schmidt. 2012. Genome streamlining and chemical defense in a coral reef symbiosis. Proc. Natl. Acad. Sci. 109: 20655–20660.

Lemaire, P. 2011. Evolutionary crossroads in developmental biology: the tunicates. Development 138: 2143–2152.

Lewandowska, A.M., D.G. Boyce, M. Hofmann, B. Matthiessen, U. Sommer and B. Worm. 2014. Effects of sea surface warming on marine plankton. Ecol. Lett. 17: 614–623.

Li, D.-X., H. Zhang, X.-H. Chen, Z.-X. Xie, Y. Zhang, S.-F. Zhang et al. 2018. Metaproteomics reveals major microbial players and their metabolic activities during the blooming period of a marine dinoflagellate *Prorocentrum donghaiense*. Environ. Microbiol. 20: 632–644.

Lima-Mendez, G., K. Faust, N. Henry, J. Decelle, S. Colin, F. Carcillo et al. 2015. Determinants of community structure in the global plankton interactome. Science 348: 1262073–1262073.

Lindeque, P.K., R.P. Harris, M.B. Jones and G.R. Smerdon. 1999. Simple molecular method to distinguish the identity of *Calanus* species (Copepoda: Calanoida) at any developmental stage. Mar. Biol. 133: 91–96.

Lindeque, P.K., R.P. Harris, M.B. Jones and G.R. Smerdon. 2004. Distribution of *Calanus* spp. as determined using a genetic identification system. Sci. Mar. 68: 121–128.

Lindeque, P.K., S.J. Hay, M.R. Heath, A. Ingvarsdottir, J. Rasmussen, G.R. Smerdon et al. 2006. Integrating conventional microscopy and molecular analysis to analyse the abundance and distribution of four *Calanus* congeners in the North Atlantic. J. Plankton Res. 28: 221–238.

Lindeque, P.K., H.E. Parry, R.A. Harmer, P.J. Somerfield and A. Atkinson 2013. Next generation sequencing reveals the hidden diversity of zooplankton assemblages. PLoS ONE 8: e81327.

Lindsay, D.J., M.M. Grossmann, J. Nishikawa, B. Bentlage and A.G. Collins. 2015. DNA barcoding of pelagic cnidarians: current status and future prospects. Bull. Plankt. Soc. Japan 62: 39–43.

Litchman, E., M.D. Ohman and T. Kiørboe. 2013. Trait-based approaches to zooplankton communities. J. Plankton Res. 35: 473–484.

López, J.L., S.L. Abalde and J. Fuentes. 2005. Proteomic approach to probe for larval proteins of the mussel *Mytilus galloprovincialis*. Mar. Biotechnol. 7: 396–404.

Machida, R.J., Y. Hashiguchi, M. Nishida and S. Nishida. 2009. Zooplankton diversity analysis through single-gene sequencing of a community sample. BMC Genomics 10: 438.

Magalhães, C., A. Martins and A.d. Santos. 2021. New approaches to study jellyfish: from autonomous apparatus to citizen science. pp. 227–251. *In*: M.A. Teodósio and A.B. Barbosa [eds.]. Zooplankton Ecology. CRC Press.

McFadden, C.S., Y. Benayahu, E. Pante, J.N. Thoma, P.A. Nevarez and S.C. France. 2011. Limitations of mitochondrial gene barcoding in Octocorallia. Mol. Ecol. Resour. 11: 19–31.

McInnes, J.C., R. Alderman, M.-A. Lea, B. Raymond, B.E. Deagle, R.A. Phillips et al. 2017. High occurrence of jellyfish predation by black-browed and Campbell albatross identified by DNA metabarcoding. Mol. Ecol. 26: 4831–4845.

Miglietta, M.P., D. Odegard, B. Faure and A. Faucci. 2015. Barcoding techniques help tracking the evolutionary history of the introduced species *Pennaria disticha* (Hydrozoa, Cnidaria). PLoS ONE 10: 1–12.

Moura, C.J., D.J. Harris, M.R. Cunha and A.D. Rogers. 2008. DNA barcoding reveals cryptic diversity in marine hydroids (Cnidaria, Hydrozoa) from coastal and deep-sea environments. Zool. Scr. 37: 93–108.

Nomura, M., A. Nakajima and K. Inaba. 2009. Proteomic profiles of embryonic development in the ascidian *Ciona intestinalis*. Dev. Biol. 325: 468–481.

Ortman, B.D., A. Bucklin, F. Pagès and M. Youngbluth. 2010. DNA barcoding the Medusozoa using mtCOI. Deep Res. Part II T 57: 2148–2156.

Ottesen, E.A., R. Marin, C.M. Preston, C.R. Young, J.P. Ryan, C.A. Scholin et al. 2011. Metatranscriptomic analysis of autonomously collected and preserved marine bacterioplankton. ISME J. 5:1881–1895.

Pan, M., A.J.A. McBeath, S.J. Hay, G.J. Pierce and C.O. Cunningham. 2008. Real-time PCR assay for detection and relative quantification of *Liocarcinus depurator* larvae from plankton samples. Mar. Biol. 153: 859–870.

Pearman, J.K. and X. Irigoien. 2015. Assessment of zooplankton community composition along a depth profile in the central red sea. PLoS ONE 10: e0133487.

Peijnenburg, K.T.C.A. and E. Goetze. 2013. High evolutionary potential of marine zooplankton. Ecol. Evol. 3: 2765–2781.

Piette, J. and P. Lemaire. 2015. Thaliaceans, the neglected pelagic relatives of Ascidians: A developmental and evolutionary enigma. Q. Rev. Biol. 90: 117–145.

Remigio, E. 2003. Testing the utility of partial COI sequences for phylogenetic estimates of gastropod relationships. Mol. Phylogenet. Evol. 29: 641–647.

Rey, A., J. Corell and N. Rodríguez-Ezpeleta. 2021. Metabarcoding to study zooplankton diversity. pp. 252–263. *In*: M.A. Teodósio and A.B. Barbosa [eds.]. Zooplankton Ecology. CRC Press.

Rombouts, I., G. Beaugrand, F. Ibañez, S. Gasparini, S. Chiba and L. Legendre. 2009. Global latitudinal variations in marine copepod diversity and environmental factors. Proc. R Soc. B Biol. Sci. 276: 3053–3062.

Salazar, G., L. Paoli, A. Alberti, J. Huerta-Cepas, H.-J. Ruscheweyh, M. Cuenca et al. 2019. Gene expression changes and community turnover differentially shape the global ocean metatranscriptome. Cell 179: 1068–1083.e21.

Santoferrara, L.F. and G.B. McManus. 2021. Diversity and biogeography as revealed by morphologies and DNA sequences: Tintinnid ciliates as an example. pp. 85–118. *In*: M.A. Teodósio and A.B. Barbosa [eds.]. Zooplankton Ecology. CRC Press.

Shearer, T.L., M.J.H. Oppen Van, S.L. Romano and G. Worheire. 2002. Slow mitochondria DNA sequence evolution in the Anthozoa. Mol. Ecol. 11: 2475–2487.

Sinniger, F., J.D. Reimer and J. Pawlowski. 2008. Potential of DNA sequences to identify Zoanthids (Cnidaria: Zoantharia). Zool. Sci. 25: 1253–1260.

Smith, J.N., G. De'ath, C. Richter, A. Cornils, J.M. Hall-Spencer and K.E. Fabricius. 2016. Ocean acidification reduces demersal zooplankton that reside in tropical coral reefs. Nat. Clim. Chang. 6: 1124–1129.

Stampar, S.N., M.M. Maronna, M.J.A. Vermeij, Fabio L.d. Silveira and A.C. Morandini. 2012. Evolutionary diversification of banded tube-dwelling anemones (Cnidaria; Ceriantharia; *Isarachnanthus*) in the Atlantic ocean. PLoS ONE 7: 41091.

Strom, S.L. 2008. Microbial ecology of ocean biogeochemistry: A community perspective. Science 320: 1043–1045.

Troedsson, C., M.E. Frischer, J.C. Nejstgaard and E.M. Thompson. 2007. Molecular quantification of differential ingestion and particle trapping rates by the appendicularian *Oikopleura dioica* as a function of prey size and shape. Limnol. Oceanogr. 52: 416–427.

Turon, X., I. Tarjuelo, S. Duran and M. Pascual. 2003. Characterising invasion processes with genetic data: an Atlantic clade of *Clavelina lepadiformis* (Ascidiacea) introduced into Mediterranean harbours. Hydrobiologia 503: 29–35.

Vadopalas, B., J.V. Bouma, C.R. Jackels and C.S. Friedman. 2006. Application of real-time PCR for simultaneous identification and quantification of larval abalone. J. Exp. Mar. Biol. Ecol. 334: 219–228.

Wang, D.-Z., Z.-X. Xie and S.-F. Zhang. 2014. Marine metaproteomics: Current status and future directions. J. Proteomics 97: 27–35.

Wang, D.-Z., L.-F. Kong, Y.-Y. Li and Z.-X. Xie. 2016. Environmental microbial community proteomics: Status, challenges and perspectives. Int. J. Mol. Sci. 17: 1275.

Wilmes, P. and P.L. Bond. 2004. The application of two-dimensional polyacrylamide gel electrophoresis and downstream analyses to a mixed community of prokaryotic microorganisms. Environ. Microbiol. 6: 911–920.

Wöhlbrand, L., B. Wemheuer, C. Feenders, H.S. Ruppersberg, C. Hinrichs, B. Blasius et al. 2017. Complementary metaproteomic approaches to assess the bacterioplankton response toward a phytoplankton spring bloom in the Southern North Sea. Front. Microbiol. 8: 442.

Yang, J., X. Zhang, W. Zhang, J. Sun, Y. Xie, Y. Zhang et al. 2017. Indigenous species barcode database improves the identification of zooplankton. PLoS ONE 12: e0185697.

Zheng, L., J. He, Y. Lin, W. Cao and W. Zhang. 2014. 16S rRNA is a better choice than COI for DNA barcoding hydrozoans in the coastal waters of China. Acta Oceanol. Sin. 33: 55–76.

Index

Printed and bound by CPI Group (UK) Ltd, Croydon, CR0 4YY

24/10/2024

01778298-0005